# Advances in horticultural soilless culture

It is widely recognised that agriculture is a significant contributor to global warming and climate change. Agriculture needs to reduce its environmental impact and adapt to current climate change whilst still feeding a growing population, i.e. become more 'climate-smart'. Burleigh Dodds Science Publishing is playing its part in achieving this by bringing together key research on making the production of the world's most important crops and livestock products more sustainable.

Based on extensive research, our publications specifically target the challenge of climate-smart agriculture. In this way we are using 'smart publishing' to help achieve climate-smart agriculture.

Burleigh Dodds Science Publishing is an independent and innovative publisher delivering high quality customer-focused agricultural science content in both print and online formats for the academic and research communities. Our aim is to build a foundation of knowledge on which researchers can build to meet the challenge of climate-smart agriculture.

For more information about Burleigh Dodds Science Publishing simply call us on +44 (0) 1223 839365, email info@bdspublishing.com or alternatively please visit our website at www.bdspublishing.com.

**Related titles:**

Robotics and automation for improving agriculture
Print (ISBN 978-1-78676-272-6); Online (ISBN 978-1-78676-274-0, 978-1-78676-275-7)

Achieving sustainable greenhouse cultivation
Print (ISBN 978-1-78676-280-1); Online (ISBN 978-1-78676-282-5, 978-1-78676-283-2)

Achieving sustainable urban agriculture
Print (ISBN 978-1-78676-316-7); Online (ISBN 978-1-78676-318-1, 978-1-78676-319-8)

Achieving sustainable cultivation of vegetables
Print (ISBN 978-1-78676-236-8); Online (ISBN 978-1-78676-238-2, 978-1-78676-239-9)

Chapters are available individually from our online bookshop:
https://shop.bdspublishing.com

BURLEIGH DODDS SERIES IN AGRICULTURAL SCIENCE
NUMBER 94

# Advances in horticultural soilless culture

Edited by Professor Nazim S. Gruda,
University of Bonn, Germany

burleigh dodds
SCIENCE PUBLISHING

Published by Burleigh Dodds Science Publishing Limited
82 High Street, Sawston, Cambridge CB22 3HJ, UK
www.bdspublishing.com

Burleigh Dodds Science Publishing, 1518 Walnut Street, Suite 900, Philadelphia, PA 19102-3406, USA

First published 2021 by Burleigh Dodds Science Publishing Limited

Cover image courtesy of Professor Nazim S. Gruda.



Library of Congress Control Number: 2020945578

British Library Cataloguing in Publication Data
A catalogue record for this book is available from the British Library

ISBN 978-1-78676-435-5 (Print)
ISBN 978-1-78676-438-6 (PDF)
ISBN 978-1-78676-437-9 (ePub)
ISSN 2059-6936 (print)
ISSN 2059-6944 (online)

DOI 10.19103/AS.2020.0076

Typeset by Deanta Global Publishing Services, Dublin, Ireland

# Contents

## Part 2 Technologies

# Series list

| Title | Series number |
| --- | --- |
| **Achieving sustainable cultivation of maize - Vol 1**<br>From improved varieties to local applications<br>Edited by: Dr Dave Watson, CGIAR Maize Research Program Manager, CIMMYT, Mexico | 001 |
| **Achieving sustainable cultivation of maize - Vol 2**<br>Cultivation techniques, pest and disease control<br>Edited by: Dr Dave Watson, CGIAR Maize Research Program Manager, CIMMYT, Mexico | 002 |
| **Achieving sustainable cultivation of rice - Vol 1**<br>Breeding for higher yield and quality<br>Edited by: Prof. Takuji Sasaki, Tokyo University of Agriculture, Japan | 003 |
| **Achieving sustainable cultivation of rice - Vol 2**<br>Cultivation, pest and disease management<br>Edited by: Prof. Takuji Sasaki, Tokyo University of Agriculture, Japan | 004 |
| **Achieving sustainable cultivation of wheat - Vol 1**<br>Breeding, quality traits, pests and diseases<br>Edited by: Prof. Peter Langridge, The University of Adelaide, Australia | 005 |
| **Achieving sustainable cultivation of wheat - Vol 2**<br>Cultivation techniques<br>Edited by: Prof. Peter Langridge, The University of Adelaide, Australia | 006 |
| **Achieving sustainable cultivation of tomatoes**<br>Edited by: Dr Autar Mattoo, USDA-ARS, USA & Prof. Avtar Handa, Purdue University, USA | 007 |
| **Achieving sustainable production of milk - Vol 1**<br>Milk composition, genetics and breeding<br>Edited by: Dr Nico van Belzen, International Dairy Federation (IDF), Belgium | 008 |
| **Achieving sustainable production of milk - Vol 2**<br>Safety, quality and sustainability<br>Edited by: Dr Nico van Belzen, International Dairy Federation (IDF), Belgium | 009 |
| **Achieving sustainable production of milk - Vol 3**<br>Dairy herd management and welfare<br>Edited by: Prof. John Webster, University of Bristol, UK | 010 |
| **Ensuring safety and quality in the production of beef - Vol 1**<br>Safety<br>Edited by: Prof. Gary Acuff, Texas A&M University, USA & Prof. James Dickson, Iowa State University, USA | 011 |
| **Ensuring safety and quality in the production of beef - Vol 2**<br>Quality<br>Edited by: Prof. Michael Dikeman, Kansas State University, USA | 012 |
| **Achieving sustainable production of poultry meat - Vol 1**<br>Safety, quality and sustainability<br>Edited by: Prof. Steven C. Ricke, University of Arkansas, USA | 013 |
| **Achieving sustainable production of poultry meat - Vol 2**<br>Breeding and nutrition<br>Edited by: Prof. Todd Applegate, University of Georgia, USA | 014 |

**Achieving sustainable greenhouse cultivation** 063
Edited by: Prof. Leo Marcelis & Dr Ep Heuvelink, Wageningen University, The Netherlands

**Achieving carbon-negative bioenergy systems from plant materials** 064
Edited by: Dr Chris Saffron, Michigan State University, USA

**Achieving sustainable cultivation of tropical fruits** 065
Edited by: Prof. Elhadi M. Yahia, Universidad Autónoma de Querétaro, Mexico

**Advances in postharvest management of horticultural produce** 066
Edited by: Prof. Chris Watkins, Cornell University, USA

**Pesticides and agriculture** 067
Profit, politics and policy
Dave Watson

**Integrated management of diseases and insect pests of tree fruit** 068
Edited by: Prof. Xiangming Xu and Dr Michelle Fountain, NIAB-EMR, UK

**Integrated management of insect pests: Current and future developments** 069
Edited by: Emeritus Prof. Marcos Kogan, Oregon State University, USA & Emeritus Prof. E. A. Heinrichs, University of Nebraska-Lincoln, USA

**Preventing food losses and waste to achieve food security and sustainability** 070
Edited by: Prof. Elhadi M. Yahia, Universidad Autónoma de Querétaro, Mexico

**Achieving sustainable management of boreal and temperate forests** 071
Edited by: Dr John Stanturf, Estonian University of Life Sciences , Estonia

**Advances in breeding of dairy cattle** 072
Edited by: Prof. Julius van der Werf, University of New England, Australia & Prof. Jennie Pryce, Agriculture Victoria and La Trobe University, Australia

**Improving gut health in poultry** 073
Edited by: Prof. Steven C. Ricke, University of Arkansas, USA

**Achieving sustainable cultivation of barley** 074
Edited by: Prof. Glen Fox, University of California-Davis, USA and The University of Queensland, Australia & Prof. Chengdao Li, Murdoch University, Australia

**Advances in crop modelling for a sustainable agriculture** 075
Edited by: Emeritus Prof. Kenneth Boote, University of Florida, USA

**Achieving sustainable crop nutrition** 076
Edited by: Prof. Zed Rengel, University of Western Australia, Australia

**Achieving sustainable urban agriculture** 077
Edited by: Prof. Johannes S. C. Wiskerke, Wageningen University, The Netherlands

**Climate change and agriculture** 078
Edited by Dr Delphine Deryng, NewClimate Institute/Integrative Research Institute on Transformations of Human-Environment Systems (IRI THESys), Humboldt-Universität zu Berlin, Germany

**Advances in poultry genetics and genomics** 079
Edited by: Prof. Samuel E. Aggrey, University of Georgia, USA; Prof. Huaijun Zhou, University of California-Davis, USA; Dr Michèle Tixier-Boichard, INRAE, France; and Prof. Douglas D. Rhoads, University of Arkansas, USA

**Achieving sustainable management of tropical forests** 080
Edited by: Prof. Jürgen Blaser, Bern University of Life Sciences, Switzerland; and Patrick D. Hardcastle, Forestry Development Specialist, UK

# Preface

Years ago, working on my PhD thesis, I was reading a book from the well-known German soil scientist, Mitscherlich. We know him among others for trial containers, under his name. In the foreword of his book 'Bodenkunde für Land- und Forstwirte' (Soil Science for Farmers and Foresters), from 1906, he wrote: 'It is completely irrelevant to our crops how the soil on which they are supposed to produce crops was once geologically created. The only thing that matters is how the soil is physically and chemically constituted when the plants grow on it.'

How right he was at that time, shows not only the history of soil science but more than a century later, the new development on growing media in soilless culture. Nowadays, we could place these words in his lips: 'It does not matter to our crops how growing media were made. The only thing that matters is how a substrate is currently constituted when plants grow in it: its physical, chemical and biological properties.' This may be the explanation as to why there are such a large number of plant substrates for horticulture on the market: from liquid to solid materials, from inert to non-inert culture media, from inorganic to organic materials, from naturally degraded to synthetically produced growing media. Regardless of whether it is peat, bark or wood fibre substrate, plants perform well, as the physical, chemical, and biological properties and plant cultivation technologies are adapted to the substrate properties used in soilless culture systems.

What are soilless culture systems? What kind of growing media and soilless culture systems do we have? Which kind of raw materials are needed to design growing media of the future? How can we analyse them? How do they perform regarding plant growth? Are they 'contemporary' and sustainable enough for the next generation? What should we do to adapt plant technologies to new rooting materials? What are developing new trends in the era of digitalisation? My colleagues, well-known scientists and specialists in their fields, respond to these, and many more questions step-by-step, chapter by chapter.

Dear colleagues, I am very proud to work with and for you! I want to thank you all for a fruitful collaboration and your excellent contributions to this book. Furthermore, my thanks go to the publishing house of Burleigh Dodds Science Publishing for providing me with the opportunity to edit this book and for their support in completing it, including during the difficult time caused by the COVID-19 pandemic.

Prof. Dr. Nazim S. Gruda,
Bonn-Berlin, Germany
September 2020

# Chapter 1

# Soilless culture systems and growing media in horticulture: an overview

*Nazim S. Gruda, University of Bonn, Germany*



## 1 Introduction

All systems of growing plants without using the soil *in situ* are called soilless culture systems. These culture procedures include systems with a solid medium, so-called aggregate systems, in which, for example, inorganic and organic substrates are used. In the English language, the term 'growing media', is widely used as a synonym for 'substrates'. We will also adopt this definition, without prejudice whether the substrates are placed in/on foils, containers, rows or thin layers.

Further, soilless culture includes systems with a liquid medium as well. These systems are also called 'hydroponics', especially when water and nutrients are mainly supplied through a nutrient solution. For both, solid or liquid systems, it is not relevant whether the nutrient solution cycles applied to plants are open or closed-loops.

## 2 Historical overview

Several civilisations have utilised soilless culture systems (SCS). Egyptian hieroglyphic records describe the growing of plants in water without soil. The hanging gardens of Babylon, one of the seven wonders of the ancient time, are another beautiful example thereof. In the eleventh century, in Mesoamerica,

http://dx.doi.org/10.19103/AS.2020.0076.20

the nomadic Mexica tribe, one of the native tribes which later formed the Aztec empire, built the City of Tenochtitlan on the shores of Lake Texcoco. Artificial islands, called chinampas, were created by fencing in the shallow lake bed with wattle. In the Andes, the ancient Aymaras built their villages on floating islands, made of totora reeds (*Schoenoplectus californicus* ssp. tatora), over the Titicaca lake at 4000 m a.s.l. People lived on fishing, and crops were grown on these islands (Gruda et al. 2016b).

Robert Boyle described the first plant-growing experiments with the roots submerged in water (Shaw 1666). Woodward (1699) showed that water alone will not suffice for plant growth but that terrestrial matter had to be added. In the middle of the nineteenth-century, water-culture began to be used for experiments on plant nutrition (Sachs 1860; Knop 1865; Hoagland and Arnon 1950). Different standard solution formulas for growing plants in water culture have been used. One from them developed by Hoagland in 1920 (Hoagland and Arnon 1950) is still in use in our time. Since the first proposal for a commercial water-culture system made by Gericke (1929), these systems were further developed in the past.

Interestingly, hydroponics has its roots in times of crises (Jones 1982; Savvas and Passam 2002). Wartime gardening was promoted as a way to increase food security in the United States during World Wars I and II. Hydroponic systems were used to some extent by the United States Army to produce vegetables in some rocky and non-arable islands in the Pacific, which were contaminated due to war operations (Jones 1982; Savvas and Passam 2002). These new gardens helped to feed both troops and civilians, and this was a significant step in establishing a commercial SCS.

More than 75 years later, based on the same principle, scientists from the University of Sheffield helped Syrian refugees at a camp in Jordan by using old mattress foam and coffee cups to produce vegetables and herbs in limited places. Thus, soilless culture made it possible to produce fresh food in a desert environment, giving people the feeling of tending to their own garden in refugees camp and helping them reuse the mountain of plastic mattresses (Morrison 2020).

***#stayathome*** was the catchphrase for the drastically changed living conditions during COVID-19. However, the crisis reveals one plus for horticulture. The lockdown of the catering industry meant that more people were cooking at home again. The crisis has also increased the demand for trustworthy, locally grown produce. Further, people enjoyed cultivating plants in their garden, sometimes by using simple hydroponic systems. According to the FAO, in the Urban Water-Food Nexus, shortening the food supply chain for urban residents and demand for locally produced food also made hydroponic systems an attractive and resilient option under COVID-19. Besides food security for the urban population, an estimated 50% rapid increase in urban water demands is expected over the next 30 years. The hydroponic and vertical farming technologies with a closed water-loop system have the advantage of enormous water and land use savings (FAO 2020). For water saving and urban gardening see some more information below.

## 3 Why soilless culture systems?

The reason for the development of soilless culture was to overcome soil-borne disease problems and other problems related to soil properties. The latest include high salt content, sodic or contaminated soils, non-arable soils with a poor soil structure and/or with a decline of soil fertility, which represent a significant proportion of cultivable land throughout the world. Moreover, soil tillage and preparation are avoided as there is no need for herbicides and frequent crop rotations. This increases the potential length of the cultivation time and yields are increased (Gruda et al. 2016b; Savvas and Gruda 2018). On the other hand, SCS promote better plant growth and earliness for vegetable crops, good quality for cut flowers and pot ornamentals in controlled environment agriculture as well as hardy nursery plants in containers. Other advantages, such as better environmental protection with closed recirculated systems and improved product quality through a precise dosage of nutrients (Schnitzler and Gruda 2002; Gruda 2009; Gruda et al. 2018) are becoming more and more crucial (Gruda and Tanny 2014; Gruda et al. 2016b). Because of a better use of energy and nutrient uptake, higher plant density, and faster plant growth, higher, reliable and predictable yields like in soil cultivation are achieved (Savvas and Gruda 2018). Finally, SCS is an accurate and excellent tool in the research work.

The majority of SCS-plants are grown in high-tech greenhouse structures with a fully automatic climate control feature. Even when there are no data about the worldwide soilless cultivation, the area has enormously increased nowadays in several countries, and it is predicted to remain a trend for the future. For instance, in Northern countries, the area of SCS has increased more than 10-fold in the last two decades. On the other hand, several container plants are cultivated in outdoor conditions. Thus, growers are progressively using SCS primarily because of high-profit margins and ten times higher yields than traditional cultivation (Hong and Gruda 2020). However, nowadays, our society is more and more interested in safety foodstuffs and horticultural goods, produced in a circular economy, including efficient waste disposal and the reduction of used resources, such as fertilisers and pesticides. All these factors make soilless culture not only a critical challenge tool of intensive agriculture worldwide but also a climate-smart production method, due to environmental concerns (Savvas and Gruda 2018; Gruda 2019), resource sustainability as well as climate changes (Gruda et al. 2019a,b).

Good sustainability indicators are the usage of resources, for example, water, fertiliser, energy, land use, for one horticultural unit. Soilless greenhouse culture leads to higher crop production for the unit area and thus to a decrease in land usage in comparison to other cultivation methods. Mugnozza et al. (2007) reported, using life cycle analysis (LCA), that soilless cultivation reduced the environmental impact, due to lower levels of fertilisers and pesticides emitted to the environment, compared to soil cultivation (Gruda 2019).

Water-saving is a significant economic and vital sustainable driver for the open field and protected cultivation horticulture, especially for countries in arid and semi-arid regions. One of the significant advantages of using SCS is water economisation. One way to measure water economisation is the water use efficiency (WUE) that is expressed in units of product weight per volume of water applied, grams or kilograms per litre (Gruda and Tanny 2014). The inverse of WUE, the product water use (PWU) provides the amount of water necessary for the production of one kilogram of fresh product. The latest is recently often used to allow a comparison of food production, expressed in litres of water used to produce one kilogram of fresh product. As shown in Tables 1 and 2, the PWU is a function of the location site, cultivation season, cultivar and irrigation used (Valenzano et al. 2008; Ntinas et al. 2017).

Table 1 shows that product water use (PWU) for tomato production in the soil is reduced from about 200 L/kg to 50–60 L/kg by using drip irrigation. A further improvement of PWU can be seen by growing tomato in a greenhouse compared to field-grown tomatoes, due to a better controlled growing condition and a lower transpiration per unit area (Nederhoff and Stanghellini 2010). Even in soil cultivation systems, with tensiometers, suitable watering volumes, and mulching, encouraging results are achieved in terms of PWU (Valenzano et al. 2008), according to Table 1 with values from 27 L/kg to 34 L/kg.

Moving from soil to SCS can further improve PWU. Generally, open-loop and closed-loop systems are different concerning PWU. However, comparing the results for an open-loop from Torrellas et al. (2012) with results for a closed-loop from Incrocci et al. (2006), we find that both, the system as well as the substrate used, are not a guarantee for good results when other conditions do not match. Beerling et al. (2014) documented the practical differences in discharges of water and nutrient solution for different growers. Theoretically, zero drainage can be achieved in open-loop systems by the adoption of modern and sophisticated techniques and technologies. However, in practice, a certain amount of leachate is usually required, even when using high-quality irrigation water (Massa et al. 2020). Excluding the results similar to open-loop systems, the best results obtained in closed-loop recirculating water systems were between 14 L/kg and 20 L/kg. This is more than 10% of the water needed from field-grown plants. Furthermore, a combination of the irrigation system with modern environmental management by the regaining and reusing of

**Table 1** Product water use efficiency (PWU) for tomato production (L $kg^{-1}$) in soil, open field and protected cultivation

| Production system | Soil | Irrigation system | PWE (L $kg^{-1}$) | Functional unit (1 kg tomato)/season or time | Country/region | Reference |
|---|---|---|---|---|---|---|
| Open field | Soil | Open loop, for example, furrow | 200±100 | 1 kg tomato, n.d. | Spain | van Kooten et al. (2004) |
| Open field | Soil | Open loop, groundwater, base/drip | 43.7 | 1 kg tomato (May-August) | Greece | Ntinas et al. (2017) |
| Open field | Soil | Open loop, surface water, base/drip | 60.0 | 1 kg tomato (April-September) | Greece | Ntinas et al. (2017) |
| Open field | Soil | Open loop, n.d. | 50.0 | 1 kg tomato, n.d. | Bundaberg, AU | Page et al. (2012) |
| Plastic greenhouse, single PE, unheated | Soil | Open loop, groundwater | 28.6 | 1 kg tomato (Winter-Spring) | Morocco | Payen et al. (2015) |
| Greenhouse, double PE/ Plexiglas, heated | Soil | Open loop, tap water, drip | 28.1 | 1 kg tomato (February-November) | Germany | Ntinas et al. (2017) |
| Greenhouse, double PE/ Plexiglas, heated | Soil | Open loop, tap water, drip | 37.0 | 1 kg tomato (February-November) | Germany | Ntinas et al. (2017) |
| Greenhouse, double F-clean, heated | Soil | Open loop, tap water, drip | 25.6 | 1 kg tomato (February-November) | Germany | Ntinas et al. (2017) |
| Plastic tunnel, two tomato fruit types | Soil, mulched | Open loop, irrigation controlled by tensiometers | 27.0 | 1 kg tomato (Winter-Spring) | Italy | Valenzano et al. (2008) |

n.d. = no data, AU = Australia.

**Table 2** Product water use (PWU) for tomato production (L $kg^{-1}$) in soilless cultures with an open and a closed-loop irrigation system

| Production system | Substrate | Irrigation system | PWU (L $kg^{-1}$) | Functional unit (1 kg tomato)/season or time | Country/region | Reference |
|---|---|---|---|---|---|---|
| Plastic tunnel, two tomato fruit types | Rockwool | Open loop, drip | 40.0 | 1 kg tomato (Winter–Spring) | Italy | Valenzano et al. (2008) |
| Low-tech greenhouse, unheated | n.d. | Open | 38.0 | 1 kg tomato, n.d. | Sydney, AU | Page et al. (2012) |
| Plastic greenhouse, single PE, heated | Rockwool | n.d., tap water, drip | 34.5 | 1 kg tomato (November–April) | Greece | Ntinas et al. (2017) |
| Plastic greenhouse, single PE, heated | Rockwool | n.d., tap water, drip | 33.5 | 1 kg tomato (November–April) | Greece | Ntinas et al. (2017) |
| Plastic greenhouse, cv. 'Floradade' | Sand | Open loop | 32.3 | 1 kg tomato (February–June) | Spain | Reina-Sanchez et al. (2005) |
| Greenhouse (glass/plastic) | Rockwool | Open/closed | 32.8 | 1 kg tomato (Winter–Spring) | France | Payen et al. (2015) |
| Plastic multi-tunnel, unheated | Perlite | Open loop | 28.8 | 1 kg tomato (Winter–Spring) | Spain | Torrellas et al. (2013) |
| Plastic multi-tunnel, unheated | Perlite | Open loop, drip | 28.8 | 1 kg tomato (52 weeks) | Spain | Torrellas et al. (2012) |
| Glasshouse Venlo, heated | Rockwool | Open loop, drip | 14.6 | 1 kg tomato (49 weeks) | Hungary | Torrellas et al. (2012) |
| Glasshouse, heated | Peat-perlite | Closed loop, sub-irrigation | 30.6 | 1 kg tomato (Spring) | Italy | Incrocci et al. (2006) |
| Glasshouse, heated | Peat-perlite | Closed loop, drip | 34.2 | 1 kg tomato (Spring) | Italy | Incrocci et al. (2006) |
| Plastic tunnel, two tomato fruit types | NFT | Closed loop | 22.2 | 1 kg tomato (Winter–Spring) | Italy | Valenzano et al. (2008) |

| | | | | | | |
|---|---|---|---|---|---|---|
| Glasshouse Venlo, heated | Rockwool | Closed loop | 20.0 | 1 kg tomato, n.d. | The Netherlands | van Kooten et al. (2004) |
| Hi-tech plastic greenhouse, heated | Rockwool | n.d. (w.a. closed system) | 18.0 | 1 kg tomato, n.d. | Ontario, CA | Dias et al. (2017) |
| Glasshouse Venlo, heated | Rockwool | Closed loop | 14.1 | 1 kg tomato (52 weeks) | The Netherlands | Torrellas et al. (2012) |
| Glasshouse Venlo, heated | Rockwool | Closed loop | 14.0 | 1 kg tomato (Winter-Spring) | Spain | Torrellas et al. (2013) |
| Closed and semi-closed glasshouses, heated | Rockwool | Closed loop | 4.0 | 1 kg tomato, n.d. | The Netherlands | van Kooten et al. (2004) |
| Closed and semi-closed glasshouses, heated | Rockwool | Closed loop | 1.5 | 1 kg tomato, n.d. | The Netherlands | van Kooten et al. (2004) |

n.d. = no data, w.a. = we assumed, NFT = nutrient film technique, AU = Australia, CA = Canada.

condensed evaporated water, using of light selective shading and evaporative cooling system, improve the PWU to 1.5–4.0 L/kg (Table 1). Even if under operational conditions, the levels of PWU are somewhat higher than this, it is clear that SCS realises a high reduction in water consumption (Gruda 2019).

On the other hand, it is objective to present also the disadvantages related to the utilisation of such systems. High investment and energy costs for the initial installation, running energy costs for high-tech greenhouses as well as increased technical skills needed to operate with such systems are some of the crucial disadvantages. High investment and high energy costs of greenhouse soilless systems have, in turn, a high impact on the production of greenhouse gas (GHG) emissions. Gruda et al. (2019a,b) recommended different measures and a new generation of greenhouses to reduce those emissions.

## 4 Range of soilless crops, cultivation places and purposes

The SCS is an intensive and sustainable tool applied for different purposes. Year-round and high-quality vegetables, cut flowers and pot ornamental plants, seedlings and transplants for indoor and outdoor production, and hard nursery plants in containers use SCS. As Hong and Gruda (2020) showed SCS is suitable also for introducing and cultivating new vegetables smoothly and efficiently, even in non-origin areas. Meanwhile, there is an increased interest in growing and producing other plants in these systems. For instance, figs (*Ficus carica* L.) and table grape (*Vitis vinifera* L.), and other small/soft fruit crops, but also fruits, such as apricots, peaches and cherries (di Lorenzo et al. 2013), or traditionally soil-grown vegetables, such as leeks (van Os et al. 2017), greens and herbs, wild vegetables (Fig. 1), and recently Cannabis are cultivated in these systems now.

For instance, we know that Cannabis is a useful plant for the extraction of fibres that are also used as a growing medium constitute (see Chapter 4, in this book). Further, we know hemp seeds, from which hemp oil is pressed. However, while fibre and grain hemp are grown in the open field, medical Cannabis is cultivated under controlled conditions in the greenhouse or climatic chamber using SCS. Cannabis plants rich in phytocannabinoid, remain significantly smaller. They grow just 1–1.5 m high and have stems of finger thickness, while fibre hemp can reach a height of 3 m and more. Moreover, the plant should branch well, to build many flowers. Only flowers from female plants are harvested. Best conditions offer completely closed systems without natural light. In order to obtain several harvests a year, plants are planted using a shortened day-night cycle and an elaborate technique. Climatic conditions have to be kept constant, and the amount of light is above average for vegetable plants. Further, according to Yep et al. (2020), root-zone-systems mediate growth, physiology, yield and potency of *Cannabis sativa*. All these conditions are better controlled

**Figure 1** Wild chicory (*Cichorium spinosum*), known as Stamnagathi, in soilless culture system at the Research Station of Agriculture, University of Athens, Greece (Photo Gruda, private collection).

in SCS that, in turn, allow the environment to remain pest-free, and for nutrients, oxygen levels and temperatures in the root zone to be well controlled. Thus, plants enjoy a faster growth, inducing high yields of excellent quality.

On the other hand, SCS is used in particular places. The integral ability of soilless culture to dissociate plant cultivation from the soil makes it possible to produce fresh vegetables in the absence of gravity. A rocket cannot transport that much food into space. Therefore, it is imperative to produce food in outer space. That is why the US space agency NASA is experimenting with hydroponic cultivation of romaine lettuce, Chinese cabbage, lettuce and mizuna, and other leave vegetables. All this is not just a material for science-fiction stories: greenhouses are also flying in satellites, and astronauts already have been able to eat lettuce grown on the international space station. Besides, a psychological effect of 'it does not feel so far from home' is achieved.

Fresh vegetables such as cucumbers are also grown in Antarctica, and nowadays it is a reality to produce fresh food in desert areas as well. In the time of the pandemic, the paradigm of food production dramatically changed into the direction of home country production, giving more impulses to SCS as an intensive production system. Therefore, the trend of soilless culture was more emphasised in some Arabic countries, such as the Bahrain, Kuwait, Oman, Qatar, Saudi Arabia and United Arab Emirates and, which are building high-tech greenhouses equipment with modern SCS.

Soilless culture is a proven tool to implement food production in megacities. Globally, the number of megacities, that is, cities with more than 10 million inhabitants, is projected to rise from 33 in 2018 to 43 in 2030 (UN 2018). This is one of the reasons why currently soilless cultures have received tremendous interest and are used in urban farming, including vertical systems (Gruda 2019). Nowadays, companies are testing vertical multiple floor modules installed in indoor environments, equipped with all needed resources for plant production, including the supply of water, nutrients, light, sensors, communication units and so on. Singapore, one of the most affluent Asian countries, has been quick to adopt indoor and rooftop farms with aims to locally produce 30% of the nation's food by 2030, realising that vertical farming can act as a safety net in times of droughts and pandemic crises (hortidaily 2020).

Because of their lightweight and sustainability in terms of resource efficiency, soilless systems are especially suitable for urban areas (Eigenbrod and Gruda 2015). Probably, the most complex and futuristic concept of urban agriculture is vertical farming (Eigenbrod and Gruda 2015). The vertical farm strategy aims to significantly increase productivity and reduce the environmental footprint within a framework of urban, indoor, climate-controlled high-rise buildings (Benke and Tomkins 2017). Cultivating vegetables vertically has a variety of advantages. However, vertical farming does not have the amount of sunlight at the same level as in a greenhouse. As light is essential for plants' photosynthesis, it is necessary to install light sources to ensure sufficient yields and high-quality products (Eigenbrod and Gruda 2015). New developments of LED lamps, concerning both light quality and lamp prices, have made it possible to increase the success of such systems in buildings.

The big sticking point is the high cost. All environmental influences such as light, temperature and water, to name just a few, must be controlled. This poses significant challenges for scientists and is also associated with very high energy costs. On the other hand, such buildings would only make sense in cities with high food requirements, and property prices are immense. Since vegetables produced should ultimately still be affordable, it seems that profitable and large-scale vegetable growing in buildings will remain a dream of the future.

However, there are still some examples around the world where it works. This is also the case in a three-storey building in Beijing, China (Fig. 2). Various plants are grown on an area of 26 000 m$^2$. Mushrooms grow in the basement, various Asian medicinal plants on the ground floor and vegetable plants in the attic with a greenhouse roof. Not only the medicinal plants, but also the mushrooms are of very high quality. Mushrooms were exported to Japan, while medicinal plants are used for the processing industry and the production of pharmaceuticals. This allows reasonable prices on average for all products, including vegetables. Vegetables can also be grown on the roofs of the buildings using this cultivation method. However, these methods should be energy-saving and suitable for producing high-quality vegetables. Furthermore, roofs have to be checked statically beforehand.

Recently the cultivation on the rooftop of several city-building typologies has been receiving attention from both researchers and practitioners, as another production kind that allows the simultaneous production of both vegetable and fish with the same water and nutrient cycle. These features make the production type very sustainable in terms of nutrients and water recycling (Eigenbrod and Gruda 2015).

**Figure 2** An example of soilless culture in a three-storey building: (a) mushrooms are grown in the basement, (b) Asian medicinal plants on the ground floor and (c) vegetable plants in the attic in contact to sunshine (Photos Gruda, private collection).

## 5 State-of-the-art: what does this book offer?

In this book, we aim to outline a wealth of recent research and all advances in the field of growing media and SCS, optimised to contribute to a sustainable, meaningful 'climate-smart' horticultural production.

Soilless culture systems can eliminate or reduce many problems associated with traditional cultivation on the soil in situ. However, advances in understanding plant root behaviour are required to understand these phenomena, because a healthy and vigorous root system is essential in SCS. The second chapter of this book addresses this topic and explains how roots take up water and nutrients explaining the implications for fertilisation in soilless systems and integrated root-zone management approach. Therefore, this chapter is a review of our understanding of root, root zone, and root-zone management in SCS. It also discusses future research trends in this area.

**Figure 3** Evaluating a growing medium means considering it from different perspectives. That includes its (i) properties, (ii) plant performance as well as (iii) sustainability concerns that in turn include economic, ecological and social aspects (Photo and design, © Gruda).

The developments of materials used within the current range of standard as well as alternatives substrates are the focus of two chapters of the book, namely Chapters 3 and 4. Here the essential requirements and the ways of assessing their suitability are addressed together within a review of renewable alternatives for peat as a finite resource and rock wool with disposal problems. Apart from inorganic materials, several organic materials such as bark, wood fibres, sawdust, coir, green compost, rice hulls, biochar, hydrochar are used as alternatives. In most cases, these components are used to optimise substrate selection/use and to build the new generation of growing media. In both chapters, after a short introduction to preparation and occurrence, growing media properties, plant performance and sustainability concerns are presented (Fig. 3). Apart from growing media properties, their ability to perform well in real growing conditions is crucial (Gruda et al. 2016a; Barrett et al. 2016; Gruda 2019). For instance, according to Knight (2012), one of the barriers to changing growing media within commercial horticulture is based on compatibility with current irrigation systems.

Environmental concerns, as well as challenges related to an anticipated global climate change, have led to a direct criticism of the use of peat and the voices to stop its use have grown louder. Therefore, efforts are being made to replace the high proportion of peat with alternative substrates. Consequently, Switzerland is looking to reduce the use of peat to 5%. Its use has reduced enormously in Austria and the UK, whereas in Germany a reduction strategy has foreseen a drastic reduction of peat, both in hobby as well as in commercial horticulture production. Reducing the use of peat in substrates is an essential contribution to reducing global greenhouse emissions. The amount of $CO_2$ equivalent produced, by use of peat in growing media, in Central Europe is comparable with that of domestic civil flight transportation per year and country. Whoever uses the chance to cultivate peat-free, can achieve advantages in plant quality, image and finally in turnover. Some farms are already producing peat-free plants because they market it under the environmental label, and the $CO_2$ footprint is a crucial criterion to do so.

Nowadays, some raw materials such as wood, and bark but also other feedstocks are not only used for production of growing media, but also for energy purposes. In the future, the tendency for competing applications will be more expanded. Both applications will support the mitigation of climate changes in a sustainable way. However, resources will more strongly compete with each other, and the possibility to extend the amount and/or the area of cultivation will be much more limited. As a consequence, we predict that prices will rise, as soon as no innovation halts (Gruda 2019).

Physical, chemical and biological properties of growing media are essential for plant growth and development. Three chapters of this book address these issues. The combinations of all these, provide the tools for the engineering of new media of the future.

Physical properties of growing media are the starting points regarding the design of new growing media. These properties can have a substantial impact on air and water availability for plant roots. However, the range of values for physical properties that growing media used are different from those in mineral soils, showing higher total pore volume. This difference makes growing media attractive (Gruda et al. 2016a). On the other hand, the volume of substrate per plant is relatively small and physical properties cannot be changed by growers during the cultivation period, at least not improved (Savvas and Gruda 2018). Apart from direct physical properties, container geometry, specifically container height, indirectly influences air/water-ratio and the hydraulic water condition in a container. Chapter 5 of this book addresses all these issues, by also emphasising the need for additional dynamic parameters, to design, optimise and guide the manufacturing of the next generations of substrates. Root growth, the microorganism activity, but also cultivation methods influence the hydraulic properties of growing media. While years ago, the static parameters were in the foreground, nowadays more and more the request has changed to analyse physical properties of growing media in situ, by including changes and performances during the plant growth and alternating water levels. In this concern, dynamic parameters are gaining more and more importance.

The most important criteria for evaluation of chemical properties are pH, cation exchange capacity and nutrient concentrations (Gruda et al. 2016a). Unlike physical, chemical properties could be adjusted at a certain level by growers at the beginning and during the cultivation period. Chapter 6 addresses these questions and new developments related to chemical properties of growing media.

On the other hand, growing media must be free from pests, pathogens and weeds, biologically stable, and non-toxic (Savvas and Gruda 2018). Moreover, since the new renewable organic materials are increasingly used as growing media, knowledge of biological characteristics has experienced considerable development. Chapter 7 addresses such new development concerning biological properties of growing media.

There is a wide assortment of typologies of SCS, ranging from simple to complicated systems. Some of them use solid inorganic and organic substrates (see Chapters 3 and 4 of this book). In contrast, some other systems use a nutrient film technique (NFT), plant-plain-hydroponics and aeroponics. Chapter 8 of this book presents the culture systems used in today's horticulture.

In SCS, improved water and fertiliser use efficiency is achieved as in conventionally soil-based cultivation methods. Recirculation and irrigation systems play an enormous role together with water quality. Sustainability and circularity can be further improved in closed and semi-closed greenhouses and by applying new cultivation methods and both greenhouse and plant growth technologies. New methods such as computational intelligence are adapted and implemented by learning a specific task from experimental work for automatic monitoring and control. These requirements are approached in Chapter 9 as well as to some degree in Chapter 10 of this book, addressing modelling of the nutrient management for SCS. As explained in Chapter 9 of this book, irrigation and fertigation are two critical inputs, to control plant development, fruit yield and the quality of produce. However, automated fertigation units have to be adapted to greenhouse environment conditions, crop types and growth stages, types of SCS and physical and hydraulic properties of individual substrates that are continually changing. The application of new approaches, such as the Internet of Things makes it possible to increase water saving to more than 30% and up to 80% in some nutrient elements. Trained deep learning models and artificial intelligence can be useful for nutrient management. Through algorithms, it is possible to predict and plan the management of nutrient solutions, reducing resource waste and offering unprecedented cultivation information (Chapter 10). However, the application of new technologies is further a question of investments and economic incentives.

For urban areas, the weight of the system is essential and could be a disadvantage, for instance, concerning rooftop installation. Chapter 11 describes the most recent innovation in hydroponic technologies for plant cultivation within cities and their adaptability to urban fabrics. Examples include rooftop farms or inside building farms through the use of artificial lighting. In these so-called plant factories, vegetables are cultivated indoors under fully controlled conditions. Computers and sensors control and measure main environmental factors that affect vegetable growth, such as lighting, carbon dioxide concentration, relative humidity and plant surface temperature (Chang et al. 2011; Eigenbrod and Gruda 2015). As the production systems are not directly dependent on soil and climate factors, cultivation can take place all year round. Thus, weather extremes are excluded and a better control of plant health can be achieved. Furthermore, food produced in urban areas usually does not have as many 'food miles' as rurally cultivated vegetables (Eigenbrod and Gruda 2015).

Gruda (2009) and Gruda et al. (2018) reported that an adaptation of cultural management to the specific cultural system of soilless culture, as well as greenhouse environmental demands (Gruda 2005), can further result in an improvement of the quality of horticultural products. Thus, SCS can serve as a tool to increase secondary plant metabolites and other relevant compounds in plant foods, and medicinal and aromatic plants. Here a compromise between yield and the quality of produce needs to be established because high yields do not automatically imply high quality. The aim is, therefore, to produce high-quality horticultural products that fit the expectations of producers, traders and consumers. Several growth factors such as climate conditions, cultural practices, including macronutrient and micronutrient concentrations, and/or the use of beneficial microorganisms could be used to induct positive stress, so-called eustress, to optimise the produce quality. The eustress used can be of physical, chemical or biological origin. Chapter 12 of this book presents the diversity of methods used and achieved results. As the authors described, the quality of horticultural produce in SCS is not only better compared to soil. The nutrient solution composition, its salt content, its pH and temperature, the fertigation management and harvesting methods make it possible to further improve the quality of produce for both vegetables and ornamental plants. Besides, vegetables could be biofortified with one or more elements promoting consumer health and/or some elements could be tailored for people with special dietary requirements.

Nearly all chapters of this book include a section, called 'case study', providing concrete and useful practical information in giving matters. Besides, Part 3 of the book, addresses advances in soilless cultivation of tomatoes (see Chapter 13), strawberries (see Chapter 14), and ornamentals (see Chapter 15) as case studies, emphasising good agricultural practices and best horticultural principles. These chapters are, therefore, a concretisation of both previous parts that lead the reader in a practical way to all details needed to grow selected crops in SCS. This part brings together the current best values of growing media, soilless culture and horticulture to create an essential reference for growers, substrate and greenhouse industry, researchers and scientists.

## 6 Conclusions and future trends

SCS are rapidly expanding and present a sector with unlimited potential for the horticulture industry. They are environmental-friendly, resource-efficient, and intensive tools for sustainable production. Future trends are very well documented in every chapter in this book for each related topic. Powered by the circular economy and a smart-climate vision, there is no doubt that the future will require not only well-designed novel growing media but also innovative technologies that will be able to recover nutrients from organic waste streams. The bio-sense of root-zone environment will further allow more cost-effective, precise and adequate real-time decision making in comparison to today methods. The coordination of possible interactions with other plant growth factors will finally contribute to yield increases by improving product quality. Further research work on these issues, but also on the accounting of GHG emissions and life-cycle assessments will most probably accompany us for the next few years and decades. It remains vital to identify and further develop sustainable materials, technologies and approaches while keeping energy and production costs low and transportation distances short.

## 7 Where to look for further information

Standard works:

The primary sources for further information are the chapters of the current book. Further, Raviv et al. (2019); Gruda et al. (2013, 2016a,b), and Savvas and Passam (2002) (see publications in the References for full details) are recommended as lectures.

Conferences and websites:

The International Society for Horticultural Science (ISHS; https://www.ishs.org/) comprises the Division Protected Cultivation and Soilless Culture that in turn includes three working groups that are primarily involved with soilless culture and growing media: (i) Substrate Analysis, (ii) Hydroponics and Aquaponics, (iii) Growing Substrates. ISHS organises some international symposia as follows:

- International Symposium on Growing Media, Soilless Cultivation and Compost Utilization in Horticulture,
- International Symposium on Soilless Culture,
- International Symposium on Growing Media, Composting and Substrate Analysis.

There are also other symposia regularly organised by the ISHS. To be mentioned are the International Symposium on Advanced Technologies and Management for Innovative Greenhouses (GreenSys), International Symposium on Horticulture in Europe (SHE) and South-Eastern Europe Symposium on Vegetables and Potatoes. They always include sessions dedicated to soilless culture and growing media. These ISHS-symposia published their proceedings in Acta Horticulturae (https://www.actahort.org/), which is a very informative source on current advances in SCS and growing media.

## 8 References

Barrett, G. E., Alexander, P. D., Robinson, J. S. and Bragg, N. C. 2016. Achieving environmentally sustainable growing media for soilless plant cultivation systems–a review. *Sci. Hort.* 212, 220–234. doi: 10.1016/j.scienta.2016.09.030.

Beerling, E. A. M., Blok, C., van der Maas, A. A. and van Os, E. A. 2014. Closing the water and nutrient cycles in soilless cultivation systems. *Acta Hortic.* 1034(1034), 49–55. doi: 10.17660/ActaHortic.2014.1034.4.

Benke, K. and Tomkins, B. 2017. Future food-production systems: vertical farming and controlled-environment agriculture. *Sustainability Sci. Pract. Policy* 13(1), 13–26, doi: 10.1080/15487733.2017.1394054.

Chang, Y. W., Lin, T. S., Wang, J. C., Chou, J. J., Liao, K. C. and Jiang, J. A. 2011. The effect of temperature distribution on the vertical cultivation in plant factories with a WSN-based environmental monitoring system. *Adv. Biomed. Eng.* 3-5, 234–240.

di Lorenzo, R., Pisciotta, A., Santamaria, P. and Scariot, V. 2013. From soil to soil-less in horticulture: quality and typicity. *Ital. J. Agron.* 8(4), 30. doi: 10.4081/ija.2013.e30.

Dias, G. M., Ayer, N. W., Khosla, S., Van Acker, R., Young, S. B., Whitney, S. and Hendricks, P. 2017. Life cycle perspectives on the sustainability of Ontario greenhouse tomato production: benchmarking and improvement opportunities. *J. Clean. Prod.* 140, 831–839.

Eigenbrod, C. and Gruda, N. 2015. Urban vegetable for food security in cities: a review. *Agron. Sustain. Dev.* 35(2), 483–498. doi: 10.1007/s13593-014-0273-y.

FAO. 2020. Home gardens/vertical farming, hydroponics and aquaponics. Available at: http://www.fao.org/land-water/overview/covid19/homegardens/en/. Accessed August 2020.

Gericke, W. F. 1929. Aquaculture, a mean of crop production. *Am. J. Botany* 16, 862.

Gruda, N. S. 2019. Increasing sustainability of growing media constituents and stand-alone substrates in soilless culture systems. *Agronomy* 9(6), 298. doi.org/10.3390/agronomy9060298.

Gruda, N. 2009. Do soilless culture systems have an influence on product quality of vegetables? J. Appl. Bot. & Food Qual. 82, 141–147. https://ojs.openagrar.de/index.php/JABFQ/article/view/2093.

Gruda, N. 2005. Impact of environmental factors on product quality of greenhouse vegetables for fresh consumption. *Crit. Rev. Plant Sci.* 24(3), 227–47. https://doi.org/10.1080/07352680591008628.

Gruda, N. and Tanny, J. 2014. Protected crops, pp. 327–405. In: *Horticulture: Plants for People and Places* (Vol. 1) Dixon, G. R. and Aldous, D. E. (Eds). Springer, Netherlands. doi: 10.1007/978-94-017-8578-5_10.

Gruda, N., Bisbis, M. B. and Tanny, J., 2019a. Impacts of protected vegetable cultivation on climate change and adaptation strategies for cleaner production – a review. *J. Clean. Prod.* 225, 324–339. https://doi.org/10.1016/j.jclepro.2019.03.295.

Gruda, N., Bisbis, M. B. and Tanny, J., 2019b. Influence of climate change on protected cultivation: impactsand sustainable adaptation strategies - a review. *J. Clean. Prod.* 225, 481–495. https://doi.org/10.1016/j.jclepro.2019.03.210.

Gruda, N., Savvas, D., Colla, G. and Rouphael, Y. 2018. Impacts of genetic material and current technologies on product quality of selected greenhouse vegetables–a review. *Europ. J. Hortic. Sci.* 83(5), 319–328. doi: 10.17660/eJHS.2018/83.5.5.

Gruda, N., Caron, J., Prasad, M. and Maher, M. J. 2016a. Growing media, pp. 1053–1058. In: *Encyclopedia of Soil Sciences* (3rd edn.) Lal, R. (Ed.). CRC Press Taylor & Francis Group, Boca Raton, FL, USA.

Gruda, N., Gianquinto, G., Tüzel, Y. and Savvas, D. 2016b. Soilless culture, p. 2804. In: *Encyclopedia of Soil Sciences* (3rd edn.) (Vols. 533–537) Lal, R. (Ed.). CRC Press Taylor & Francis Group.

Gruda, N., Qaryouti, M. M. and Leonardi, C. 2013. Growing media, pp. 271–302. In: *Good Agricultural Practices for Greenhouse Vegetable Crops–Principles for Mediterranean Climate Areas*. (Plant Production and Protection Paper 217). Food and Agriculture Organization, Rome, Italy.

Hoagland, D. R. and Arnon, D. I. 1950. *The Water-Culture Method for Growing Plants without Soil*. California Agricultural Experiment Station, Circular 347. Berkeley: College of Agriculture, University of California.

Hong, J. and Gruda, N. 2020. The potential of introduction of Asian vegetables in Europe. *Horticulturae* 6(3), 38. https://doi.org/10.3390/horticulturae6030038.

hortidaily 2020. Vertical Farming: The answer to global food security? https://www.hortidaily.com/article/9225447/vertical-farming-the-answer-to-global-food-security/. Accessed on 11. December 2020.

Incrocci, L., Malorgio, F., Della Bartola, A. and Pardossi, A. 2006. The influence of drip irrigation or subirrigation on tomato grown in closed-loop substrate culture with saline water. *Sci. Hortic.* 107(4), 365–372.

Jones, J. B. Jr. 1982. Hydroponics: its history and use in plant nutrition studies. *Nut. R.* 5(8), 1003–1030. doi: 10.1080/01904168209363035.

Knight, A. 2012. Towards sustainable growing media. Chairman's report and roadmap, sustainable growing media task force. https://assets.publishing.service.gov.uk/government/uploads/system/uploads/attachment_data/file/221019/pb13867-towards-sustainable-growing-media.pdf.

Knop, W. 1865. Quantitative Untersuchungen über die Ernährungsprozesse der Pflanzen. *Landwirtsch. Vers. Stn.* 7, 93–107.

Massa, D., Magán, J. J., Montesano, F. F. and Tzortzakis, N. 2020. Minimizing water and nutrient losses from soilless cropping in southern Europe. *Agric. Water Manag.* 241. doi: 10.1016/j.agwat.2020.106395.

Morrison, R. 2020. Feed the World with Foam! Old and Discarded Mattresses Are Being Used to Replace Soil and Grow Vegetables for Refugees in Desert Environments. *Mailonline*.

Mugnozza, G. S., Russo, G. and De Lucia Zeller, B. 2007. LCA methodology application in flower protected cultivation. *Acta Hortic* 761, 625–632.

Nederhoff, E. and Stanghellini, C. 2010. Water use efficiency of tomatoes–in greenhouses and hydroponics. *Pract. Hydroponics Greenh.*, 115, 53–59.

Ntinas, G. K., Neumair, M., Tsadilas, C. D. and Meyer, J. 2017. Carbon footprint and cumulative energy demand of greenhouse and open-field tomato cultivation systems under Southern and Central European climatic conditions. *J. Clean. Prod.* 142, 3617–3626.

Page, G., Ridoutt, B. and Bellotti, B. 2012. Carbon and water footprint tradeoffs in fresh tomato production. *J. Clean. Prod.* 32, 219–226.

Payen, S., Basset-Mens, C. and Perret, S. 2015. LCA of local and imported tomato: an energy and water trade-off. *J. Clean. Prod.* 87, 139–148.

Raviv, M., Lieth, J. H. and Bar-Tal, A. 2019. *Soilless Culture: Theory and Practice*. Academic Press, 712 pp. ISBN: 9780444636966.

Reina-Sanchez, A., Romero-Aranda, R. and Cuartero, R. 2005. Plant water uptake and water use efficiency of greenhouse tomato cultivars irrigated with saline water. *Agric. Water Manag.* 78(1-2), 54–66.

Sachs, J. 1860. Berichte über die physiologische Tätigkeit an der Versuchsstation in Tharandt. IV. Vegetationsversuche mit Ausschluss des Bodens über die Nährstoffe und sonstigen Ernährungsbedingungen von Mais, Bohnen und anderen Pflanzen. *Landwirtsch. Vers. Stn.* 2, 219–68.

Savvas, D. and Gruda, N. 2018. Application of soilless culture technologies in the modern greenhouse industry - A review. *Eur. J. Hortic. Sci.* 83(5), 280–293. https://doi.org/10.17660/eJHS.2018/83.5.2.

Savvas, D. and Passam, H. 2002. *Hydroponic Production of Vegetables and Ornamentals*, pp. 463. Embryo Publications, Athens, Greece.

Schnitzler, W. H. and Gruda, N. 2002. Hydroponics and product quality. In *Hydroponic Production of Vegetables and Ornamentals*, D. Savvas, and H. C. Passam (Eds). Embryo Publications, Athens, Greece, pp. 373–411.

Shaw, P. 1725. *The Philosophical Works of the Honourable Robert Boyle*. W. Inuys and J. Inuys, London, Vol. 1, pp. 248–249 (originally from Boyle's Origin of form and qualities according to the corpuscular philosophy, 1666).

Torrellas, M., Antón, A., López, J. C., Baeza, E. J., Parra, J. P., Muñoz, P. and Montero, J. I. 2012. LCA of a tomato crop in a multi-tunnel greenhouse in Almeria. *Int. J. Life Cycle Assess.* 17(7), 863–875.

Torrellas, M., Antón, A. and Montero, J. I. 2013. An environmental impact calculator for greenhouse production systems. *J. Envir. Manag.* 118C, 186–195. doi: 10.1016/j.jenvman.2013.01.011.

UN. 2018. The world's cities in 2018. Available at: https://www.un.org/en/events/citiesday/assets/pdf/the_worlds_cities_in_2018_data_booklet.pdf. Accessed 31 August 2020.

Valenzano, V., Parente, A., Serio, F. and Santamaria, P. 2008. Effect of growing system and cultivar on yield and water-use efficiency of greenhouse-grown tomato. *J. Hortic. Sci. Biotechnol.* 83(1), 71–75.

van Kooten, O., Heuvelink, E. and Stanghellini, C. 2004. Nutrient supply in soilless culture: on-demand strategies. *Acta Hortic.* 659(659), 533–540. doi: 10.17660/ActaHortic.2004.659.69.

van Os, E. A., van Weel, P. A., Wilms, J. A. M., Bruins, M. A., Verhoeven, J. and van Wijk, K. 2017. Hydroponically grown leek: slowly growing to maturity. *Acta Hortic.* 1176(1176), 79–86. doi: 10.17660/ActaHortic.2017.1176.11. Available at: www.hortidaily.com/article/9225447/vertical-farming-the-answer-to-global-food-security/. Accessed 13 July 2020.

Woodward, J. 1699. Thoughts and experiments on vegetation. *Philos. Trans. R. Soc. Lond.* 21, 193–227.

Yep, B., Gale, N. V. and Zheng, Y. 2020. Comparing hydroponic and aquaponic rootzones on the growth of two drug-type cannabis sativa L. Cultivars during the flowering stage. *Indust. Crops Prod.* 157. doi: 10.1016/j.indcrop.2020.112881.

# Part 1

## Materials

# Chapter 2

## Advances in understanding plant root behaviour and rootzone management in soilless culture systems

*Youbin Zheng, University of Guelph, Canada*



## 1 Introduction

When a plant is growing in soil, roots are essential for (1) anchoring and physical support, (2) taking up and conducting water and nutrients, and (3) storage of food. When a plant is grown in a soilless culture system (SCS), the anchoring may be less important, for example, in solution culture and aeroponics, and the primary function of the root is water and nutrient uptake and conduction. When a plant is grown in soil in the open field, water and nutrient can be limited, which is why the majority of root-related research has been focussed on how to get plant roots to be more efficient in reaching and taking up water and nutrient, and on how to reduce water and nutrient stresses (see Section 9 'Where to look for further information'). In SCSs, water and nutrient are continuously supplied targeting at the optimal levels; therefore, water and nutrient limitation is of less concern than overly supplied. In soil, roots can have penetration concerns when the soil is too hard, for example, in hard clay soil, SCS do not have this

http://dx.doi.org/10.19103/AS.2020.0076.01

issue. However, since most of the SCSs use containers, the container size and shape can create certain restrictions for root growth and development (Volder and Van Iersel, 2019).

When growing in soil, plant rhizosphere is the narrow region of soil that is directly influenced by root exudates and micro-organisms. Roots can play an essential role in converting some of the mineral nutrients in the rhizosphere from unavailable forms to available forms for plant uptake. However, when the plant is growing in a SCS, water/nutrient solution is frequently added to the growing medium, or moving around the root surface in solution culture or aeroponics. Therefore, one would speculate that the rhizosphere is much narrower than that in soil, potentially making it less important for plant nutrient uptake or much more challenging to study. Instead, considering the whole rootzone is a more appropriate approach for a SCS. The rootzone is defined as the environment where roots reside (Zheng, 2020). In a SCS, the rootzone can be the environment within the container containing growing media, or the nutrient solution in direct contact with the roots in the case of solution culture or aeroponics. The rootzone is affected by root exudate and micro-organisms as well; however, the significance can, in some situations, be less than the influence from the supplied nutrient solution.

In soilless systems, a healthy root system is essential for water and nutrient uptake. The majority of crops grown in soilless systems are for the aboveground parts, either as food, for example, fruit vegetables, leaf greens, or as ornamentals, for example, cut flowers, potted ornamentals, except for crops producing roots as the ultimate goal, such as *Echinacea* as medicine (Zheng et al., 2006) and sweet potato as food (Sakamoto and Suzuki, 2018). It is well-known that roots grown in different systems, in soil or different SCSs, can have different morphologies and hydraulic conductivities (Graves, 1992). Therefore, root-related research for a SCS should have different emphases than for soil-grown plants. However, so far, the vast majority of the root-related research has been focussed on plants grown in soil either naturally or for agricultural purpose, even there are some studies that used SCSs for easy root observation and measurements. Therefore, this chapter will discuss the advances in understanding roots in all the systems and emphasise on how to apply the relevant knowledge to soilless systems, and it will also discuss how to manage the rootzone for achieving different objectives of plant cultivation in soilless systems.

## 2 How a root takes up water and nutrients

When a plant is growing in a soilless system, the main driving forces for water flow from the rootzone through the plant to the atmosphere are the differences (1) in water vapour concentrations between leaf and air, (2) in pressure potentials in the xylem conduits, (3) in water potentials of the living cells in the root, and (4) in water potentials between the rootzone and the root. In soil, quite often there is high negative water potential which can limit water uptake by the root; however, in the rootzone of a SCS, negative water potential is not commonly high enough to lead to limit root water uptake. Water should be readily available in the nutrient solution surrounding the root surfaces, such as in NFT or in large pores of a growing medium, as free water.

When water is taken up from the rootzone to the root and moved up to the leaves, the nutrients dissolved in water are also taken up and transported to the other parts of the plant. In the rootzone, nutrients move to the root surface either by bulk flow or by diffusion. Nutrients are carried in water flow from the rootzone to the root in bulk flow, and by diffusion from a high concentration region in the rootzone to a low concentration region close to the root surface. One would speculate that bulk flow is dominant in solution culture and aeroponics, and at the time right after each fertigation event in a system using growing media, diffusion becomes important only when the growing medium's water content becomes very low.

Plant nutrient uptake has three characteristics: (1) selective, (2) accumulative, and (3) genotype dependent. It is important to keep these characteristics in mind when operating a SCS, since these systems require a constant supply of water and nutrients to the plant rootzone. Increasingly, the nutrient solution in soilless systems is encouraged to be recycled and reused for both environmental and economic reasons (Zheng, 2018). When the nutrient solution has been reused, certain ions can accumulate, and certain ions can be depleted over time, even when a balanced feeding solution is initially used (Savvas and Gruda, 2018). This is due to the selective nutrient uptake characteristics of roots. For example, if KCl is used to make nutrient solution, plants typically take up more $K^+$ than $Cl^-$. Eventually, $Cl^-$ will accumulate in the solution to a detrimental level if the solution is not amended. Current water treatment technologies are not effective in removing only $Cl^-$ without removing other ions from a nutrient solution; therefore, eventually, the $Cl^-$-loaded nutrient solution has to be discharged into the environment. It is recommended to limit the additions of ions that are not taken up and used by plants in large amount, and these ions can include $Na^+$, $Cl^-$, and $SO_4^{2-}$.

Different plant species and even the same species at different development stages may have different nutrient uptake and requirements. It is important to provide different recipes for different plant species at their different development stages (Savvas and Gruda, 2018). However, currently, there is limited research on nutrient uptake of different species at their respective development stages (Albornoz and Lieth, 2016). The mismatch between ion ratio of the supplied nutrient solution and the ratio of uptake by the plant can cause nutrient solution to be short-lived, with imbalanced nutrients and need to be discharged after a short-term of use. Root can take up sufficient nutrient from low concentration solution (accumulative characteristic) to sustain normal plant growth, as long as the nutrient is continuously supplied without any local depletion at the root surface. Water and nutrient delivery methods can take advantage of this characteristic for different purposes. This is particularly true in NFT and aeroponic systems (Adams, 1992). In soilless cultivation, high concentration nutrient solution often causes disorders (e.g. salt stress, root and stem rot) and reduces plant growth and productivity. For example, stem rot is often observed in potted gerbera production when plants are sub-irrigated with high-concentration nutrient solution. Zheng et al. (2004) explored whether lower nutrient concentrations can be utilised. Potted gerberas were fertigated with different concentrations of nutrient solution, namely 100%, with a total N of 12.7 mM, 50%, 25%, or 10% of full strength in a sub-irrigation system. The results demonstrate that the commonly used full-strength nutrient concentration can be reduced to 50% without affecting leaf tissue N, P, K, Ca, and Mg concentrations and plant performance. Besides, the full-strength nutrient solution causes high salt accumulation in the upper section of the growing medium in the container, which may harm plant health.

In SCS, root nutrient uptake can generally be described by the Michaelis-Menten equation:

$$I = \left(I_{max} \times C\right) / \left(Km + C\right)$$

where I is the ion uptake rate, $I_{max}$ is the maximum ion uptake rate, C is the concentration of the ion in the nutrient solution, and Km is the Michaelis constant.

The uptake rate increases almost linearly with the increase of nutrient concentration when the concentration is low, and gradually the increase rate reduces to zero (uptake rate reach $I_{max}$) when the concentration is reaching a certain level. As reviewed by White (2012), the root nutrient uptake consumes considerable amount of energy of the plant, especially when nutrient ions, such as $K^+$, $NH_4^+$, $NO_3^-$, and $SO_4^{2-}$, are in high concentration in the rootzone. It can consume energy for useless cycling of ions across the plasma membrane of root cells. It is well-known that at high nutrient supply, plant can have the least nutrient-use efficiency. Therefore, it is important to determine the optimal nutrient application rate for soilless plant production based on the production system and plant species. Even though this topic is important, it is often either ignored or there is a lack of funding, which has resulted in a limited number of publications on a limited number of crops, for example, cannabis and some container-grown ornamental plants, in this aspect (Caplan et al., 2017a,b; Zheng et al., 2010, 2018).

## 3 How environmental factors influence root water and nutrient uptake

Many environmental factors can influence root water and nutrient uptake. The following will briefly discuss some of the significant relevant environmental factors.

### *3.1 Vapour pressure deficit*

One of the major driving forces for root to take up water and nutrient is the vapour pressure difference between the leaf and the air, the vapour pressure deficit (VPD). When VPD is high, water is readily transpired from the leaf to the air which can pull water with dissolved nutrients from the rootzone to the plant root and up to the leaves. However, when VPD is too high, and water cannot flow quick enough from the rootzone to the root, plant leaves start to close their stomates. When VPD is low (high humidity), root uptake of nutrients can be negatively affected as well. For example, as reviewed by Adams (1992), when growing tomato in a high relative humidity (>90%) environment, leaf Ca content dramatically decreased. VPD can influence water and nutrient uptake, and at the same time it can also affect plant photosynthesis, diseases and insects, including pollinators (Shamshiri et al., 2017); therefore, when setting VPD range in a soilless production system, it should take a balanced approach.

The growing root-tip region is much more active in water uptake than the non-growing parts of the root; nutrients can be taken up either through the growing tip region or the entire surface of the root, depending on the plant species and nutrients. That is why it is important not to damage the root tips when transplanting, or to avoid a high VPD environment for the first few days right after transplanting if the root tips are disturbed. Due to this reason and other logistic considerations, modern SCS usually produce transplants using different plugs.

### *3.2 Lighting*

Lighting can affect water and nutrient uptake in different ways. For example, both light quality and intensity can (1) control the opening of stomates to increase or decrease the driving force for root water, and nutrient uptake, (2) affect photosynthesis; therefore, the energy and material supply for root growth, water and nutrient uptake, (3) affect root initiation and growth.

It is well established that up to a limit, plant water and nutrient uptake increases with an increase of light intensity. For example, Adams (1992) demonstrated that in low light-intensity environments (e.g. winter months in the northern hemisphere), root water and some nutrient (e.g. N, K) uptake has a positive correlation with light intensity. However, research on how different light qualities (i.e. spectrum and spectra combinations) affect plant root growth, water and nutrient uptake has only started to be active recently as light-emitting diodes (LED) are becoming economically feasible to be used in commercial plant cultivation. Barickman et al. (2020) grew baby green Chinese kale (*Brassica oleracea* var. *alboglabra*) in a SCS under either fluorescence light or LEDs with different ratios from blue (B) to red (R) (1:9, 2:8 or 4:6). They found that the shoot tissue P, S, K, Ca, and Mg concentrations increase under the LEDs comparing to those under the fluorescence light. However, for the root tissue concentration, K increases, Mg decreases under LEDs, but there is no light treatment effect on Ca or P uptake, especially when half-strength Hoagland solution was used. Clavijo-Herrera et al. (2018) grew lettuce in soilless system under LEDs with different B and R ratios (0:10, 7:93, 26:74; 42:58; 66:34; or 100:0) and a broad-spectrum (B19: green (G) 43: R38) with the same photosynthetic photon flux density (PPFD) of 270 ± 5 $\mu mol \cdot m^{-2} \cdot s^{-1}$, and found that water-use efficiency generally decreases in response to blue light. However, nitrogen uptake was not affected by light quality. They indicate that the decreased water-use efficiency is due to increased stomatal conductance in response to blue light. Pinho et al. (2017) grew lettuce under HPS or four different LEDs with different combinations of red, blue and far-red spectra, and they found that, in general, the LEDs are better than the HPS in promoting the growth and nutrient uptake of plants. LEDs with a combination of the far-red, deep-red and blue have the highest growth and nutrient uptake indices. Zhao et al. (2019) grow *Larix principis-rupprechtii* seedlings under different fertilisation levels, and two LED spectra (R, G, and B ratios). They found that there are light and nutrient interaction effects on root dry weight, uptake of N and P, and the nutrient utilisation efficiencies.

Light can also influence the root initiation and growth, as demonstrated in some vegetative propagation research. It is generally accepted that there is an optimal light intensity for different species; however, optimal light quality depends on species, culture environment, light intensities and so on (Christiaens et al., 2016). For example, Morini et al. (1990) used rooted 'Leccino' olive cuttings under different colours (white, red, yellow, green, and blue) of fluorescence lights, and found that cuttings under yellow light have higher percentage rooting and longer root length than those under the other lighting treatments. Olschowski et al. (2016) showed that *Calibrachoa* cuttings rooted in soilless plugs have higher root dry weight under the high pressure sodium (HPS) light and the LED with a mix of white, blue, and red than those under the red-, blue-, or white-only LEDs. Also, the plants generally have higher root dry weight and total root length under higher light intensity (PPFD of 80 $\mu mol\ m^{-2}\ s^{-1}$) comparing to their counterparts with lower light intensity (40 $\mu mol\ m^{-2}\ s^{-1}$).

Since different research used either different plant species or different light spectra combinations, based on the limited available literature, it is currently difficult to generalise which spectrum or spectra combination is the best in promoting root initiation, root growth, root water and nutrient uptake. What is clear is that both light quality and intensity can have an effect on water and nutrient uptake, and the effect is species, light and environment dependent. More research is needed to determine the optimal light intensity and quality for different economically important species in SCS in the aspects of root growth, water and nutrient uptake.

### 3.3 Air and rootzone temperature

Different plants require different temperature ranges for growth and development, and also may have different optimal temperatures for water and nutrient uptake. How air temperature affects plant water and nutrient uptake has been discussed in many publications (Giri et al., 2017; Gruda, 2005); this chapter will mainly focus on rootzone temperature effects. Adams (1989) demonstrated that increasing root temperature of tomato plants in a soilless system from 16°C to 28°C, increases water, N, P, K, Ca, Mg uptake by 8–33%, however, without a change in N/K ratio. Tindall et al. (1990) grew tomato plants under six rootzone temperatures (10°C, 15.6°C, 21.1°C, 26.7°C, 32.2°C, and 37.8°C) with the same ambient air temperature of 21.1°C, and found that nutrient uptake for all nutrient elements generally peak at 26.7°C. Temperature has significant effects on the uptake of all mineral elements except B, Fe, and Mo which do not respond to temperature. They also found that the plant growth and water uptake peak at 25°C. By growing cucumber in hydroponic system with rootzone temperatures of 12°C, 20°C, 28°C, or 36°C, Daskalaki and Burrage (1998) found that water uptake and plant growth are highest at 28°C. All measured nutrient (N, P, K, Ca, and Mg) uptake increase rapidly from 12°C to 20°C. From 28°C to 36°C the uptake of P and Ca continue to increase, but there is little effect on the uptake of N, K, and Mg. Yan et al. (2012) grew *Cucumis sativus* L. in hydroponic system with the rootzone temperature at either 10°C or 20°C. The authors found that the low rootzone temperature reduces N, P, and K translocation to the shoots and more nutrients accumulate in roots than at higher rootzone temperature. The higher rootzone temperature also increases plant biomass and mineral nutrient uptake. When air temperature is low, rootzone heating can improve root nutrient uptake and tomato yield (Kawasaki et al., 2014), and this could be an alternative way to heat SCS rather than heating the air. Al-Rawahy et al. (2019) grew cucumber in hydroponic system at rootzone temperatures of 22°C, 25°C, 28°C, and 36°C, and found that cucumber growth and yields are the best at 22°C and 25°C. Research also demonstrates that when roots are exposed to a high temperature for short term, it increases water and nutrient uptake through a decrease in water viscosity. However, when exposed to a high temperature for long term, it reduces water and nutrient uptake (Affan et al., 2005). As demonstrated by Giri et al. (2017), severe short-term heat stress in tomato plants can also decrease total root protein concentration and levels of nutrient-uptake and nutrient-assimilation proteins.

It is necessary to determine the optimal rootzone temperature for different economically important species in SCS. However, many species currently still do not have this type of data available in the scientific literature.

## 3.4 Rootzone pH

Rootzone pH can affect nutrient availability, micro-organism and root growth, which therefore affects root water and nutrient uptake. Different plant species can also affect rootzone pH differently, which in turn, influences root nutrient uptake. Many nutrients, such as phosphorus, iron, manganese, and calcium, have different forms under different pH conditions. Some of the forms are readily available for plant uptake, and some of them are not. It is generally recommended for most of the plant species in soilless system to keep the rootzone pH at 5.5–6.5 since most of the mineral nutrient elements are relatively available at this pH range. Most of the plant species can grow well without any nutrient deficiency or toxicity when a balanced and correct nutrient rate is supplied through nutrient solution. Outside of this pH range, many plants can have nutrient disorder issues. However, when the same species of plants grow in soil, the pH range can be much wider than in soilless production system, which indicates that certain nutrient elements can be available in soil even when the pH is higher than 6.5. For example, phosphorus has the highest availability at pH below 6.2 in soilless media; however, its availability is the highest between pH 6.2 and pH 8 in mineral soil. There is hardly any explanation in the literature on why the optimal pH ranges for nutrient availability in soilless media is around 5.5–6.5, and it is much wider in mineral soils. One may speculate that when the plant is growing in soil, the rhizosphere pH is majorly controlled by root exudates (especially organic acids) and stays acidic (see Section 4), even the bulk soil pH stays high. However, in a soilless system, the rhizosphere is much thinner due to the continual supply of nutrient solution (e.g. many times a day in a greenhouse tomato production system when rock wool is used as growing medium) to the rootzone. Therefore the rootzone pH can be dominantly affected by the supplied nutrient solution and the pH range of the nutrient solution should be between 5.5 and 6.5. However, this is not to say that plant root does not affect rootzone pH in soilless systems.

Different plant species affect rootzone-pH differently. For example, Huang et al. (2001) conducted a soilless trial with seedlings and found that pansy (*Viola × wittrockiana*), petunia (*Petunia × hybrida*), and vinca (*Catharanthus roseus*) raise rootzone pH, but celosia (*Celosia cristata*), tomato (*Lycopersicon esculentum*), and zinnia (*Zinnia elegans*) lower rootzone pH. Johnson et al. (2013) grew seedlings in soilless medium. The authors found that plant species vary in their pH effect, in the order from acidic to basic: geranium (*Pelargonium × hortorum*), Impatiens (*Impatiens wallerana*), and Petunia (*Petunia × hybrida*). Savvas et al. (2006) found that lettuce can increase rootzone pH. Zheng et al. (2006) grew *Echinacea purpurea* and *E. angustifolia* using three different soilless growing substrates and three $NO_3^-/NH_4^+$ ratios and found the rootzone pH remains stable regardless of the species, medium and $NO_3^-/NH_4^+$ ratio. The differences of different species in affecting rootzone-pH can mainly be contributed to their different preference in $NH_4^+$ versus $NO_3^-$ uptake (Dickson et al., 2016; Savvas et al., 2003, 2006). When roots take up $NH_4^+$, they excrete the same equivalent of $H^+$, and when $NO_3^-$ is taken up by roots, they excrete the same equivalent of $OH^-$ or $HCO_3^-$. Therefore, they either decrease or increase the rootzone pH. Rootzone pH, therefore, is not only controlled by the pH of the feeding nutrient solution, but also depending on the $NH_4^+/NO_3^-$ ratio, and the total ionic concentration of the feeding nutrient solution (Zheng et al., 2004, 2006). When potted gerbera (*Gerbera jamesonii*; Zheng et al., 2004) and potted miniature rose (*Rosa chineersis minima;* Zheng et al., 2006) were sub-irrigated with different concentrations of the same nutrient solution for the same crop species, rootzone-pH increases when the nutrient concentration is lower and more stable or decreases when the concentration is higher.

## 3.5 Rootzone oxygen

Adequate rootzone oxygen level is essential for root formation, water and nutrient uptake, growth, and health. Rootzone oxygen level is mainly controlled by the diffusing speed of atmospheric oxygen to the rootzone, and it can also be determined by factors such as the root and microbial respiration, nutrient (Suhl et al., 2019), and temperature. Generally it is not possible to simply predict rootzone oxygen by either rootzone temperature or water content as demonstrated by some of our measurements in greenhouse with plants grown in soilless media under hot summer conditions in Southwest Ontario, Canada (see Fig. 1 as an example). In situ dissolved oxygen (DO), temperature, and water content sensors were inserted into coir slabs with tomato plants. Data were collected once every 15 min over two 24-h periods, one in the middle and one in the end of July, in two greenhouses. Both rootzone water content and temperature were negatively correlated with DO level; however, neither water content nor temperature could be used to predict rootzone oxygen level. Rootzone DO levels were between 5 ppm and 10 ppm for a majority of the times; however, it dropped to <2.5 ppm from time to time as well. DO that is too low or too high can affect root health and growth (Zheng et al., 2007).

As mentioned above, nutrient uptake needs energy and root needs oxygen for respiration to generate some of the energy for the uptake. A rootzone deficient in DO can negatively affect nutrient uptake. Morard et al. (2000) grew

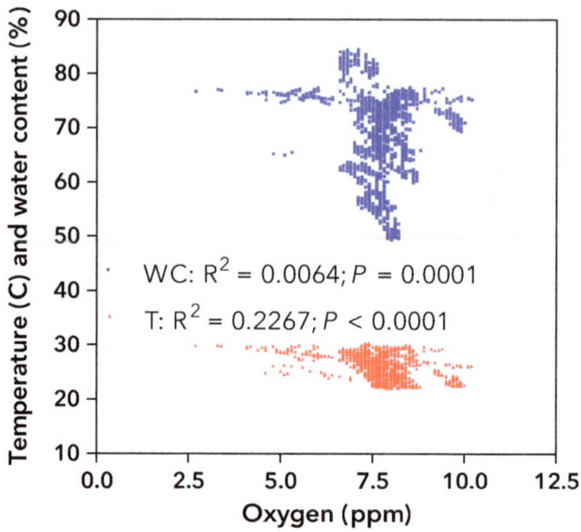


**Figure 1** Relationships between rootzone dissolved oxygen level and temperature (T), or water content (WC) in coir slabs with tomato plants. Each data point is the result of an in situ measurement of either water content or temperature against an in situ measurement of dissolved oxygen. The measurements were conducted over two summer days in two different greenhouses in Southwest Ontario, Canada. $R^2$ is the coefficient of determination. Source: Zheng, unpublished.

tomato plants in a SCS with adequate DO or with DO-deficiency conditions and found that the water uptake is reduced by 20–30% at 48 h from the onset of oxygen deficiency treatment. Rootzone oxygen deficiency also reduces or stops almost all the macronutrient uptake, including $H_2PO_4^-$, $SO_4^{2-}$, $K^+$, $Ca^{2+}$, $Mg^{2+}$, except $NO_3^-$ uptake, which is not affected, at 10 h from the onset of oxygen deficiency treatment. The critical low DO levels are different depending on plant species. For example, Goto et al. (1996) grew lettuce (*Lactuca sativa* L.) in hydroponic system with a DO of 2.1, 4.2, 8.4 or 16.8 mg·L$^{-1}$, and found that there were no treatment effects on the fresh and dry weight of shoots or roots. The authors concluded that the critical low DO level for vigorous lettuce growth is below 2.1 mg·L$^{-1}$. Rootzone oxygen deficiency can cause root diseases such as Pythium root rot, which in turn affects root water and nutrient uptake. Chérif et al. (1997) grew tomato plants in a soilless system, and found that when the rootzone is short of oxygen (with a DO < 3 mg·L$^{-1}$), roots are more susceptible to inoculated *Pythium* F707 comparing with a rootzone with higher DO (>5 mg·L$^{-1}$).

Dissolved oxygen is also essential for root formation and root growth. For example, Soffer and Burger's (1988) rooted cuttings of *Fictis benjamins* L. and *Chrvsanthemum × morifolium* in aero-hydroponics found that lowering DO concentration increases the time required to form adventitious roots, reduces rooting percentages, numbers of roots formed per cutting, and root lengths, for both species. Delay and reduction in root growth can affect water and nutrient uptake for cuttings in the propagation stage.

## 4 Root exudates and allelopathy

Plant roots release about 11% net fixed carbon or 27% of carbon allocated to roots is released to the rootzone (Vives-Peris et al., 2020). About 15–60% of carbon allocated to the roots is used by roots for respiration and released as $CO_2$ (Lambers et al., 2002). Root exudates can include low-molecular weight compounds, primary and secondary metabolites, and high-molecular weight compounds such as mucilage and proteins. Strehmel et al. (2014) detected 103 compounds of root exudates from *Arabidopsis thaliana* grown in a hydroponic system. Common root exudates include sugar, amino acids and other organic acids, nucleosides, flavonoids, glucosinolates, phenolics, salicylic and jasmonic acid catabolites, enzymes and vitamins (Strehmel et al., 2014; Oburgera and Jones, 2018). The types and amount of exudates depend on plant species, ages, rootzone micro-organisms, and the growing environment (Vives-Peris et al., 2020). Root exudates can improve plant resistance to adverse conditions by attracting beneficial microbiota, chelating toxic compounds in the rootzone, changing rootzone pH, or increase the availability of certain nutrient elements (Erro et al., 2010; Vives-Peris et al., 2020).

On the other hand, root exudates from most of the species have inhibitory effects (Narwal, 2004) except for a few species. These exudates can inhibit seed germination, growth or development of the same species (autotoxicity) or other species. This kind of chemicals is called allelochemicals. Allelochemicals can cause different stresses to plants, such as oxidative stress (Cruz-Ortega et al., 2007). They can cause injury to the root, reduce water and nutrient uptake and eventually reduce photosynthesis and retard plant growth (Lyu and Blum, 1990; Blum and Gerig, 2005; Lee et al., 2006; Hussain and Reigosa, 2011; Talukder et al., 2020). Soilless production systems quite often reuse nutrient solution as discussed above. Plant stresses caused by allelochemicals in this type of system have been reported in quite a few studies in the past two decades (Kitazawa et al., 2005; Asao et al., 2004, 2007; Asaduzzaman et al., 2012; Talukder et al., 2020).

The negative effect response and sensitivity of different species to another species' allelochemicals is species-dependent in hydroponic systems as demonstrated by Cheng and Xu (2013). Research to identify which allelochemicals from one species can have a negative effect on another species is important in soilless production in order to avoid reusing the same nutrient solution for those not compatible plant species in the same system. It is also important to develop and select technologies to alleviate the stresses caused by allelochemicals in soilless systems. So far, there are a few technologies that have been explored with some success in alleviating the stresses caused by allelochemicals in soilless production systems. These technologies include the use of micro-organisms to degrade (Asao et al., 2004), active charcoal or Amberlite XAD-4 to adsorb (Lee et al., 2006; Talukder et al., 2020), electro-degradation to degrade (Asaduzzman et al., 2012; Talukder et al., 2020) and slow sand filtration, membrane filtration, photo-catalytic treatment, and heat treatment to remove allelochemicals (Hosseinzadeh et al., 2017). More research needs to investigate the critical levels of accumulated root exudates above which there will be negative impacts for different species, so that soilless cultivators are able to decide when and what technologies to use for extending the nutrient solution life for reuse.

## 5 Beneficial micro-organism versus root nutrient uptake and plant pathogen

It has been well-known for many years that plant roots have a symbiotic relationship with micro-organisms in soil. The rootzone microbial community is not as diverse in soilless systems as they are in soil, but they are still important for plant health. Rootzone microbes provide a wide range of services and benefits to the plant, and in return, the plant provides the microbial community with photosynthates and other metabolites. Rootzone microbes include beneficial micro-organisms and plant pathogens. Beneficial micro-organisms can improve root nutrient uptake and suppress plant pathogens.

Rootzone microbes can improve root water and nutrient uptake by increasing root and substrate interaction surfaces, producing organic chemicals to convert nutrients from insoluble forms to available forms. In soil, the way in which mycorrhiza and plant growth-promoting rhizobacteria (PGPB) improve root water and nutrient uptake and promote plant growth has been extensively studied and documented (Baum et al., 2015). However, in soilless systems their benefits are less studied and also have mixed results, with many studies showing no positive effects. For example, Maboko et al. (2013) inoculated four arbuscular mycorrhizal species (*Glomus etunicatum*, *Paraglomus occultum*, *Glomus clarum*, and *Glomus mosseae*) to tomato plants grown in coir or saw dust at both seedling and transplant stages, and found that there is no positive effect on yield or quality. Cwala et al. (2010) inoculated arbuscular mycorrhizal fungi to soilless-grown tomato plants and found that the inoculation does not improve plant growth or plant tissue micro- and macro-nutrient contents, except copper. However, there are also studies showing positive effects. For example, Unal et al. (2020) showed that inoculating mycorrhiza fungus to strawberry (*Fragaria × ananassa*) roots grown in perlite increases total and marketable yields and water-use efficiency. It is generally believed that when there is enough phosphate in the rootzone, mycorrhiza may not be able to inoculate roots as successfully as in phosphate-deficient environment (Kowalska and Konieczny, 2019). Also, mycorrhiza help roots take up nutrients, especially phosphate, and in return, there is a carbohydrate cost to the plant. One would speculate that mycorrhiza does not consistently show benefits to plants in soilless systems due to the abundant nutrients normally provided to the rootzone. The abundant nutrients may reduce the success rate of mycorrhiza inoculation, and when the inoculation is successful, the plant has to provide mycorrhiza with photosynthate. So there is a cost-benefit balance in play. More research is needed to investigate under what circumstances mycorrhiza can have a positive effect and whether it pays to use mycorrhiza in soilless systems.

Another active research area is using beneficial micro-organisms in the rootzone to suppress plant pathogens in soilless systems. This is especially important in production systems that recycle nutrient solution (Zheng, 2019). When the rootzone, especially organic materials such as coir and peat are used as growing media, is infected by plant pathogens, it may not be effective to use water-disinfecting technologies such as UV or ozone to control these pathogens. Many recent studies have demonstrated that inoculating beneficial micro-organisms into the rootzone can suppress certain plant pathogens (Tian and Zheng, 2013a,b; Gravel et al., 2009, 2015; Vallance et al., 2011). Readers can also find more detailed discussion on this topic in a book chapter recently authored by Alsanius and Wohanka (2019). The modes of action for beneficial micro-organisms in suppressing pathogens can include (Köhl et al., 2019): (1) resistance or prime-enhanced resistance against infections by a pathogen, (2) competition for nutrients and space, and (3) interacting directly with the pathogen by hyperparasitism or antibiosis.

## 6 Rootzone management for plant health and productivity

A healthy plant root system is essential for water and nutrient uptake, and also important for the whole plant's health and productivity. In the rootzone of a soilless system, the main factors that can affect plant health include water, nutrients, dissolved oxygen, temperature, and biotic factors such as micro-organisms. All these factors are interconnected as well; changing one can affect the others. For example, when there is too much water in the rootzone, it can cause oxygen deficiency and eventually cause root diseases such as Pythium root rot. Different plant species, plant size, and microclimate can result in different water uptake.

When irrigating, it is important to consider the physical properties of the growing media (e.g. water holding capacity and air porosity), the container size and shape. Container size (Volder and Van Iersel, 2019) and other factors in the rootzone can limit plant growth and productivity. Container size can not only affect rootzone water and nutrient availability, it can also induce root signals (e.g. hormone) to regulate both root and shoot growth (Di Benedetto, 2011). Container size can dramatically affect plant growth as reviewed by Poorter et al. (2012). By conducting a meta-analysis on 65 studies using different pot sizes for plant growth research, the authors found that a 100% increase in pot size, generally results in 43% increase in plant biomass. However, to enhance plant growth, one cannot merely use larger pots. Ideally, container size needs to match plant size and the growing environment. When the plant is small, but the container is too large, it can result in too much water retained in the growing media for a prolonged period, which can restrict oxygen diffusing into the rootzone and cause plant to suffer from oxygen deficiency (Fig. 2; Zheng, 2020). On the other hand, with the same container volume, different shapes can have different effects on plant growth (Liptay and Edwards, 1994), which may be caused by the differences in water and air dynamics in the rootzone.

However, since all the elements (e.g. water, nutrient) can easily be controlled, there are many advantages of using SCS. For example, one can easily control rootzone salt concentration to improve plant product quality, or easily manage water stress to enhance certain second metabolite productions. One of our recent studies grew *Cannabis sativa* in a 11-litre pot containing organic

*Photo A* *Photo B*

**Figure 2** Basil (*Ocimum basilicum*) plant performs well when transplanted to the right size container, a smaller pot (left) compared with those transplanted to containers which are too big for the plants, in larger pots (right) (Photo A; in a peat and perlite mix). Photo B shows a close look of the typical waterlogging injury symptoms observed on plants grown in the larger pots in Photo A. The plants were transplanted at the same size and date. Adapted from Zheng, 2020.

soilless substrate, with a single drought-stress event during the flowering stage which resulted in an increase of the concentrations of tetrahydrocannabinolic acid (THCA) and cannabidiolic acid (CBDA) in the inflorescences by 12% and 13%, respectively, and an increase in the yield per unit growing area of THCA, CBDA, THC and CBD by 43%, 47%, 50%, and 67%, respectively, compared to the control (Caplan et al., 2019).

Growing media, container size, and shape can all affect rootzone water status, nutrient availability, and DO (Zheng, 2020), and also affect root growth and distribution (Gruda and Schnitzler, 1999; Gruda et al., 2000), and eventually affect root water and nutrient uptake. Research by looking at one single factor, such as medium water content, quite often has limited practical significance, even it is essential to do so. Therefore, it would be appropriate to take an integrated approach to manage plant rootzone. Integrated rootzone management (IRM) is important for both research purposes and practical plant cultivation (Zheng, 2020). The IRM considers all of the interconnected elements of the rootzone (e.g. physical, chemical and biological) and how changing any one element can affect the others (Zheng, 2019). Since the rootzone is a very complicated and interconnected system, to create an optimal rootzone for plant cultivation, Zheng (2019) advocates horticulturists to collaborate with engineers to develop artificial intelligence systems by utilising different advanced sensors. For example, there are technologies which can continuously take root images in situ. Research may correlate these images with root water and nutrient uptake data to investigate how root morphology, growth and distribution affect water and nutrient uptake, and also develop feedback control systems to fertigate plants in soilless systems.

For soilless systems, there is limited research on how root physical restriction affects plant growth, whether there is an optimal root size, that is, larger than which can waste photosynthates and reduce aboveground growth, smaller than which may not be able to adequately supply the plant with water and nutrients.

## 7 Summary

A healthy and vigorous root system is essential for plant water and nutrient uptake, health and productivity in soilless systems. Water and nutrient uptake in the roots can be affected by many factors: air and rootzone temperature, VPD, light, DO, and pH. Root exudate and rootzone micro-organisms also play an important role in plant health. In soilless systems, understanding root nutrient uptake characteristics, allelopathy and rootzone micro-organisms are important in supplying and maintaining the right quality nutrient solution that can be reused for a longer term without compromising plant health. To create and maintain a rootzone with adequate water, nutrients, and oxygen, within an appropriate temperature range, and without abiotic or biotic stressors, an IRM approach is the ultimate choice.

## 8 Future trends in research

There is some progress in recent decades in our understanding of root and rootzone management in soilless system. There are many areas that need further research. Future research could include:

1. Developing and utilising artificial intelligence systems for IRM.
2. Determining whether it is necessary to have a large root system for plants in soilless system, and whether there is an optimal root size for different species at different development stages.
3. Enhancing the quality of plant product by manipulating the rootzone (e.g. apply controlled stresses) in soilless culture, since the rootzone is easier to manipulate in soilless systems compared to soil.
4. Determining the critical levels of accumulated allelochemicals in the rootzone above which there will be negative impacts for different species.

## 9 Where to look for further information

### 9.1 Books

1. *Plant Roots: The Hidden Half*. 3rd ed. Waisel, Y., Eshel, A., and Kafkafi, U. (eds), CRC Press, 2002.
2. *Plant Roots: The Hidden Half*. 4th ed. Eshel, A. and Beeckman, T. (eds), CRC Press, 2013.
3. *Marschner's Mineral Nutrition of Higher Plants*. 3rd ed. Marschner, P. (eds), Academic Press, 2012.

### 9.2 Articles

1. Hosseinzadeh, S., Verheust, Y., Bonarrigo, G. and Van Hulle, S. (2017). 'Closed hydroponic systems: operational parameters, root exudates occurrence and related water treatment'. *Rev Environ Sci Biotechnol.*, 16, 59–79. DOI 10.1007/s11157-016-9418-6.
2. Volder, A. and Van Iersel, M. (2019). 'Root growth, physiology, and potential impact of soilless culture on their functioning'. In: Raviv, M., Lieth, H. and Bar-Tal, A. (Eds), *Soilless Culture: Theory and Practice* (2nd ed.). Elsevier, pp. 15–32. Elsevier, eBook ISBN: 9780444636973. DOI 10.1016/B978-0-444-63696-6.00002-5.
3. Vives-Peris, V., de Ollas, C., Gómez-Cadenas, A. and Pérez-Clemente, R. M. (2020). 'Root exudates: from plant to rhizosphere and beyond'. *Plant Cell Rep.*, 39, 3–17.
4. Zheng, Y. (2020). 'Integrated root-zone management for successful soilless culture'. *Acta Hortic.*, 1273, 1–8. DOI 10.17660/ActaHortic.2020.1273.1.

### 9.3 Conferences and websites

The International Society for Horticultural Science (ISHS; https://www.ishs.org/) organises some international symposia such as: (1) International Symposium on Growing Media, Soilless Cultivation, and Compost Utilization in Horticulture, occurring on average once every two years, and (2) International Symposium on Soilless Culture. There are also other symposia regularly organised by the ISHS such as Greensys. All of these symposia discuss the science and technologies related to soilless production, and some discussions can include root and rootzone management-related topics. These symposia normally result in proceedings published through Acta Horticulturae (https://www.actahort.org/), which is a great source for current advances in root, rootzone management in soilless systems.

## 10 References

Adams, P. 1989. Some effects of root temperature on the growth and nutrient uptake of tomato in NFT. *Proc. 7th. Int. Cong. Soilless Culture*, Flevohof 1988, Wageningen, 73–82.

Adams, P. 1992. Plant nutrition in hydroponics. *Acta Hortic.* 323, 289–305.

Affan, F. F. M., Kitano, M., Yasutake, D., Wajima, T. and Yasunaga, T. 2005. Study on root absorption responding to environmental stress by using hydroponic systems. *Phyton Ann. Rei Bot.* 45, 223–228.

Albornoz, F. and Lieth, J. H. 2016. Daily macronutrient uptake patterns in relation to plant age in hydroponic lettuce. *J. Plant Nutr.* 39(10), 1357–1364. DOI: 10.1080/01904167.2015.1109110.

Al-Rawahy, M. S., Al-Rawahy, S. A., Al-Mulla, Y. A. and Nadaf, S. K. 2019. Effect of cooling root-zone temperature on growth, yield and nutrient uptake in cucumber grown in hydroponic system during summer season in cooled greenhouse. *J. Agr. Sci.* 11(1), 47–60.

Alsanius, B. W. and Wohanka, W. 2019. Root zone microbiology of soilless cropping systems. In: Raviv, M., Lieth, H. and Bar-Tal, A. (Eds), *Soilless Culture: Theory and Practice* (2nd edn.). Elsevier, pp. 149–194. eBook ISBN: 9780444636973.

Asaduzzaman, M., Kobayashi, Y., Isogami, K., Tokura, M., Tokumasa, K. and Asao ,T. 2012. Growth and yield recovery in strawberry plants under autotoxicity through electro-degradation. *Eur. J. Hortic. Sci.* 77, 58–67.

Asao, T., Kitazawa, H., Tomita, K., Suyama, K., Yamamoto, H., Hosoki, T. and Pramanik, M. H. R. 2004. Mitigation of cucumber autotoxicity in hydroponic culture using microbial strain. *Sci. Hortic.* 99(3–4), 207–214.

Asao, T., Kitazawa, H., Ushio, K., Sueda, Y., Ban, T. and Pramanik, M. H. R. 2007. Autotoxicity in some ornamentals with means to overcome it. *Hortscience* 42, 1346–1350.

Barickman, T. C., Kopsell, D. A., Sams, C. E. and Morrow, R. C. 2020. Sole-source LED lighting and fertility impact shoot and root tissue mineral elements in Chinese kale (*Brassica oleracea* var. *alboglabra*). *Horticulturae* 6(3). DOI: 10.20944/preprints202001.0178.v1.

Baum, C., El-Tohamy, W. and Gruda, N. 2015. Increasing the productivity and product quality of vegetable crops using arbuscular mycorrhizal fungi: a review. *Sci. Hortic.* 187, 131–141.

Blum, U. and Gerig, T. M. 2005. Relationships between phenolic acid concentrations, transpiration, water utilization, leaf area expansion, and uptake of phenolic acids: nutrient culture studies. *J. Chem. Ecolog.* 31(8), 1907–1932.

Caplan, D., Dixon, M. and Zheng, Y. 2017a. Optimal rate of organic fertilizer during the vegetative-stage for cannabis grown in two coir-based substrates. *Hortscience* 52(9), 1307–1312.

Caplan, D., Dixon, M. and Zheng, Y. 2017b. Optimal rate of organic fertilizer during the flowering-stage for cannabis grown in two coir-based substrates. *Hortscience* 52(12), 1796–1803.

Caplan, D., Dixon, M. and Zheng, Y. 2019. Increasing inflorescence dry weight and cannabinoid content in medical cannabis using controlled drought stress. *Hortscience* 54(5), 964–969.

Cheng, Z. and Xu, P. 2013. Lily (*Lilium* spp.) root exudates exhibit different allelopathies on four vegetable crops. *Acta Agr. Scand. B* 63(2), 169–175. DOI: 10.1080/09064710.2012.734323.

Chérif, M., Tirilly, Y. and Bélanger, R. R. 1997. Effect of oxygen concentration on plant growth, lipidperoxidation, and receptivity of tomato roots to pythium under hydroponic conditions. *Eur. J. Plant Path.* 103(3), 255–264.

Christiaens, A., Gobin, B. and Van Labeke, M. C. 2016. Light quality and adventitious rooting: a mini-review. *Acta Hortic.* 1134, 385–394. DOI: 10.17660/ActaHortic.2016.1134.50.

Clavijo-Herrera, J., van Santen, E. and Gómez, C. 2018. 'Growth, water-use efficiency, stomatal conductance, and nitrogen uptake of two lettuce cultivars grown under different percentages of blue and red light. *Horticulturae* 4(3), 16. DOI: 10.3390/horticulturae4030016.

Cruz-Ortega, R., Lara-Núñez, A. and Anaya, A. L. 2007. Allelochemical stress can trigger oxidative damage in receptor plants: mode of action of phytotoxicity. *Plant Signal. Behav.* 2(4), 269–270.

Cwala, Y., Laubscher, C. P., Ndakidemi, P. A. and Meyer, A. H. 2010. Mycorrhizal root colonisation and the subsequent host plant response of soil less grown tomato plants in the presence and absence of the mycorrhizal stimulant, Mycotech. *Afr. J. Microbiol. Res.* 4, 414–419.

Daskalaki, A. and Burrage, S. W. 1998. Solution temperature and the uptake of water and nutrients by cucumber (*Cucumis sativus* L.). *Acta Hortic.* 458, 317–322.

Di Benedetto, A. 2011. Root restriction and post-transplant effects for bedding pot plants. In: Aquino, J. C. (Ed.), *Ornamental Plants: Types, Cultivation and Nutrition*, pp. 47–79. ISBN 978-1-61761-736-2.

Dickson, R. W., Fisher, P. R., Argo, W. R., Jacques, D. J., Sartain, J. B., Trenholm, L. E. and Yeagera, T. H. 2016. Solution ammonium: nitrate ratio and cation/anion uptake affect acidity or basicity with floriculture species in hydroponics. *Sci. Hortic.* 200, 36–44.

Erro, J., Zamarreño, A. M. and García-Mina, J. M. 2010. Ability of various water-insoluble fertilizers to supply available phosphorus in hydroponics to plant species with diverse phosphorus-acquisition efficiency: involvement of organic acid accumulation in plant tissues and root exudates. *J. Plant Nutr. Soil Sci.* 173, 772–777. DOI: 10.1002/jpln.200900263.

Giri, A., Heckathorn, S., Mishra, S. and Krause, C. 2017. Heat stress decreases levels of nutrient-uptake and -assimilation proteins in tomato roots. *Plants* 6(1). DOI: 10.3390/plants6010006.

Goto, E., Both, A. J., Albright, L. D., Langhans, R. W. and Leed, A. R. 1996. Effect of dissolved oxygen concentration on lettuce growth in floating hydroponics. *Acta Hortic.* 440, 205–210.

Gravel, V., Ménard, C. and Dorais, M. 2009. Pythium root rot and growth responses of organically grown geranium plants to beneficial microorganisms. *Hortscience* 44(6), 1622–1627.

Gravel, V., Dorais, M., Dey, D. and Vandenberg, G. 2015. Fish effluents promote root growth and suppress fungal diseases in tomato transplants. *Can. J. Plant Sci.* 95(2), 427–436. DOI: 10.4141/cjps-2014-315.

Graves, W. R. 1992. Influence of hydroponic culture method on morphology and hydraulic conductivity of roots of honey locust. *Tree Physiol.* 11(2), 205–211.

Gruda, N. 2005. Impact of environmental factors on product quality of greenhouse vegetables for resh consumption. *Crit. Rev. Plant Sci.* 24(3), 227–247. DOI: 10.1080/07352680591008628.

Gruda, N. and Schnitzler, W. H. 1999. Influence of wood fiber substrates and application rates on the growth of tomato transplants. *Adv. Hort. Sci.* 13, 20–24.

Gruda, N., Tucher, S. V. and Schnitzler, W. H. 2000. N-immobilization of wood fiber substrates in the production of tomato transplants (*Lycopersicon lycopersicum* (L.) Karst. ex Farw.). *J. Appl. Bot.* 74, 32–37 (In German).

Hosseinzadeh, S., Verheust, Y., Bonarrigo, G. and Van Hulle, S. 2017. Closed hydroponic systems: operational parameters, root exudates occurrence and related water treatment. *Rev. Environ. Sci. Biotechnol.* 16(1), 59–79. DOI: 10.1007/s11157-016-9418-6.

Huang, J., Nelson, P. V. and Lee, J. 2001. Seedling effect on root substrate pH. *J. Plant Nutr.* 24, 1133–1147.

Hussain, M. I. and Reigosa, M. J. 2011. Allelochemical stress inhibits growth, leaf water relations, PSII photochemistry, non-photochemical fluorescence quenching, and heat energy dissipation in three C3 perennial species. *J. Exp. Bot.* 62(13), 4533–4545.

Johnson, C. N., Fisher, P. R., Huang, J., Yeager, T. H., Obreza, T. A., Vetanovetz, R. P., Argo, W. R. and Bishko, A. J. 2013. Effect of fertilizer potential acidity and nitrogen form on the pH response in a peat-based substrate with three floriculture species. *Sci. Hortic.* 162, 135–143.

Kawasaki, Y., Matsuo, S., Kanayama, Y. and Kanahama, K. 2014. Effect of root-zone heating on root growth and activity, nutrient uptake, and fruit yield of tomato at low air temperatures. *J. Japan. Soc. Hort. Sci.* 83(4), 295–301. https://doi.org/10.2503/jjshs1.MI-001.

Kitazawa, H., Asao, T., Ban, T., Pramanik, M. H. R. and Hosoki, T. 2005. Autotoxicity of root exudates from strawberry in hydroponic culture. *J. Hortic. Sci. Biotech.* 80(6), 677–680.

Köhl, J., Kolnaar, R. and Ravensberg, W. J. 2019. Mode of action of microbial biological control agents against plant diseases: relevance beyond efficacy. *Front. Plant Sci.* 10, 845. DOI: 10.3389/fpls.2019.00845.

Kowalska, I. and Konieczny, A. 2019. Effect of mycorrhiza on yield and quality of lettuce grown on medium with different levels of phosphorus and selenium. *Agr. Food Sci.* 28(2), 84–92.

Lambers, H., Atkin, O. K. and Millenaar, F. F. 2002. Respiratory patterns in roots in relation to their functioning. In: Waisel, Y., Eshel, A. and Kafkafi, U. (Eds), *Plant Roots: The Hidden Half* (3rd edn.). CRC Press, Marcel Dekker, New York, pp. 521–552.

Lee, J. G., Lee, B. Y. and Lee, H. J. 2006. Accumulation of phytotoxic organic acids in reused nutrient solution during hydroponic cultivation of lettuce (*Lactuca sativa* L.). *Sci. Hortic.* 110(2), 119–128.

Liptay, A. and Edwards, D. 1994. Tomato seedling growth in response to variation in root container shape. *Hortscience* 29(6), 633–635.

Lyu, S. W. and Blum, U. 1990. Effects of ferulic acid, an allelopathic compound, on net P, K, and water uptake by cucumber seedlings in a split root system. *J. Chem. Ecol.* 16, 2429–2439.

Maboko, M. M., Bertling, I. and Du Plooy, C. P. 2013. Arbuscular mycorrhiza has limited effects on yield and quality of tomatoes grown under soilless cultivation. *Acta Agr. Scand. B* 63(3), 261–270, DOI: 10.1080/09064710.2012.755219.

Morard, P., Lacoste, L. and Silvestre, J. 2000. Effect of oxygen deficiency on uptake of water and mineral nutrients by tomato plants in soilless culture. *J. Plant Nutr.* 23(8), 1063–1078.

Morini, S., Loreti, F. and Sciutti, R. 1990. Effect of light quality on rooting of 'Leccino' olive cuttings. *Acta Hortic.* 286, 73–76. http://dx.doi.org/10.17660/ActaHortic.1990.286.11.

Narwal, S. S. 2004. *Allelopathy in Crop Production*. Pawan Kumar Scientific Publisher, India.

Oburger, E. and Jones, D. L. 2018. Sampling root exudates–mission impossible? *Rhizosphere* 6, 116–133.

Olschowski, S., Geiger, E.-M., Herrmann, J. V., Sander, G. and Grüneberg, H. 2016 Effects of red, blue, and white LED irradiation on root and shoot development of Calibrachoa cuttings in comparison to high pressure sodium lamps. *Acta Hortic.* 1134, 245–250. DOI: 10.17660/ActaHortic.2016.1134.33.

Pinho, P., Jokinen, K. and Halonen, L. 2017. The influence of the LED light spectrum on the growth and nutrient uptake of hydroponically grown lettuce. *Lighting Res. Technol.* 49(7), 866–881.

Poorter, H., Bühler, J., van Dusschoten, D., Climent, J. and Postma, J. A. 2012. Pot size matters: a meta-analysis of the effects of rooting volume on plant growth. *Funct. Plant Biol.* 39(11), 839–850. http://dx.doi.org/10.1071/FP12049.

Sakamoto, M., and Suzuki, T. 2018. Effect of pot volume on the growth of sweetpotato cultivated in the new hydroponic system. *Sustain. Agr. Res.* 7, 137–145.

Savvas, D. and Gruda, N. 2018. Application of soilless culture technologies in the modern greenhouse industry - a review. *Eur. J. Hortic. Sci.* 83(5), 280–293. doi.org/10.17660/eJHS.2018/83.5.2.

Savvas, D., Karagianni, V., Kotsiras, A., Demopoulos, V., Karkamisi, I. and Pakou, P. 2003. Interactions between ammonium and pH of the nutrient solution supplied to gerbera (*Gerbera jamesonii*) grown in pumice. *Plant Soil* 254(2), 393–402.

Savvas, D., Passam, H. C., Olympios, C., Nasi, E., Moustaka, E., Mantzos, N. and Barouchas, P. 2006. Effects of ammonium nitrogen on lettuce grown on pumice in a closed hydroponic system. *Hortscience* 41(7), 1667–1673.

Shamshiri, R. R., Jones, J. W., Thorp, K. R., Ahmad, D., Man, H. C. and Taheri, S. 2017. Review of optimum temperature, humidity, and vapour pressure deficit for microclimate evaluation and control in greenhouse cultivation of tomato: a review. *Int. Agrophys.* 32(2), 287-302. DOI: 10.1515/intag-2017-0005.

Soffer, H. and Burger, D. W. 1988. Effects of dissolved oxygen concentrations in aero-hydroponics on the formation and growth of adventitious toots. *J. Amer. Soc. Hort. Sci.* 113, 218-221.

Strehmel, N., Böttcher, C., Schmidt, S. and Scheel, D. 2014. Profiling of secondary metabolites in root exudates of Arabidopsis thaliana. *PhytoChemistry* 108, 35-46.

Suhl, J., Oppedijk, B., Baganz, D., Kloas, W., Schmidt, U. and van Duijn, B. 2019. Oxygen consumption in recirculating nutrient film technique in aquaponics. *Sci. Hortic.* 255, 281-291.

Talukder, M. R., Asaduzzaman, M., Ueno, M., Tanaka, H. and Asao, T. 2020. Alleviation of allelochemical stress induced growth inhibition and oxidative damage in lettuce under closed hydroponics through electro-degradation. *Hort. Sci. (Prague)* 47(1), 53-68.

Tian, X. and Zheng, Y. 2013a. Compost teas and reused greenhouse nutrient solution suppress several plant pathogens in vitro. *Hortscience* 48(4), 510-512. DOI:10.21273/HORTSCI.48.4.510.

Tian, X. and Zheng, Y. 2013b. Evaluation of biological control agents for Fusarium wilt in *Hiemalis begonia*. *Can. J. Plant Pathol.* 35(3), 363-370. DOI:10.1080/07060661.2013.812580.

Tindall, J. A., Mills, H. A. and Radcliffe, D. E. 1990. The effect of root zone temperature on nutrient uptake of tomato. *J. Plant Nutr.* 13(8), 939-956.

Unal, N., Tuzel, Y., Tuzel, I. H. and Oztekin, G. B. 2020. Effects of mycorrhiza and irrigation programs on strawberry production in substrate culture. *Acta Hortic.* 1273, 177-182. DOI: 10.17660/ActaHortic.2020.1273.24.

Vallance, J., Déniel, F., Floch, G. L., Guérin-Dubrana, L., Blancard, D. and Rey, P. 2011. Pathogenic and beneficial microorganisms in soilless cultures. In: Lichtfouse, E., Hamelin, M., Navarrete, M. and Debaeke, P. (Eds), *Sustainable Agriculture*, Volume 2. Springer, Dordrecht, pp. 711-726. https://doi.org/10.1007/978-94-007-0394-0_31.

Vives-Peris, V., de Ollas, C., Gómez-Cadenas, A. and Pérez-Clemente, R. M. 2020. Root exudates: from plant to rhizosphere and beyond. *Plant Cell Rep.* 39(1), 3-17.

Volder, A. and Van Iersel, M. 2019. Root growth, physiology, and potential impact of soilless culture on their functioning. In: Raviv, M., Lieth, H. and Bar-Tal, A. (Eds), *Soilless Culture: Theory and Practice* (2nd edn.). Elsevier, pp. 15-32. eBook ISBN: 9780444636973. DOI: 10.1016/B978-0-444-63696-6.00002-5.

White, P. J. 2012. Ion uptake mechanisms of individual cells and roots: short-distance transport. In: Marschner, P. (Ed), *Marschner's Mineral Nutrition of Higher Plants* (3rd edn.). Academic Press, London, pp. 7-47.

Yan, Q., Duan, Z., Mao, J., Li, X. and Dong, F. 2012. Effects of root-zone temperature and N, P, and K supplies on nutrient uptake of cucumber (*Cucumis sativus* L.) seedlings in hydroponics. *Soil Sci. Plant Nutr.* 58(6), 707-717. DOI: 10.1080/00380768.2012.733925.

Zhao, J., Chen, X., Wei, H., Lv, J., Chen, C., Liu, X., Wen, Q. and Jia, L. 2019. Nutrient uptake and utilization in Prince Rupprecht's larch (*Larix principis- rupprechtii* Mayr.) seedlings exposed to a combination of light-emitting diode spectra and exponential fertilization. *Soil Sci. Plant Nutr.* 65(4), 358-368. DOI: 10.1080/00380768.2019.1631715.

Zheng, Y. 2018. Current nutrient management practices and technologies used in North American greenhouse and nursery industries. *Acta Hortic.* 1227, 435–442.

Zheng, Y. 2019. Developments in growing substrates for soilless cultivation. In: Marcelis, L. F. M. and Heuvelink, E. (Eds), *Achieve Sustainable Greenhouse Cultivation*. Burleigh Dodds Science Publishing, Cambridge, UK, pp. 225–240.

Zheng, Y. 2020. Integrated root-zone management for successful soilless culture. *Acta Hortic.* 1273, 1–8. DOI: 10.17660/ActaHortic.2020.1273.1.

Zheng, Y., Graham, T., Richard, S. and Dixon, M. 2004. Potted gerbera production in a subirrigation system using low-concentration nutrient solutions. *Hort. Sci.* 39(6), 1283–1286.

Zheng, Y., Dixon, M. and Saxena, P. 2006. Greenhouse production of Echinacea purpurea (L.) and E. angustifolia using different growing media, $NO_3^-/NH_4^+$ ratios and watering regimes. *Can. J. Plant Sci.* 86(3), 809–815.

Zheng, Y., Wang, L. and Dixon, M. 2007. An upper limit for elevated root zone dissolved oxygen concentration for tomato. *Sci. Hortic.* 113(2), 162–165.

Zheng, Y., Cayanan, D. F. and Dixon, M. 2010. Optimum feeding nutrient solution concentration for greenhouse potted miniature rose production in a recirculating subirrigation system. *Hortscience* 45(9), 1378–1383.

Zheng, Y., Clark, M. J., Agro, E., Bowser, D., MacDougall, H. and Schiestel, K. 2018. Fertilization guide for container-grown nursery crops. Available at: https://legacy.horttrades.com/assets/1532525007.Fertilization_Guide_20180622.pdf.

# Chapter 3

## Developments in inorganic materials, synthetic organic materials and peat in soilless culture systems

*Jeb S. Fields, Louisiana State University Agricultural Center, USA; and Nazim S. Gruda, University of Bonn, Germany*



## 1 Introduction

Across the world, soilless culture is utilised in a broad spectrum of production systems, from small volume precision greenhouse vegetable and floriculture containers to large nursery trees, and from systems such as hydroponics utilising the nutrient film technique (NFT) to long-trough grown strawberries. In all these production systems, the one similarity is that components and materials have been carefully selected and engineered to fit the system. Many soilless culture systems (SCS) use a solid growing medium. Growing media have been used traditionally mostly for plant propagation, bedding and pot plant production. However, this range of use has expanded to include the total production of many food and ornamental crops, especially high-value ones, grown under protection in greenhouses (Gruda et al. 2016a,b). The growing medium takes the role of soil. It provides anchorage for the root system, a supply of water and nutrients for the plant and adequate aeration

http://dx.doi.org/10.19103/AS.2020.0076.02

in the root area. Therefore, the growing medium must be free of pests and toxic substances, have a low soluble salt content, be uniform and have a high content of available water. At the same time, it must be well aerated and drained. However, availability of materials, cost, texture, pH-consistency and, more recently, environmental concerns must be considered (Gruda et al. 2013; Gruda 2019).

Soilless substrates, a technical term given to growing media, require specific properties for optimal function. The primary functions are:

1 serving as an anchor point for roots;
2 retaining sufficient moisture and serve as a repository for nutrients for plant uptake;
3 providing ample room for root growth and air exchange; and
4 providing a substrate for beneficial microfauna populations to thrive.

As such, we tend to consider specific properties when assessing the potential of a material to be developed as a substrate component. These properties are categorised as:

- physical;
- chemical; and
- biological.

Most characterisation properties are interlinked between multiple categories.

Physical properties encompass the physical structure of a substrate and are often characterised by static physical and dynamic hydraulic properties. Into the first group belong properties that characterise a growing medium such as material density, and particle size but also properties give insights to the air and water relationships of the substrate, the associated mass or weight of the material, processing and material breakdown. Some commonly measured static physical properties are the maximum water-holding capacity or container capacity, and minimum air-filled porosity or air space. Substrates are traditionally designed with these static physical properties in mind, to be very forgiving to growers who, in the past, were often not faced with similar ecological and economic issues associated with high water usage, which we face today.

There is now a greater focus on the water-holding properties of substrates. As a plant receives irrigation, the water content changes within the substrate over time. Similarly, after the irrigation event ends, the water slowly drains away over time. We also know that the water within the substrate does not just move upwards and downwards throughout the profile or in-and-out of the container. Instead, there is lateral movement and redistribution continually occurring within the substrate, and there is channelling, with both water and nutrient gradients changing within the profile of the container.

Properties that consider these changing or dynamic measures are termed dynamic properties. The hydraulic properties of substrates provide insight as to water retention and distribution, drainage, hydraulic gradients and water fluxes within the substrate. The primary difference is that physical properties are often static, or very slow to change (i.e. total porosity), and hydraulic properties are dynamic, fluctuating with wetting and drying cycles (i.e. hydraulic conductivity). Using dynamic properties to develop soilless substrates, growers have been able to produce more sustainable container substrates.

One such case involves utilising substrate hydraulic conductivity to engineer a substrate that allows plants access to higher proportions of the water held, allowing fewer or reduced volume irrigations. Improved 'delivery of water' to the roots allows plants to access water from previously inaccessible pore spaces in the container. By ensuring that higher proportions of water are available to plants, we can reduce irrigation requirements. Low water irrigation systems, when paired with the properly optimised substrates, have been shown to not only reduce water and fertility requirements but also encourage increased growth rates, resulting in reaching marketability sooner. Static and dynamic physical properties of substrates are discussed elsewhere in this book.

Chemical properties refer to the substrate's chemistry and influence how nutrients are managed. The most basic chemical property is substrate pH, which is of the utmost importance as pH regulates nutrient availability. Cation exchange capacity and anion exchange capacity are also primary chemical properties that dictate the potential consideration of a material as a substrate component. Solute transport or the measure of the movement and retardation of nutrient flux through the substrate can be considered both a physical and chemical property. The chemical properties of growing media are discussed in Chapter 6 in this book.

For substrate science, biological properties are a somewhat new frontier. Biological properties can refer to the organic structure of the materials (i.e. lignin or cellulose content), presence and change in micro-colonies within the substrate, as well as the interaction of these properties with microorganism and all changes in biological level. Biological properties referring to the microflora and fauna within the substrate can both be beneficial or harmful. Often substrate materials are chosen for their sterility or ability to be sterilised. However, many substrate materials benefit from an ageing process which involves the breakdown of carbon. This carbon breakdown results from microbial activity. The processes that are involved in substrate breakdown, degradation and composting can all be considered biological as well as physical changes. For more information, see Chapter 7 on this topic in this book.

It is evident that these properties are densely interlinked, and often factors that influence one set of characteristics will have an influence on multiple or all characteristics. For instance, the water content can directly influence physical, that is, hydraulic, but in some degree also the chemical, and biological properties of growing medium. The lack of water can lead to the hydrophobicity of materials, a reduction in microbial activity and can affect a substrate's chemistry. In contrast, overhydration can lead to anaerobic conditions, reducing nutrient movement, altering substrate water potential, and result in changes to biological communities and biological functions. External abiotic stimuli such as temperature, light concentration, humidity, moisture tend to lend themselves to influence substrate properties across the board.

## 2 The development of solid growing media

It is essential to differentiate between the growing media components and growing media additives. Components are materials that are used to constitute a specific horticultural substrate. They are added in significant volumes; sometimes even used as stand-alone substrates. Components are often incorporated or introduced to a substrate to provide significant changes in physical, chemical or biological properties. Moreover, components will have a lasting effect on the substrate system.

On the contrary, growing media additives are added in small amounts and usually on a weight basis. Apart from fertilisers, lime, wetting and/or binding agents, plant growth regulators, hydrogels, and sometimes dying stuff are used as additives. They often have a limited duration of effect in the container system, with influence diminishing over time. A substrate composite is an overall substrate in its entirety. While composites can be single components, it is more likely that a composite will be made of multiple components and additives. For example, one of the most common substrates used throughout the world for plant production is 'peatlite'. Peatlite is a composite as it is a blend of two primary components 'peat' and 'perlite'. These components are blended to achieve targeted physical properties optimal for growing conditions. Often peatlite composites will have additives, such as fertiliser and lime, to add nutrition and balance pH in the composite. However, depending on the production system, either peat or perlite can be used as a stand-alone substrate. As such, it is crucial to keep in mind that growing media selection and development must be considered for each production system.

Often production systems are a result of current infrastructure, crops or production goals, and generally are chosen before the selection of substrate. As such, it is feasible to develop, engineer, or prefer substrates to fit the needs of the system. Growing media selection is a unique process for every system; however, often substrates are selected for physical, chemical and biological properties, in that specific order. The physical structure of the growing medium defines the air: water dynamics of the substrate system, and is the most difficult to modify post-processing. Chemical and biological properties are more easily manageable through the use of additives in most cases. As such, substrate selection and development primarily focus on the physical properties of components and the resulting physical properties of composites after blending and processing.

Essential properties to consider when choosing, developing or engineering substrates have been discussed at length in the literature. Gruda et al. (2013) have compiled a list of general requirements in aggregate materials for growing media. These include:

- High pore volume with favourable pore size distribution.
- Sufficient air supply, even with a high-water content.
- Suitable drainage properties.
- High water capillarity.
- Good re-wettability after being dried out.
- Structural stability over a long cultivation period.
- Favourable levels of plant-available nutrients and other salts.
- High cation exchange capacity.
- Manageable nutrient dynamics (low definition and release of nutrients).
- Favourable pH value.
- Buffering against pH shift.
- Free from pests and pathogens.
- Free from plant-damaging and growth-inhibiting substances.
- Free from weed seeds and expansive parts of plants.
- Long-term storability.
- Uniformity, always representing the same quality.

We will briefly review these critical factors.

*High pore volume with favourable pore size distribution.* All necessary functions of a growing medium occur in the pore space. Water, air, nutrients are all held and retained in pores. Moreover, root growth exists only in pore spaces. The structure of a growing medium is primarily designed to create optimal pore space. As such, having ample pore space is necessary for a substrate.

*Sufficient air supply, even with a high-water content.* This meant that no matter the conditions, an adequate volume of air would remain within the substrate. Much effort has been devoted to help growers in providing sustainable air and water balance. The results are countless variations of soilless substrates used in productions throughout the industry.

*Suitable drainage properties.* Drainage is a significant factor to consider when selecting or designing soilless substrates. One of the significant issues associated with production in containers is known as the 'container effect' (Bilderback 1980). What does it mean? When growers first started using containers to produce plants, they noticed that the water did not drain well in mineral soil that was used to fill the containers. This is in part a result of the small particles of the soil, creating a pore structure that is unable to overcome the effect of the height of the container. The relatively short height of the container, compared to the depth of the soil profile, does not provide sufficient gravitational force to overcome the suction of the pore structure, inhibiting proper drainage. To counteract the 'container effect', soilless substrates were initially designed to allow for ample drainage, regardless of the volume of water that was applied.

*High water capillarity.* Water movement within the substrate is vital for maintaining a beneficial hydraulic gradient within the substrate system. Many growing systems utilise some sort of sub-irrigation (i.e. capillary mats and ebb and flow systems). High capillarity ensures that the substrate can have sufficient moisture throughout. Hoskins et al. (2014) observed water profile movement in overhead irrigation settings. Water moves quickly down through the profile of the substrate until it reaches the bottom. Water begins to leach, while simultaneously, some water is pooled and retained in the bottom layer of the substrate. Slowly, as more water is added, the profile will backfill through capillary rise.

*Good re-wettability after being dried out.* Growing media will inevitably dry below the desired water content at some point. A perfect example is peat. This leads to issues with hydrophobicity and water repellency. It is crucial that substrates can overcome hydrophobicity from drying during or before plant production.

*Structural stability over a long cultivation period.* Some crops can take multiple years to grow to a marketable state fully. As such, soilless substrates utilised in these systems must not degrade over time.

*Favourable levels of plant-available nutrients and other salts.* Nutrients in growing media are provided in accordance with plant species. This is the reason for the diversity of container growing media. When a nutrient solution is used, plant-available nutrients are delivered through the nutrient solution, and the growing media serve to anchor the plants.

*High cation exchange capacity.* Cation exchange capacity (CEC) indicates the ability of the substrate to adsorb mineral nutrients and retain cations for plant uptake. However, in comparison to soils, the impact of CEC on fertiliser and pH management in soilless culture growing media is not so important. The precise distribution of nutrient solutions reduces the need for a strong buffer capacity of the growing medium.

*Manageable nutrient dynamics (low definition and release of nutrients).* Just like water, substrate nutrients are not static. For more information, see Chapter 6.

*Favourable pH value.* While pH is adjustable post-processing, components with favourable pH are more desirable due to reduction of additives needed, and thus reduction of cost associated with growing media use.

*Buffering against pH shift.* As for nutrients, the pH-buffer capacity of a growing medium has a significant impact in pot plants that are not directly supplied with nutrient solution.

*Free from pests and pathogens.* Pests and pathogens can be a significant economic hindrance for a grower. Pest management is often one of the most significant financial and labour costs associated with agricultural production. Therefore, soilless substrate materials must be free from pests and pathogens. Frequently, inert materials can easily be sterilised, but organic materials that are not carefully handled can have issues with introducing new pests or outbreaks into an operation.

*Free from plant-damaging and growth-inhibiting substances.* Similar to freedom from pests, anything that can retard growth or inhibit healthy crop production will be very detrimental to a grower. Substrate materials must not contain any phytotoxic materials that can adversely affect crop quality and production timing. This can sometimes be the case with treated materials, materials with too high salt concentrations, or those derived from industrial applications which are not properly treated.

*Free from weed seeds and expansive parts of plants.* Substrate materials must be adequately processed and weed-free when incorporated into a new production system. Poorly processed materials can have weeds present, which will allow for the expansion of plant materials into the rest of the operation and prove costly to manage.

*Long-term storability.* Substrates must be stored, often for relatively long periods before use. Materials that degrade too quickly are unsuitable, e.g. for crops growing over a long period.

*Uniformity, always representing the same quality.* As growing operations are very repeatable, substrates must be uniform. Non-uniformity between substrate mixes can compromise growing operations. As the substrate influences all aspects of production, irregularity within the substrate will affect the entire production line, reducing efficiency and productivity associated with plant production.

## 3 Substrate materials

Substrate materials can be classified into two main groups: inorganic and organic. Inorganic materials can be (Gruda et al. 2013):

- natural: such as sand and tuff;
- processed: such as perlite and vermiculite; and
- synthetic: such as mineral wool.

Some examples are shown in Fig. 1.

In the same way, organic materials can be classified as:

- natural: such as peat; and
- processed: such as wood fibres and bark.

These are also shown in Fig. 1. Substrate components can also be classified as fibrous, such as coir, or granular, such as perlite. Moreover, components can also be classified as active, such as compost, or inert, such as mineral wool (Gruda et al. 2013). Here we provide a brief description of standard and traditionally utilised inorganic and organic components. The remaining sections in the chapter will focus primarily on inorganic, organic synthetic and the primary organic substrate utilised, peat. Alternative organic components will be discussed in Chapter 4.

The first sections focus on inorganic materials for growing media. In comparison to organic raw materials, mineral materials are used in much smaller quantities in growing media. For instance, for EU countries, they

**Figure 1** Some materials used as growing media. From left to right: Rock wool, polyurethane foam, expanded shale, volcanic material, open porous clay granulate, expanded clay, perlite, black peat, coarse wood fibres, fine wood fibres, vermiculite and light peat (Photo: Gruda, private collection).

comprise 2 131 000 m$^3$ 6.2% of the total amount of all constituents. This total includes pre-shaped mineral wool (Schmilewski 2017). The main reasons for this are normally high bulk density, with issues of low water holding capacity and problems associated with mechanical processing. Mineral substrate components are primarily from natural sources that are processed granules. In some cases, a unique preparation process is needed. This is especially true with expanded clay, perlite or vermiculite. Other materials are residues from industrial processes, i.e. sand, or recycling materials, i.e. crushed bricks.

Mineral materials can be used as a stand-alone substrate or as components in composite substrates, mixed with other inorganic and organic raw materials. As stand-alone substrates, chemically inert materials such as mineral wool, perlite or expanded clay are used. An important area of application is SCS with recirculating nutrient solution, in which the nutrient supply can be precisely controlled. Different methods are used for this purpose. A detailed overview is given in Chapter 8 of this book.

As previously discussed, by selecting suitable starting materials as components for growing media, their properties can be influenced in a targeted manner. In the case of sand, it is possible to affect increases in water storage capacity, improvement in re-wettability and especially increases in nutrient buffering. For the other materials, it is essentially the air capacity and structural stability that can be optimised.

## 4 Natural inorganic substrate materials

### *4.1 Sand*

Sand is a naturally occurring mineral aggregate that is classified based on its particle diameters. The International Society of Soil Sciences categorises sand as particles ranging between 0.02 mm and 2.0 mm. While proper sand is a mineral material that has been broken down from parent materials over time, sand is often considered any mineral aggregate between these particle diameters. In horticultural production, multiple different grades of sand are used. These are often separated by smaller particle diameter ranges within the sand range or composed of a specific mineral compound (Table 1).

*Properties.* Sand is a mineral aggregate formed naturally from a variety of parent materials. However, it is common to consider sand to be primarily composed of quartz, which has a particle density of roughly 2.65 g cm$^{-3}$. As a result, the bulk density of sand is relatively high, relative to other organic and inorganic materials used as soilless culture growing media. Sand is often incorporated into other substrates to increase weight (to reduce tipping) and to retain water. In soils, high concentrations of sand tend to

**Table 1** Main advantages and disadvantages of peat and rock wool, and some inorganic and synthetic organic growing media constituents

| Material | Origin | Advantages | Disadvantages |
|---|---|---|---|
| Sand | Natural with particles of 0.02–2.0 mm | Relatively inexpensive, good drainage ability. | Low nutrient and water-holding capacity, heavy volume-weight (1400–1800 kg $m^{-3}$), low TPS (40–50% V/V). |
| Pumice | Light silicate mineral of volcanic origin | Volume weight (450–670 kg $m^{-3}$), good TPS (70–85% V/V), cheap and long-lasting, environmental-friendly. | High transport costs, pH may initially be high: pH 7.5-8.5. |
| Tuff | Pyroclastic volcanic material | TPS of 60–80%, stable material | Relatively heavy volume weight (800–1500 kg $cm^{-3}$), at pH below five there may be a risk due to Al and Mn toxicity |
| Vermiculite | $Mg^{++}$, $Al^{+++}$ and $Fe^{++}$ silicates which are heated to 1500 °C and then sieved to give various size grades | Light volume weight (70–100 kg $m^{-3}$), high nutrient holding ability, good water holding ability, suitable pH buffering capacity, good aeration: TPS (70–80% V/V). | Compacts when too wet, energy-consuming product, expensive. |
| Perlite | Siliceous volcanic mineral sieved and heated to 1000 °C | Sterile, neutral in pH (6.5–7.5), no decay, TPS (50–75% V/V). | Volume weight (700–1000 kg $m^{-3}$), low nutrient capacity, energy-consuming product, expensive. |
| Expanded clay granules | Pelletising and firing in a kiln at temperatures of 1200 °C | Light volume weight (280-630 kg $m^{-3}$), chemically inert and are neutral with a pH of approximately 7.0, resistant to compression. | Low amount of easily available water, Phosphate fixation may occur on the Al oxide fraction of the heat-treated clays |
| Zeolite | Crystalline aluminium silicates granular mineral, usually formed by metamorphism of volcanic rocks. | Could be used as a slow-release material for $K^+$, $Mg^{++}$, and $NH_4^+$ ions. | Heavy volume-weight (1900–2300 g $cm^{-3}$), the price of zeolite could high. |

| | | | |
|---|---|---|---|
| Mineral wool | Melted silicates at 1200–1600 °C | Light volume weight, high total pore space (92–98% V/V), ease of handling, totally inert, nutrition can be carefully controlled. | Disposal problems, energy consumed during manufacture, may be irritating to the operator skin. |
| Polyurethane | Synthetic organic polymer, produced from mineral oil mixed with glycol. | Light volume weight (78–1190 kg $m^{-3}$), high TPS (95%), chemically inert, and an almost neutral pH (6.5–7.0). | Not decomposed by microbial enzymes, e.g. cannot be composted after use. |
| Peat | Natural anaerobically processed plant residues, | Physical stability, good air and water holding capacity: TPS (85–97% V/V), light volume weight (60–200 kg $m^{-3}$), low and easy to adjusted pH, low nutrient content, low microbial activity. | Finite resource, environmental concerns and contribution to $CO_2$ release, drying leads to substrate hydro-repellence, may be strongly acidic. |

Adapted, based on Gruda et al. (2016b), and Savvas and Gruda (2018).
TPS = total pore space, V = volume.

improve drainage, as sand is one of the more abundant constituents of field soil. However, in soilless culture, sand is one of the finer particles and tends to reduce drainage.

*Performance*. Gravel and sand were used in older SCS to improve aeration. They have tended to be replaced by lighter materials such as mineral wool in more modern operations.

*Sustainability concerns*. As sand is a naturally occurring abundant resource, there are few known environmental concerns associated with using sand in soilless culture.

## 4.2 Pumice

Pumice is created in the course of volcanic eruptions in which 800–1,200°C hot magma is ejected. The high-pressure rock melt, coming from the earth's interior is decompressed, which leads to a substantial expansion of different gases. The material usually formed from silicic lavas has a certain degree of porosity–high-porosity materials as called pumice, in contrast to the lower porosity of lava. Depending on the origin and purpose of use, the raw material is washed, broken and fractionated (Table 1).

*Properties*: The composition of the minerals varies greatly depending on their origin. The more fine-pores there are, the lighter the material. Pumice shows a yellow-white or grey-white colour. Pumice is one of the lightest porous materials of natural origin, and is used as a growing media. Pumice has a low volume weight of 0.45–0.67, sometimes to 0.80 g $cm^{-3}$ and a TPS of 70–85% (Boertje 1995). According to Maucieri et al. (2019), the structure of pumice tends to deteriorate fairly quickly, due to the natural breaking up of the particles. Pumice, added to peat, increases the drainage and aeration of the substrate. Pumice has a neutral pH that is only weakly buffered and is therefore not very stable. The contents of water-soluble salts, as well as nitrogen and phosphate, are meagre. Potassium is available in 200–400 mg $K_2O$ per litre (Amberger-Ochsenbauer and Meinken 2020). Despite the large internal surface area, pumice only has a low cation exchange capacity that contributes little to plant nutrition but does not decrease the availability of fertiliser nutrients (Handreck and Black 2005).

*Performance*: Pumice is used both in its pure form and in mixtures with other mineral or organic components. For use as a horticultural substrate, pumice blocks are ground into small particles that range from fine dust to 8 mm (Gizas and Savvas 2007). Using pumice as a growing medium requires low investment, and it is easily applicable in existing growing systems (Gruda et al. 2013). It is used as a substrate for fruit vegetables, such as tomato, cucumber, pepper, and cut flowers. Pumice is also used in landscaping, interior greening and smaller quantities, for producing container plants. Pérez-Urrestarazu et al. (2019) found that pumice showed a slightly better performance in terms of green cover and biomass production as perlite, and expanded clay when used for living walls. Gizas and Savvas (2007) investigated the effect of particle size of pumice on plant growth and yield and found that gypsophila and cucumber gave the highest yields when grown in pots filled with the two fine pumice grades (0–2 mm and 0–5 mm). In contrast, the yield obtained from plants grown in bags was reduced regardless of the particle size range. In contrast, lettuce, and to higher degree roses, exhibited a weaker response to the different pumice grades and growing systems.

*Sustainability concerns*: Pumice is friendly to the environment because no costly production processes are involved (Boertje 1995). Pumice can be used for many years, so it produces relatively little substrate waste (Gruda et al. 2013). European deposits are present in Turkey, Greece, Italy, Iceland and Tenerife. Pumice can also be found in Japan, New Zealand, Russia and the United States (Raviv et al. 2002). High transport costs limit its use in areas that do not have local deposits.

## 4.3 Tuff

Tuff is a pyroclastic volcanic material, characterised by a high total pore space and high surface area (Table 1). Lavas with a percentage of silica higher than 65% contain more Si and Al, whereas those with a silica percentage of less than 50% contain higher amounts of Fe, Mn, Ca, and Mg and present a dark colour (Bar-Tal et al. 2019).

*Properties*: Tuff has a volume weight of 0.8–1.5 g $cm^{-3}$, and total porosity of 60–80% (Gruda et al. 2013; Bar-Tal et al. 2019), and a water-retention capacity between 15% and 25% by volume (Kipp et al. 2001). The physical and chemical properties and even the colour of tuff, black, red and yellow, are determined mainly by its mineralogical composition and weathering stages, as well as the grinding and sieving processes (Raviv et al. 2002). Al-Ajlouni et al. (2017) found differences in properties of tuff by investigating five grades of the particle-size substrate, consisting of granulated tuff volcanic material: 0–2, 0–4, 0–8, 2–4, and 4–8 mm.

*Performance*: Volcanic tuff is widely used in many Mediterranean countries for growing vegetable crops, such as tomatoes, cucumbers and bell peppers, and cut flower plants. Tomatoes (*Lycopersicum esculentum* L.) grown in volcanic tuff had a higher marketable yield than those grown in perlite and soilless sand substrates (Al-Ajmi et al. 2009). Al-Ajlouni et al. (2017) found that using 0–4 mm tuff improved growth and flower quality of

Asiatic hybrid lily grown under soilless culture. Introducing plants to black or even red tuff, after equilibration with an acidic electrolyte (pH below 5), may be risky due to Al and Mn toxicity (Bar-Tal et al. 2019).

*Sustainability concerns*: Tuff can be used for many years because it is a stable material. According to Raviv et al. (1998), growing plants in tuff increases the accumulation of organic matter. Soluble organic compounds are adsorbed by tuff surfaces, changing its surface charge (Silber and Raviv 2001) and enhancing the yield of cut roses (*Rosa x hybrida*) (Raviv et al. 1998).

## 5 Processed inorganic substrate materials

### 5.1 Perlite

Perlite is one of the most commonly used aggregate soilless culture substrates (Evans 2004). It is an aluminosilicate rock that is formed by volcanic activity, and then mined, ground to desired particle size and heated to around 1,000°C (Nelson 2012; Gruda et al. 2013) (Table 1). At these temperatures, perlite expands to 4–20 times its original volume, producing a lightweight material with high porosity and structurally stable form (Bunt 1988; Grillas et al. 2001; Gruda et al. 2013; Bar-Tal et al. 2019).

*Properties*. Perlite has a closed structure, which means it has little to no internal porosity. Therefore, water is not absorbed inside the mineral, thus increasing the substrate's drainage. Perlite, like most aggregates, is added to a substrate to influence physical properties. Its most significant role in most substrates is to increase aeration which, in turn, will often increase drainage in as well as improve wettability, capillary movement, and gas exchange (Bunt 1988). Perlite helps reduce compaction of a substrate. However, this also means that perlite does not break down readily (Bunt 1988). Perlite is sterile and has a neutral pH, so it will not affect pH or crop fertility. The properties of perlite mean it does not contribute to the ability to hold nutrients (Bunt 1988). Perlite has a CEC between 25 cmol $kg^{-1}$ and 35 cmol $kg^{-1}$, which is very low relative to other substrate components (Doğan and Alkan 2012).

*Performance*. Perlite can be used alone or mixed with other substrates for greenhouse plant production (Gruda et al. 2013; Gruda 2019). It is used in many soilless growing systems around the world due to its relative stability. In Europe, perlite is well established, mainly in the Mediterranean region. In SCS it is used for the production of fruit vegetables and cut flowers, sometimes mixed with other materials. In container crops, perlite is usually mixed with organic materials. As an inert material, perlite does not degrade in the system, nor does it influence a system chemically or biologically.

*Sustainability concerns.* Because perlite is produced from a mined rock and not renewable, and it does not break down, it is not considered sustainable or eco-friendly (Drotleff 2007). When dried, perlite has been seen to be potentially harmful as dust, acting as an irritant to the eyes, skin and lungs, though there are no confirmed long-term effects (Whitcomb, 1984). However, perlite is not generally considered a dangerous material for use as a growing media. Another issue with perlite is its high cost which reflects its resource-intensive production methods, including mining, heating, and transportation (Evans and Gachukia 2007). Generally, perlite is an expensive material for growing media. Due to perlite's lack of compressibility and relatively high bulk density (0.17 g $cm^{-3}$), shipping costs are generally high.

## 5.2 Vermiculite

Vermiculite is the name given to hydrated laminar aluminium-iron magnesium silicate minerals that are heated and expanded in a furnace at approximately 1000°C. Vermiculite consists of a clay mineral which, as with all clays, initially exists in a layered structure. As the vermiculite is expanded, these layers separate and provide increased water holding capacity, and create a 'sponge-like' texture. Vermiculite improves water holding capacity, conditions the substrate, and helps with retaining nutrients for plant uptake. Fine to coarse grade vermiculite is often incorporated into peat or peat-like substrate to increase water and nutrient retention. However, vermiculite can also be utilised as a sole component in some settings, especially in SCS production, cutting propagation and seed germination.

*Properties.* Vermiculite is a very lightweight inorganic mineral (bulk density ranging from 0.7 to 1.0 g/$cm^3$). It has good aeration with a total pore space of 70–80% v/v. As with many inorganic minerals, vermiculite is graded during processing and sold in individual grades, to allow growers to meet specific needs. Vermiculite is a neutral component (pH of 7.0), and may slightly increase the pH of acidic substrates. Vermiculite also retains a fair amount of nutrition, with an average CEC of approximately 20 cmol/kg.

*Performance.* Vermiculite is often an incorporated component of soilless substrates, especially in seedling and greenhouse production, used to provide water and nutrient retention. Vermiculite acts as a buffer for water as the system dries, releasing water to plants as needed. Vermiculite does not break down nor decay. However, cleaning processes can destroy vermiculite structure and as such it is not considered to be reusable for multiple applications. Vermiculite is commonly used as a standard for rooting cuttings and germinating seeds. Vermiculite regulates water content retaining a desirable level of moisture for propagation. Vermiculite can also be utilised by home gardeners to store bulbs. The absorptive capacity of vermiculite removes moisture from the surface of the bulbs by preventing rot.

*Sustainability concerns.* In the 1990s links between some vermiculite sources and asbestos contamination became public knowledge. However, that was due to a single source (a mine near Libby, Montana) which had natural deposit asbestos. That mine has since been closed; however, the news damaged the industry. Since then, vermiculite is re-gaining in popularity. Disposal of vermiculite poses no known environmental risks.

## 5.3 Expanded clay granules

In soil science clay is defined as soil particles with a size of less than 0.002 mm. Clays mainly consist of clay minerals that have been created by chemical weathering of primary minerals. In general, clay additions in soilless substrates retain relatively significant quantities of water. However, expanded clays have been developed to retain minimal water during hydroponic production. Clays do provide added nutrient retention in soilless substrates. Expanded clays bring added nutrient retention, albeit significantly reduced from non-processed clays, but without the high water retention. The most common type of expanded clay is light expanded clay aggregates (LECA). These LECA are formed by pelletising clays and firing in a kiln at temperatures of 1200°C (Boudaghpour and Hashemi 2008). The gases released in the firing process expand, creating micropores within the structures that remain once the beads cool and harden (Jones et al. 1959).

*Properties.* The diameter of the expanded clays determines much of the physical properties of expanded clays. Depending on the source, expanded clays used in SCS can range from 4 to 16 mm. Expanded clays are chemically inert and are neutral with a pH of approximately 7.0.

*Performance.* Expanded clays can be used as an aggregate in a substrate composite or as a stand-alone substrate in SCS. Minimal degradation occurs over time so that they may be used in long-term crop production. Moreover, the ability to dry out between irrigations and the sterile nature of expanded clays results in few pest issues. This rapid drainage can require careful attention as crops can dry out quickly with little water reserves in expanded clay substrates. It is crucial to soak expanded clays before use to limit this risk.

*Sustainability concerns.* There are few environmental concerns associated with expanded clays. Some concern has been expressed about potential strip mining associated with harvesting the clays. Expanded clays do need to be washed before use to remove any dust and between production cycles. Due to the lack of degradation, expanded clays can be washed and sterilised after production cycles and reused in the next crop. The easy recycling of expanded clays and the relatively low costs make expanded clays one of the most economically efficient inorganic substrate components for SCS.

## 5.4 Zeolite

Natural zeolite is a crystalline aluminium silicate granular mineral. Due to their ion exchange, adsorption, hydration/dehydration, and catalysis properties, zeolites are widely used in agriculture and numerous industries for the removal of pollutants from waste and drinking water (Bar-Tal et al. 2019).

*Properties*: Zeolite possesses a relatively high water-holding capacity and pore space. Due to stable cavity structures, the mineral has a large internal surface and a high cation exchange capacity. Zeolites possess a relatively high volume-weight (1.9–2.3 g $cm^{-3}$) and are therefore used in substrate mixtures to increase the stability of container crops; however, they are also used as single growing media (Gruda et al. 2013). The proportions of cations zeolites contain vary considerably in different zeolites found around the world (Handreck and Black 2005). The different climatic conditions of the respective locations and the different nature of the volcanic deposits result in different properties. Savvas et al. (2004) reported that, during the initial wetting of the substrates with nutrient solution, most of the K was absorbed on the surface of the zeolite; as a result, the K concentration was sharply reduced in the solutions drained from substrates with those constituents comprising zeolite.

$NH_4^+$ ions can be absorbed in the same way. This property makes zeolite a substrate of great interest as it firstly absorbs and then slowly releases K+ and NH4+ ions, while it is not able to absorb Cl− and Na+, which are hazardous to plants (Maucieri et al. 2019). According to Broschat (2001), zeolite can save water-soluble nutrients and prevents cation leaching. It promotes leaching of nitrate and potassium (Kimberly and Nelson 1997). Leggo (2000) and Zwingmann et al. (2009) showed that zeolite can reduce losses of ammonium ($NH_4^+$) and increase efficiency in its absorption and later slow release. In the same way (Savvas et al. 2004), reported a reduction of potassium leaching and generally a balance between macronutrient cations using zeolite. Using zeolite in sand mixtures offers potential in countries where sand is abundant (Al-Ajmi et al. 2009). Zeolite has also been reported to protect plants against toxicity (e.g. from ammonium–Handreck and Black 2005) or from heavy metals (Kapetanios and Loizidou 1992).

*Performance*: As a stand-alone substrate and in mixtures with organic growing media, zeolite can be used for the production of vegetables, cut flowers and container plants. In a study with tomatoes, Savvas et al. (2004) reported the highest yields when plants were grown in the zeolite, followed by treatment using zeolite in a substrate mixture. The excellent performance of the plants grown in zeolite was due to higher cation exchange capacity, enabling a more efficient buffering of excess ammonium and Mg concentrations in the root environment. Moreover, zeolite was capable of absorbing excess Mg, resulting in more balanced macronutrient cation ratios in the root environment. Gül et al. (2005) described the use of zeolites as a component of growing media for cultivation of crisp-head lettuce. Mixtures with zeolite are used to produce high-quality, structurally stable growing media. For instance, adding zeolite to perlite improved physicochemical properties of the substrate, enhanced water retention capacity and led to an increase in yield of greenhouse cucumber grown under water limited conditions (Gholamhoseini et al. 2018). Zeolites are marketed in formulations which differ in the N and P content and which can be used in seed sowing, for the rooting of cuttings or during the cultivation phase (Pickering et al. 2002). In addition to the slow-releasing capacity for $K^+$ and $NH_4^-$, Eghtedary-Naeini et al. (2016), used enriched zeolite as a slow release of iron fertiliser for soilless greenhouse cultivation of cucumber, resulting in an increase in yield and plant vigour.

*Sustainability concerns*: Since the physical properties of zeolite do not change much, zeolite can be used several times. Depending on the origin and quality, the price of zeolite can be high. Zeolites are currently increasingly used in interior greening cultivation.

## 6 Mineral wool as a substrate

The term 'mineral wool' includes two different materials: stone or rock wool and glass wool. For stone wool, the stone is selected; for glass wool glass is chosen as raw material. Both are used as a thermal and acoustic insulating material in the construction industry. Mineral wool was accidentally discovered as a plant substrate. A staff member of a construction company in Denmark saw sprouted seeds on a wetted piece of stone wool. That led to the use of rock wool in horticulture. The first experiments were carried out at universities in Denmark and the Netherlands in the second half of the 1960s. It quickly became apparent that cultivation using this material made possible a more precise level of watering and nutrient supply control. The development of recirculated systems using mineral wool made it possible to operate in a more environmentally friendly way as well as save input costs. The first horticultural companies then introduced this cultivation system for soilless culture of cut roses, gerbera, tomatoes and other fruit vegetables.

Stone wool is manufactured by heating a mixture of three natural raw materials: 60% diabase (a form of basalt rock, dolerite), 20% limestone, and 20% coke. These materials are fused at high temperature, with the coke acting as fuel in a blast furnace through which air is forced to raise the temperature to 1600°C. The molten mixture is then spun at high speed into thin fibres of about 0.5 mm diameter, which are then cooled by an airstream. For horticultural use, wetting agents are added to improve water absorption. It is then pressed into blocks or slabs of various sizes (Smith 1987).

Glass wool is made by melting quartz sand in electric furnaces at 1200°C. The process of production of glass wool fibres and slabs is similar to that of stone wool. In general, both stone wool and glass wool (mineral wool) are sterile, easy to manage, and consistent in performance.

*Properties*: Rock wool is a lightweight material with a low volume weight of about 0.05–0.2 g $cm^3$ and a very high total pore volume of 92–98%, depending on the producer (Smith 1987). The water holding capacity is also relatively high. The water buffer capacity is low, and a steep gradient in water content occurs from the top to the bottom of the slab following irrigation and free drainage (Bar-Tal et al. 2019). The air volume is low at the bottom, just 4% at the height of 1 cm above the base, while the upper layers are dry. The recommended height of a slab for optimal water to air ratio is 7.5–10 cm (Smith 1987).

There are no plant-available nutrients in rock wool. It is a chemically inert material. This means that it has neither a nutrient sorption capacity nor pH buffering and is therefore well suited for soilless culture processes with continuous nutrient solution supply. The pH-values of industry produced rock wool slabs and blocks are today around 6–6.5. Below pH 5.0, stone wool starts to dissolve; hence low pH values should be avoided (Bar-Tal et al. 2019).

*Performance*: Polythene-wrapped rock wool has become the dominant soilless culture system for greenhouse vegetables and roses worldwide and especially in Europe (Gruda et al. 2016a). Growers grow seedling and transplant them into blocks of different sizes or smaller round plugs made of rock wool (Gruda et al. 2013). Small blocks are put into containers and used for the cultivation of mother plants. Stone wool is also used as a growing medium for 'green walls'. Before planting, it is advisable to saturate rock wool with nutrient solution to ensure adequate moistening and thus proper water distribution. Stone wool should not be allowed to dry out, as it is difficult to wet again. The setup of a rock wool growing system is simple: rock wool slabs are placed in the rows, holes for plants are cut in the plastic surrounding the pieces, and the slabs are filled with solution. After soaking for about 24 h, the transplants are placed on the slabs with drainage slits cut at the bottom (Gruda et al. 2013). A complete nutrient solution is supplied through the irrigation system to feed plants with nutrients (Fig. 2) (Gruda 2019).

Due to the inert nature of the material, culture control in rock wool is straightforward: the buffering capacity is limited to the amount of nutrients and water held within the pore space of the medium. On the other hand, it is 'unforgiving' to management errors, because it lacks a buffering capacity for nutrients, pH and water (Bar-Tal et al. 2019).

*Sustainability concerns.* Problems in disposing of mineral wool have led to criticism of its current usage, even if it can be partially reused after appropriate recycling. After removing the covering film, the material is shredded. It is then used for the production of new rock wool or as an admixture in some industrial processes (Bussell and Mckennie 2004). In a life cycle assessment of growing media, Quantis (2012) reported that mineral wool had the highest negative impacts. Purchase costs and freight costs are also relatively high.

**Figure 2** Tomato production in soilless culture with rock wool as a growing medium. Transplants in rock wool cubes, shortly before the start of greenhouse cultivation (Photo: Gruda, private collection).

# 7 Synthetic organic materials as substrates

## 7.1 Polyurethane

Polyurethane is a synthetic organic polymer, produced from mineral oil mixed with glycol.

*Properties*: Polyurethane is a very lightweight material with a low particle density. It is chemically inert, has an almost neutral pH (6.5–7.0), does not contain useful nutrients available to plants and does not undergo decomposition (Kipp et al. 2001). The porosity of polyurethane is very high at 95%, while the available water within the matric potential range from -10 to -100 cm $H_2O$ amounts only to 2% of its volume (Bar-Tal et al. 2019). In a four-year trial, Benoit and Ceustermans (1995) observed a shift of the pore size distribution to smaller pores in recycled polyurethane substrate. The roots left in mats increased water retention and decreased the air volume.

*Performance*: The use of polyurethane is limited, even if the polyurethane foam slabs could be used for the production of vegetables and cut ornamentals in the same way as mineral wool. The latest types of polyurethane foams are not phytotoxic. They are used for the conservation of fresh-cut flowers and may be used as support matrices for hydroponics and most recently for vertical farming. Hardgrave (1995) and Komosa et al. (2020) reported equal or better productivity and fruit quality in reused rather than new polyurethane foam slabs for cucumbers and tomatoes, respectively. According to Komosa et al. (2020), the marketable and total greenhouse tomato yields obtained in reused one- or two-year-old polyurethane foam slabs were respectively 14.7–15.9% and 14.2–19.7% higher than in new polyurethane slabs. Such beneficial effects on crop performance might be attributed to the improved water-holding capacity due to the presence of root materials within the recycled slabs (Bar-Tal et al. 2019) and/or the increased amount of mineral element such $N\text{-}NO_3$, Ca in the slabs (Komosa et al. 2020)

*Sustainability concerns*: Polyurethane is a sterile product that can be steam cleaned between crops for several growing cycles and used for 10–15 years (Benoit and Ceustermans 1995). However, it is resistant to acids and is not decomposed by microbial enzymes, hence cannot be composted after use (Bar-Tal et al. 2019). Although the waste can be reused for the production of polyurethane slabs for construction or maybe incinerated (Bar-Tal et al. 2019), natural materials are preferred. This may be one reason why use of polyurethane is limited (Schmilewski 2017).

### *7.2 Other synthetic organic materials*

Apart from polyurethane, polystyrene and polyester fleece are some of commonly used synthetic organic materials in soilless culture. Information concerning these materials may be found in Bar-Tal et al. (2019).

## 8 Peat as a substrate material

Peat is formed by partial decomposition of Sphagnum, other mosses and sedges at an acidic milieu and in the absence of oxygen (Gruda et al. 2016b). Peatlands occur over a wide range of climates. A significant proportion occurs in the higher middle latitudes (50–60°N) in Canada, Russia and Finland. These peat deposits developed in the post-glacial period after the recession of the ice sheets 14 000–11 000 years ago (Maher et al. 2008). According to Schmilewski (2009, 2017), over 75% of growing media used annually in Europe's horticultural industry is based on peat. The use of peat as a central component of growing media is due to its relatively low costs, its excellent chemical, biological and physical properties with low nutrient content, low pH, high water-holding capacity, high air space and its light weight (Gruda et al. 2016b; Savvas and Gruda 2018).

In the peat bogs, we can differentiate between younger, little to moderately decomposed 'white peat' from the upper layers and the older and more decomposed 'black peat', also known as 'humic peat'. Differences between different peat are related to variation in the degree of decomposition, plant species and the place of origin, which are affected by variation in climate and water quality (Handreck and Black 2005; Raviv et al. 2002). The Von Post (1937) scale gives a more precise classification, indicating degrees of decomposition or humusity (H) from H1 to H10. The peat for use in horticulture is classified as light peat (H1-H3), dark peat (H4–H6) and black peat (H7–H10). Dark, older peat is less prone to biodegradation during cropping compared to light younger peats (Prasad and Maher 2000). According to Rydin and Jeglum (2013), Sphagnum mosses do not possess lignin of the type found in plants. The proportion of pores decreases with increasing decomposition, whereas the percentage of solid substance increases. Thus, the water holding capacity increases and the air capacity decreases.

*Properties*: Peat has a very stable structure, is easy to store and, after use, easy to reuse or recycle (Gruda et al. 2013). It has a high volume of pores (85–97%), good air and water capacity, and a high cation exchange capacity (Table 1). Besides decomposition, the extraction process, the effects of frost and the degree of crushing during processing influence the physical properties of peat. Peat possesses both small dead-end-pores as well as large interparticle macropores that vary depending on decomposition grade. It has low volume-weight. The light weight of peat can be an advantage related to transportation costs.

The nutrient content and pH of peat substrates are easy to control and allow adjustment to the plant requirements, because of the initial very low content of soluble nutrients and extremely low pH with values of 2.5–3.5. The higher the degree of composition, the higher the pH value of the peat. The low pH could be adjusted by lime. Besides the pH-regulation, dolomitic limestone provides calcium and magnesium. When ground limestone is used, Mg should be added separately. Micronutrients are usually added to peat-based growing media.

Generally, peat is free from pollutants and pathogens. The decomposition of peat is very slow, even after the liming and the fertilisation process. Therefore, the nitrogen immobilisation is not significant for peat, despite a relatively wide C:N ratio. Sagging or the volume reduction cultivation is low, in comparison with other organic materials. The resistance to decomposition may be linked to Sphagnum, a pectin-like polysaccharide in the cell walls of sphagnum mosses (Hajek et al. 2011). Younger peat has more hemicelluloses and celluloses. The loss of volume under greenhouse conditions is a little faster than more decomposed white peat (Amberger-Ochsenbauer and Meinken 2020).

The population of microorganisms in peat is low. Thus, the antagonistic effects against soil-borne pathogens, such as Pythium or Phytophthora *spp.*, are low. This disadvantage could be improved by mixed peat with materials like compost that possess a suppressive ability.

The poor rewetting ability after drying out is one of the few disadvantages when using peat as a growing media. This is one of the reasons for using clay in peat substrates. Clay has previously been used in self-mixed substrates. Nowadays, substrate companies usually use wetting agents, such as non-ionic ethylene oxide derivatives. Mixtures with suitable hydrophilic raw materials are also used to counteract this property e.g. coir which is also considered as an excellent material to be mixed with peat.

*Performance*: The use of peat for vegetable and cut flower production in soilless culture is limited. However, seedlings and transplants of horticultural plants in nurseries and ornamental pot plants are grown predominantly in organic substrates based on peat (Savvas and Gruda 2018). Commercial applications of peat include lawn and garden soil amendments, potting soils, and turf maintenance on golf courses (Apodaca 2018). These applications have made peat the most widely used growing medium and substrate component in horticulture (Savvas and Gruda 2018).

*Environmental concerns*: Peat is a slow-growth resource. Its extraction can lead to a permanent loss or degradation of wetland habitats - some of our most ecologically significant and biologically diverse environments, which are of considerable conservation, landscape and cultural significance (Morrish and Hofstede 2000). The regrowth rate of peat bogs occurs very

> slowly with around 1 mm per year. Due to its slow formation, peat use in horticulture cannot be sustainable. The use of peat in horticulture also contributes significantly to the release of carbon dioxide. In the long-term, peatlands are one of the largest stores of organic carbon and are an important provider of ecosystem services (UNEP 2019; Gruda 2019). They provide carbon (C) storage, regulation of local water quality and hydrological conditions, including flood protection (Gruda 2012a; Barrett et al. 2016; Kern et al. 2017).

Poorer performance of some non-peat based growing media has been expressed by some gardeners. This has created resistance to using different peat-reduced or peat-free growing media (Knight 2012). Some more information on peat reduction work in Europe is presented in chapter 4 'Developments in alternative organic substrates for soilless culture' by Gruda and Bragg.

## 9 Summary and future trends

Soilless substrates utilised in traditional hydroponics are often inorganic or synthetic materials, as opposed to organic substrate components utilised in other forms of soilless culture. As growers seek more precision production applications, more operations are shifting to SCS to improve resource management. The standard substrate components utilised in soilless plant production have been well researched and engineered to fit into specific operations. Understanding the relationship between the substrate, water and fertiliser in a container and knowing the movement within is the key to designing not only substrates but irrigation and fertility systems that will work for the specific crop that will then allow for continued beneficial improvements in soilless culture and the container horticulture industry.

This chapter has reviewed standard inorganic and organic substrate components and how these components are developed, engineered and processed. The selection of growing media and substrate constitute is based on the properties of the material used, plant performance and most recently on sustainability concerns. Sustainability is based on three well-known criteria including not only economic aspects, such as cost and availability, but also ecological and societal considerations. We strongly believe that ecological and societal considerations will be more and more critical requirements for substrate choice in the future. The new customer is looking for new alternatives and novel substrates with a better life cycle assessment. As we improve agricultural practices, new optimised substrate materials should be developed.

In response to these pressures, soilless substrate research is continuously evolving to develop more resource-efficient and sustainable production practices. Globally, sustainability is becoming extremely important for the continued success of the horticultural industry. Water, mineral nutrients and organic materials are often the centre of attention. As such, besides the study of new materials for growing media, there is significant research pursuing more efficient use of water and fertilisers in soilless media, as well as how to manage substrate systems.

## 10 Where to look for further information

Books and book chapter: Standard introductions to the subject are contributions by Raviv et al. (2002), Bar-Tal et al. (2019), and Gruda et al. (2013, 2016a,b) (see publications in the References for full details).

Articles: Gruda (2012, 2019), and Savvas and Gruda (2018) give recent developments to alternative inorganic and synthetic materials worldwide through overviews (see these contributions in the References for full details).

Conferences and websites: The International Society for Horticultural Science (ISHS; https://www.ishs.org/) organises different international symposia such as the International Symposium on Growing Media, Soilless Cultivation, and Compost Utilization in Horticulture, every two years.

## 11 References

Al-Ajlouni, M. G., Ayad, J. Y. and Othman, Y. A. 2017. Particle size of volcanic tuff improves shoot growth and flower quality of Asiatic hybrid lily using soilless culture. *hortte* 27(2), 223–227. doi:10.21273/HORTTECH03585-16.

Al-Ajmi, A., Al-Karaki, G. and Othman, Y. 2009. Effect of different substrates on fruit yield and quality of cherry tomato grown under recalculating soilless system. *Acta Hortic.* 2(807), 491–494.

Amberger-Ochsenbauer, S. and Meinken, E. 2020. *Torf und Alternative Substratausgangsstoffe*. Zarbock GmbH & Co. KG, Frankfurt am Main, Germany.

Apodaca, L. E. 2018. Peat (advance release). In: *Minerals Yearbook* (vol. 2016). United States Department of the Interior, Washington, DC; United States Geological Survey, Reston, VA.

Barrett, G. E., Alexander, P. D., Robinson, J. S. and Bragg, N. C. 2016. Achieving environmentally sustainable growing media for soilless plant cultivation systems-A review. *Sci. Hortic.* 212, 220–234.

Bar-Tal, A., Saha, U. K., Raviv, M. and Tuller, M. 2019. Chapter 7. *Inorganic and Synthetic Organic Components of Soilless Culture and Potting Mixtures. Soilless Culture–Theories and Practice* (2nd edn.). Elsevier Science, pp. 259–301. doi: 10.1016/B978-0-444-63696-6.00007-4.

Benoit, F. and Ceustermans, N. 1995. A decade of research on ecologically sound substrates. *Acta Hortic.* 408, 17–30. doi: 10.17660/ActaHortic.1995.408.2.

Bilderback, T. E. 1980. Container soils and soilless media. In: *North Carolina Nursery Crop Production Manual* (vol. 9) Bonaminio, V. P. (Ed.). North Carolina State Extension, pp. 1-12.

Boertje, G. A. 1995. Chemical and physical characteristics of pumice as a growing medium. *Acta Hortic.* 401, 85–88.

Boudaghpour, S. and Hashemi, S. 2008. A study on light expanded clay aggregate (LECA) in a geotechnical view and its application on greenhouse greenroof cultivation. *Int. J. Geo.* 4, 59–63.

Broschat, T. K. 2001. Substrate nutrient retention and growth of container-grown plants in clinoptilolitic zeolite-amended substrates. *Horttech* 11(1), 75–78.

Bunt, A. C. 1988. *Media and Mixes for Container-Grown Plants*. Springer Verlag. ISBN 978-94-011-7904-1.

Bussell, W. T. and Mckennie, S. 2004. Rockwool in horticulture, and its importance and sustainable use in New Zealand. *N. Z. J. Crop Hortic. Sci.* 32(1), 29–37. doi:10.1080/01140671.2004.9514277.

Doğan, M. and Alkan, M. 2012. Some physicochemical properties of perlite as an adsorbent. *Fresen. Environ. Bull.* 13(3B), 252–257.

Drotleff, L. 2007. *Mixing Up Organics*. s.l.: Greenhouse Grower.

Eghtedary-Naeini, A., Golabadi, M. and Hoodaji, M. 2016. Using enriched zeolite as a slow release iron fertilizer for soilless greenhouse cultivation of cucumber. *J. Plant Nutr.* 39(4), 523–530. doi:10.1080/01904167.2015.1109123.

Evans, M. 2004. Ground bovine bone as a perlite alternative in horticultural substrates. *Horttechnology*. doi:10.21273/HORTTECH.14.2.0171.

Evans, M. R. and Gachukia, M. M. 2007. Physical properties of sphagnum peat-based root substrates amended with perlite or parboiled fresh rice hulls. *HortTechnology* 17, 312–315.

Gholamhoseini, M., Habibzadeh, F., Ataei, R., Hemmati, P. and Ebrahimian, E. 2018. Zeolite and hydrogel improve yield of greenhouse cucumber in soil-less medium under water limitation. *Rhizosphere* 6, 7–10. https://doi.org/10.1016/j.rhisph.2018.01.006.

Gizas, G. and Savvas, D. 2007. Particle size and hydraulic properties of pumice affect growth and yield of greenhouse crops in soilless culture. *Hortscience* 42(5), 1274–1280. https://doi.org/10.21273/HORTSCI.42.5.1274.

Grillas, S., Lucas, M., Bardopoulous, E., Sarafopoulos, S. and Voulgari, M. 2001. Perlite based soilless culture system: current commercial applications and prospects. *Acta Hortic.* 548(548), 105–114.

Gruda, N. 2012. Current and future perspective of growing media in Europe. *Acta Hortic.* 960(960), 37–43. https://doi.org/10.17660/ActaHortic.2012.960.3.

Gruda, N. S. 2019. Increasing sustainability of growing media constituents and stand-alone substrates in soilless culture systems. *Agronomy* 9(6), 298. doi.org/10.3390/agronomy9060298.

Gruda, N., Gianquinto, G., Tüzel, Y. and Savvas, D. 2016a. Culture soil-less. In: *Encyclopedia of Soil Sciences* (3rd edn.) Lal, R. (Ed.). CRC Press, Taylor & Francis Group, pp. 533-537. doi:10.1081/E-ESS3-120053777.

Gruda, N., Caron, J., Prasad, M. and Maher, M. J. 2016b. Growing media. In: *Encyclopedia of Soil Sciences* (3rd edn.) Lal, R. (Ed.). CRC Press, Taylor & Francis Group, pp. 1053-1058. https://doi.org/10.1081/EESS3-120053784.

Gruda, N., Qaryouti, M. M. and Leonardi, C. 2013. Growing media. In: *Good Agricultural Practices for Greenhouse Vegetable Crops–Principles for Mediterranean Climate Areas*, Plant Production and Protection Paper 217, FAO, Rome, Italy, pp. 271-302.

Gruda, N. and Bragg, N. (2021). Chapter 4. Developments in alternative organic materials for growing media in soilless culture systems. In: *Advances in Horticultural Soilless Culture*, Gruda, N. (Ed.). Burleigh Dodds Science Publishing Limited, Cambridge, UK.

Gül, A., Erogul, D. and Ongun, A. R. 2005. Comparison of the use of zeolite and perlite as substrate for crisp-head lettuce. *Sci. Hortic.* 106(4), 464-471.

Hajek, T., Ballance, S., Limpens, J., Zijlstraa, M. and Verhoeven, J. T. A. 2011. Cell-wall polysaccharides play an important role in decay resistance of Sphagnum and actively depressed decomposition in vitro. *Biogeochemistry* 103(1-3), 45-57.

Handreck, K. A. and Black, N. D. 2005. *Growing Media for Ornamental Plants and Turf*. A UNSW Press Book, Sydney.

Hardgrave, M. 1995. An evaluation of polyurethane foam as a reusable substrate for hydroponic cucumber production. *Acta Hortic.* 401, 201-208. doi:10.17660/ActaHortic.1995.401.24.

Hoskins, T. C., Owen, J. S., Jr. and Niemiera, A. X. 2014. Water movement through a pine-bark substrate during irrigation. *horts* 49(11), 1432-1436.

Jones, T. R., Hirsch, T. J. and Stephenson, H. K. 1959. *The Physical Properties of Structural Quality Lightweight Aggregate Concrete*. Texas Transporatation Institute, Texas.

Kapetanios, E. G. and Loizidou, M. 1992. Heavy metal removal by zeolite in tomato cultivation using compost. *Acta Hortic.* 302, 63-74. doi:10.17660/ActaHortic.1992.302.6.

Kern, J., Tammeorg, P., Shanskiy, M., Sakrabani, R., Knicker, H., Kammann, C., Tuhkanen, E.-M., Smidt, G., Prasad, M., Tiilikkala, K., Sohi, S., Gascó, G., Steiner, C. and Glaser, B. 2017. Synergistic use of peat and charred material in growing media–an option to reduce the pressure on peatlands? *J. Environ. Eng. Landsc. Manag.* 25(2), 160-174.

Kimberly, A. W. and Nelson, P. V. 1997. Using precharged zeolite as a source of potassium and phosphate in a soilless container medium during potted chrysanthemum production. *J. Amer. Soc. Hort. Sci.* 122, 703-708.

Kipp, J. A., Wever, G. and de Kreij, C. 2001. *International Substrate Manual*. Elsevier, Amsterdam.

Knight, A. 2012. Towards sustainable growing media. Chairman's chairman's report and roadmap, sustainable growing media task force. Available at: https://assets.publishing.service.gov.uk/government/uploads/system/uploads/attachment_data/file/221019/pb13867-towards-sustainable-growing-media.pdf.

Komosa, A., Kleiber, T., Komosa, A., Markiewicz, B., Mieloszyk, E. and Mieloch, M. 2020. Yield and nutrient status of greenhouse tomato (*Lycopersicon esculentum* Mill.) grown in new and re-used rockwool, polyurethane, NFT and aeroponics. *J. Elem.* 25(2), 523-536. doi:10.5601/jelem.2019.24.3.1894

Leggo, P. J. 2000. An investigation of plant growth in an organo-zeolitic substrate and its ecological significance. *Plant Soil* 219(1/2), 135-146.

Maher, M., Prasad, M. and Raviv, M. 2008. Organic soilless media components. In *Soilless Culture: Theory and Practice* (2nd ed.) Raviv, M. and Lieth, J. H., (Eds). Andre Gerhard Wole, Amsterdam, The Netherlands, pp. 459-504.

Maucieri, C., Nicoletto, C., Os, E., Anseeuw, D., Havermaet, R. V., and Junge, R. 2019 Hydroponic technologies. In: *Aquaponics Food Production Systems*, Goddek, S., Joyce, A., Kotzen, B. and Burnell, G. (Eds). Springer, Cham. https://doi.org/10.1007/978-3-030-15943-6_4.

Morrish, R. and Hofstede, H. 2000. *Alternatives to Peat*. Environmental Science Report #00/1. Murdoch University, Perth, Western Australia, p. 18.

Nelson, P. V. 2012. Root substrate. In: *Greenhouse Operation and Management* Nelson, P. V. (Eds), 7th edition. Prentice Hall, Upper Saddle River, NJ, pp. 161–194.

Pérez-Urrestarazu, L., Fernández-Cañero, R., Campos-Navarro, P., Sousa-Ortega, C. and Egea, G. 2019. Assessment of perlite, expanded clay and pumice as substrates for living walls. *Scientia Horticul.* 254, 48–54. https://doi.org/10.1016/j.scienta.2019.04.078.

Pickering, H. W., Menzies, N. W. and Hunter, M. N. 2002. Zeolite/rock phosphate–a novel slow release phosphorus fertiliser for potted plant production. *Sci. Hortic.* 94(3–4), 333–343.

Prasad, M. and Maher, M. J. 2000. Horticultural properties of peat as related to degree of decomposition. Sustaining our Peatlands, Proceeding of the 11th International Peat Congress, pp. 523–527.

QUANTIS. 2012. Comparative life cycle assessment of horticultural growing media based on peat and other growing media constituents 2012. Available at: http://epagma.eu/evidence-based (accessed on 02 May 2020).

Raviv, M., Silber, A. and Medina, Sh. 1998. The effect of medium disinfestation on cut rose productivity and on some chemical properties of tuff. *Sci. Hortic.* 74, 285–293.

Raviv, M., Wallach, R., Silber, A. and Bar-Tal, A. 2002. Substrates and their analysis. In: *Hydroponic Production of Vegetables and Ornamentals* Savvas, D. and Passam, H. (Eds). Embryo Publications, Athens, pp. 25–102.

Rydin, H. and Jeglum, J. K. 2013. *The Biology of Peatlands* (2nd edn.). Oxford University Press, Oxford, UK.

Savvas, D. and Gruda, N. 2018. Application of soilless culture technologies in the modern greenhouse industry–a review. *Europ. J. Hortic. Sci.* 83(5), 280–293. https://doi.org/10.17660/eJHS.2018/83.5.2.

Savvas, D., Samantouros, K., Paralemos, D., Vlachakos, G., Stamnatakis, M. and Vassilatos, C. 2004. Yield and nutrient status in the root environment of tomatoes (*Lycopersicon esculentum*) grown on chemically active and inactive inorganic substrates. *Acta Hortic.* 644(644), 377–383.

Schmilewski, G. 2009. Growing medium constituents used in the E.U. *Acta Hortic.* 819(819), 33–46. https://doi.org/10.17660/ActaHortic.2009.819.3.

Schmilewski, G. 2017. Growing media constituents used in the EU in 2013. *Acta Hortic.* 1168(1168), 85–92. doi: 10.17660/ActaHortic.2017.1168.12.

Silber, A. and Raviv, M. 2001. Disinfestation effects on the chemical properties of tuff. *Acta Hortic.* 554, 41–50. doi:10.17660/ActaHortic.2001.554.2.

Smith, D. L. 1987. *Rockwool in Horticulture*. Grower Books, London, UK.

UNEP. 2019. *Frontiers 2018/19 Emerging Issues of Environmental Concern*. United Nations Environment Programme, Nairobi, Kenya.

von Post, L. 1937. The geographical survey of Irish bogs. *Ir. Natl J.* 6, 210–227.

Whitcomb, C. E. 1984. *Plant Production in Containers*. Lacebark Publications, Stillwater, OK, 638 p.

Zwingmann, N., Singh, B., Mackinnon, I. D. R. and Gilkes, R. J. 2009. Zeolite from alkali modified kaolin increases $NH_4^+$ retention by sandy soil: column experiments. *Appl. Clay Sci.* 46(1), 7–12.

# Chapter 4

# Developments in alternative organic materials for growing media in soilless culture systems

*Nazim S. Gruda, University of Bonn, Germany; and Neil Bragg, Substrate Associates Ltd, United Kingdom*



## 1 Introduction

The use of organic growing media is not new and was practiced in ancient civilisations. Since then soil, manure and composted materials have often been used in mixtures. The effort to develop standard growing media on a scientific basis began with the work of Lawrence and Newell at the John Innes Horticultural Institution in the UK towards the end of 1930s. They developed John-Innes compost, based on the use of sterilised loam, peat and sand with added fertilisers. In Germany, Fruhstorfer developed peat-clay-substrates called 'Einheitserde' at the beginning of 1950s. In the 1950s Baker at the University of California developed UC loam-less composts based on mixes of peat and sand as suitable loam was challenging to obtain. Later Puustjarvi in Finland and Penningsfeld in Germany reported the use of growing media based exclusively on peat. Parallel to these developments, production began to shift from researchers and growers to specialised industrial substrate companies (Gruda et al. 2016b).

http://dx.doi.org/10.19103/AS.2020.0076.03

In 1990 a review produced for the Horticultural Development Company in the UK identified upwards of 30 materials which might contribute to peat replacement (Bragg 1990). At the time of writing this chapter, issues such as cost and availability of materials have limited choice primarily from six to eight main substrate constituents. According to Schmilewski (2017), in the EU in 2013, 10.8% of all reported constituents used in soilless culture systems (SCS), equalling 3 719 000 $m^3$, were organic materials other than peat. In this organic category, coir materials, wood fibres and bark were the main constituents used. About 7.9% of all reported constituents for the manufacture of growing media was compost, amounting to 2 748 000 $m^3$.

Organic raw materials come from different sources. They could be classified in materials coming from:

1 natural degradation, for example, peat;
2 residues of manufacturing processes, for example, wood, tree bark and coconut fibres and coir;
3 waste and composted materials, for example, green compost; and
4 materials especially cultivated for soilless culture, for example, Sphagnum and miscanthus.

In this chapter, we will attempt to distil the available information on alternative organic materials which have emerged as the contenders for peat dilution or replacement.

## 2 Organic raw materials as residues of manufacturing processes: wood-based materials

Wood fibres, wood chips and sawdust are wood-based materials and renewable resources from the lumber and paper industry. All these products are characterised by low water retention, good air content and high-saturated hydraulic conductivity (Gruda et al. 2016b). Wood fibres are the most used material. Wood chips are shredded between 2 mm and 15 mm. Wood chips and sawdust are usually used in the proportion of 20–30% in mixtures with other substrate components. Wood chips give the growing media better aeration and structural stabilisation. The substrates may be used in containers but also for plant cultivation in ebb/flow systems.

Fresh sawdust from hardwood negatively affects plant growth because of phytotoxins (Maas and Adamson 1982). Handreck and Black (2010) reported a rapid decomposition of whitewood sawdust in pots, with a volume loss of up to 50% in one year. Slumping (loss of volume) and a significant loss of air porosity were also observed. Starck et al. (1991) found a low content of total and soluble nitrogen in leaves of carnation plants grown in sawdust in comparison to peat or mixtures of peat and sawdust. High applications of nitrogen fertiliser are required to balance N immobilisation in sawdust (Handreck and Black 2010). Another possibility is to compost it (Gruda et al. 2013).

Wood chips and wood fibre were first used as materials in Europe in the 1980s. The first information on the use of non-composted wood raw materials for growing media comes from France in 1984 when a company managed to bring an industrially produced substrate onto the market (Lemaire et al. 1989). At the beginning of the 1990s, wood fibres were investigated in Germany, for example, the work at Weihenstephan with ornamentals and vegetable plants. At the beginning of this century, wood fibre substrates also started to be used in the United States (Jackson et al. 2009). Wood fibres are now gaining in importance as a substitute for peat. In Germany, for example, several substrate manufacturers have recently invested in equipment to produce wood-fibre products for the horticultural sector.

Wood fibres that are suitable for use in growing media can be produced from untreated wood which contains little or no bark. Usually, the content of bark is less than 5%. As materials, softwood residues from the wood and paper industry, such as spruce, pine and sometimes Douglas fir are mainly used. Depending on the initial material, they could sometimes contain phytotoxins that may affect the plant growth at the beginning of cultivation. Gruda et al. (2009) recommend pre-treatment with substrate washing to deal with this problem.

According to Jackson and Bragg (2016), the term wood fibre is ill-defined and is taken to include any of the following:

- Hammer-milled wood chip: a purely mechanical process caused by the hammers in and around the screen around the mill;
- mechanically extruded wood chip: where the chips are contra-screwed between two rotating screws, resulting in a certain amount of heat being applied as a result of friction; and
- steam-heated wood chip: where the chips are steam-heated under increasing pressure and then released under pressure between contra-rotating plates to produce varying lengths of fibre.

Each of the processes produces usable materials. However, they do differ in the end properties and potential use within growing media mixes. Twin-screw extruders work with two screws running against each other. Due to frictional pressures, the material is heated up to 70°C for a short time. When disc refiners are used, the wood is shredded between grinding discs. This takes place either at atmospheric pressure or under increased pressure. Some manufactures use hammer mills to prepare wood fibres, sometimes utilising a whole tree substrate (Jackson et al. 2009). The quality of wood fibres has substantially improved in the last 20–30 years to present a homogenous fibre structure that, in turn, depends on the design and capabilities of the equipment used and adjustable process conditions. The machine settings make it possible to influence the fibre structure and thus the physical properties of the wood fibres, such as water retention and water-holding capacity (Gruda and Schnitzler 2000a,b, 2004a).

*Properties*: Wood fibres have a high air capacity. The total pore space of wood fibre substrates is similar to that of peat substrates and is over 90%. The volume weight is generally 0.083–1.50 or higher, depending on particle size and substrate compression (Table 1) (Gruda and Schnitzler 2004a). Wood-fibre substrates are characterised by low water retention with less-easily available water and water-buffering capacity compared to peat-based substrates, good air content and high saturated hydraulic conductivity (Gruda and Schnitzler 2004a). These characteristics lead to better aeration of substrate mixtures and promote their water permeability. A reduction of particle size, an increase of volume weight and an increase of irrigation frequency is recommended, to increase matric potential (Gruda and Schnitzler 2000a,b, 2004a). Excellent correlations were detected between the high percentage of particles <1 mm and maximum water-holding capacity and, therefore, plant growth (Gruda and Schnitzler 2006).

Macronutrient and micronutrient concentrations are deficient, except for K in some materials. Wood-fibre substrates are generally free of pathogens and weed seeds unless poorly stored. Due to the wide C/N ratio and the large surface area, untreated wood fibres are subject to a high level of microbial degradation and N-immobilisation. Gruda et al. (2000) found rapid nitrogen immobilisation in untreated wood fibres. N immobilisation can be 300–400 mg of nitrogen per litre of wood fibres, and sometimes more. This situation leads to nitrogen deficiency in the plants without corrective measures. Some companies have 'impregnated' wood fibre materials during their manufacture by adding slow-release nitrogen fertiliser. Gruda and Schnitzler (1999) and Gruda et al. (2000) found that N-immobilization could be effectively counterbalanced, even in 100% wood fibre substrates, with impregnation and additional N fertilizing during cultivation. In N-treated wood fibre, immobilisation was similar to white peat. Impregnation with a slow-release N-source provided enough N for initial microbial fixation. Generally, the amount of fertiliser corresponds to the N-withdrawal by the plant. However, general recommendations for the additional amount of fertiliser are challenging because the initial $N_{min}$ content of substrates varies enormously (Handreck and Black 2010; Gruda et al. 2000). Different cultivation systems also make general recommendations difficult.

Sulphur deficiency can also be a problem when using a substrate with wood-fibre components. It is therefore recommended to fertilise the plant with

**Table 1** Main advantages and disadvantages of alternative organic materials used as growing media constituents

| Material | Origin | Advantages | Disadvantages |
|---|---|---|---|
| Wood chips, wood fibres, sawdust and other wood by-products | By-products of wood manufacture | Good air content, and good TPS (>90% V/V), average volume weight (100–150 kg $m^{-3}$ and higher) | Variability, extremely high C:N ratio, high N-immobilisation, low water retention, increasing cost since used as an alternative to fuel and in landscaping |
| Bark (well-aged or composted) | By-product or waste of wood manufacture | Good air content and water-holding capacity, good TPS (75–90% V/V), sub-acid-neutral pH (5–7), average volume weight (320–750 kg $m^{-3}$), long-lasting | High variability, need time to reduce C:N ratio and terpene concentrations, increasing cost since used as an alternative to fuel and in landscaping |
| Coir | A by-product of fibre coconut processing | Physical stability, lightweight (65–110 kg $m^{-3}$), good air content TPS (85–96% V/V) and water-holding capacity, subacid to neutral pH (5–6.8) | May contain high salt levels, energy consumption during transport |
| Rice hulls | Residue of rice processing | Good TPS (89–93%), degradable processes are slow | Low water-holding capacity, available only in limited geographical areas |
| Xylit | By-product of brown coal mining | Good air capacity, low levels of soluble nutrients, low salt content, high cation exchange capacity, favourable pH-value, and low nitrogen immobilisation | Low water-holding capacity, the future availability is politically uncertain |
| Green compost | Composted plant residues | Good source of potassium and micronutrients, suppression of diseases, functional moisture-holding capacity, urban waste reduction | Variable in composition, high volume weight (600–950 kg $m^{-3}$), may contain excess salt, need time to be prepared, depending on material sometimes becomes easily waterlogged |
| Anaerobic digestate | Fermentation of residue for biogas production | Good air content, some fertiliser benefit due to some nutrients in the physical residue | Low easily available water, nutrient and salt-rich, continued degradation leading to slumping in use as GM |

*(Continued)*

**Table 1** (*Continued*)

| Material | Origin | Advantages | Disadvantages |
|---|---|---|---|
| Biochar | Solid material derived from biomass pyrolysis | Production is energy neutral, helps with carbon sequestration, biologically very stable. Possess a high TPC and water-holding capacity and high retention capacity for nutrients | Properties vary dependent on feedstock (biochar), relatively high production costs, often a high pH, can be dusty |
| Hydrochar | Material derived from biomass hydrolysis | Production is energy neutral, helps with carbon sequestration, wet material can be used. Possess a low EC and pH-value | Properties vary depending on the material used, high N-immobilisation, presence of some phytotoxic substances, relatively high production costs |
| *Sphagnum moss* | May be grown on the surface of re-wetted peat soils | Good TPS, air content, and water-holding capacity, acidic pH-value that can easily be adjusted | Contain herbal or grass seeds, poorly re-wettable; investment costs are still apparently high |
| Miscanthus | C4 perennial biomass crop | Low weight and high air capacity, low salt and nutrient contents | Low water-holding capacity, high nitrogen immobilisation, care needed in handling the feedstock, very sharp leaf mid-ribs which easily could cut the skin |

Modelled with a few modifications after Gruda et al. (2016b), and Savvas and Gruda (2018). TPS = total pore space.

additional calcium sulphate to avoid chlorosis and/or necrosis of the youngest leaves or to stimulate plant growth. Depending on the material used, the drying process may also cause a pH reduction. The species, age and origin of trees, including the site location and soil and, of course fibre processing and storage of the material, may also influence the pH value of the material.

Warm and humid conditions in the greenhouse, in combination with the loose structure of wood fibres, as well as water and nutrient supply, contribute to a high decomposition rate, and volume loss also called 'slumping'. The longer the plant growth, the higher the degree of slumping. Slumping is coupled with a reduction of macropores and an increase of micropores. Biological processes thus influence the physical properties of substrates, affecting plant development. To keep slumping within limits, Gruda and Schnitzler (2004a,b) recommend a relatively firm potting, that is, compression of wood-fibre substrates in the pots.

*Performance*: With a supply of nutrient solution in a soilless culture system (SCS), it is possible to use wood fibres as a stand-alone growing medium. Substrates based on wood fibre are already used in ornamental crop growing, nursery gardens and in vegetable growing. Because of high total pore space and air capacity, plants develop excellent roots in wood-fibre substrates. Investigations with ornamental plants (Steinbacher and Amberger-Ochsenbauer 1997; Gerber et al. 1999) and vegetable seedlings and transplants (Gruda and Schnitzler 2000a,b) showed that increasing the wood-fibre content of a substrate decreases the ability of plants to hold water and, accordingly, the frequency of irrigation should be increased. However, the total water quantity needed for irrigation is no higher than using a pure peat substrate. Wood fibres are further used to optimise the physical properties of other material components, that is, reducing volume weight, increasing air space and improving re-wetting capacity (Gruda and Schnitzler 2006; Barrett et al. 2016).

Because of their low content of available nutrients and other soluble salts, as well as their low weight, wood fibres are ideal for mixtures with nutrient-rich and heavy materials such as compost. The good buffering of pH can be an advantage when using irrigation water with low alkalinity and acidic fertilisers. In contrast, the use of carbonate-rich irrigation water can lead to increased pH value (Amberger-Ochsenbauer and Meinken 2020). As mentioned before, there is a risk of N immobilisation. Additional nitrogen supply is required, preferably with ammonium-based fertilisers. Sometimes, to avoid damage in young leaves or to avoid reduced plant growth, additional fertilising with calcium sulphate is recommended. In a similarly way to bark, wood chips are also used as organic mulch to reduce soil temperature fluctuations, water evaporation and suppress weeds (Gruda 2007, 2008), sometimes covering the container surface (Fig. 1).

*Environmental and economic concerns*: Wood fibres come from locally available and renewable sources and are, therefore, a sustainable material which can be used even in organic production. Worldwide competition in the wood products industry also influences the prices of the substrate. In recent years, the energy crisis has made the situation even more critical as wood is used as renewable fuel material. While the use of wood as an energy source is not a new phenomenon, the impact of its use as a biomass energy source has increased significantly in recent years (Gruda et al. 2013).

## 3 Organic raw materials as residues of manufacturing processes: bark, coir, rice hulls and xylit

### *3.1 Bark*

Bark is a significant component of growing media, particularly in areas where peat is scarce or expensive due to transportation cost (Gruda 2012, 2019).

**Figure 1** Wood chips used to cover the container surface, improving aesthetic and functionality by reducing substrate temperature fluctuations, water evaporation and suppress weeds.

It is a waste product from forestry. It started to be used as growing media in horticulture when the trunks of the trees were debarked mechanically in centralised production systems. Bark is usually stripped from trees, milled and screened into various sizes. Various by-products have been developed, for example, pine bark from the western Mediterranean. Some types of bark are used exclusively as mulch and do not count towards substrate components (Amberger-Ochsenbauer and Meinken 2020).

The production process influences bark quality. In order to meet the performance requirements of growers, growing media manufacturers usually undertake some secondary processing, often composting or ageing the material (Barrett et al. 2016). In Germany, for example, a composted bark or a so-called Rindenhums (bark humus) is made from predominantly native crushed and fractionated bark with or without nutrient additives. The common spruce *Picea abies* and, to a lesser extent, the forest pine *Pinus sylvestris* are used as sources for bark. They first undergo an aerobic decomposition process. Microbiological and biochemical processes break down growth-inhibiting ingredients such as resins, phenols and tannic acids. About one kilogram of nitrogen per cubic metre is added to the shredded raw bark, mostly in the form of urea, which is kept sufficiently moist and turned several times to create optimal conditions for high microbial activity. The activity of the microorganisms should reach temperatures of 60–70°C (Amberger-Ochsenbauer and Meinken 2020). Composted bark is mostly used in a grain size of 0–10 mm or 0–20 mm.

Pine bark from *Pinus pinaster*, which consists of relatively hard particles, is not subjected to microbial degradation and is therefore used without composting. The stable structure, the exceptionally high air capacity and the lack of any significant amounts of nutrients except moderate amounts of potassium made it suitable for use as growing media for orchids. The pH of around four sometimes makes it necessary to add lime (Amberger-Ochsenbauer and Meinken 2020). In some countries well-aged bark is used as a plant substrate. Aged bark is stockpiled and weathered for several months after production to promote biological stability and remove phytotoxic volatiles such as terpenes (Naasz et al. 2009; Gaches et al. 2011; Barrett et al. 2016).

*Properties*: Pine bark is not produced specifically for use in growing media and tends to have variable physical, chemical and biological properties (Barrett et al. 2016). However, after processing bark shows generally a very high air capacity. Air-filled porosity is high while water-holding capacity is generally low (Table 1). Bark is therefore used in mixtures with peat to improve both air and water-holding capacity. Bark generally increases the volume weight of a peat-based mixture. Polymers in bark include lignin, polysaccharides, monosaccharides and disaccharides. Microorganisms can assimilate free monosaccharides and disaccharides and degrade polysaccharide components, resulting in a high N-immobilisation rate. N-immobilisation is controlled through different manufacture processes.

Some fresh bark types may contain toxins, including high levels of monoterpenes and phenols, which may prove harmful to plants. Factors such as tree species, age, time of harvesting, soil type and geographical region affect phytotoxicity (Raviv et al. 2002). Recent studies have shown that hydrothermal treatments are in reducing phytotoxicity in industrial bark. After this treatment, bark maintains a very high air content that is beneficial when added to commercial peat-based substrates (Chemetova et al. 2018). Composted bark contains some plant nutrients. It is therefore recommended that it is mixed with nutrient-poor materials such as peat and clay. High manganese content, especially at low pH, could also be a source of potential phytotoxicity (Maher and Thomson 1991), for instance, in spruce (*Picea abies*) bark (Solbraa and Selmer-Olsen 1981). The authors recognised the potential problem with bark taken from different locations and developed a method for extraction and characterisation of potentially harmful levels of soluble Mn. The regulation of pH through liming can also improve bark as growing media.

*Performance*: Bark is used as a stand-alone substrate in the production of orchids and as a constituent in growing media in pot ornamentals (Gruda 2019). White spruce and fir bark alone, or mixed with low-grade peat, showed good potential for greenhouse tomato production as an environmentally sound alternative to rock wool (Allaire et al. 2005). Pine bark can also be successfully used as a stand-alone substrate for soilless cultivation of vegetables such as bell pepper, cucumbers, and muskmelons (Shaw et al. 2004; Rodriguez et al. 2006; Cantliffe et al. 2007), and as a bark-based material for strawberry production (Depardieu et al. 2016). Pine bark provides a root zone structure that does not inhibit water drainage. Plant roots in a container thus have sufficient airspace and water-holding capacity for growth. When the bark is well composted, it can comprise up to 50% of a substrate (Maher and Thomson 1991). It can be used for container trees and salt-tolerant pot plants. Irrigation intervals have to be adapted to counterbalance the relatively low water-storage capacity. The high content of manganese can induce visible chlorosis of plants and lead to iron deficiency (Ortega et al. 1996). Amberger-Ochsenbauer and Meinken (2020) recommend avoiding pH values below 6 to limit the availability of manganese and thus avoid the risk of induced iron deficiency. Additional iron supplements in the form of chelates also reduce this risk.

*Environmental and economic concerns*: The price of bark depends on availability, the distance from sources of bark (affecting transportation costs) and type of processing. Economic analysis has found that pine bark is nearly one-eighth the cost of perlite and can be re-used for several consecutive crops, resulting in reduced production costs and higher profits. However, bark can become a limited resource due to the changing nature of the timber industry and the fact that bark is a useful energy source (Owen et al. 2008). There are also restrictions on trade in bark between countries due to phytosanitary concerns, particularly invasive pests such as the pinewood nematode and the spruce larch beetle. Bark exports from Southern European countries, for example, need sterilising before an export licence can be granted.

## *3.2 Coconut materials (coir)*

Coir comes from the fruit of the coconut palm, *Cocos nucifera*. Coir is mainly imported from Sri Lanka and India, but also from Vietnam, the Philippines and Mexico. The skin of the ripe fruit consists of the outer leathery and smooth exocarp, the underlying spongy-fibre mesocarp and, finally, the inner, relatively hard endocarp.

The mesocarp contains a dense network of fibres. The long fibres of coir extracted from the coconut husk are used in the manufacture of industrial products, for example, mats or ropes. Traditionally the dust and small fibres were left behind and accumulated as a waste product. From the late 1980s this material has been used as a growing medium (Maher et al. 2008). Coir fibres, coir pith (sometimes mistakenly referred to in the literature as coco peat) and coir chips are some of the most abundant plant-derived organic waste materials in many tropical and subtropical countries which can be used as a renewable by-product. The use of coir in growing media has increased significantly since 2000, particularly in Europe, but also in the western United States (Carlile et al. 2015; Gruda 2019).

Husks are sometimes soaked in water to soften them and facilitate grinding (Maher et al. 2008). Before the processing, the raw material can be stored on paved surfaces for about six months to stabilise its physical properties. During this time phytotoxic substances and rapidly decomposing organic components are broken down. Alternatively, to meet the ever-increasing market demand for coir, fresh green husks are collected and shredded and milled to separate the 'green' fibres and coir pith. The latter is then loaded into trucks and hose washed with water or solutions of calcium nitrate. However, sometimes only some of these steps are carried out. The material is then laid out on pads to dry in the sun with occasional turning to assist drying, before being compressed into bricks for transport. Coir fibres usually are bound in bales to facilitate transportation (Salvador et al. 2005).

*Properties*: Coir fibres have a pore volume of 85–96% by volume, increasing air capacity when added to growing (Table 1). On the other hand, coir pith has high water capacity and easily available water. High lignin content contributes to low microbial degradability and nitrogen immobilisation (Rencoret et al. 2013). Since coir contains more lignin and less cellulose than peat, it is more resistant to microbial breakdown and may shrink less (Robbins and Evans 2010). Low microbial decomposition makes this material suitable as a stand-alone substrate for long-term cultivation with high demands on air capacity, as well as a constituent in substrate mixtures. All coconut materials are easily re-wettable, which improves the water absorption of substrate mixtures and water distribution in the growing medium.

Coir pH values are 5–6.8. The difference to peat is that coir has little buffering capacity against pH changes. Fertiliser application becomes the controlling factor affecting pH. The initial electric conductivity (EC) of coir may be high. Sequential washing is therefore recommended to reduce EC. Leaching of nitrogen is marginally higher than in peat when comparing materials of similar particle size (Noguera et al. 2000; Gruda et al. 2013).

*Performance*: Coir is used pure, as a stand-alone substrate, or in mixtures with other components. It is particularly suitable for long-term cultivation, especially orchids and container trees, but is also used for ebb/flow and closed cultivation systems. Together with coir pith, coir fibres are also used as bag cultures in so-called growbag systems for berry crop, vegetable and cut flower production. Coir needs to be pre-moistened before use. Mariotti et al. (2020) have recommended the use of coir as an alternative substrate to grow Quercus species seedlings. The right degree of fertilisation can offset coir chemical deficiencies. Excellent physical and chemical properties make coir a suitable growing medium for herbs and, more recently, cannabis.

*Sustainably concerns*: The use of coir, especially of coir pith, is to be seen as environmental-friendly. However, coir production represents the use of an end-of-life product from coconut production. There is therefore a need to take into account the environmental impact of coconut production. Coir is also a finite resource determined by the demand for coconuts as a food crop. Significant disadvantages include the high demand for freshwater in those countries where coir is washed – this needs serious consideration in areas where potable water is already limited. Since production is limited to tropical areas, transportation is another issue, though this is reduced by coir being light in weight (Gruda 2012, 2019; Barrett et al. 2016). Transportation costs can be further reduced by drying and compression of the material before transportation. Sometimes the washing takes place in the countries of use. Another social issue is child labour and poor labour rights in countries of origin. There is also domestic use of coir in Asian countries as a fuel for cooking. Product certification is needed to show that growers and producers have taken account of environmental and social issues. However, as environmental issues have grown in importance, the relative expense of coir compared with peat has become less of a constraint in its uptake (Barrett et al. 2016).

## 3.3 Rice hulls

The hulls of rice are the outermost layer of the rice (*Oryza sativa* L.) grain, a residue of rice processing. Rice hulls have been used to a limited extent as growing media components since the 1980s. Usually, the hulls from the 'parboiled process' are suitable for this purpose, because the germination of weeds and rice is inhibited. Besides, hulls from this manufacturing process have more or less consistent quality. Rice hulls are light and have a high total pore volume of 89–93% (Gómez and Robbins 2011), but a low water capacity (Table 1). They are, therefore, suitable for improving the aeration and drainage ability of substrates. Due to the content of lignin and silica, the medium of rice hulls is stable, and the degradable processes are slow. Several studies showed that rice hulls are used to cultivate different plants, such as seedlings of *Pinus* (Tsakaldimi 2006), *Gerbera* (Barreto and Jagtap 2006), *Spiraea x bumalda* (Gómez and Robbins 2011) and *Cyclamen* (Bonaguro et al. 2017). Besides, aged, composted and pyrolysed rice hulls are used as growing media components in some countries.

### 3.4 *Xylit - the brown coal fibre*

When brown coal is extracted, xylitol (trade name Xylit) also occurs. Xylitol is inferior due to the lower calorific value and is usually separated. It is a preliminary stage of brown coal or lignite and consists of not fully carbonised over 10 million year-old plant materials, mainly trees and in which the wood structure can still be recognised (Kaever 2004). The largest lignite reserves in the world, corresponding to reserves of xylitol are Germany, Russia and America. The fibres are mechanically shredded, screened and cleaned from brown coal particles. Xylitol has a high air capacity and low water capacity. Besides, it has low levels of soluble nutrients, low salt content and a favourable pH-value. These chemical properties are similar to peat. The high exchange capacity can be an advantage (Table 1).

In contrast to wood fibre, the nitrogen immobilisation is low, and the structural stability is high. Kamp and Emino (1983) reported that a peat-vermiculite potting medium mixed with 25% and 33% lignite by volume, did not alter the growth parameters of *Pelargonium*, but reduced the growth of *Ficus*, *Ardisia*, *Euphorbia* and *Chrysanthemum* plants. Dysko et al. (2015) found that lignite was on a par with rock wool for the marketable and total yield of tomatoes, cultivated in SCS. Since the material is a by-product of brown coal mining, the future availability of xylit is politically uncertain in some European countries.

## 4 Waste and composted materials other than bark

Waste is an environmental problem and an economic loss with a direct and indirect impact on climate change, air pollution and ecosystems. Recycling in the form of potting media provides a potential solution in reducing waste amount and in increasing the proportion of peat-free material in plant substrates (Gruda 2012, 2019; Barrett et al. 2016). Abad et al. (2001) differentiate between urban, marine, agricultural, forest, animal, industrial and food waste. Compost is a bioresource that includes all organic matter that has undergone thermophilic, aerobic decomposition (Raviv et al. 2002; Gruda et al. 2013). Depending on the starting material and production processes, several kinds of composts can be produced. These include urban and municipal solid wastes; animal manure; grape, olive and other food processing wastes; bark; sewage sludge or paper waste; greenhouse waste, pruning waste, spent mushroom compost and green waste (Table 2).

Generally, we can differentiate between various types of compost:

- self-produced compost;
- bio-waste compost; and
- green-waste compost.

**Table 2** Waste materials used for compost, which, in turn, is used as a plant substrate on its own or in mixture with other materials

| Feedstock waste | Use as a growing medium for the production of | Reference |
|---|---|---|
| Animal manures | Pot plant production, landscape nurseries, vegetables and cut flowers production | Raviv et al. (2004) |
| Broccoli | Lettuce | Santos et al. (2016) |
| Chestnut | Lettuce | Santos et al. (2016) |
| Coconut coir dust | Gerbera | Riaz et al. (2015) |
| Coffee pulp | Tomato seedling | Berecha et al. (2011) |
| Corn cobs | *Anthurium* | Suo et al. (2011) |
| Cotton gin | *Azalea* | Cole et al. (2005) |
| Dredged sediment co-composted with green waste | Ornamental plants | Mattei et al. (2017) |
| Farmyard manure | Gerbera | Riaz et al. (2015) |
| Grapefruit with coir or vermiculite | Seedlings of lemon basil | El-Mahrouk et al. (2017) |
| Grapes | Lettuce | Santos et al. (2016) |
| Green waste and sewage sludge | Ornamental bedding plant | Grigatti et al. (2007) |
| Green/pruning; green/pruning wastes compost, vermicompost and slumgum compost | Rosemary, Leyland cypress, lettuce, onion, petunia and pansy | Morales-Corts et al. (2014) |
| Olive mill[a], olive[b] | Melon, cress and tomato plants[a]; lettuces[b] | Aviani et al. (2010)[a], Santos et al. (2016)[b] |
| Plant leaves | Gerbera | Riaz et al. (2015) |
| Posidonia residues | Tomato[a], lettuce seedlings[b], melon and tomato seedlings[c], pot basil[d], pot sea fennel[e] | Castaldi and Melis (2004)[a], Mininni et al. (2012)[b], Mininni et al. (2013)[c], Mininni et al. (2015)[d], Montesano et al (2018)[e] |
| Pruning wastes; pruning waste and municipal solid, or sewage sludge | Ryegrass and cypress[a], Pistacia (nursery)[b] | Benito et al. (2005)[a], Ostos et al. (2008)[b] |
| Sewage sludge | Ornamental conifer plants | Hernández-Apaolaza et al. (2005) |
| Slumgum compost | Rosemary, Leyland cypress, lettuce, onion, petunia and pansy | Morales-Corts et al. (2014) |
| Spent mushroom | Ryegrass and cypress | Benito et al. (2005) |
| Urban solid wastes | Tomato transplant | Diaz-Perez and Camacho-Ferre (2010) |

According to Gruda (2019). Superscript reference (e.g. [a,b,c]) link feedstock waste, and growing media with the corresponding literature, applicable only within rows, not columns.

Self-produced composts came from organic residues of the farm, such as vegetable waste and substrate residues. They have traditionally been an essential component of growing media in the past. Their importance decreased significantly with the development of industrially produced growing media. However, they are still used in developing countries. Self-produced compost is used to a small extent in developed countries as well, contributing to creating more circular horticultural production. Since it is a spontaneous process involving non-homogenous materials, there are strong fluctuations in the levels of soluble nutrients (Amberger-Ochsenbauer and Meinken 2020). It is therefore recommended that material is analysed and pasteurised before it is used as a medium to grow plants.

Bio-waste compost is produced under aerobic conditions in closed decomposition facilities from kitchen and garden waste. Bio-waste compost is not homogeneous, has high salt levels, and therefore, is not recommended for use by the growing media industry.

In contrast, compost from green materials is widely used in Europe, primarily in media used for the hobby market. Green waste compost is made from greenhouse vegetables, nursery shrubs, branches, plant trimmings, leaves, and grass from gardens, public green spaces and other landscapes. The woody material is usually chopped and mixed with the remaining green residues and gathered together in clamps. Adequate moisture, nutrients, oxygen and pH-value are essential prerequisites to promote the microorganisms that enable the aerobic rotting process. These microbial degradation processes raise the temperature in the clamps to over 60°C. These temperatures lead to a pasteurised composted material that prevents weed seeds from germinating. Non-mature compost can immobilise a significant amount of N but, once stabilised, the compost acts as a slow-release fertiliser (Raviv et al. 2002). However, even though the nutrients and salinity of green-waste composts can fluctuate, high-quality composts can be achieved by selecting the right raw materials and by optimally managing the composting process (Massa et al. 2018). Amberger-Ochsenbauer and Meinken (2020) suggest mixing woody with herbaceous material, nutrient-poor with nutrient-rich as well as moist with dry material. Adequate moistening and/or addition of nitrogen is recommended. There is a need for new technologies to facilitate the production of compost (Massa et al. 2018). One possible alternative to conventional composting is to use the action of worms to break down organic waste materials to produce vermicompost (Gruda 2019).

*Properties*: Compost is not a consistent product. Physical, chemical and biological properties vary, depending on the raw materials used and the production processes. A mature and stable material is recommended for growing medium use. Generally, composts are relatively heavy (600–950 g/L fresh wt compared to peat-based substrates at 200 g/L fresh wt). Their use usually increases the volume weight of a substrate mixture and reduces the air capacity. It is therefore recommended to add other organic or mineral materials with a high air capacity, such as wood fibres or vermiculite. Green-waste compost contains valuable nutrients such as phosphorus, potassium and trace elements, and microorganisms in the compost may also have suppressive properties against soil-borne diseases (Table 1). Composts can also be used as bio-fertilisers and bio-stimulants (Gruda 2019).

Depending on the origin of the starting material, the limiting factor regarding the use of composted green waste is its high concentrations of EC, potassium (K) and calcium carbonate ($CaCO_3$). High amounts of free calcium carbonate can lead to pH values above 8. Composts should, therefore, contain less than 10% by mass of $CaCO_3$. Carbonate can be partially neutralised by adding elemental sulphur. Sulphuric acid is then formed in the substrate by sulphur-oxidising bacteria (Amberger-Ochsenbauer and Meinken 2020). Maher and Prasad (2001) have suggested the use of up to 50% compost as a component in a growing medium but not as a stand-alone substrate. There can also be problems from plant and human pathogens as well as weed contamination if the composting process is not conducted correctly, that is, if the temperature and the time of exposure are not sufficient (Gruda et al. 2016b). Moreover, compost has low (5–10%) carbon efficiency which is reflected in material mass and volume reduction and a relatively high pH (Gruda 2019). The quality assurance of green compost established should be regularly analysed by testing for weed seeds, nutrient content and pollutants.

Particle-size distribution and fertility are superior in vermicompost-based media than in the conventional compost media. However, it should be noted that vermicompost materials are not pasteurised and weed seed may still be viable. In experiments carried out by Gong et al. (2018), vermicompost medium showed an approx. 2.2-time higher fineness index than compost-based media. Earthworms increase the number of small particles by ingesting, mixing, grinding and then egesting organic material (Garg et al. 2012). The nutrient level is also higher and heavy metal concentration is lower in vermicompost (Lazcano et al. 2008; Wang et al. 2013; Gong et al. 2018).

*Performance*: As can be seen in Table 2, different nursery, ornamental and vegetable plants can grow into compost substrates. Compost, when mixed into growing media, is a rooting medium, and an essential source of nutrients such as nitrogen (N), phosphorus (P), potassium (K) and microelements. Substrate mixtures containing compost require adjusted fertigation to take account of nutrients supplied by the compost (Vandecasteele et al. 2018a). The raw material is of crucial importance. Massa et al. (2018) reported that compost derived from mixed raw material negatively influenced *Pelargonium* plant nutrition and photosynthesis, thus significantly reducing plant biomass accumulation and quality. Compost is a suitable component for growing media used in pot and balcony plants, perennials and shrubs that require a high amount of nutrients. In contrast, it is not suitable for salt-sensitive plant species.

*Sustainability concerns*: The use of waste as a composting material can be of dual benefit (Gruda 2019). An example is the removal and disposal of large volumes of plant biomasses of Posidonia, a damaging marine phanerogam endemic in the Mediterranean (Mininni et al. 2012). After composting the material can become a component for plant substrates (Table 2). High volume weight in association with transportation costs can reduce the economic viability of compost and increase its environmental impact (Mininni et al. 2012). From a commercial perspective, the most critical problems are contamination of the waste stream, principally from herbicides but also foreign bodies (e.g. glass) and human pathogens (Barrett et al. 2016). Microparticles and nanoparticles of plastic are a more recent problem. Compost therefore needs appropriate quality standards and procedure protocols.

### 4.1 Anaerobic digestate

Some composts from the anaerobic fermentation of residue from biogas production have recently been tested for use as growing medium components. Fresh residues contain some phytotoxic intermediates and can have high phosphate content and an unstable nitrogen balance. Due to the presence of a high proportion of animal raw materials, these materials should be pasteurised. Dry fermentation usually leads to nutrient and salt-rich composts that cannot be used as growing media materials. In the case of wet fermentation, approximately 10% of dry matter, a high proportion of the soluble constituents is removed with the process water (Amberger-Ochsenbauer and Meinken 2020). Ridout and Tripepi (2011) reported pH-values as high as 7, and EC-values as high as 5.0 dS $m^{-1}$ (Table 1), mainly caused from high concentrations of potassium (K), phosphorus (P), sodium (Na) and chloride (Cl). Composts may also contain high concentrations of nitrogen, primarily in ammonium nitrogen form. The addition of 20% dry solid-phase digestate to peat-based substrates increased air capacity and decreased easily available water content, and have been successfully used in greenhouse experiments with pot plants, such as *Petunia, Impatiens* and *Pelargonium* (Dubský et al. 2019).

## 5 Biochar and hydrochar

Biochar is a form of charcoal which is manufactured from organic matter by heating in an anoxic situation (pyrolysis). Relatively dry materials, such as wood-chips, sawdust and other wood materials are used to produce biochar, though coir is also used.

In contrast to biochar production, the hydrothermal carbonisation (HTC) process requires moderate temperatures and pressures. Water is also needed to accelerate the reaction. Any raw organic material with relatively high water content such as general waste, as well as residues from biogas plants, vegetable material or sewage sludge, can be used. The final product is similar in structure to biochar produced by pyrolysis (Gruda 2012; Gruda et al. 2013). However, the properties are different, and this is reflected in plant performance as well.

*Properties*: Biochar and HTC char have physical and chemical properties that are variable, depending on the raw material used and the carbonisation technique (Gruda 2019). Biochar has high stability, low volume weight, with good aeration and water-holding capacity. The total pore space of biochar can be 90–95% by volume and is similar to peat. Due to the large surface area, biochar possesses a high nutrient exchange capacity (Nemati et al. 2014), especially for phosphorus and nitrogen, reducing nutrient leaching (Table 1). Kim et al. (2017) suggested using rice hull-derived biochar to increase cation exchange capacity. EC and pH values are low in both peat and HTC but are higher in biochar (Kern et al. 2017). Biochars from nutrient-rich raw materials may be too salty (Kaudal et al. 2015) and alkaline (Blok et al. 2017).

Prasad et al. (2019) stated the need for particle size information, as fractions from the same biochar can have different levels of total extractable nutrients and pH levels. This is needed if, for instance, the biochar is used to replace or reduce lime application rates of a peat-biochar mixture. Generally, the properties of HTC are not comparable with biochar in their use as growing media. Future issues to resolve include mould formation, presence of pesticides and odour (Gruda 2012). Another issue is high N-immobilisation in hydrochar as well as the presence of some phytotoxic substances (Table 1) (Neumaier et al. 2017).

*Performance*: Several studies of the application of biochar and hydrochar as growing media consider them as a peat replacement (Gruda et al. 2019a,b). However, research on their effects on plant growth is not consistent and sometimes contradictory. Nobile et al. (2020) studied six different biochars for basil (*Ocimum basilicum*), lettuce (*Lactuca sativa* var. *crispa*) and pansy (*Viola wittrockiana*). They found that physical and chemical properties of biochars had significant positive impacts on plant growth. Álvarez et al. (2018) showed that it is possible to grow container plants of geranium (*Pelargonium peltatum*) and petunia (*Petunia* × *hybrida*) using a peat-based substrate mixed with 20–25% biochar and/or vermicompost by volume. Plants in these substrates showed a similar or enhanced physiological response to those grown under control using a commercial peat-based substrate. Tian et al. (2012) and Conversa et al. (2015) reported that biochar results in the same plant growth as in peat-based standard substrates for *Calathea* and *Pelargonium zonale*. Dumroese et al. (2018) reported that biochar could successfully be used in substrate mixtures in nursery production of *Pinus ponderosa* seedlings, while Frenkel et al. (2017) reported on effects from the use of biochar in reducing plant diseases. In contrast, HTC char does not always perform as well as established growing media components. High N-immobilisation and the presence of some phytotoxic substances reduced growth of potted basil, even in mixtures of only 30% by volume HTC (Neumaier et al. 2017).

*Sustainability concerns*: Both pyrolysis and HTC have potential for the production of sustainable $CO_2$-neutral energy from biomass. However, the carbon efficiency of pyrolysis is nearly 50%, whereas the carbon efficiency of HTC is 80–100%. Carbon efficiency, usually used in energy calculations, is the amount of carbon of biomass contained in the end product compared to the amount before the treatment (Gruda 2012). The high-energy input in the manufacturing process of biochar is a weakness in terms of environmental impact. However, carbon storage is higher than $CO_2$ emissions from the production of biochar which means that, in the long term, biochar can store carbon (Steiner and Harttung 2014). Levesque et al. (2018) evaluated the effects of additional biochar at a rate of 15% (v/v) to a peat-based substrate. They found that the biochar addition increased C, decreased N availability in fertigated peat-based growing media, and mitigated carbon dioxide, methane, and nitrous oxide emissions. However, biochar production is still relatively expensive.

## 6 Cultivated organic raw materials

### *6.1 Sphagnum moss*

*Sphagnum* moss is a plant that may be grown on the surface of re-wetted peat soils. Agriculture on rewetted peatlands is called 'paludiculture'. It provides a sustainable alternative to drainage-based agriculture. One form of paludiculture is the cultivation of *Sphagnum*, which can be used as a raw material for growing media (Temminka et al. 2017). The cultivation of *Sphagnum* ('Sphagnum farming'), the technology of using *Sphagnum* biomass in industrial substrate production, and the use of growing media based on *Sphagnum* biomass in professional horticulture, have been extensively studied in recent years (Gaudig et al. 2014). New areas have been cultivated in the last twenty years, primarily in Canada, Finland, Latvia and Germany. Experiments showed that a high and stable water level is the most decisive factor for *Sphagnum* growth. From the species used, *Sphagnum palustre* is one of the most promising species for both growing media and Sphagnum farming (Gaudig et al. 2014). Nevertheless, the opinions of the scientists on this topic differ widely. While Gaudig et al. (2014, 2018) enthusiastically speak for a replacement of peat, Amberger-Ochsenbauer and Meinken (2020) reported a series of questions that have not yet been sufficiently clarified.

*Properties*: The horticultural properties are similar to those of white peat (Gaudig et al. 2008). *Sphagnum* has high pore space, air capacity and water-holding capacity (Table 1). These favourable physical properties result from the structure of the sphagnum leaflets, which consist of two different cell types. Thanks to the hyaline cells (hyalocytes, water cells), peat mosses can absorb 15–20 times their dry mass of water. Even in the event of drought, the structure as a whole is retained (Amberger-Ochsenbauer and Meinken 2020) after drying out entirely; peat moss is initially poorly re-wettable and similar to peat. The extremely low nutrient and salt contents, except for potassium, are an advantage. The pH-value is strongly acidic and similar to peat. The stable nitrogen balance also makes peat moss a suitable raw material for growing media. Nevertheless, some batches of Sphagnum have led to less than optimal results in plant experiments. Allelopathic effects between mosses and crop plants could be a possible reason. The presence of herbal or grass seeds can also be disadvantageous. Sphagnum is therefore often gamma radiated or treated with steam vapour before use to kill any germinating seeds it may contain (Gaudig et al. 2014).

*Performance*: The quality of orchid roots has been found to be excellent in pure *Sphagnum* (Emmel 2008). Together with peat substrates, *Sphagnum* is suitable for cultivation of *Tagetes* (Emmel 2008), *Pelargonium* and *Petunia* (Jobin et al. 2014). Plant substrates with up to 50% by volume of *Sphagnum* performed well with different horticultural species, including ornamental, vegetables, herbs, shrub and trees (Gaudig et al. 2018). *Sphagnum* has been used even in press pots for production of vegetable seedlings without loss of quality or press pot stability when black peat was replaced with 25% *Sphagnum* biomass (Emmel 2017). *Sphagnum* biomass production and use in substrates is likely continued to develop (Gruda et al. 2013; Jobin et al. 2014).

*Sustainable concerns*: Studies from the University of Greifswald, Germany have shown that *Sphagnum* farming might provide a sustainable, high-quality alternative to fossil peat as a raw material for growing media. *Sphagnum* farming is a climate-friendly and sustainable land use option for abandoned cutover bogs and degraded bog grassland (Gaudig et al. 2014, 2018). From an economic point of view, investment costs are still high, requiring new technologies to compete with peat production (Gaudig et al. 2018).

## *6.2 Giant miscanthus*

*Miscanthus* × *giganteus*, a hybrid of *M. sinensis* and *M. sacchariflorus* is a promising C4 perennial biomass crop. It has high yields, low input demand, excellent environmental performance, multiple biomass-use options and the potential to grow on land considered marginal for food production (Muylle et al. 2015; Lewandowski et al. 2016, Debode et al. 2018). The straw is usually used in the construction, automotive and pulp industries. However, it can be shredded or processed with an extruder, as for wood fibre production, and then used as a raw, locally sourced material for growing media.

*Properties*: The straw of miscanthus has a low weight and high air capacity (Table 1). In contrast, the water-holding capacity is relatively low. The salt and nutrient contents are low, as is the pH and nutrient buffering. Due to the high C/N ratio, the straw of miscanthus is subject to nitrogen immobilisation (Vandecasteele et al. 2018b). Nitrogen in the form of ammonium sulphate or urea is sometimes added.

*Performance*: Miscanthus is used as a stand-alone substrate in SCS (Fig. 2), or as a component for growing media mixtures. Frequent water supply in small quantities can compensate for low water-holding capacity. Due to high air capacity, miscanthus straw is suitable for cultivation where there are high air demands in the root area. A continuous supply with nutrients, starting with the beginning of plant culture, is needed to offset N-immobilisation. Debode et al. (2018) reported that pre-colonisation of straw with *Trichoderma* increased post-harvest disease suppressiveness against *B. cinerea*. The high silicon content of miscanthus could have a positive phytosanitary effect (Woli et al. 2011).

*Sustainability concerns*: Miscanthus is a renewable material that can be locally cultivated on land considered as marginal for food production. However, production costs are higher than wood-fibre substrates. Whether the harvest of biofuel crops for alternative use as growing media is economically viable remains unclear (Barrett et al. 2016).

Switchgrass (*Panicum virgatum* L.) (Altland and Krause 2009; Altland 2010), giant reed (*Arundo donax* L.) (Andreu-Rodriguez et al. 2013), reed canary grass (*Phalaris arundinacea* L.) straw (Kuisma et al. 2014), willow (*Salix* spp.) (Altland and Krause 2009), flax (*Linum usitatissimum*) and hemp shives (*Cannabis sativa*) (Vandecasteele et al. 2018b) have been used in plant production as sole substrates or in mixtures with other materials (Gruda 2019). Flax shives

(a) (b)

**Figure 2** Strawberries cultivated in a stand-alone *Miscanthus × giganteus* substrate in a soilless culture system (Photo: Gruda 2019, private collection).

possess a low weight, a high total pore volume with a high air capacity, and are, therefore, suitable to be mixed with heavy growing media components.

## 7 Case study: peat use reduction in European countries

Peat is a limited resource and peatlands are the most important long-term carbon sinks in the terrestrial biosphere. They only cover about 3% of the earth's land area but may store nearly to two times more of organic carbon as all the forest areas in the world (Yu et al. 2010; Weissert and Disney 2013; Gruda 2019, 2020). However, when peat bogs are drained and extracted, that is, used in agriculture and forestry, they no longer act as carbon sinks. Degraded peatlands contribute disproportionately to global greenhouse gas emissions, with approximately 25% of all $CO_2$ emissions from the land-use sector (Bonn et al. 2014). According to Quantis (2012) and Stucki (2020), peat has the highest impact on 'climate change' of all commonly used substrate materials. Peatlands are also associated with other important ecosystem services such as water quality and flood prevention (Gruda 2019; Gruda et al. 2019a,b).

European countries are actively discussing whether further peat harvesting should be stopped. Peat extraction for non-energy use in Europe is approximately 30 million $m^3$. Voluntary measures to reduce peat use in growing media are part of the approach to reduce the need for peat extraction. Some countries such as Austria, Norway, The Netherlands, Switzerland and the UK have set the goal of reducing peat use. Currently, in the UK peat represents around 53% of all substrates and soil conditioners (Bragg 2020). In 2019 a declaration of intent was signed in Switzerland for commercial horticultural products, setting a target to reduce peat to a maximum of 50% of substrates by 2025 and 5% by 2030 (Carlen 2020). Recently, Norway set a national target to phase out the use of peat in amateur growing media by 2025 and professional horticulture by 2030 (Anonymous, 2020).

Germany is one of the most significant players in the peat market, importing almost a third of its peat, mainly from the Baltic countries, while a good third of the total production of ready-to-be-used growing media is exported to other countries. The Federal Ministry of Food and Agriculture in Germany has also developed a strategy to significantly reduce peat use, including discussions with the horticultural sector (Gruda 2020). As a result, peat has been replaced in such sectors as soil conditioners, the use of substrates for SCS production of tomatoes, peppers, cucumbers, lettuce and cut flowers, to a large extent in pot herb production and even in the area of ornamental pot plants, such as *Gaultheria*, *Poinsettia* and *Calluna*. Meanwhile, companies have started to market peat-reduced and peat-free substrates for hobby growing (Fig. 3).

However, challenging will remain as protected SCS expand into new areas such as berry fruit cultivation, where they offer the opportunity to avoid soil

(a) (b)

**Figure 3** Peat-reduced (a) and peat-free (b) substrates offered in the German market. (a) 50% peat substrate; (b) 100% peat-free substrate.

diseases, reduce vulnerability to crop damage from rain and hail storms in the early summer and provide greater ease of picking. Due to the low pH needs of berry crops, peat is the major growing media used though coconut products are being introduced. Lowering of pH is required when alkaline growing media are used, which, for instance, can be achieved by adding different sulphur-based acidifying agents.

Looking more generally, cost remains a key issue, particularly in countries in Northern Europe with closer proximity to peat extraction areas where peat products are cheaper. According to Knight (2012), the transition to sustainable, protected cultivation systems needs to be economically viable. If inexpensive and better-quality growing media were available, they will be used in a large scale. Research also suggests that growers and gardeners tend to favour the types of growing media that they are accustomed to and know how to manage. Hence inertia is also a barrier to change. Even the change from John Innes formulations to peat in the 1960s did not occur overnight (Knight 2012).

The search for sustainable alternatives to peat will continue. Physical, chemical and biological characteristics of substrates influence plant performance. Working with alternative substrates requires changes to cultivation practices. Some practical considerations are:

- Fine-grained substrates with small particle sizes should be used for germination, direct sowing and young plant cultivation in small bales or multi-pot plates. For other plants that need large pots, relatively coarse substrates or substrate mixtures could be used. Fine-grained substrate has a higher water capacity, but a coarse-grained substrate is advantageous for later development of plants.
- Those substrates that have lower water retention, for example, wood fibres, should be moistened well before being used. In addition, they need to be irrigated more often. Wetting agents and non-ionic surfactants may be applied to help with re-wetting.
- To avoid slumping, try to add more substrate to the pot or container from the beginning by compressing it.
- The container size and shape have a significant impact on cultivation success.
- In order to account for N immobilisation, supply plants with the appropriate nutrients straight away.
- Note the pH at the beginning of the culture, and then check regularly as with other substrates.

Further progress requires action not only from researchers and producers but also legislators across regions such as the EU-wide, both to limit further peat extraction and use as well as to facilitate the market for alternative substrate materials that are environmental-friendly. The reduction of using peat is a challenge, but it is also a considerable opportunity for horticulture (Gruda 2020).

## 8 Summary and future trends

Peat will remain a significant constituent of growing media in the near and mid future. Similarly, rock wool will be the primary substrate used in SCS based on nutrient solutions, although coir will be a competitive substrate (Gruda 2012). However, there is increasing pressure on both growing media manufacturers and horticulturalists to significantly reduce their reliance on peat as a constituent of substrates.

There has been a search for alternative materials which could either reduce peat content of mixes and/or replace the use of peat entirely. In this chapter, we attempted to distil down the available information on those materials which have emerged as the major contenders for peat dilution or replacement. The list of alternatives is driven in part by local availability of materials, for example, while rice hulls have been investigated in areas such as Italy and the Southern states of the United States. Away from these geographic areas, there is no local supply and hence the use in substrates becomes limited by availability and cost.

A big issue beyond the availability of specific alternatives identified is quality. Physical, chemical and biological properties of organic materials are generally very different from peats. The unique processes that have led to peat formation have created specific properties which are extremely difficult to find in many alternatives. For instance, the water-holding capacity of peat(s) compared to wood fibres is entirely different. It does not mean that the latter cannot be used but does require a management shift in irrigation and fertilisation applications to adjust for the new material in use. Biologically many of the alternatives are potentially far more active than peats, and this is demonstrated in the frequency of N-immobilisation and the need for either pre-treatment of the alternative or the constant addition of additional N during the use of the material in order to prevent the growing plant being starved of available N. Water-holding capacity and availability of nitrogen during the use of the alternatives is a matter of understanding of the material's inherent properties and adjusting management appropriately during their use. Finally, consideration must be given to any phytosanitary issues associated with the use of some of the alternatives identified. For instance, there are areas of Europe where tree barks are available, but specific pests have become endemic to the point that the movement of materials is significantly restricted. So far, the emergence of a single material to replace peat has not materialised. The horticultural industry is becoming increasingly aware that the future of growing media will be based on blends of a number of components which will need to be fully characterised and analysed to ensure that their properties are managed by the industry. The preferred components will be renewable and locally produced materials with a better life-cycle assessment that takes into consideration sustainability concerns.

As noted, the trend in the field of peat-alternative growing media is towards the use of natural, local resources as well as renewable raw materials (Gruda 2012, 2019). Particularly in the industrialised countries, the re-use of wastes has become common. However, availability alone is not sufficient. A substrate should also be a standardised and growth-promoting product (Gruda et al. 1997). The improvement and the quality assurance of peat substitutes and their combination will, therefore, continue to be of high importance in the future. With the spectrum of substrates offered in horticulture, performance and suitability for plant cultivation must also include adjustment of production technologies and cultivation guidelines, especially irrigation and fertilisation processes. Only properly adapted guidelines will realise the advantages of substrates for successful cultivation (Gruda 2009). However, with changes in how plants are irrigated to meet reduced water availability, different growing media properties may better suit these new systems. Future labour availability and technology, such as robotics, are also likely to transform the sector and the systems designed to deal with this should be built around the use of sustainable growing media (Knight 2012).

Recently, more and more substrate manufacturers include on the packaging materials reference to peat-free and peat-reduced products. Some advertise the greenhouse gas (GHG) savings of their products compared to peat. However, since the manufacturers do not use a uniform GHG calculation method (e.g. peat reference value, reference scenario, end-of-life emissions), the values cannot be compared. Proper certification will help to resolve this and other challenges in optimising the use of alternative substrates.

## 9 Where to look for further information

Standard introductions to the subject are contributions by Raviv et al. (2002), Maher et al. (2008), and Gruda et al. (2013, 2016a,b) (see publications in the References for full details). Barrett et al. (2016), Gruda (2012, 2019), and Savvas and Gruda (2018) further give recent developments to alternative organic materials worldwide through overviews (see these contributions in the References for full details).

The websites of ministries and authorities, such as www.gov.uk/government/organisations/department-for-environment-food-rural-affairs, www.bmel.de/DE/themen/landwirtschaft/pflanzenbau/gartenbau/torf.html, www.bafu.admin.ch deliver updated information for peat-substitute growing media in the UK, Germany and Switzerland, respectively.

For wood products and bark, information could be found at: www.forestresearch.gov.uk.

## 10 References

Abad, M., Noguera, P. and Burés, S. National inventory of organic wastes for use as growing media for ornamental potted plant production: case study in Spain. *Bioresour. Technol.* 2001 77(2), 197–200.

Allaire, S. E., Caron, J., Menard, C. and Dorais, M. Potential replacements for rockwool as growing substrate for greenhouse tomato. *Can. J. Soil Sci.* 2005 85(1), 67–74.

Altland, J. Use of processed biofuel crops for nursery substrates. *J. Environ. Hortic.* 2010 28(3), 129–134.

Altland, J. E. and Krause, C. Use of switchgrass as a nursery container substrate. *Hortscience* 2009 44(7), 1861–1865.

Álvarez, J. M., Pasian, C., Lal, R., López, R., Díaz, M. J. and Fernández, M. Morpho-physiological plant quality when biochar and vermicompost are used as growing media replacement in urban horticulture. *Urban. For. Urban Green* 2018 34, 175–180.

Amberger-Ochsenbauer, S. and Meinken, E. *Torf und alternative Substratausgangsstoffe*. Zarbock GmbH & Co. KG: Frankfurt am Main, Germany, 2020.

Andreu-Rodriguez, J., Medina, E., Ferrandez-Garcia, M. T., Ferrandez-Villena, M., Ferrandez-Garcia, C. E., Paredes, C., Bustamante, M. A. and Moreno-Caselles, J. Agricultural and industrial valorization of *Arundo donax* L. *Commun. Soil Sci. Plant Anal.* 2013 44(1-4), 598-609.

Anonymous 2020. https://www.compostnetwork.info/policy/from-waste/compost-digestate-markets/. Accessed on 31. November 2020.

Aviani, I., Laor, Y., Medina, Sh, Krassnovsky, A. and Raviv, M. Co-composting of solid and liquid olive mill wastes: management aspects and the horticultural value of the resulting composts. *Bioresour. Technol.* 2010 101(17), 6699-6706.

Barreto, M. S. and Jagtap, K. B. Assessment of substrates for economical production of gerbera (*Gerbera jamesonii* Bolus ex Hooker F.) flowers under protected cultivation. *J. Ornam. Hortic.* 2006 9(2), 136-138.

Barrett, G. E., Alexander, P. D., Robinson, J. S. and Bragg, N. C. Achieving environmentally sustainable growing media for soilless plant cultivation systems–a review. *Sci. Hortic.* 2016 212, 220-234.

Benito, M., Masaguer, A., De Antonio, R. and Moliner, A. Use of pruning waste compost as a component in soilless growing media. *Bioresour. Technol.* 2005 96(5), 597-603.

Berecha, G., Lemessa, F. and Wakjira, M. Exploring the suitability of coffee pulp compost as growth media substitute in greenhouse production. *Int. J. Agri. Res.* 2011 6(3), 255-267.

Blok, C., Van Der Salm, C., Hofland-Zijlstra, J., Streminska, M., Eveleens, B., Regelink, I., Fryda, L. and Visser, R. Biochar for horticultural rooting media improvement: evaluation of biochar from gasification and slow pyrolysis. *Agronomy* 2017 7(1), 6. doi:10.3390/agronomy7010006.

Bonaguro, J. E., Coletto, L. and Zanin, G. Environmental and agronomic performance of fresh rice hulls used as growing medium component for *Cyclamen persicum* L. pot plants. *J. Clean. Prod.* 2017 142, 2125-2132. doi:10.1016/j.jclepro.2016.11.071.

Bonn, A., Reed, M. S., Evans, C. D., Joosten, H., Bain, C., Farmer, J., Emmer, I., Couwenberg, J., Moxey, A., Artz, R., Tanneberger, F., von Unger, M., Smyth, M. and Birnie, D. Investing in nature: developing eco-system service markets for peatland restoration. *Ecosyst. Serv.* 2014 9, 54-65.

Bragg, N. Peat and its alternatives, HDC (now AHDB-Horticulture UK), 1990. Available at: www.ahdb.co.uk.

Bragg, N. The responsible sourcing of components for growing media–a UK perspective. Conference "Torfminderung", 18. und 19. February 2020, BMEL Berlin, 2020, pp. 14-15.

Cantliffe, D. J., Shaw, N. L., Saha, S. K. and Gruda, N. Greenhouse cooling for production of peppers under hot-humid summer conditions in a high-roof passively-ventilated greenhouse. *Acta Hortic.* 2007 761(761), 41-48.

Carlen, C. Torfausstiegskonzept–Forschung und Umsetzung in der Schweiz. Tagung „Torfminderung", 18. und 19. Februar 2020, BMEL Berlin, 2020, 13.

Carlile, W. R., Cattivello, C. and Zaccheo, P. Organic growing media: constituents and properties. *Vadose Zone J.* 2015 14(6), 1-13.

Castaldi, P. and Melis, P. Growth and yield characteristics and heavy metal content on tomatoes grown in different growing media. *Commun. Soil Sci. Plant* 2004 35(1-2), 85-98.

Chemetova, C., Fabião, A., Gominho, J. and Ribeiroa, H. Range analysis of Eucalyptus globulus bark low-temperature hydrothermal treatment to produce a new component for growing media industry. *Waste Manag.* 2018 79, 1–7.

Cole, D. M., Sibley, J. L., Blythe, E. K., Eakes, D. J. and Tilt, K. M. Effect of cotton gin compost on substrate properties and growth of azalea under differing irrigation regimes in a greenhouse setting. *hortTechnology* 2005 15(1), 145–148.

Conversa, G., Bonasia, A., Lazzizera, C. and Elia, A. Influence of biochar, mycorrhizal inoculation, and fertiliser rate on growth and flowering of Pelargonium (*Pelargonium zonale* L.) plants. *Front. Plant Sci.* 2015 6, 1–11.

Debode, J., De Tender, C., Cremelie, P., Lee, A. S., Kyndt, T., Muylle, H., De Swaef, T. and Vandecasteele, B. Trichoderma-inoculated miscanthus straw can replace peat in strawberry cultivation, with beneficial effects on disease control. *Front. Plant Sci.* 2018 9, 213. https://doi.org/10.3389/fpls.2018.00213.

Depardieu, C., Prémont, V., Boily, C. and Caron, J. Sawdust and bark-based substrates for soilless strawberry production: irrigation and electrical conductivity management. *PLoS ONE* 2016 11(4), e0154104.

Diaz-Perez, M. and Camacho-Ferre, F. Effect of composts in substrates on the growth of tomato transplants. *hortTechnology* 2010 20(2), 361–367.

Dubský, M., Chaloupková, Š., Kaplan, L., Vondráčková, S. and Tlustoš, P. Use of solid phase of digestate for production of growing horticultural substrates. *Hort. Sci. (Prague)* 2019 46, 34–42. https://doi.org/10.17221/221/2016-HORTSCI.

Dumroese, R., Pinto, J., Heiskanen, J., Tervahauta, A., Mcburney, K., Page-Dumroese, D. and Englund, K. Biochar can be a suitable replacement for Sphagnum Peat in nursery production of Pinus ponderosa seedlings. *Forests* 2018 9(5), 232. doi:10.3390/f9050232.

Dysko, J., Kaniszewski, S. and Kowalczyk, W. Lignite as a new medium in soilless cultivation of tomato. *J. Elementol.* 2015 20(3), 559–569.

El-Mahrouk, M. E., Dewir, Y. H. Utilization of grape fruit waste-based substrates for seed germination and seedling growth of lemon basil. *hortTechnology* 2017 27(4), 523–529.

Emmel, M. Growing ornamental plants in Sphagnum biomass. *Acta Hortic.* 2008 779(779), 173–178.

Emmel, M. Sphagnum instead of peat pots–how much is possible? *(in German) Gemüse* 2017 11, 18–21.

Frenkel, O., Jaiswal, A. K., Elad, Y., Lew, B. and Graber, E. R. The effect of biochar on plant diseases: what should we learn while designing biochar substrates? *J. Environ. Engg. Landsc. Manag.* 2017 25(2), 105–113. doi:10.3846/16486897.2017.1307202.

Gaches, W. G., Fain, G. B., Eakes, D. J., Gilliam, C. H. and Sibley, J. L. Comparison of aged and fresh WholeTree as a substrate component for production of greenhouse-grown annuals. *J. Environ. Hortic.* 2011 29(1), 39–44.

Garg, V. K., Suthar, S. and Yadav, A. Management of food industry waste employing vermicomposting technology. *Bioresour. Technol.* 2012 126, 437–443.

Gaudig, G., Joosten, H. and Kamermann, D. Growing media: promises of Sphagnum biomass. *Acta Hortic.* 2008 779(165), 165–172.

Gaudig, G., Fengler, F., Krebs, M., Prager, A., Schulz, J., Wichmann, S., et al. Sphagnum farming in Germany–a review of progress. *Mires Peat* 2014 13, 1–11.

Gaudig, G., Krebs, M., Joosten, H. Sphagnum farming from species selection to the production of growing media: a review. *Mires Peat* 2018 20 (2017/18), 1–30. Available at: http://www.mires-and-peat.net/, ISSN 1819-754X, Article 13, doi:10.19189/MaP.2018.OMB.340.

Gerber, T., Steinbacher, F. F. and Hauser, B. Holzfasersubstrat zur Kultur von Pelargo-nium-Zonale-Hybriden - biophysikalische und pflanzenbauliche Unter-suchung. *Angew. Bot.* 1999 73, 217-221.

Gómez, C. and Robbins, J. Pine bark substrates amended with parboiled rice hulls: physical properties and growth of container-grown during long-term nursery production. *horts* 2011 46(5), 784–790. doi:10.21273/HORTSCI.46.5.784.

Gong, X., Li, S., Sun, X., Wang, L., Cai, L., Zhang, J. and Wei, L. Green waste compost and vermicompost as peat substitutes in growing media for geranium (Pelargonium zonale L.) and calendula (Calendula officinalis L.). *Sci. Hortic.* 2018 236, 186–191.

Grigatti, M., Giorgioni, M. E. and Ciavatta, C. Compost-based growing media: influence on growth and nutrient use of bedding plants. *Bioresour. Technol.* 2007 98(18), 3526–3534.

Gruda, N. Weed suppression in vegetable crops using wood fibre mulch (In German). *Rep. Agriculture* 2007 85, 329–334.

Gruda, N. The effect of wood fiber mulch on water retention, soil temperature and growth of vegetable plants. *J. Sustain. Agric.* 2008 32(4), 629–643.

Gruda, N. Do soilless culture systems have an influence on product quality of vegetables? *J. Appl. Bot. Food Qual.* 2009 82, 141–147. Available at: https://ojs.openagrar.de/index.php/JABFQ/article/view/2093.

Gruda, N. Current and future perspective of growing media in Europe. *Acta Hortic.* 2012 960(960), 37–43. https://doi.org/10.17660/ ActaHortic.2012.960.3.

Gruda, N. S. Increasing sustainability of growing media constituents and stand-alone substrates in soilless culture systems. *Agronomy* 2019 9(6), 298. doi.org/10.3390/agronomy9060298.

Gruda, N. Einführung–Torfminderungsstrategie der Bundesregierung. Tagung "Torfminderung" 18. und 19. Februar 2020, BMEL Berlin, 2020, 5–6.

Gruda, N. and Schnitzler, W. H. Influence of wood fiber substrates and application rates on the growth of tomato transplants. *Adv. Hort. Sci.* 1999 13(1), 20–24.

Gruda, N. and Schnitzler, W. H. The effect of water supply on bio-morphological and plant-physiological parameters of tomato transplants cultivated in wood fiber substrate (in German). *J. Appl. Bot.* 2000a 74, 233–239.

Gruda, N. and Schnitzler, W. H. The effect of water supply of seedlings, cultivated in different substrates and raising systems on the bio-morphological and plant-physiological parameters of lettuce (In German). *J. Appl. Bot.* 2000b 74(5–6), 240–247.

Gruda, N. and Schnitzler, W. H. Suitability of wood fiber substrates for production of vegetable transplants. I. Physical properties of wood fiber substrates. *Sci. Hortic.* 2004a 100, 309–322. https://doi.org/10.1016/j.scienta.2003.10.001.

Gruda, N. and Schnitzler, W. H. Suitability of wood fiber substrates for production of vegetable transplants. II. The effect of wood fiber substrates and their volume weights on the growth of tomato transplants. *Sci. Horticul.* 2004b 100(1–4), 333–340. doi:10.1016/j.scienta.2003.09.004.

Gruda, N. and Schnitzler, W. H. Wood fiber substrates as a peat alternative for vegetable production. *Eur. J. Wood Prod.* 2006 64, 347–350.

Gruda, N., Michalsky, F. and Schnitzler, W. H. Substrate für Gemüsejungpflanzen. *Dtsch. Gartenbau* 1997 51(37), 1985–1988.

Gruda, N., Tucher, S. V. and Schnitzler, W. H. N-immobilization of wood fiber substrates in the production of tomato transplants (*Lycopersicon lycopersicum* (L.) Karst. ex. Farw.). *J. Appl. Bot.* 2000 74, 32–37.

Gruda, N., Rau, B. and Wright, R. D. Laboratory bioassay and greenhouse evaluation of a pine tree substrate used as a container substrate. *Europ. J. Hortic. Sci.* 2009 74, 73–78.

Gruda, N., Qaryouti, M. M. and Leonardi, C. Growing media. In: *Good Agricultural Practices for Greenhouse Vegetable Crops–Principles for Mediterranean Climate Areas (Rome, Italy: FAO), Plant Production and Protection*, 2013, pp. 271–302, Paper 217.

Gruda, N., Gianquinto, G., Tüzel, Y. and Savvas, D. Culture soil-less. In: *Encyclopedia of Soil Sciences* (3rd edn.), Lal, R. (Ed.). CRC Press, Taylor & Francis Group: Boca Raton, Florida, 2016a, pp. 533–537. doi:10.1081/E-ESS3-120053777.

Gruda, N., Caron, J., Prasad, M. and Maher, M. J. Growing media. In: *Encyclopedia of Soil Sciences* (3rd edn.), Lal, R. (Ed.). CRC Press, Taylor & Francis Group: Boca Raton, Florida, 2016b, pp. 1053–1058. https://doi.org/10.1081/EESS3-120053784.

Gruda, N., Bisbis, M. B. and Tanny, J. Impacts of protected vegetable cultivation on climate change and adaptation strategies for cleaner production–a review. *J. Clean. Prod.* 2019a 225, 324–339. https://doi.org/10.1016/j.jclepro.2019.03.295.

Gruda, N., Bisbis, M. B. and Tanny, J. Influence of climate change on protected cultivation: impacts and sustainable adaptation strategies–a review. *J. Clean. Prod.* 2019b 225, 481–495, https://doi.org/10.1016/j.jclepro.2019.03.210.

Handreck, K. A. and Black, N. D. *Growing Media for* Ornamental Plants *and* Turf. UNSW Press, 2010.

Hernández-Apaolaza, L., Gascó, A. M., Gascó, J. M. and Guerrero, F. Reuse of waste materials as growing media for ornamental plants. *Bioresour. Technol.* 2005 96(1), 125–131.

Jackson, B. E. and Bragg, N. C. Wood components: a step towards a sustainable growing media. *FCI- Magazine* Issue September, 2016, pp. 30–31.

Jackson, B. E., Wright, R. D. and Seiler, J. R. Changes in chemical and physical properties of pine tree substrate and pine bark during long-term nursery production. *Hortscience* 2009 44(3), 791–799.

Jobin, P., Caron, J. and Rochefort, L. Developing new potting mixes with Sphagnum fibers. *Can. J. Soil Sci.* 2014 94(5), 585–593.

Kaever, M. 2004. *Die nicht erneuerbaren Energieträger zwischen Rur und Maas*. LIT Verlag: Münster.

Kamp, M. and Emino, E. R. Growth response of container-grown plants in potting media amended with lignite. *Acta Hortic.* 1983 133(133), 191–198. https://doi.org/10.17660/ActaHortic.1983.133.21.

Kaudal, B. B., Chen, D., Madhavan, D. B., Downie, A. and Weatherley, A. Pyrolysis of urban waste streams: their potential use as horticultural media. *J. Anal. Appl. Pyrol.* 2015 112, 105–112, DOI:10.1016/j.jaap.2015.02.011.

Kern, J., Tammeorg, P., Shanskiy, M., Sakrabani, R., Knicker, H., Kammann, C., Tuhkanen, E.-M., Smidt, G., Prasad, M., Tiilikkala, K., Sohi, S., Gascó, G., Steiner, C. and Glaser, B. Synergistic use of peat and charred material in growing media-an option to reduce the pressure on peatlands? *J. Environ. Eng. Landsc. Manag.* 2017 25(2), 160–174.

Kim, H. S., Kim, K. R., Yang, J. E., Ok, Y. S., Kim,W. I., Kunhikrishnan, A. and Kim, K.-H. Amelioration of horticultural growing media properties through rice hull biochar incorporation. *Waste Biomass Valorization* 2017 8(2), 483–492.

Knight, A. Towards sustainable growing media. Chairman's report and roadmap, sustainable growing media task force, 2012. Available at: https://assets.publishing.service.gov.uk/government/uploads/system/uploads/attachment_data/file/221019/pb13867-towards-sustainable-growing-media.pdf.

Kuisma, E., Palonen, P. and Yli-Hallab, M. Reed canary grass straw as a substrate in soilless cultivation of strawberry. *Sci. Hortic.* 2014 178, 217–223.

Lazcano, C., Gómez-Brandón, M. and Domínguez, J. Comparison of the effectiveness of composting and vermicomposting for the biological stabilization of cattle manure. *Chemosphere* 2008 72(7), 1013–1019.

Lemaire, F., Dartigues, A. and Rivière, L. M. Physical and chemical characteristics of a ligno-cellulosic material. *Acta Hortic.* 1989 238(238), 9–22.

Levesque, V., Rochette, P., Ziadi, N., Dorais, M. and Antoun, H. Mitigation of CO2, CH4 and N2O from a fertigated horticultural growing medium amended with biochars and a compost. *Appl. Soil Ecol.* 2018 126, 129–139.

Lewandowski, I., Clifton-Brown, J., Trindade, L. M., van der Linden, G. C., Schwarz, K. U., Müller-Sämann, K., Anisimov, A., Chen, C. L., Dolstra, O., Donnison, I. S., Farrar, K., Fonteyne, S., Harding, G., Hastings, A., Huxley, L. M., Iqbal, Y., Khokhlov, N., Kiesel, A., Lootens, P., Meyer, H., Mos, M., Muylle, H., Nunn, C., Özgüven, M., Roldán-Ruiz, I., Schüle, H., Tarakanov, I., van der Weijde, T., Wagner, M., Xi, Q. and Kalinina, O. Progress on optimising Miscanthus biomass production for the European bioeconomy: results of the EU FP7 project OPTIMISC. *Front. Plant Sci.* 2016 7, 1620. doi: 10.3389/fpls.2016.01620.

Maas, E. F. and Adamson, R. M.. *Artificial Media in Horticulture–Their Formulation and Fertilisation*. Agriculture Canada, 1982. Publication 1726/E.

Maher, M., Prasad, M. and Raviv, M. Organic soilless media components. In: *Soilless Culture: Theory and Practice*, Raviv, M. and Lieth, J. H. (Eds). Academic Press: San Diego, 2008, pp. 459–504.

Maher, M. J. and Prasad, M. The effect of N source on the composting of green waste and its properties as a component of a peat growing medium. In: *Proceedings of the International Conference Orbit 2001 on Biological Processing of Waste*, Seville, May 9–12, 2001; Spanish Waste Club, Orbit Association: Seville, 2001, pp. 299–306.

Maher, M. J. and Thomson, D.. Growth and manganese content of tomato (Lycopersicon Esculentum) seedlings grown in sitka spruce (Picea Sitchensis (bong.) Carr.) Bark Substrate. *Scientia Horticul.* 1991 48, 223–231.

Mariotti, B., Martini, S., Raddi, S., Tani, A., Jacobs, D. F., Oliet, J. A. and Maltoni, A. Coconut coir as a sustainable nursery growing media for seedling production of the ecologically diverse Quercus species. *Forests* 2020 11(5), 522.

Massa, D., Malorgio, F., Lazzereschi, S., Carmassi, G., Prisa, D. and Burchi, G. Evaluation of two green composts for peat substitution in geranium (*Pelargonium zonale* L.) cultivation: Effect on plant growth, quality, nutrition, and photosynthesis. *Sci. Hortic.* 2018, 228, 213–221.

Mattei, P., Pastorelli, R., Rami, G., Mocali, S., Giagnoni, L., Gonnelli, C. and Renella, G. Evaluation of dredged sediment co-composted with green waste as plant growing media assessed by eco-toxicological tests, plant growth and microbial community structure. *J. Hazard. Mater.* 2017 333, 144–153.

Mininni, C., Bustamante, M. A., Medina, E., Montesano, F., Paredes, C., Pérez-Espinosa, A., Moral, R. and Santamaria, P. Evaluation of Posidonia seaweed-based compost as a substrate for melon and tomato seedling production. *J. Hortic. Sci. Biotechnol.* 2013 88(3), 345–351.

Mininni, C., Grassi, F., Traversa, A., Cocozza, C., Parente, A., Miano, T. and Santamaria, P. *Posidonia oceanica* (L.) based compost as substrate for potted basil production. *J. Sci. Food Agric.* 2015 95(10), 2041–2046.

Mininni, C., Santamaria, P., Abdelrahman, H. M., Cocozza, C., Miano, T., Montesano, F. and Parente, A. 2020. Posidonia-based Compost as a Peat Substitute for Lettuce Transplant Production. *HortScience* 47(10), 1438–1444. Doi: 10.21273/HORTSCI.47.10.1438.

Montesano, F., Gattullo, C., Parente, A., Terzano, R. and Renna, M. Cultivation of potted sea fennel, an emerging Mediterranean halophyte, using a renewable seaweed-based material as a peat substitute. *Agriculture* 2018 8(7), 96.

Morales-Corts, M. R., Gómez-Sánchez, M. Á and Pérez-Sánchez, R. Evaluation of green/pruning wastes compost and vermicompost, slumgum compost and their mixes as growing media for horticultural production. *Sci. Hortic.* 2014 172, 155–160.

Muylle, H., Van Hulle, S., De Vliegher, A., Baert, J., Van Bockstaele, E. and Roldán-Ruiz, I. Yield and energy balance of annual and perennial lignocellulosic crops for bio-refinery use: a 4-year field experiment in Belgium. *Eur. J. Agron.* 2015 63, 62–70. doi: 10.1016/j.eja.2014.11.001.

Naasz, R., Caron, J., Legault, J. and Pichette, A. Efficiency factors for bark substrates biostability, aeration, or phytotoxicity. *Soil Sci. Soc. Am. J.* 2009 73(3), 780–791.

Nemati, M. R., Simard, F., Fortin, J.-P. and Beaudoin, J. Potential use of biochar in growing media. *Vadose Zone J.* 2014 14, 6. https://doi.org/10.2136/vzj2014.06.0074.

Neumaier, D., Lohr, D., Voßeler, R., Girmann, S., Kolbinger, S. and Meinken, E. Hydrochars as peat substitute in growing media for organically grown potted herbs. *Acta Hortic.* 2017 1168(1168), 377–386.

Nobile, C., Denier, J. and Houben, D. Linking biochar properties to biomass of basil, lettuce and pansy cultivated in growing media. *Sci. Horticul.* 2020 261, https://doi.org/10.1016/j.scienta.2019.109001.

Noguera, P., Abad, M., Noguera, V., Puchades, R. and Maquieira, A. Coconut coir waste, a new and viable ecologically-friendly peat substitute. *Acta Hortic.* 2000 517(517), 279–286.

Ortega, M. C., Moreno, M. T., Ordovás, J. and Aguado, M. T. Behavior of different horticultural species in phytotoxicity bioassays of bark substrates. *Sci. Horticul.* 1996 66(1–2), 125–132.

Ostos, J. C., López-Garrido, R., Murillo, J. M. and López, R. Substitution of peat for municipal solid waste- and sewage sludge-based composts in nursery growing media: effects on growth and nutrition of the native shrub *Pistacia lentiscus* L. *Bioresour. Technol.* 2008 99(6), 1793–1800.

Owen Jr., J. S., Warren, S. L., Bilderback, T. E., Cassel, D. K. and Albano, J. P. Physical properties of pine bark substrate amended with industrial mineral aggregate. *Acta Hort.* 2008 779, 131–138.

Prasad, M., Chrysargyris, A., McDaniel, N., Kavanagh, A., Gruda, N. S. and Tzortzakis, N. Plant nutrient availability and pH of biochars and their fractions, with the possible use as a component in a growing media. *Agronomy* 2019 10(1). doi:10.3390/agronomy10010010.

QUANTIS. Comparative life cycle assessment of horticultural growing media based on peat and other growing media constituents 2012. Available at: http://epagma.eu/evidence-based (accessed on 2 Mai 2020).

Raviv, M., Medina, S., Krasnovsky, A. and Ziadna, H. Organic matter and nitrogen conservation in dairy manure composting for organic agriculture. *Compost Sci. Util.* 2004,12, 6-10.

Raviv, M., Wallach, R., Silber, A. and Bar-Tal, A. Substrates and their analysis. In: *Hydroponic Production of Vegetables and Ornamentals*, Savvas, D. and Passam, H. (Eds). Embrio publications: Athens, 2002, pp. 25-102.

Rencoret, J. J., Ralph, G., Marques, A., Gutierrez, A., Martinez, A. and del Rio, J. C. Structural characterisation of lignin isolated from coconut (Cocos nucifera) coir fibers. *J. Agric. Food Chem.* 2013 61(10), 2343-2445.

Riaz, A., Younis, A., Ghani, I., Tariq, U. and Ahsan, M. Agricultural waste as growing media component for the growth and flowering of *Gerbera jamesonii* cv. hybrid mix. *Int. J. Recycl. Org. Waste Agric.* 2015 4(3), 197-204.

Ridout, M. E. and Tripepi, R. R. Initial chemical and physical properties of potting mixes amended with anaerobically digested cattle biosolids. *Acta Hortic.* 2011 891, 167-172.

Robbins, J. A. and Evans, M. R. Growing media for container production in a greenhouse or nursery. Part I (Components and Mixes). Greenhouse and Nursery Series, University of Arkansas, 2010. Available at: http://www.uaex.edu.

Rodriguez, J. C., Cantliffe, D. J., Shaw, N. L. and Karchi, Z. Soilless media and containers for greenhouse production of 'Galia' type muskmelon. *Hort.Sci.* 2006 41(5), 1200-1205.

Salvador, E. D., Haugen, L. E. and Gislerød, H. R. Compressed coir as substrate in ornamental plants growing - Part I: Physical properties. *Acta Hort.* 2005 683, 215-222.

Santos, F. T., Goufo, P., Santos, C., Botelho, D., Fonseca, J., Queirós, A., Costa, M. S. and Trindade, H. Comparison of five agro-industrial waste-based composts as growing media for lettuce: effect on yield, phenolic compounds and vitamin C. *Food Chem.* 2016 209, 293-301.

Savvas, D. and Gruda, N. Application of soilless culture technologies in the modern greenhouse industry–a review. *Europ. J. Hortic. Sci.* 2018 83(5), 280-293. doi.org/10.17660/eJHS.2018/83.5.2.

Schmilewski, G. Growing media constituents used in the EU in 2013. *Acta Hortic.* 2017 1168(1168), 85-92. doi:10.17660/ActaHortic.2017.1168.12.

Shaw, N. L., Cantliffe, D. J., Funes, J. and Shine, C. Successful beit alpha cucumber production in the greenhouse using pine bark as an alternative soilless media. *HortTechnology* 2004 14(2), 289-294.

Solbraa, K. and Selmer-Olsen, A. R. Manganese Toxicity - in particular when growing plants in bark compost. *Acta Agric. Scand.* 1981 31(1), 29-39.

Starck, J. R., Lukaszuk, K. and Maciejewski, M. Effect of fertiliser nitrogen and potassium upon yield and quality of carnations grown in peat and sawdust. *Acta Hortic.* 1991 294(294), 289-296.

Steinbacher, F. and Amberger-Ochsenbauer, S. Mehr Blüten bei Holzfaser-Torf-Substrat. *DeGa* 1997 51(20), 1138-1141.

Steiner, C. and Harttung, T. Biochar as a growing media additive and peat substitute. *Solid Earth* 2014 5(2), 995-999. doi:10.5194/se-5-995-2014.

Stucki, M. Torf und Ersatzstoffe im Ökobilanzvergleich. Tagung "Torfminderung" 18. und 19. Februar, BMEL Berlin, 2020, 26–27.

Suo, L. N., Sun, X. Y. and Li, S. Y. Use of organic agricultural wastes as growing media for the production of *Anthurium andraeanum* 'Pink Lady'. *J. Hortic. Sci. Biotechnol.* 2011 86(4), 366–370.

Temminka, R. J. M., Fritz, C., van Dijk, G., Hensgens, G., Lamers, L. P. M., Krebs, M., Gaudig, G. and Joosten, H. Sphagnum farming in a eutrophic world: the importance of optimal nutrient stoichiometry. *Ecol. Eng.* 2017 98, 196–205.

Tian, Y., Sun, X., Li, S., Wang, H., Wang, L., Cao, J. and Zhang, L. Biochar made from green waste as peat substitute in growth media for *Calathea rotundifola* cv. 'Fasciata'. *Sci. Hortic.* 2012 143, 15–18.

Tsakaldimi, M. Kenaf (*Hibiscus cannabinus* L.) core and rice hulls as components of container media for growing *Pinus halepensis* M. seedlings. *Bioresour. Technol.* 2006 97(14), 1631–1639. doi:10.1016/j.biortech.2005.07.027.

Vandecasteele, B., Debode, J., Willekens, K. and Van Delm, T. Recycling of P and K in circular horticulture through compost application in sustainable growing media for fertigated strawberry cultivation. *Eur. J. Agron.* 2018a 96, 131–145.

Vandecasteele, B., Muylle, H., De Windt, I., Van Acker, J., Ameloot, N., Moreaux, K., Coucke, P. and Debode, J. Plant fibers for renewable growing media: potential of defibration, acidification or inoculation with biocontrol fungi to reduce the N drawdown and plant pathogens. *J. Cleaner Prod.* 2018b 203, 1143–1154, https://doi.org/10.1016/j.jclepro.2018.08.167.

Wang, L. M., Zhang, Y. M., Lian, J. J., Chao, J. Y., Gao, Y. X., Yang, F. and Zhang, L. Y. Impact of fly ash and phosphatic rock on metal stabilisation and bioavailability during sewage sludge vermicomposting. *Bioresour. Technol.* 2013 136, 281–287.

Weissert, L. F. and Disney, M. Carbon storage in peatlands: a case study on the Isle of Man. *Geoderma* 2013 204–205, 111–119.

Woli, K. P., David, M. B., Tsai, J., Voigt, T. B., Darmody, R. G. and Mitchell, C. A. Evaluating silicon concentrations in biofuel feedstock crops Miscanthus and switchgrass. *Biomass Bioenergy* 2011 35(7), 2807–2813. https://doi.org/10.1016/j.biombioe.2011.03.007.

Yu, Z. C., Loisel, J., Brosseau, D. P., Beilman, D. W. and Hunt, S. J. Global peatland dynamics since the last glacial maximum. *Geophys. Res. Lett.* 2010 37(13), n/a–n/a.

# Chapter 5

# Understanding and optimizing the physical properties of growing media for soilless cultivation

*Jean Caron, Université Laval, Canada; and Jean-Charles Michel, L'Institut Agro, France*



## 1 Introduction

The science behind growing media design and manufacturing is in constant evolution because new practices are being introduced, and essential components used in the formulation of soilless mixes are also continually changing in properties, availability and price. This chapter addresses the critical basic physical parameters associated with their design and manufacturing.

In terms of mineral soil physics, earlier design of growing media focussed on issues such as container size, growing media constituents and irrigation type for a limited number of crops. Since then, the design of growing media has faced additional challenges. The first is to optimize fertigation management to reduce pesticide and nutrient leaching to meet environmental regulations and cope with increasing social awareness about environmental issues. The second constraint is the need to find peat alternatives to reduce the carbon footprint in manufacturing soilless media. This has resulted in increasing

http://dx.doi.org/10.19103/AS.2020.0076.04

amounts of organic green waste, wood-based products and by-products and compost of various origins being used in the manufacturing process. A third challenge is the evolution of irrigation systems from overhead irrigation to drip irrigation, pulse misting, and ebb and flow irrigation strategies. Finally, a fourth type of constraint is the result of changes in container size and shape introduced in horticulture, green roof technologies, and urban agriculture and the introduction of new crops such as soilless strawberries in the field (Cormier et al., 2020) or under shelters (Depardieu et al., 2018).

An in-depth understanding of water and air dynamics in the soil-plant-atmosphere system is needed to design and use growing media to meet this broader range of challenges. Root water uptake in container-grown plants depends on the balance between existing atmospheric conditions (radiation, temperature, relative humidity, etc.) and water availability to plant roots at a given moment itself controlled by the hydraulic properties of the growing media at that time. The same applies to oxygen uptake by actively growing roots and microbes.

While this understanding is of fundamental importance in field production, additional knowledge is required in soilless culture, due to the specific production constraints brought by pot and container production systems. Relative to field conditions, water, gas and nutrient bio-availability fluctuate faster because of the limited rooting volume. This requires the quantification of dynamic processes aimed at ensuring a constant supply and elimination of water and gas to actively transpiring plants. Hence, new concepts must be introduced to cope with such highly dynamic systems like pulse irrigation, capillary mat (Caron et al., 2005a,b), tension reservoir (Caron et al., 2020) and aeration devices (Lemay et al., 2012) as they can increase productivity relative to open-field production.

In agricultural soils facing aeration limitation due to compaction and poor drainage or in poorly designed or compacted substrates, maximum plant productivity is usually achieved by maintaining soil water potential and substrate water content within a quite restricted irrigation range (Fig. 1). Improvement of substrate aeration through better design (de Boodt and Verdonck, 1972) has allowed growers to achieve higher productivity. However, because of this increased aeration, the substrate can be used across a wider range of soil water potentials than in low-aeration growing media. This has resulted in a characteristic type II plant productivity response curve to soil water potential (Fig. 1).

The more recent development of continuously fed irrigation systems (capillary mat, timer or computer control highly frequent irrigation) has led to the observation of a type III plant productivity response curve to soil water potential (Caron et al., 1998). In such system, the substrate is irrigated constantly through a capillary mat system providing a nearly constant water source. The high productivity of a very wet growing medium (close to zero soil

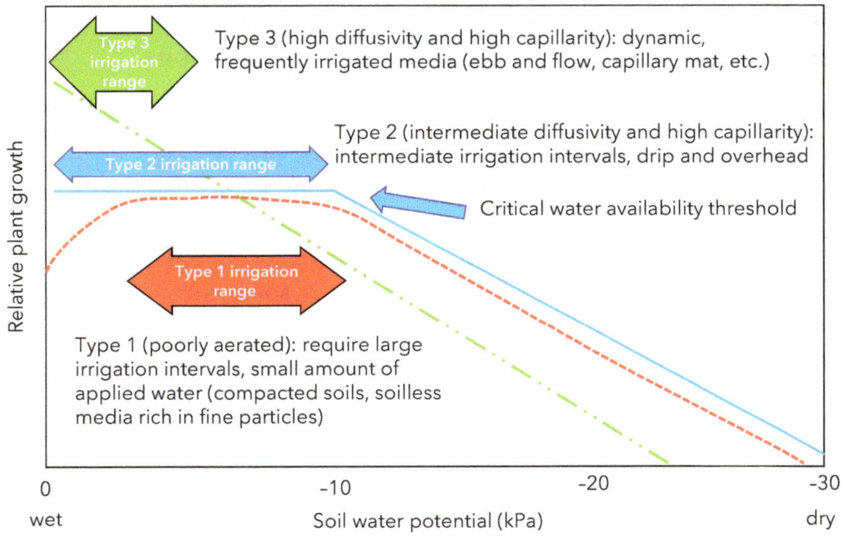


**Figure 1** Theoretical relationship between soil water potential and plant activity with three contrasting growing media types.

water potential) is reached by using growing media with high diffusivity and high capillarity (Caron et al., 2004, 2005a; Michel and Kerloch, 2017). Hence, designing substrate requires these additional concepts to be understood and implemented (Caron et al., 2015a,b).

Substrates used in horticulture often have a high proportion of organic matter of different origins, which may differ very much in terms of physical, chemical and biological properties. Relative to mineral soils, they typically have a low weight and a low bulk density. In terms of soil physics, the organic components are chosen for their high water retention, high aeration and low mechanical impedance, which then make them suitable for use as a growing medium and appropriate as physical support for plant rooting. However, growing media high in organic matter can be very fragile and of low biostability. Hence, despite suitable initial physical properties, the organic components used are sensitive to processing and harvesting, stockpiling, bagging, loosening, and mixing steps (Caron et al., 2015c). On top of the above factors, organic matter once mixed and used within a substrate can undergo rapid decomposition, and the initial properties of the growing media will evolve. Because of all of this, it is important to carry out follow-up analyses of the properties of the media to anticipate potential growth problems beyond their initial characterization.

The next two sections aim at defining those physical parameters of substrates critical to understanding and measuring them, and then to identify the key factors explaining their specific physical behaviour.

## 2 Parameters used to describe the physical properties of substrates and their measurement

### 2.1 Static parameters

The different components and their proportional use in preparing growing media are chosen to give a substrate the ability to provide water to the root system without cutting off its oxygen supply. Most of the time, the characterization of substrates' physical properties is usually based on analysing the distribution of different fluid (air, water) and solid (substrate) phases in relation to an equilibrium soil water potential of the substrate critical to plant growth. This critical equilibrium soil water potential is usually the container capacity for aeration (the equivalent of field capacity). A second critical water potential is usually defined at −10 kPa, a soilless water potential acknowledged to limit growth (de Boodt and Verdonck, 1972). This soil water potential is usually considered as the one below which water will not be available for growth without plant stress. That is why they are referred to as static parameters.

On this basis, important static parameter guidelines for growing media have been defined (Gabriels, 1994; European procedure EN 13041, 2000) and are still the basis for substrate characterization (European procedure EN 13041, 2000). Those parameters are:

- total porosity, corresponding to the proportion of the total bulk volume available for storing water and/or air;
- air-filled porosity, corresponding to the proportion of air that has replaced water initially occupying coarser porosity that has rapidly drained (usually water potentials < −1 kPa);
- pore size distribution: obtained from the water desorption curve measured from the volumetric water content release from saturation to the −10 kPa range, corresponding to the proportion of water relative to the bulk volume retained in the substrate that can be extracted by roots, usually defined for a range of potentials from −1 to −10 kPa and also called gravitational water (de Boodt and Verdonck, 1972);
- available water, referring to the proportion of water released between −1 and −10 kPa, itself subdivided into:
  - easily available water, referring to the proportion of water released between container capacity (−1 kPa) and −5 kPa;
  - water buffering capacity, referring to the proportion of water released between −5 and −10 kPa, and that enables the physiological adaptation of the plant to the increase in water potential (de Boodt and Verdonck, 1972);
  - unavailable water, referring to the proportion of water stored below −10 kPa.

With respect to aeration, air-filled porosity is one commonly used static parameter criterion which must be around 0.20–0.30 $cm^3\ cm^{-3}$ of air content (at −1 kPa of matric water potential) for optimal growth. Authors report potential problems at values classically below 0.10–0.15 $cm^3\ cm^{-3}$ air content. Naasz et al. (2009) have reported that air-filled porosity should be even higher than 0.30 $cm^3\ cm^{-3}$ for a highly respiring substrate made of non-composted bark (Fig. 2). Available water is another criterion, which is defined as that water between −1 and −10 kPa and which should be about 0.20–0.30 $cm^3\ cm^{-3}$ for a good substrate. These parameters can be obtained from interpreting the water desorption curve (Fig. 3). The figure illustrates an optimal water desorption curve and the curve of the same severely compacted growing media, indicating how critical the applied pressure at potting may impact on the aeration as well as the water retention property of a same growing medium, given the highly sensitive structure of growing media. Table 1 quantifies how compaction for a same substrate may drop air-filled porosity and increase available water of a same substrate, quantifying also how critical the compaction process at potting may influence aeration and available water of the potted substrate (see below).

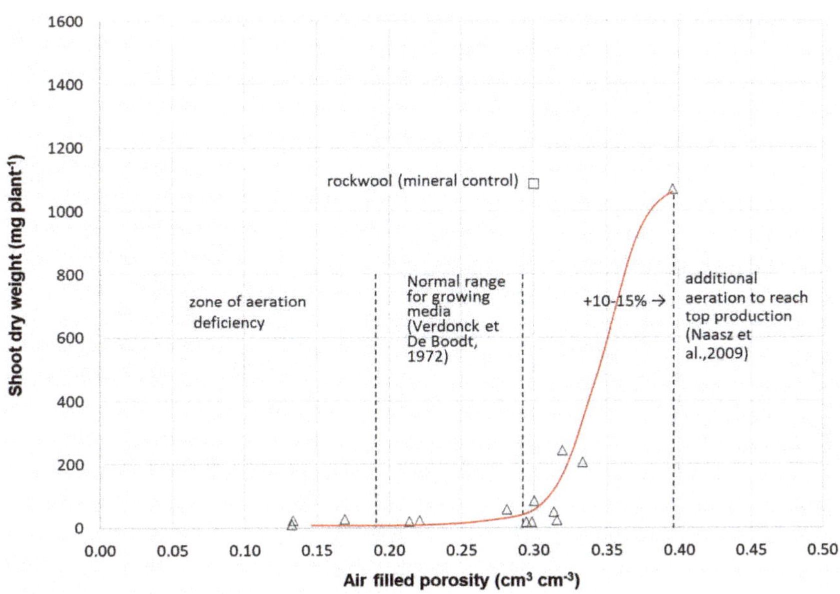


**Figure 2** Relationship between air-filled porosity for young tomato plants in organic (non-composted) bark/peat substrates and mineral (rock wool) growing media. Data are taken from Naasz et al. (2009).

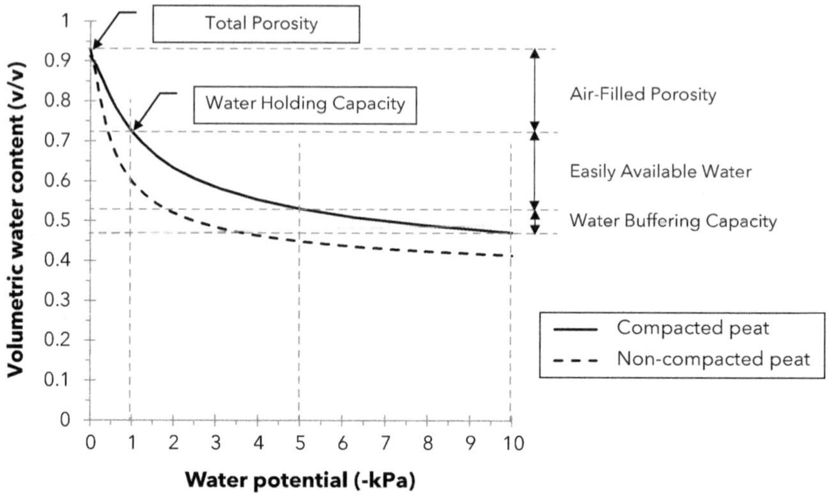


**Figure 3** Water retention curves for a compacted- and a non-compacted white- (or blond) milled peat, with graphical identification of the main physical parameters for compacted peat.

**Table 1** Main physical parameters calculated from their water retention curves

| | Compacted peat | Non-compacted peat |
|---|---|---|
| Total porosity (TP) v/v | 0.91 | 0.93 |
| Air-filled porosity (AFP) v/v | 0.19 | 0.33 |
| Water holding capacity (WHC) v/v | 0.72 | 0.60 |
| Easily available water (EAW) v/v | 0.20 | 0.15 |
| Water buffering capacity (WBC) v/v | 0.06 | 0.03 |

## 2.2 Dynamic parameters

The above parameters are still highly relevant for quality control purposes in growing media manufacturing. Initially derived from the pioneering work of de Boodt and Verdonck (1972), they corresponded to growing conditions used in greenhouse 20-cm high beds that were top irrigated. The principle was to manage the growing bed as a reservoir of water and air that should be maintained within a given range of soil matric potential for optimal aeration and water availability, typically from −1 to −10 kPa (Fig. 1). Because of the risk of low aeration at high water potential maintained in 20-cm bed and small containers, they required air-filled porosity of the substrate to be high enough and the substrate to have a typical type II response curve (Fig. 1).

However, it appears that knowledge of the available water content in the medium or even of its water potential under equilibrium conditions is not always enough to account for typical response commonly observed in soilless cultures. Indeed, signs of water stress in the case of potted plants at relatively high apparent water potentials (−5 kPa) have been reported for *Prunus xcistena* (Caron et al., 1998), white spruce, (Boudreault et al., 2014); *Viburnum* spp. and *Ligustrum* spp. (Caron et al., 2005a; Jobin et al., 2004), a matric potential considerably higher than the −10 kPa of de Boodt and Verdonck (1972). This difference can be explained by strong resistance to hydraulic transfers at the interface between the growing media and the roots (Orlander and Due, 1986). For this reason, a great deal of research has been devoted to the study of water and air dynamics in relation to plant development. The improvement of techniques and models since the 1990s has allowed researchers to improve and better understand water transfer processes in dry substrate zones to anticipate plant water stress. This has also helped in avoiding oxygen stress in wet substrates. Finally, improved models and techniques have helped linking these fluid transfer capacities to the physiological requirements of the plant to define thresholds for the bioavailability of water and oxygen in the growing media (Wallach et al., 1992; Allaire et al., 1996; Naasz et al., 2005; Caron et al., 2013).

Rekika et al. (2014) studied water availability in muck (organic) soils have based their model on the unsaturated hydraulic conductivity k(h) of the soil, described by Eq. 1:

$$k(h) = K_s e^{-\alpha h} \tag{1}$$

where the soil saturated hydraulic conductivity $K_s$ and $\alpha$ (Table 2), a parameter describing the shape of the unsaturated hydraulic conductivity, decreases as soil water potential (*h*) decreases.

These authors have derived a dynamic equilibrium model relating the dynamic properties for water transfer of the soil and the evapotranspiration demand of the plant rooted to a depth *z*. It shows that the equilibrium of water from the plant to the soil would be maintained up to a critical threshold, which would vary according to $K_s$ and $\alpha$. Using such a model in growing media, Fig. 4 indicates that for a substrate with an $\alpha$ around 0.03 cm, water should be available from container capacity down to about −6 kPa (Caron et al., 2005a). Caron et al. (1998) studying that availability showed the threshold to vary between −5 kPa and −15 kPa, agreeing with de Boodt and Verdonck (1972) who suggested beginning of stress about −5 kPa. This would be consistent with a type II substrate (Fig. 1). This may not apply in very coarse and loose growing media, made to have high gas diffusion, where a type III behaviour is expected. Boudreault et al. (2014) reported a typical type III response curve (Fig. 4), consistent with soil hydraulic parameters with an $\alpha$ value typical of coarse media set at 0.30 $cm^{-1}$ (Caron et al., 2005a). Such a high $\alpha$ value is characteristic of low bulk-density substrates, reported in Boudreault et al. (2014).

**Table 2** Criterion used to evaluate growth limitations in growing media

| Parameters | Aeration | | Water | | | |
|---|---|---|---|---|---|---|
| | Air-filled porosity at −1 kPa | Gas diffusivity | Available water (from −1 to −10 kPa) | *Ks* | α | Wettability (θ) |
| units | $cm^3\ cm^{-3}$ | | $cm^3\ cm^{-3}$ | $cm\ s^{-1}$ | $kPa^{-1}$ | ° (degrees) |
| static | 0.15–0.40[a,b,c] | | 0.20–0.30[b] | | | <90[h] |
| dynamic | | 0.010–0.015[d] | | 0.08[f] | | <87[i] |
| | | 0.003–0.005[e] | | | 9.6[g] | |

*Ks* and α are unsaturated hydraulic conductivity curve parameter, and θ is the contact angle.
[a] From Wever (1994).
[b] de Boodt and Verdonck (1972).
[c] Naasz et al. (2008).
[d] From Allaire et al. (1996) for −0.8 kPa in 5, 9 and 14 L nursery container.
[e] From Boudreault et al. (2014) for −0.6 kPa in tree seedlings.
[f] Mustin (1987).
[g] Caron et al. (2005a).
[h] From Michel et al. (2001).
[i] From Michel (2015).

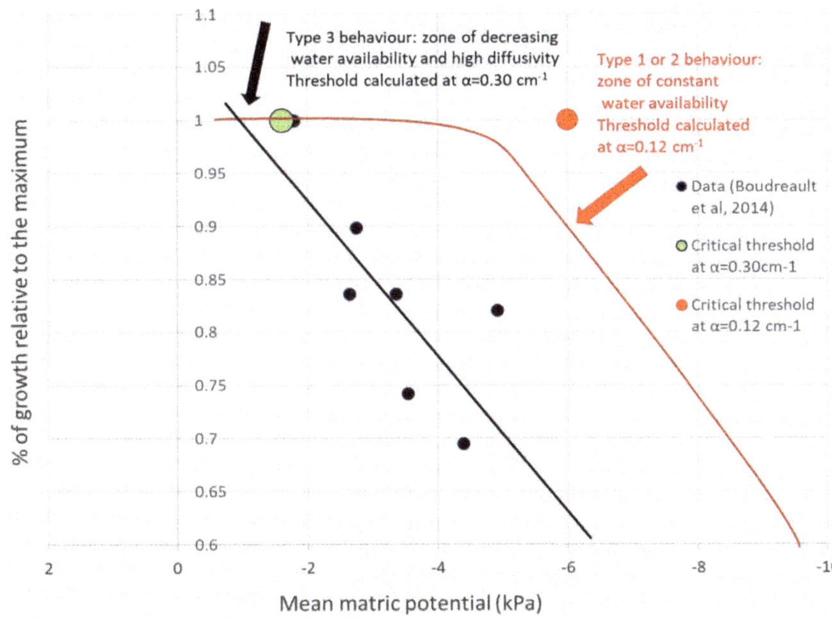


**Figure 4** Relationship between the mean matric potential over the growth period and the percentage of maximum tree seedlings growth (data obtained from Boudreault et al., 2014).

As opposed to water flow, the dynamics of air or one of its components ($CO_2$, $O_2$, etc.) in the substrates has received limited attention. Most studies dealt with the concept of substrate aeration based on the volumetric air content value at −1 kPa (de Boodt and Verdonck, 1972; Wallach et al., 1992) based on the fact that gas exchange (mainly $O_2$ and $CO_2$) does occur through the gaseous phase. However, studies undertaken to correlate these aeration criteria with the quality and the growth of potted plants have turned out to be more (Ouimet et al., 1990) or less (Brown and Emino, 1981; Karlovich and Fonteno, 1986) significant.

Glinski and Lipiec (1990), Allaire et al. (1996), and Caron et al. (2010) suggested that similar to water flow, the evaluation of gas transfer properties (that is, their relative diffusivity) is more relevant for describing gases availability to the respiring root than air-filled porosity. This would be because gas exchange will occur not only through the air-filled pore space, but requires those air-filled pores to be connected, a second parameter influencing gas diffusion in most equations (Caron and Nkongolo, 2004). One additional reason is the fact that organic growing media do have high respiration requirements, first due to microbial decomposition taking place, and second due to a high respiration activity of a densely rooted volume. This implies that storage of oxygen will rapidly become depleted (Naasz et al., 2009). Adequate respiration will then rely more on the rate of dynamic exchange of gases than on the static storage of gases itself within the substrate. Figure 5 plots the different relationship between gas diffusivity and plant performances obtained in such studies.

Hence, additional growth limitations have also been identified and linked to dynamic gas and water exchanges, namely gas relative diffusivity and unsaturated hydraulic conductivity, all summarized in Table 2.

With respect to measuring dynamic parameters (gas diffusivity and unsaturated hydraulic conductivity), such parameters should be characterized with appropriate methodologies adapted to different container geometries to be of any use (Caron et al, 2002, 2010, 2013, 2015c). Indeed, their values are strongly affected by container geometry (Fonteno, 1988). Limited disturbance is also a requirement for such measurements to be representative of all prior operations, and also to account for the evolution of the substrate in the container itself (Paquet et al., 1993; Caron et al., 2010).

Some studies have reported possible ways to measure the gas diffusion rate (Paul and Lee, 1976; Bunt, 1991) or the gas diffusion coefficient (Gislerød, 1982; King and Smith, 1987) in different types of substrates. Other authors, such as Allaire et al. (1994, 1996), Allaire-Leung et al. (1999), Caron and Nkongolo (2004) and Caron et al. (2010), developed methodologies capable of effectively measuring gas exchange characteristics in potted substrates and revealed their high correlation with plant growth, contrary to the traditional substrate aeration indicator (air content at −1 kPa) used in parallel.

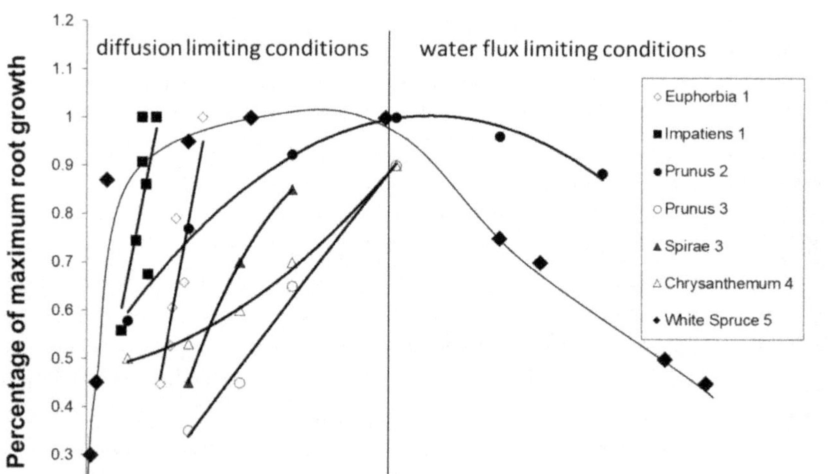


**Figure 5** Relationship between gas diffusivity and root growth for different experiments [1]Caron et al. (2010), [2]Allaire et al. (1996), [3]Nkongolo and Caron (2006b), [4]Nkongolo and Caron (2006a), [5]Boudreault et al. (2014).

Many studies, including da Silva et al. (1993, 1995), Laurén and Heiskanen (1997), Caron et al. (1998), Holden (2001) and Otten et al. (1999) have provided data on unsaturated hydraulic conductivity of pure or mixed substrate samples, estimated in the laboratory using transient methods (instantaneous profile) or steady-state methods, and so on that generally require relatively precise and accurate instruments. Moreover, several authors have developed techniques adapted to the constraints imposed by the studied materials and the containers to directly measure these properties during growth (Allaire et al., 1994; Heiskanen et al., 1996; Caron and Elrick, 2005a,b). The latter, in particular, developed an experimental procedure that makes it possible to directly measure unsaturated hydraulic conductivity in containers.

Wallach et al. (1992), da Silva et al. (1993) and Naasz et al. (2005) showed that hydraulic conductivity of various horticultural substrates, such as peat, composted bark, agricultural waste compost and so on radically decreases by several orders of magnitude within a small range of water potential. This could rapidly lead to very low water flows in the substrate and therefore limit root uptake. Additional work also reported such limitation restricting plant growth on capillary mats (Caron et al., 2005a).

Caron and Elrick (2005b) proposed norms for the shape of the unsaturated hydraulic conductivity curve (Fig. 6, Table 2) based on results of various experiments. These norms were proposed for substrates used on capillary rise operated systems. A model was derived for its predictive use (Caron et al., 2005b). Alternatively, models like Hydrus 1-D and 2-D can be successfully used, provided hysteresis is considered (Naasz et al., 2008) and that the Durner-Van Genuchten function is used (Caron et al., 2015c, 2020).

Finally, recent finding have suggested that water flow in substrates, as in peat, takes place as a bimodal (two distinct pore size distribution for large and fine pores) that should be taken into account when describing water flow (Weber et al., 2017), solutes transport (Hoag and Price, 1997; Rezanezhad et al., 2012; Caron et al., 2015b) and gas transport (Caron and Nkongolo, 2004).

## 3 Factors explaining substrate physical behaviour: particle size

The range of granulometry and particle size (sand, silt and clay) classically used in civil and agricultural engineering and soil science for describing the texture of mineral soils and their expected engineering behaviour finds limited relevance in growing media science. While particle size distribution may give more or less precise indications of the expected suitable physical properties (Bunt, 1983, 1988; Verdonck, 2004; Abad et al., 2005), particle shape is also of importance. Some impermeable components (like bark) may create barriers to gas diffusion

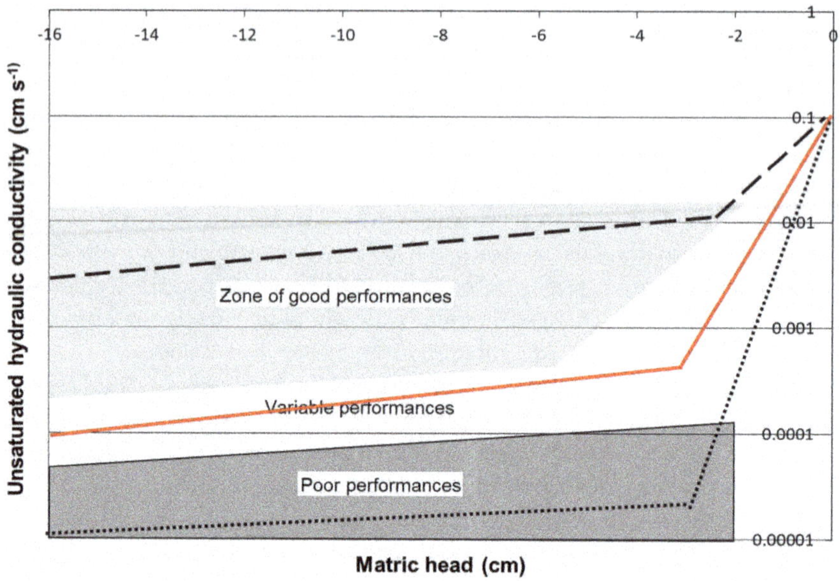


**Figure 6** Shapes for the unsaturated hydraulic conductivity-water potential curves. The red line represents minimal requirement for good performances in ebb and flow and capillary mat systems for proper capillary rise (redrawn from Caron and Elrick, 2005b).

(Nkongolo and Caron, 1999) or liquid water movement under unsaturated conditions (Caron et al., 2005a). Developments have been recently carried out (Bartley, 2019) and are still in progress (Michel, personal communication) about particle size and shape analysis using specific tools based on dynamic image analysis or X-ray microtomography.

### *3.1 Aeration*

Initially, several authors, including Carlson (1983), Bernier and Gonzalez (1995) and Puutsjärvi (1982), attempted to associate the physical properties of growing media to their particle size distribution. However, later studies have shown that this approach was insufficient for different reasons. First, organic components are fragile to sieving (Dinel and Lévesque, 1976) and as highlighted above, sieving, loosening of compacted raw material, mixing operations and automated potting procedures may result in alteration of initial fibre size, with improved or degraded properties (Heiskanen et al., 1996). Second, some studies have shown that the relationship may depend on the parameters chosen for characterizing proper growth indices. For example, if air-filled porosity is used to diagnose aeration limitations, Verhagen and Schmilewski (1997) have linked reduced aeration to particles < 1 mm, while Caron et al. (2005b) have linked it to particles of diameter <0.25 mm. (Fig. 7). In contrast, when using gas exchange (gas diffusivity) as a key aeration property, too coarse impermeable particles (>4 mm) may have a detrimental effect on that property (Nkongolo and Caron, 2006a,b; Caron et al., 2005a, 2010) despite no change in air-filled porosity (Fig. 8). Third, results are somewhat inconsistent as sometimes no relation with plant growth can be found (Boudreault et al., 2014). Despite this, an optimal growing media may be made of a combination of less than 4 mm material with a coarse fraction in between 2 mm and 4 mm, but not larger. Indeed, coarser fragments are impermeable to water and gas exchange and may limit aeration (Jobin et al., 2014). A fair proportion of intermediate particles, likely between 0.25 mm and 2 mm must also be present in an optimized growing media. This has been observed, as substrates made of biostable material less than 0.25 mm or even 0.10 mm often experience aeration and drainage problems.

However, because there is a growing interest in using non-composted material to reduce the cost associated with the composting step itself (Lemay et al., 2012), aeration limitations may result from this early higher microbial respiration in some non-composted material, competing with plant respiration, despite an air-filled porosity of at least 0.20 $cm^3\ cm^{-3}$ (Naasz et al., 2009). While this may not be an issue with biologically stable material like coir and peat, Naasz et al. (2009) have suggested that the aeration problem might be solved by increasing air-filled porosity above 0.30 $cm^3\ cm^{-3}$ (Fig. 2) in some non-composted bark to compensate for oxygen limitations. This may also

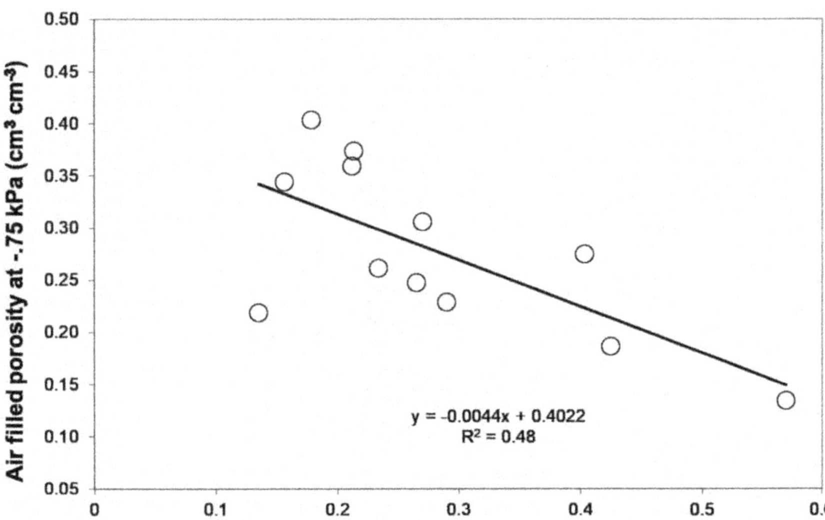


**Figure 7** Relationship between the proportion of fines less than 0.8 mm and air-filled porosity at −0.75 kPa of 12 different growing media (data taken from Caron et al., 2005a).

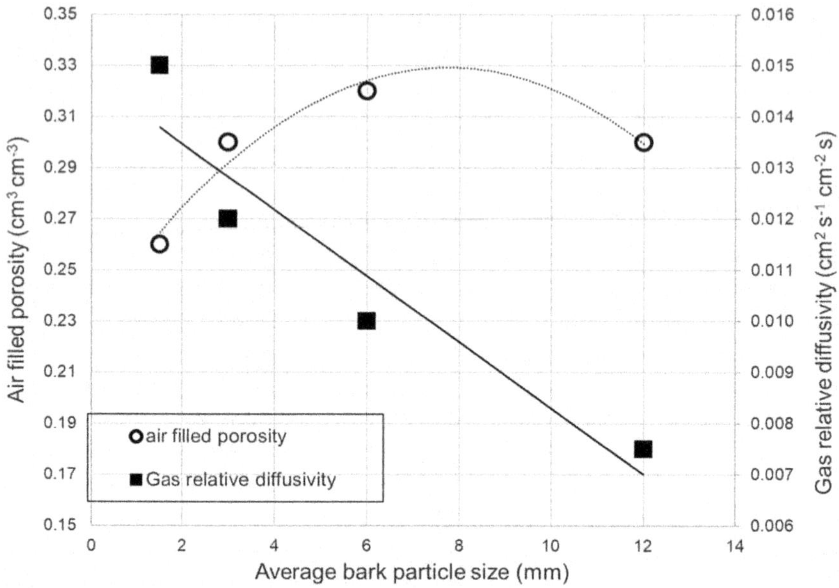


**Figure 8** Relationship between bark particle size, air-filled porosity and gas diffusivity in 50% peat/50% bark (volume per volume) growing media (data from Nkongolo and Caron, 1999).

occur with fresh sawdust material, where macronutrients' immobilization has been suggested to limit their use in the substrate. In the case of possible competition for oxygen, the processing of sawdust is one route commonly used to cause a net reduction in *N* immobilization and likely a reduction in microbial activity through elimination of free sugars involving in the immobilization and microbial respiration processes (Tišler et al., 1986). Naasz (2007, personal communication) has tested convective devices to increase oxygenation and accelerate $CO_2$ elimination in growing greenhouse tomatoes with the same yield as in the control. Using a passive (diffusive) aeration system with constant water availability, Lemay et al. (2012) grew greenhouse tomatoes in a sawdust-peat mix relative with the same yield as the rockwood control. Both studies suggest therefore that aeration limitation contributes to apparent nutrient immobilization.

### *3.2 Capillary rise*

Despite the pioneering work on capillary rise undertaken by Wallach et al. (1992), Otten (1994) and Naasz et al. (2005, 2008), water availability studies have focussed on the water reservoir itself. However, growing media are now used in various containers with varying geometry and irrigated by different systems, such as micro-sprayers, drip lines and overhead sprinklers (now potentially automated). Growing media can also be irrigated through a bottom-up process, like ebb and flow or use of capillary mats, for example. In these cases, while the reservoir itself is of importance, the rate of exchange of water between the source and the sink will be dominant (like aeration). This is related to the unsaturated hydraulic conductivity of the growing media with the process driven by capillary rise. In such a case, this property will be defined by the unsaturated hydraulic conductivity curve (Fig. 6) and can be described by the Gardner relationship (see Eq. 1) for a range of soil matric potential (Caron and Elrick, 2005a,b).

Moreover, this property can be described for a drainage process as well as a rewetting process and the shape of the curve will differ significantly, a phenomenon called hysteresis (Wever et al., 1996). Because of the considerable difference in the shape of the drainage or the rewetting unsaturated hydraulic conductivity curve, particularly in coarse growing media, this difference should be taken into account. Indeed, particular attention should be paid to the proportion of coarse fraction as it may negatively affect the capillary rise properties of the growing media, causing an increase in the magnitude of hysteresis (Caron et al., 2005a). Hence, too high a proportion of coarse fragments has also resulted in poor capillary rise properties, and a too low a volume of water being transferred to the actively growing plant, and thus imposed temporary water stresses to the crop and caused subsequent yield drops (Caron et al., 2005a; Caron and Elrick, 2005b). The problem has been solved by increasing the proportion of fine material (generally less than 0.25–1 mm depending on authors) in substrates up to 50–60% sphagnum peat (v/v), 30% of fines being typically too low (Caron et al., 2004, 2005a).

## 4 Factors explaining substrate physical behaviour: structure (solid-phase organization)

### *4.1 Volume weight*

Growing media are generally fragile and easy to compact. Through years of practice, growers have been able to determine an approximate volume weight, also called 'bulk density' to achieve their maximum yield with the same growing media of similar composition from year to year. For simulating greenhouse operation conditions, some laboratory methods include a compacting step of samples to reach a volume weight similar to that found in the container, like applying a weight once the substrate is in the measuring cylinder or applying a suction at the bottom of the filled cylinder (Gabriels, 1994; Morel et al., 2000; Fonteno and Harden, 2003). The compacting process under greenhouse operation is a critical step, as a too low or too high bulk density at the beginning of growth has been observed to reduce rooting into the substrate, with consequent poor growth of up to 50% less root development, a problem solved by managing the compaction at potting (Gruda and Schnitzler, 2004a,b). As mentioned above, such a problem can be diagnosed by characterizing the substrate properties once potted (Caron et al., 2008, 2010, 2015c; Cannavo and Michel, 2013; Heiskanen et al., 1996) as highlighted above.

### *4.2 Container geometry*

After irrigation and drainage in a growing medium, free water often may appear at the bottom layer of the container as extra water is applied to cause leaching. The presence of this water table at the base of the container affects the water potential distribution throughout the profile. Since the average potential applied at the top surface of the potted substrate will be in direct connection with the water table, this will result in different suction applied on the substrate at the end of drainage, because the higher the container, the higher the weight of the substrate above applied to the bottom of the container. Additionally, the higher the container, the lower the water potential at the top will be, and the more significant the effect of retreating water menisci throughout the whole substrate in bringing particles closer to one another, leading to an increased bulk density of the substrate with depth (Raviv et al., 2002). Hence, different sizes and heights of container will affect the in-situ material properties, thus air and water storage (Bilderback and Fonteno, 1987; Raviv et al., 2002).

When this modification of the matric potential within the substrate and bulk density is associated with depth, the result is a general increase of air-filled porosity with container height and a decrease in available water (Fig. 9). Equilibrium values of air and volumetric water content in the container will, therefore, be a function of container height and geometry. Therefore, the volumetric air content in the container will be characteristic of the height. Its effects on average potential, but also the compaction process, results from the container height. This is an additional reason for which the concept of field capacity used in agriculture is of little relevance here and rather replaced by container capacity itself depending on container geometry (Bilderback and Fonteno, 1987; Caron et al., 2015a). It is why also the characterization of properties should be performed in containers in diagnosing potential growth problems.

## 5 Factors explaining substrate physical behaviour: change in physical properties

As written in a previous section, the analysis of the hydraulic properties of growing media is generally carried out before use. However, many factors can

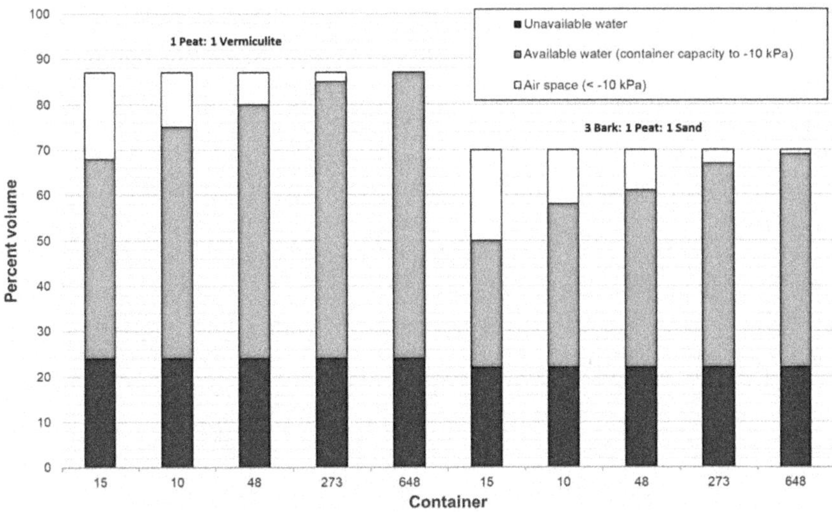


**Figure 9** Air-filled porosity and available water of two different substrates in containers of different sizes (from Bilderback and Fonteno, 1987). 15 = 15 cm pot, 10 = 10 cm pot, 48 = 48 cells/tray plug cells, 273 = 273 cells/tray plug cells, 648 = 648 cells/try waffle plug cells.

be significantly modified at potting and throughout plant growth. Solid phase organization (material deformation), change in material hydrophobia (decrease in water affinity for the solid phase) and soil structure change associated with the growth of the root system and microbial composition takes place. Changes of growing media properties during their use are mainly observed through their indirect effects on plant development, except in a few cases where some indirect physical and/or chemical and/or biological parameter of the medium has been measured over time (during cultivation or not).

## 5.1 Shrinkage and stability of growing media

Gradual decomposition of the organic constituents of the medium may lead to the shrinkage of the substrate with a decreased air-filled porosity and an increased bulk density. Hence, high biostability is looked for when testing peat alternatives (Morel et al., 2000). Peat moss is generally considered as a relatively stable substrate for a greenhouse crop cycle (4–12 months). Peat types vary in their stability level though. Coir is considered as of or even more stable than peat and Prasad (1997) reported a volume reduction of coir of 24% over 24 months, which is the characteristic of a relatively stable material, at least for greenhouse crop use for less than a year. Typically, some compost types may be prone to excessive shrinkage and deterioration of their physical properties may occur rapidly, that is less than from 3 to 6 months (Chen et al., 2002). However, some made of dairy manure (Jeong et al., 2011) and aged pine bark (Jackson et al., 2009) were found of adequate stability. Besides, wood fibre substrates are reported to have a high shrinkage rate. Therefore, Gruda and Schnitzler (2004a and b) recommend increasing substrate compression by filling the pots to reduce this effect.

The rapid consumption of water by roots results in frequent irrigation cycles during growth. As a consequence, growing media may exhibit more hysteresis phenomena with regard to water retention, and successive drying/wetting cycles of the substrate during cultivation can also lead to its shrinkage and then compaction (Qi et al., 2011; Kerloch and Michel, 2015), with a negative impact in terms of aeration properties (Langerud, 1986; Heiskanen et al., 1996; Michel and Kerloch, 2017) that could cut off the supply of oxygen to the roots during growth. This loss in rooting total substrate volume is mainly observed in fibrous and in fine materials, such as peat, wood fibres, and so on. Michel (2019) demonstrated that irrigation management (i.e. threshold for triggering irrigation) largely influences both water and air content and flow within substrates.

## 5.2 Wettability and hysteresis

The ability of organic matter used in growing media to acquire a hydrophobic character depends on its degree of decomposition and initial water content (Michel et al., 2001, 2017; Michel, 2015). The hydrophobic character then affects its physical and hydraulic properties by inducing rewetting difficulties as well as non-uniform wetting (Valat et al., 1991; Michel et al., 2001, 2017) and unstable wetting front (Dekker et al., 2000).

Wettability of peat has been a long-term, well-known limitation, solved by adding a wetting agent, sometimes clay (20–40 kg per cubic meter in the mix), coir (Michel et al., 2020) or wood fibre in peat-based mixes. Most materials are incorrectly considered hydrophilic so the importance of wettability issue tends to be of lesser importance in material other than peat, where it is a critical issue. Even if the risk of hydrophobicity in organic materials other than peat is low, some organic substrate constituents can also exhibit a variable affinity for water (Michel, 2015). A classification based on their surface properties and their reversibility to rewetting has been recently proposed, defining four types with increasing risk of hydrophobicity (Michel et al., 2017). Type 1 remained hydrophilic and recovered their water retention after drying, whatever the intensity of drying, whereas types 2, 3 and 4 showed gradual changes from hydrophilicity to hydrophobicity and their ability to capture water after drying decreased with the intensity of drying. Wettability gradually changes during the drying and rewetting processes and should be considered as one of the main parameters explaining hysteresis phenomena in water retention (Naasz et al., 2008) and unsaturated hydraulic conductivity as well (Naasz et al., 2005).

Kerloch and Michel (2015) and Michel and Kerloch (2017) used wettability to partly explain some changes in physical properties of different substrates, such as peat, but also bark, coir, wood fibre and peat-based mixes, during plant growth. Michel (2019) reported a decrease in wettability over time, limiting the increase in water retention and especially creating preferential air and water flows within the substrate. Hence, wettability and the associated hysteresis phenomenon have a strong influence on the air and water exchange processes (Otten, 1994; Wever et al., 1996; Michel and Kerloch, 2017; Michel, 2019). According to Naasz et al. (2005) and Michel (2019), such processes should be integrated into decision support models used in optimizing irrigation management.

The magnitude of the hysteresis effect is indirectly associated with the presence of fine material and the uniformity of the pores, and the fact that they are angular-disconnected pores rather than tubular pores (Hoag and Price, 1997; Caron and Nkongolo, 2004; Caron et al., 2005a). One of the consequences of a strong hysteresis effect, associated with too coarse fragments and lower wettability upon rewetting, is a poor capillary rise, which will become critical in ebb and flow system but even in drip irrigation systems (Wallach et al., 1992; Otten, 1994; Lemay et al., 2012; Caron et al., 2005a; Michel et al., 2017).

## *5.3 The root system and its development*

The root system is affected by the physical constraints imposed by the substrate and by the fact that the plant must have a sufficient volume of the substrate to provide the quantities of water, air and minerals necessary to its growth. As a result, several factors affect the potential growth of the root system. A first factor is the plant species: root colonization is very different between plants with shallow rooting compared to those with taproots (Lemaire et al., 2003), a factor which becomes increasingly complex in the case of woody perennials over time. A second factor is the size-related characteristics of the container (volume, depth, diameter), where increasing volume and height has a positive impact on the growth and distribution of plant shoots and roots (Biran and Eliassaf, 1980; Keever and Cobb, 1987). A final factor is the type of substrates and its properties, particularly the granulometry and solid phase and pore size distribution.

In certain cases, the roots may take up a large part of the empty spaces in the substrate (sometimes more than the volume of the solid, that is, substrate phase) and lead to a steep decrease in the total porosity where water and air are easily available to the plant. In other cases, the structuring effect of the roots can have a beneficial impact on the substrates' physical stability by facilitating the formation of aggregates and by reducing compaction during growth (Rivière, 1991). It can sometimes even improve gas and/or water transfers by increasing the effectiveness of the substrate's pore network (Allaire-Leung et al., 1999; Cannavo and Michel, 2013; Michel, 2019). Michel (2019) reported a higher wettability as a function of root content, in relation to the higher water retention in the substrates; an effect may be associated with mucigel influence on the rhizosphere water desorption curve (Carminati et al., 2010).

## 6 Case study: designing a substrate for production of forest tree seedlings

Previous studies have illustrated the need to include dynamic measurements of growing media, ideally directly in the container (Allaire et al., 1996; Nkongolo and Caron, 2006a,b, Caron et al., 2010). More recently, a comprehensive case study has been undertaken on the development of growing media for forest tree seedlings in Canada (Boudreault et al., 2014). This study provides a holistic approach to illustrate the importance of dynamic parameters in designing a growing medium and illustrates the key message of this chapter. A similar case study for potted plants can also be found in Jobin et al. (2014).

In forest nurseries a significant number of young trees are rejected because of insufficient root development and poor root plug cohesion. It appears to be quite common in forestry industry to base substrate design and purchase on the ratio from coarse to fine particle sizes (Carlson, 1983). Boudreault et al. (2014) conducted a 2-year experiment to determine the effects of substrate physico-chemical properties on seedling growth. Seedlings were grown in different organic substrates composed of peat with some mineral components like perlite and vermiculite. Such mixes created an extensive range of static and dynamic physical properties. Although the peat used in the substrate was from the same bog, the particle size distributions differed from year to year. Substrates were also voluntarily more compacted on the second year. The experiment was repeated on two consecutive years under plastic tunnels with boom irrigation as required.

On top of measuring various seedling growth parameters and nutrient content, physical properties were measured initially, directly in the container and at the end of each growing season once the roots had affected physical properties of the growing media. The parameters measured included static parameters, such as bulk density, available water, air-filled porosity, as well as dynamic parameters, such as saturated hydraulic conductivity, gas diffusivity and pore effectiveness coefficient. Moreover, the static and dynamic characteristics were used to simulate oxygen profile in the container and the availability of water. Substrate properties had significant effects on growth parameters in both years.

### 6.1 Dynamics of water availability

As described in Fig. 4, the study used the model developed by Rekika et al. (2014) with the saturated hydraulic conductivity of the substrate used in the case study as well as the $\alpha$ parameter (Eq. 1) obtained from Caron et al. (2005a) for $\alpha$ = 0.30 (coarse medium) and $\alpha$ = 0.03 (fine medium). The model showed that the plants probably suffered from water stress at very high soil matric potential, probably because water transfer was too slow (Boudreault et al., 2014). Indeed, there was a significant decrease in seedling height and the average matric potential of the different substrates during the growing season even if mean potentials were all above −5 kPa. The alpha value of $\alpha = 0.30\ cm^{-1}$, consistent with coarse, low volume weight substrate (Caron et al., 2005a) led to the conclusion that water transfer should start limiting crop growth at about −2 kPa (Fig. 4). The conclusion was consistent with previous work (Caron et al., 1998, 2005a) and a type III substrate behaviour (Fig. 1).

Data indicated no correlation with mean weight diameter overall, a summary parameter reflecting particle size distribution (Table 3). Decreasing plant performance was related to increasing air-filled porosity, increasing saturated hydraulic conductivity and decreasing bulk density. This relationship is consistent with the conclusion that unsaturated hydraulic conductivity limited the growth during the season (Fig. 4). With respect to static parameters, no relationship was found with available water (Table 3), likely because the threshold point of availability would not be representative of the critical threshold of −10 kPa below which water becomes very difficult to extract (Jobin et al., 2004; Caron et al., 1998). Negative dominant contribution of air-filled

**Table 3** Raw correlation between some physical parameters and root dry mass (RDM) and shoot dry mass (SDM) in the case study of Boudreault et al. (2014)

| | 2008 ($n$ = 26) | | 2009 ($n$ = 22) | |
|---|---|---|---|---|
| Property[a] | RDM | SDM | RDM | SDM |
| Physical | *Static parameters* | | | |
| MWD | −0.12 | −0.07 | −0.68** | −0.72** |
| $\rho_b$ | 0.53** | 0.63** | −0.08 | 0.07 |
| AW | 0.35 | 0.26 | −0.20 | −0.19 |
| | *Dynamic parameters* | | | |
| $K_s$ | −0.56** | −0.62** | 0.39 | 0.30 |
| $\theta_a$ | −0.65** | −0.71** | −0.08 | −0.19 |
| Y | 0.02 | 0.01 | 0.71** | 0.72** |
| $D_sD_o^{-1}$ | −0.46* | −0.44* | 0.55** | 0.42* |

[a] ** and * indicate correlation coefficient significant at p= 0.01 and 0.05, respectively.

**Table 4** Stepwise multiple regression for the two years of the case study (Boudreault et al., 2014) relating static and dynamic to plant growth parameters

| Growth variable[a] | Standard estimate | Property[b] | Partial $R^2$ | Standard estimate | Property[c] | Partial $R^2$ | Total $R^2$ |
|---|---|---|---|---|---|---|---|
| | 2008 ($n$ = 26) | | | | | | |
| RDM | −0.64 $\theta_a$ | | 0.42 | 0.35 | EC | 0.12 | 0.54 |
| SDM | −0.49 $\theta_a$ | | 0.52 | −0.36 | $K_s$ | 0.08 | 0.60 |
| | 2009 ($n$ = 22) | | | | | | |
| RDM | 0.54 $\gamma$ | | 0.54 | −0.43 | MWD | 0.15 | 0.69 |
| SDM | 0.75 $\gamma$ | | | | | | 0.56 |

[a] RDM = root dry mass; SDM = shoot dry mass.
[b] $\theta_a$ = air-filled porosity; $\rho_b$ = bulk density; $\gamma$ = coefficient of pore effectiveness.
[c] EC = electrical conductivity; $K_s$ = saturated hydraulic conductivity; MWD = mean weight diameter.
All regression coefficient are signficant at p=0.05

porosity suggested excess macropores and limited water transfer (Table 4, Fig. 5), as highlighted above.

### 6.2 Dynamics of gas exchange

The year after seedlings were planted in the substrate, when substrates were compacted high enough to have a low air-filled porosity, a high bulk density and a low gas diffusivity, plant parameters were linked positively to pore effectiveness for gas transport. They were very weakly related to particle size distribution for the growth parameters on one year only (Tables 3 and 4).

As before in Allaire et al. (1996), Nkongolo and Caron (2006a,b) and Caron et al. (2010), dynamic properties explained the growth performance observed after two years and were consistent with oxygen and water flow limitations modelled for the container geometry used. The multiple stepwise regression indicated that pore effectiveness coefficient dominated the regression for the 2009 data, consistent with aeration limiting water transfer, as this parameter was negatively correlated to air-filled porosity. Hence, this case study suggests again that, when designing and using substrates at the nursery scale, more attention should be paid to aeration properties after potting and water dynamic availability, rather than to the coarse/fine particle ratio, and to dynamic more than static parameters.

## 7 Summary and future trends

Initially, most 'recipes' proposed by substrate manufacturers refer to using different particle size distribution to infer substrate growth performances once potted. Thereafter, physical limitations appeared linked to aeration and water retention problems, and were diagnosed using static property indices like air-filled porosity and available water, characteristics of the medium at a given equilibrium soil water potential. But because of increasing irrigation and management systems with frequent adjustment of water potential and air content in substrates, dynamic properties linked to the exchange of fluids have become more important. Hence, gas diffusivity and substrate respiration indices appear relevant in guiding the manufacturing of growing media. Wettability and hydraulic conductivity appear to be additional properties, particularly in very coarse substrate products. Other limitations appear linked to volume weight and structural stability, in case of substrates components with low biological stability.

Efficient design of substrates will therefore require the use of key static and dynamic parameters and of appropriately chosen methods to measure and diagnose them. Additionally, given the high variability in container configuration, it becomes critical to adapt these parameters to container geometry and to production system constraints. Design criteria and the expected substrate performance will depend on the particular container geometry chosen. Pot size, shape and hole distribution are, therefore, to be considered in combination with the chosen substrate along with the irrigation strategy for optimal productivity. Finally, given the change in growing media properties, such characteristics should be obtained ideally from the substrates once used in the production system. This will allow a better understanding and diagnosis of growth limitation. This will also help in achieving gains in productivity when this information can be used in computerized and automatized irrigation and fertilization management tools, whether sensors, software or models.

Recent research results highlighted in this chapter have thus indicated the need for additional dynamic parameters to guide the manufacturing of the next generations of substrates. Recent developments have been initiated (Bartley, 2019) and are currently in progress to characterize particle size and shape and pore and solid phase organization with the aim of predicting physical properties and to design new optimized substrates. This will have implications for the parameters to be measured, the operating conditions under which they should be measured and, finally, the frequency with which they should be measured. New norms should therefore be developed to take into account all these changes. Such norms should be related not only to static but also to dynamic parameters related to flux, gradient and consumption, as air-filled porosity and available water are shown to be incomplete. More specifically, these norms should move to incorporate gas diffusivity, substrate respiration rate and saturated and unsaturated hydraulic conductivity among others and take into account the context of the application (Caron et al., 2013). Recent evidence of the bimodal nature of pore size distribution in substrates is growing and should also be taken into account in the characterization of growing media (Weber et al., 2017; Caron et al., 2005b; Caron and Nkongolo, 2004; Rezanezhad et al., 2012).

The gain in precision in the measurement of such parameters and the associated potential gain in productivity will in turn boost the development of new management technology. Powerful computer models will be part of this technology and will be increasingly used to manage greenhouse and nursery irrigation and in situ operations. Indeed, computer models are already part of the design process (Fields et al., 2020; Caron et al., 2020) and are expected to increase in importance with the development and implementation of precision irrigation technologies (Michel and Kerloch, 2017; Depardieu et al., 2018).

## 8 Acknowledgements

The authors thank France Chabot for help in preparing the text and Genevieve Montminy for help with some of the figures.

## 9 Where to look for further information

For readers interested in more detailed basic definitions of concepts and methodology raised in this chapter, extensive descriptions were made recently by Raviv et al (2019). They described in different chapters the basic physical properties of soilless using approaches taken in classical soil physics, with some illustrative examples of papers using these parameters from the growing media research field. They also emphasized the methodologies associated with the characterization of these physical properties, for guiding the manufacturing process, with most of them being developed by International Society of Horticultural Science (ISHS) and by Comité européen de normalisation (CEN). However, the authors emphasize mainly static properties, such as bulk density, particle size distribution, water retention, hydrophobicity and container capacity, though with some limited coverage of dynamic properties like saturated and unsaturated hydraulic conductivity and oxygen diffusion among other factors.

Given what was presented above, complementary data should be gathered on dynamic parameters for research on growing media. Readers are referred to Caron et al. (2008) and Caron et al. (2015a) for such description. Data collected on changes in substrate properties indicate potential productivity benefits (Caron et al., 2010; Cannavo and Michel, 2013; Kerloch and Michel, 2015; Michel and Kerloch, 2017). The future will see increasing control systems that will be able to take into account change in the physical properties and adapt the management tools with it. Readers and researchers should therefore look at methodologies capable of measuring properties in various sets of configurations and operational conditions. Examples of such approaches are given by Paquet et al. (1993), Caron et al. (2002, 2010), Allaire et al. (1994) and Michel (2019).

Finally, we recommend using computer models in order to develop improved management approaches and more performing growing media while reducing the needs for extensive experimental development. In the field of soil physics, *Hydrus* is one of the widely used software that should be investigated for use by future substrate physicists (Fields et al., 2020; Caron et al., 2020).

## 10 References

Abad, M., Fornes, F., Carrión, C., Noguera, V., Noguera, P., Maquieira, A. and Puchades, R. 2005. Physical properties of various coconut coir dusts compared to peat. *HortScience* 40(7):2138–2144.

Allaire, S. E., Caron, J. and Gallichand, J. 1994. Measuring the saturated hydraulic conductivity of peat substrates in nursery containers. *Canadian Journal of Soil Science* 74(4):431–437.

Allaire, S. E., Caron, J., Duchesne, I., Parent, L. E. and Rioux, J. A. 1996. Air-filled porosity, gas relative diffusivity and tortuosity: indices of *Prunus* x *cystena* sp.

growth in peat substrates. *Journal of the American Society for Horticultural Science* 121(2):236-242.

Allaire-Leung, S. E., Caron, J. and Parent, L. E. 1999. Changes in physical properties of peat substrates during plant growxth. *Canadian Journal of Soil Science* 79(1):137-139.

Bartley, P. C. 2019. Multidimensional characterization of horticultural substrates. PhD thesis. North Carolina State University, USA.

Bernier, P. Y. and Gonzalez, A. 1995. Effects of the physical properties of Sphagnum peat on the nursery growth of containerized *Picea mariana* and *Picea glauca* seedlings. *Scandinavian Journal of Forest Research* 10(1-4):176-183.

Bilderback, T. E. and Fonteno, W. C. 1987. Effects of container geometry and media physical properties on air and water volumes in containers. *Journal of Environmental Horticulture* 5(4):180-182.

Biran, I. and Eliassaf, A. 1980. The effect of container shape on the development of roots and canopy of woody plants. *Scientia Horticulturae* 12(2):183-193.

Boudreault, S., Pepin, S., Caron, J., Lamhamedi, M. S. and Paiement, I. 2014. Substrate aeration properties and growth of containerized white spruce: A case study. *Vadose Zone Journal* 13:3-18.

Brown, O. D. R. and Emino, E. R. 1981. Response of container-grown plants to six consumer growing media. *Horticulturalscience* 16:78-80.

Bunt, A. C. 1983. Physical properties of mixtures of peats and minerals of different particle size and bulk density for potting substrates. *Acta Horticulturae* 150(150):143-154.

Bunt, A. C. 1988. *Media and Mixes for Container-Grown Plants* (2nd edition). Unwin Hyman, London.

Bunt, A. C. 1991. The relationship of oxygen diffusion rate to the air-filled porosity of potting substrates. *Acta Horticulturae* 294(294):215-224.

Cannavo, P. and Michel, J.-C. 2013. Peat particle size effects on spatial root distribution, and changes on hydraulic and aeration properties. *Scientia Horticulturae* 151:11-21.

Carlson, L. W. 1983. Guidelines for rearing containerized conifer seedlings in the Prairie Provinces. Northern Forest Research Centre, Edmonton (Alberta), Information Report NOR-X-214E.

Carminati, A., Moradi, A. B., Vetterlein, D., Vontobel, P., Lehmann, E., Weller, U., Vogel, H. J. and Oswald, S. E. 2010. Dynamics of soil water content in the rhizosphere. *Plant and Soil* 332(1-2):163-176.

Caron, J. and Nkongolo, N. V. 2004. Assessing gas diffusion coefficients in growing media from in situ water flow and storage measurements. *Vadose Zone Journal* 3:300-311.

Caron, J. and Elrick, D. E. 2005a. Measuring the unsaturated hydraulic conductivity of growing media with a tension disc. *Soil Science Society of America Journal* 69(3):783-793.

Caron, J. and Elrick, D. E. 2005b. Defining critical capillary rise properties for growing media: model and methodology. *Acta Horticulturae* 779(779):149-154.

Caron, J., Xu, H. L., Bernier, P. Y., Duchesne, I. and Tardif, P. 1998. Water availability in three artificial substrates during Prunus xcistena growth: variable threshold values. *Journal of the American Society for Horticultural Science* 123(5):931-936.

Caron, J., Rivière, L. M., Charpentier, S., Renault, P. and Michel, J. C. 2002. Using TDR to estimate hydraulic conductivity and air entry in growing media and sand. *Soil Science Society of America Journal* 66(2):373-383.

Caron, J., Beeson, R., Haydu, J. and Boudreau, J. 2004. Saving water with Sphagnum peat in nursery growing media. *Acta Horticulturae* 664(664):119-124.

Caron, J., Elrick, D. E., Beeson, R. and Boudreau, J. 2005a. Defining critical capillary rise properties for growing media in nurseries. *Soil Science Society of America Journal* 69(3):783–793.

Caron, J., Rivière, L.-M. and Guillemain, G. 2005b. Gas diffusion and air-filled porosity: effect of some oversize fragments in growing media. *Canadian Journal of Soil Science* 85(1):57–65.

Caron, J., Parent, L. E., Elrick, D. E. and Naasz, R. 2008. Physical properties of organic soils and growing media: water and air storage and flow dynamics. In: Carter, M. R. and Gregorich, E. G. (Eds) *Soil Sampling and Methods of Analysis*. CRC Press, Boca Raton, FL. 885–912.

Caron, J., Morel, P., Rivière, L. M. and Guillemain, G. 2010. Identifying appropriate methodology to diagnose aeration limitations with large peat and bark particles in growing media. *Canadian Journal of Soil Science* 90(3):481–494.

Caron, J., Pepin, S. and Périard, Y. 2013. Physics of growing media in a green future. *International Symposium on Growing Media and Soilless Cultivation* 1034:309–317.

Caron, J., Heinse, R. and Charpentier, S. 2015a. Organic materials used in agriculture, horticulture, reconstructed soils, and filtering applications. *Vadose Zone Journal* 14(6):6.

Caron, J., Létourneau, G. and Fortin, J. 2015b. Electrical conductivity breakthrough experiment and immobile water estimation in organic substrates: is R = 1 a realistic assumption? *Vadose Zone Journal* 14(9):9.

Caron, J., Price, J. S. and Rochefort, L. 2015c. Physical properties of organic soil: adapting mineral soil concepts to horticultural growing media and histosol characterization. *Vadose Zone Journal* 14(6) :6.

Caron, J., Brisson, D. and Dansereau, B. 2020. Comparing water availability of injected polymers and tension reservoirs in growing media used in hanging baskets. *Acta Horticulturae* (in press).

Chen, J. J., McConnell, D. B., Robinson, C. A., Caldwell, R. D. and Huang, Y. F. 2002. Production and interior performances of tropical ornamental foliage plants grown in container substrates amended with composts. *Compost Science and Utilization* 10(3):217 -225.

Cormier, J., Depardieu, C., Letourneau, G., Boily, C., Gallichand, J. and Caron, J. 2020. Tensiometer-based irrigation scheduling and water use efficiency of field-grown strawberries. *Agronomy Journal* 112(4):2581–2597. doi.org/10.1002/agj2.20205.

da Silva, F. F., Wallach, R. and Chen, Y. 1993. Hydraulic properties of sphagnum peat moss and tuff (scoria) and their potential effects on water availability. *Plant and Soil* 154(1):119–126.

da Silva, F. F., Wallach, R. and Chen, Y. 1995. Hydraulic properties of rockwool slabs used as substrates in horticulture. *Acta Horticulturae* 401(401):71–76.

de Boodt, M. and Verdonck, O. 1972. The physical properties of the substrate in horticulture. *Acta Horticulturae* 26(26):37–44.

Dekker, L. W., Ritsema, C. J. and Oostindie, K. 2000. Wettability and wetting rate of Sphagnum peat and turf on dune sand effected by surfactant treatments. *Proceedings of the 11th International Peat Congress*, 6–12 August 2000, Quebec City, Canada II, pp. 566–574.

Depardieu, C., Watters, N., Gendron, L., Boily, C., Pépin, S. and Caron, J. 2018. High productivity of soilless strawberry cultivation under rain shelters. *Scientia Horticulturae* 232:127–138. doi.org/10.1016/j.scienta.2017.12.056.

Dinel, H. and Lévesque, M. 1976. A simple technique for the granulometric analysis of peat in aqueous medium. *Canadian Journal of Soil Science* 56:119–120.

Fields, J. S., Owen, J. S., Stewart, R. D., Heitman, J. L. and Caron, J. 2020. Modeling water fluxes through containerized soilless substrates using HYDRUS. *Vadose Zone Journal* 19:e20031. https://doi.org/10.1002/vzj2.20031.

Fonteno, W. C. 1988. An approach to modeling air and water status of horticultural substrates. *Acta Horticulturae* 238(238):67–74.

Fonteno, W. C. and Harden, C. T. 2003. *Procedures for Determining Physical Properties of Horticultural Substrates Using the NCSU Porometer*. Horticultural Substrates Laboratory, North Carolina State University, Rayleigh.

Gabriels, R. 1994. Standardization of growing media analysis and evaluation: CEN/ISO/ISHS. *Acta Horticulturae International Symposium on Growing Media & Plant Nutrition in Horticulture* 401:555–558.

Gislerød, H. R. 1982. Physical conditions of propagation media and their influence on the rooting of cuttings. *Plant and Soil* 69(3):445–456.

Glinski, J. and Lipiec, J. 1990. *Soil Physical Conditions and Plant Roots*. CRC Press, Boca Raton, FL. 250 pp.

Gruda, N. and Schnitzler, W. H. 2004a. Suitability of wood fiber substrate for production of vegetable transplants: I. Physical properties of wood fiber substrates. *Scientia Horticulturae* 100(1–4):309–322.

Gruda, N. and Schnitzler, W. H. 2004b. Suitability of wood fiber substrates for production of vegetable transplants II: The effect of wood fiber substrates and their volume weights on the growth of tomato transplants. *Scientia Horticulturae* 100(1):333-340.

Heiskanen, J., Tervo, L. and Heinonen, J. 1996. Effects of mechanical container-filling methods on texture and water retention of peat growth media. *Scandinavian Journal of Forest Research* 11(1–4):351–355.

Hoag, R. S. and Price, J. S. 1997. The effects of matrix diffusion on solute transport and retardation in undisturbed peat in laboratory columns. *Journal of Contaminant Hydrology* 28(3):193–205.

Holden, N. M. 2001. A note on the determination of unsaturated hydraulic conductivity of moss peat using time domain reflectrometry. *Irish Journal of Agricultural and Food Research* 40:277–282.

Jackson, B. E., Wright, R. D. and Seiler, J. R. 2009. Changes in chemical and physical properties of pine tree substrate and pine bark during long-term nursery crop production. *Hortscience* 44(3):791–799.

Jeong, K. Y., Nelson, P. V., Frantz, J. and Brinton, W. 2011. Impact of composted dairy manure on pH management and physical properties of soilless substrate. *Acta Horticulturae* 891:173–180.

Jobin, P., Caron, J., Bernier, P. Y. and Dansereau, B. 2004. Impact of two hydrophilic polymers on the physical properties of three substrates and the growth of Petunia hybrida 'Brillliant pink'. *Journal of the American Society for Horticultural Science* 129(3):449–457.

Jobin, P., Caron, J. and Rochefort, L. 2014. Developing new potting mixes with Sphagnum fibers. *Canadian Journal of Soil Science* 94(5):585–593.

Karlovich, P. T. and Fonteno, W. C. 1986. Effect of soil moisture tension and soil water content on the growth of chrysanthemum in 3 container media. *Journal of the American Society for Horticultural Science* 111(2): 191–195.

Keever, G. J. and Cobb, G. S. 1987. Effects of container volume and fertility rate on growth of two woody ornamentals. *Horticulturalscience* 22(5):891–893.

Kerloch, E. and Michel, J. -C. 2015. Pore tortuosity and wettability as main characteristics of the evolution of hydraulic properties of organic growing media during cultivation. *Vadose Zone Journal* 14(6). doi.org/10.2136/vzj2014.11.0162.

King, J. A. and Smith, K. A. 1987. Gaseous diffusion through peat. *Journal of Soil Science* 38(2):173–177.

Langerud, B. R. 1986. A simple in situ method for the characterization of porosity in growth media. *Plant and Soil* 93(3):413–425.

Laurén, A. and Heiskanen, J. 1997. Physical properties of the mor layer in a Scots pine stand I. Hydraulic conductivity. *Canadian Journal of Soil Science* 77(4):627–634.

Lemaire, F., Dartigues, A., Rivière, L.-M., Charpentier, S. and Morel, P. 2003. *Culture en pots et conteneurs, principes agronomiques et applications* (2ème édition). INRA Éditions, Paris, France.

Lemay, I., Caron, J., Dorais, M. and Pepin, S. 2012. Defining irrigation set points based on substrate properties for variable irrigation and constant matric potential devices in greenhouse tomato. *Horticulturalscience* 47(8):1141–1152.

Michel, J. C. 2015. Wettability of organic growing media used in horticulture: a review. *Vadose Zone Journal* 14(6). doi.org/10.2136/VZJ2014.09.0124.

Michel, J. C. 2019. Relevant hydraulic properties of peat-based mixes for explaining root development according to irrigation strategies. *Acta Horticulturae* 1266:353–360. doi.org/10.17660/ActaHortic.2019.1266.49.

Michel, J. C. and Kerloch, E. 2017. Evolution of hydraulic properties and wettability of organic growing media during cultivation according to irrigation strategies. *Scientia Horticulturae* 217:28–35.

Michel, J. C., Rivière, L. M. and Bellon-Fontaine, M. N. 2001. Measurement of the wettability of organic materials in relation to water content by the capillary rise method. *European Journal of Soil Science* 52(3):459–467.

Michel, J. C., Jackson, B. E. and Fonteno, W. C. 2017. Classification of organic substrates' wettability from contact angle measurements and hydration efficiency tests. *Acta Horticulturae* 1168(1168):199–206. doi.org/10.17660/ActaHortic.2017.1168.26.

Michel, J. C., Jackson, B. E. and Fonteno, W. C. 2020. The use of coir for reducing risks of substrates hydrophobicity. *Acta Horticulturae* (in press).

Morel, P., Poncet, L. and Rivière, L. M. 2000. *Les supports de culture horticoles: les matériaux complémentaires et alternatifs à la tourbe*. Editions Quae, Versailles, France.

Mustin, M. 1987. *Le compost. Gestion de la matière organique*. Dubusc, Paris.

Naasz, R., Michel, J. -C. and Charpentier, S. 2005. Measuring hysteretic hydraulic properties of peat and pine bark using a transient method. *Soil Science Society of America Journal* 69(1):13–22.

Naasz, R. 2007. Evaluation of convective aeration devices for greenhouse tomato production. Unpublished post-doctoral internal report. Soil science and agri-Food engineering department, Laval University, Québec, Canada.

Naasz, R., Michel, J.-C. and Charpentier, S. 2008. Water repellency of organic growing media and its influence on hysteretic water retention properties. *European Journal of Soil Science* 59(2):156–165.

Naasz, R., Caron, J., Legault, J. and Pichette, A. 2009. Efficiency factors for bark substrates: biostability, aeration or phytotoxicity. *Soil Science Society of America Journal* 73(3):780–791.

Nkongolo, N. V. and Caron, J. 1999. Bark particle sizes and the modification of the physical properties of peat substrates. *Canadian Journal of Soil Science* 79(1):111–116.

Nkongolo, N. V. and Caron, J. 2006a. Pore space organization and plant response in peat substrates: II. Dendrathemum morifolium Ramat. *Scientific Research and Essays* 1:93–102.

Nkongolo, N. V. and Caron, J. 2006b. Pore space organization and plant response in peat substrates: I. Prunus× cistena and Spiraea japonica. *Scientific Research and Essays* 1:77–89.

Orlander, G. and Due, K. 1986. Location of hydraulic resistance in the soil-plant patway in seedlings of Pinus sylvestris L. grown in peat. *Canadian Journal of Forest Research* 16(1):115–123.

Otten, W. 1994. Dynamics of water and nutrients for potted plants induced by flooded bench fertigation: experiments and simulation. Ph.D. Thesis. Wageningen Agricultural University, Wageningen, The Netherlands.

Otten, W., Raats, P. A. C., Baas, R., Challa, H. and Kabat, P. 1999. Spatial and temporal dynamics of water in the root environment of potted plants on a flooded bench fertigation system. *NJAS – Wageningen Journal of Life Sciences* 47(1):51–65.

Ouimet, R., Charbonneau, J., Parent, L. E., Blain, J., Joyal, P. and Gosselin, A. 1990. The effects of peat substrate composition and grow bag volume on the productivity of greenhouse tomato. *Canadian Journal of Plant Science* 70(2):585–590.

Paquet, J. M., Caron, J. and Banton, O. 1993. In-situ determination of the water desorption characteristics of peat substrates. *Canadian Journal of Soil Science* 73(3):329–339.

Paul, J. L. and Lee, C. I. 1976. Relation between growth of Chrysanthemum and aeration of various container media. *Journal of the American Society for Horticultural Science* 101:500–503.

Prasad, M. 1997. Physical, chemical and biological properties of coir dust. *Acta Horticulturae* 450(450):21–30.

Puutsjärvi, V. 1982. *Peat and Plant Yearbook 1981–1982*. Peat Research Institute, Helsinki, Finland.

Qi, G., Michel, J. C., Boivin, P. and Charpentier, S. 2011. A laboratory device for continual measurement of water retention and shrink/swell properties during drying/wetting cycles. *HortScience* 46(9):1298–1302.

Raviv, M., Wallach, R., Silber, A. and Bar-Tal, A. 2002. Substrates and their analysis. In: Savvas, D. and Passam, H. (Ed.) *Hydroponic Production of Vegetables and Ornamentals*. Embryo Publications, Athens. 25–105.

Raviv, M., Lieth, J. H. and Bar-Tal, A. (Eds). 2019. *Soilless Culture: Theory and Practice*. Elsevier.

Rekika, D., Caron, J., Rancourt, G. T., Lafond, J. A., Gumiere, S. J., Jenni, S. and Gosselin, A. 2014. Optimal irrigation for onion and celery production and spinach seed germination in histosols. *Agronomy Journal* 106(3):981–994.

Rezanezhad, F., Price, J. S. and Craig, J. R. 2012. The effects of dual porosity on transport and retardation in peat: a laboratory experiment. *Canadian Journal of Soil Science* 92(5):723–732.

Rivière, L. M. 1991. Contribution à l'étude du fonctionnement hydrique du systeme substrat-plante en culture hors sol. Mémoire d'Habilité à diriger la recherche (HDR), Institut National de la recherche agronomique (INRA), France, 135.

Tišler, V., Galla, E. and Pulkkinen, E. 1986. Fractionation of hot water extract from Picea abies Karst. Bark. *Holz als Roh- und Werkstoff* 44(11):427–431.

Valat, B., Jouany, C. and Riviere, L. M. 1991. Characterization of the wetting properties of air-dried peats and composts. *Soil Science* 152(2):100–107.

Verdonck, O. 2004. The influence of the particle size on the physical properties of growing media. *Acta Horticulturae* 644:99–101.

Verhagen, J. B. G. M. and Schmilewski, G. 1997. Particle size distribution to qualify milled peat; a prediction of air content of ultimate mixtures. In Schmilewski, G. (Ed.) *Peat in Horticulture*. International Dutch National Committee of the International Peat Society, Eelde, the Netherlands, 53–56.

Wallach, R., da Silva, F. F. and Chen, Y. 1992. Hydraulic characteristics of tuff (scoria) used as a container medium. *Journal of the American Society for Horticultural Science* 117(3):415–421.

Weber, T. K. D., Iden, S. C. and Durner, W. 2017. Unsaturated hydraulic properties of Sphagnum moss and peat reveal trimodal pore-size distributions. *Water Resources Research* 53(1):415–434.

Wever, G. 1994. Physical analysis of peat and peat-based growing media. *Acta Horticulturae* 401(401):561–568.

Wever, G., Leeuwen, A. V. and Meer, M. C. 1996. Saturation rate and hysteresis of substrates. *Acta Horticulturae* 450:287–296.

# Chapter 6

## Understanding and optimising the chemical properties of growing media for soilless cultivation

*Patrizia Zaccheo and Laura Crippa, University of Milan, Italy; and Francesco Giuffrida, University of Catania, Italy*



### 1 Introduction

Plants in soilless culture systems (SCS) generally develop their roots in a limited space in a container. The root environment must therefore offer the best growth conditions, including adequate and balanced concentrations of nutrients. Chemical properties of growing media - particularly cation exchange capacity (CEC) and pH - play a crucial role in determining availability and uptake of nutrients. CEC exerts a regulative influence on nutrients by storing and releasing them, and pH is a crucial determinant of the availability of essential macroelements and microelements, and for the potential solubilisation of toxic compounds that, in turn, also affect CEC.

Electric conductivity (EC) measures the salinity of growing media. When in excess amounts, soluble salts released from the solid substrates affect nutrient solution composition, at least in the first stages of cultivation. Growing media with high salt content need appropriate management of nutrient solution composition in order to keep substrate conditions suitable for plants, avoiding salt stress and nutrient imbalances.

http://dx.doi.org/10.19103/AS.2020.0076.05

For both CEC and EC, respective analysis of the ion composition of the exchange complex and soluble ions is of paramount importance for the management of growing media. This is particularly the case for organic substrates, which require a different approach compared to mineral substrates. Their chemical characteristics, that is, buffering power and exchange capacity, lead to interactions with the outer environment which alter their properties.

## 2 Cation exchange capacity (CEC): role and effect on nutrient solutions

The majority of organic and some mineral constituents of growing media carry electrical charges on their outer surfaces. These charges attract and hold nutrients and non-nutrients dissolved in the solution. CEC is the sum of negative charges held by solids. Anion exchange capacity (AEC) is the sum of positive charges. Both are expressed as $cmol_c\ kg^{-1}$ or $cmol_c\ L^{-1}$ (or meq 100 $g^{-1}$ in older texts–not a SI unit). Due to the predominance of acidic functional groups in many substrate components, CEC is usually much higher than AEC; consequently, cations may be attracted and held on solid surfaces, buffering the acidity of the solution. Elements in anionic forms are retained to a lesser extent. CECs of growing media and their components are usually expressed on a weight basis. Nevertheless, since the roots are in contact with a limited volume of a medium in a pot, it is more useful to measure by volume, especially when comparing constituents with different bulk densities.

In general, inorganic components of growing media possess both a structural or permanent charge and a variable charge, depending on the solution pH. The first is generated by the replacement of one atom by another of similar size in a crystal lattice (heterovalent isomorphous substitution). Variable charges come from the protonation or deprotonation of hydroxyl-groups (-OH) at the edges or surface of the mineral (Sposito, 1989; Sumner and Miller, 1996).

Table 1 lists the CEC for key components of growing media, calculated on a volume basis. The highest CEC of mineral materials used as substrates is associated with zeolites such as clinoptilolite. Their extremely high adsorption capability and high cost may restrict their use as a component in a substrate blend (Bar-Tal et al., 2019). At the other extreme, tuff and pumice have low CECs, especially when used in relatively big particles. Perlite and mineral wool are in practical terms inert materials: due to thermal treatment, they do not show any detectable buffering power and adsorption capability. With inorganic components, it is often particle size that determines the CEC, due to the increase in available surface area from a larger number of smaller particles. Another factor affecting variation in CEC is the degree of chemical weathering, highlighted in tuff by Silber et al. (1994).

**Table 1** Cation exchange capacity (CEC) and pH of selected organic and mineral components of growing media, expressed on a volume basis (Carlile et al., 2015, 2019; Silber, 2019; Maucieri et al., 2019; Landis, 1990)

| | CEC ($cmol_c L^{-1}$) | pH range |
|---|---|---|
| *Organic costituents* | | |
| Peat | 7–25 | 3.4–4.4[a] |
| Coir | 2–15 | 4.8–6.7 |
| Bark | 10–74 | 4.0–6.0[b] |
| Wood fibre | 20–30 | 3.8–6.6 |
| Compost | 15–20 | 7.5–8 |
| *Mineral costituents* | | |
| Perlite | 0.1–0.4 | 6.0–8.0 |
| Tuff | 10–60 | 6.6–7.8[c] |
| Clinoptilolite | 400–800 | |
| Sand | 30–38 | 6.4–7.9 |
| Vermiculite | 8–15 | 6–7.6 |
| Pumice | - | 6.3 |
| Mineral wool | - | 5.2–7.8 |

[a] Fen, sedge and reed-derived peats may have much higher pH (up to 7.5) (Robbins and Evans, 2006).
[b] Highest values for composted bark.
[c] Measured in 1M KCl.

Organic components only have variable charges, generated from the ionisation of carboxylic acid and phenolic OH. As a consequence, CEC is more strictly related to substrate pH. On a weight basis, peats show relatively high CEC, from 150 $cmol_c kg^{-1}$ to 250 $cmol_c kg^{-1}$ (Puustjarvi and Robertson, 1975; Carlile et al., 2015). CEC varies depending on degree of humification and biological composition (Rippy and Nelson, 2007) and particle size (Preston et al., 1989). Coir has a lower CEC than peat, with values ranging from 35 $cmol_c kg^{-1}$ to 130 $cmol_c kg^{-1}$ (Handreck and Black, 1994; Abad et al., 2002; Enzo et al., 2001). Pine bark CEC have been recorded at 58–74 $cmol_c kg^{-1}$ (Daniels and Wright, 1988). In a more recent study, three out of four batches of pine bark showed lower values, from 21.3 $cmol_c kg^{-1}$ to 29.5 $cmol_c kg^{-1}$ with CEC dependent on particle-size distribution (Altland et al., 2014). Bark aging and composting increases CEC (Handreck and Black, 1994). The CEC of wood fibres has been reported at 22–28 $cmol_c kg^{-1}$ by Domeño et al. (2010) and in the range from 7.4 $cmol_c kg^{-1}$to 14.3 $cmol_c kg^{-1}$ by Dumont et al. (2014). Composts are usually very stable. In a recent study (Li et al., 2018) a wide range of CEC values, from 3.3 $cmol_c kg^{-1}$ to 95 $cmol_c kg^{-1}$ were recorded in 74 composts from the United States and Canada. The CEC depend on the nature of source materials. Also, the length of composting process can lead to higher values (Inbar et al., 1989).

Base saturation represents the percentage of CEC occupied by basic cations other than $H^+$ and $Al^{+++}$. It differs significantly between substrate components, depending on the nature and origin of inorganic and organic constituents and is strongly correlated with pH. Peat is reported to have a variation in base saturation ranging from 15% to 71%, with calcium accounting for the major proportion. A high base saturation indicates a low requirement for lime to buffer acidity. Base saturation depends on peat origin, composition of Sphagnum species, amount of sedges and degree of decomposition (Puustjarvi and Robertson, 1975; Rippy and Nelson, 2007).

In a study of the adsorption complex of 17 samples of coir from different sources, Verhagen (1999) found that, among the adsorbed cations, $K^+$, $Na^+$ and $Mg^{++}$ are more abundant than $Ca^{++}$ with high variability, probably depending on the quality of waters used for washing. Base saturation of inorganic components occurs from dissolution of minerals. CEC of Italian pumice, for example, resulted in 100% base saturation with the predominance of $Ca^{++}$ and $K^+$, followed by $Mg^{++}$ and $Na^+$ (Crippa et al., 2019).

Mineral and organic-based growing media may affect the composition of the nutrient solution, especially at the beginning of the cultivation period, because of the release and/or retainment of cationic nutrients from/on their exchange complexes. These processes have a direct effect not only on the concentration of single nutrients but also on the relative proportion of different nutrients. Any nutrient imbalance may affect plant growth and tolerance to abiotic and biotic stresses (Huber et al., 2012). Different approaches for dealing with these effects are discussed below, based on practical calculation and experience as well as research conducted under laboratory and glasshouse conditions.

Practical calculation of the effects of cations is based on Lemaire et al. (1989) who developed the work of André (1981, 1985) to maintain $Ca^{++}$ and $Mg^{++}$ balance between solid and liquid phases of Sphagnum peat after neutralisation. Pre-plant fertilisation can be set for $K^+$ and $NH_4^+$. Figure 1 shows the calculation, based on the different adsorption affinities of cations, to predict the type of liming materials and the composition of pre-plant fertilisation to be supplied to the limed peat. This method takes into account peat pH, CEC and base saturation so as to minimise the effect of peat-based growing media on nutrient solution composition.

Practical experiences in soilless culture are related to the use of coir as a stand-alone substrate to replace mineral wool and perlite, especially for horticultural crops such as tomato and strawberry. One of the most common cultivation problems arises from unwashed and undusted coir. During initial contact, coir release $K^+$, $Na^+$, $Cl^-$ into the nutrient solution. Moreover, a darkening of the nutrient solution may occur, due to the release of tannins. This complicates UV disinfection during recirculation of the nutrient solution.

(a) **How to calculate the amount and type of liming materials**

i. Assuming a lime requirement (from the neutralization curve) of 5 g $L^{-1}$ pure $CaCO_3$ to bring peat pH to 5.5, corresponding to 10 $cmol_c\ L^{-1}$ $H^+$ displaced.

ii. Assuming the following nutrient solution (NS) composition:

$Ca^{2+}$ = 0.46 $cmol_c\ L^{-1}$; $Mg^{2+}$ = 0.2 $cmol_c\ L^{-1}$; $K^+$ = 0.8 $cmol_c\ L^{-1}$; $NH_4^+$ = 0.1 $cmol_c\ L^{-1}$.

To keep unchanged the ratio between $Ca^{2+}$ and $Mg^{2+}$ in the NS, the proper ratios among Ca- and Mg-based liming materials must be assessed.

According to the equilibrium exchange rate calculated by André (1981) for a white peat, to maintain the NS $Ca^{2+}/Mg^{2+} = 2.3$, corresponding to $Ca^{2+}/(Ca^{2+}+Mg^{2+}) = 0.7$, the ratio between the sorbed $\overline{Ca^{2+}}/(\overline{Ca^{2+}}+\overline{Mg^{2+}})$ must be kept around *0.8*.

If we assume that 10 cmol $L^{-1}$ $H^+$ are displaced by $Ca^{2+}$ and $Mg^{2+}$:

$\overline{Ca^{2+}}/(\overline{Ca^{2+}}+\overline{Mg^{2+}}) = Ca^{2+}/10 = 0.8$ that is $Ca^{2+}$ = 8 $cmol_c\ L^{-1}$ and $Mg^{2+}$ = 2 $cmol_c\ L^{-1}$

**corresponding to 1.84 g $L^{-1}$ of pure dolomite and 3 g $L^{-1}$ of pure lime**

(b) **How to formulate pre-plant fertilization**

i. Assuming the following amount of sorbed nutrients on CEC of the limed peat

$\overline{Ca^{2+}}$ = 8 $cmol_c\ L^{-1}$ (from lime and dolomite); $\overline{Mg^{2+}}$ = 2 $cmol_c\ L^{-1}$ (from dolomite); $\overline{K^+}$ = 0.05 $cmol_c\ L^{-1}$ (from analysis); $\overline{NH_4^+}$ = 0. 02 $cmol_c\ L^{-1}$ (from analysis).

According to the following experimental derived relationship (André, 1985):

$$(\overline{K^+}+\overline{NH_4^+})/(\overline{Ca^{2+}}+\overline{Mg^{2+}}) = 0.08\ (K^+ + NH_4^+)/(Ca^{2+} + Mg^{2+})$$

$(\overline{K^+}+\overline{NH_4^+})/10 = 0.08\ (0.8+0.1)/(0.46+0.2)$

$\overline{K^+}+\overline{NH_4^+}$ = 1.09 $cmol_c\ L^{-1}$

ii. Assuming that $(\overline{K^+}/\overline{NH_4^+}) = (K^+/NH_4^+)$ (André, *1985):*

$\overline{K^+}/(\overline{K^+}+\overline{NH_4^+}) = K^+/(K^+ + NH_4^+) = 0.8/(0.8 + 0.1) = 0.89$

$\overline{K^+} = 0.89 \times (\overline{K^+}+\overline{NH_4^+}) = 0.89 \times 1.09 = 0.97$ $cmol_c\ L^{-1}$

$\overline{NH_4^+} = 1.09 - 0.97 = 0.12$ $cmol_c\ L^{-1}$

by subtracting the amount present in peat:

$\overline{K^+} = 0.97 - 0.05 = 0.92$ $cmol_c\ L^{-1}$ and $\overline{NH_4^+} = 0.12 - 0.02 = 0.1$ $cmol_c\ L^{-1}$

**corresponding to 0.36 g $L^{-1}$ of K and 0.014 g $L^{-1}$ of N-$NH_4$**

**Figure 1** Example calculations of liming materials and pre-plant fertilization composition in a peat-based growing medium (modified from Lemaire, 1995).

The high-water retention induced by dust increases water content and reduces water drainage, complicating nutrient replacement and rapidly increasing EC. Abundant flushes are needed before planting until drainage EC equals drip EC, avoiding seedlings getting stressed and in some cases poisoned by the high levels of salts. When using washed coir, initial flushing only lasts a few days to achieve low salt levels. When not buffered, the coir shows an excess of $K^+$ and $Na^+$ that need to be replaced by an excess of $Ca^{++}$ in the first 'exchange' recipe used for hydration. Practical experience suggests that a nutrient solution can be used with +2 mmol $L^{-1}$ $Ca^{++}$ and −3 mmol $L^{-1}$ $K^+$. It takes about 2–4 weeks of

**Table 2** Nutrient composition of drainage waters from coco coir grow bags (volume 18 L) after irrigation with exchange nutrient solution (7 L day$^{-1}$)

| | K | Na | Ca | Mg | $NO_3$ | Cl | $SO_4$ | $PO_4$ | Fe |
|---|---|---|---|---|---|---|---|---|---|
| | mmol L$^{-1}$ | | | | | | | | µmol L$^{-1}$ |
| 14 DAT | 7.3 | 3.2 | 9.0 | 3.5 | 24.2 | 4.7 | 2.6 | 1.6 | 19.6 |
| 28 DAT | 6.1 | 2.5 | 9.4 | 3.9 | 21.7 | 2.6 | 2.9 | 1.4 | 12.9 |
| $14_{DAT}$-$28_{DAT}$ | −1.2 | −0.7 | +0.4 | +0.4 | −2.5 | −2.1 | +0.3 | −0.2 | −6.7 |

DAT = Days after tomato transplanting.

irrigation to normalise excess cations (Table 2). In the case of sensitive crops, it is advisable to pre-flush only with Ca or to buy pre-treated buffered product (Di Pierro, Ageon Ltd., personal comm.).

Studies conducted under laboratory and glasshouse conditions refer to the ability of biochar and pumice to interact with nutrients. Biochar is increasingly used as a partial peat substitute. Pumice, either alone or in combination with other materials, is widely used in soilless culture in the Mediterranean region. When added to peat-based growing media, biochars, in particular those obtained from pyrolysis/gasification of woody feedstocks, may influence their chemical and physical properties. In particular, biochars buffer the acidity of peat and significantly influence nutrient availability. Biochar can retain cations and anions (Yao et al., 2012; Clough et al., 2013). It can also increase mineral availability, mainly potassium (Chan and Xu, 2009; Blok et al., 2017; Prasad et al., 2020). These effects occur mainly in the initial phase of cultivation but persist over time.

In a pot experiment carried out with basil (Bedussi et al., 2017), spruce and poplar biochar were found to enhance the removal of $NH_4$-N and $NO_3$-N from the nutrient solution, though they remained available for plant uptake (Fig. 2). Both biochars increased $K^+$ in the solution (Fig. 2) and promoted early release of $K^+$ when mixed with peat, whether or not there was initial potassium fertilisation. The $CaCl_2$/DTPA (CAT) extractable potassium (Fig. 3) measured in the substrates 150 days after cyclamen transplant showed that biochar preserves potassium at the beginning of cultivation. Biochar is able to store potassium fertiliser in a plant-available pool, acting as a sink and a source of plant-available potassium (Zaccheo et al., 2019). These results suggest reducing the amount of potassium in the base dressing before the beginning of fertigation. These findings are consistent with the results obtained by Altland and Locke (2012) in leaching experiments with peat:perlite substrates amended with biochar. To make the results of these studies transferable to real-world conditions, we need to know more about variability in composition and properties of biochars from different feedstocks/processes.

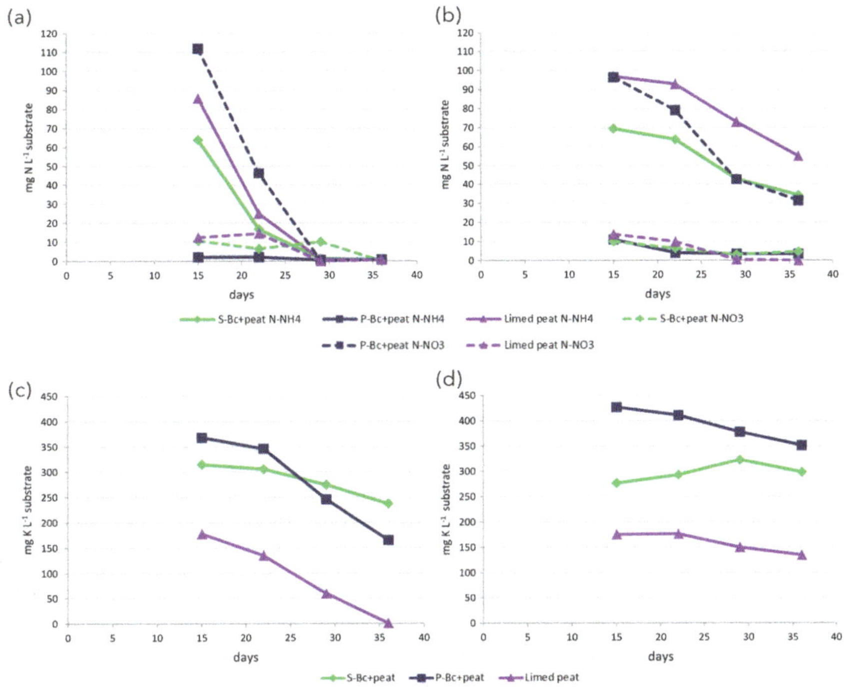


**Figure 2** Nutrient solution composition in limed white peat (limed peat), acid white peat mixed with sprucewood biochar (S-Bc+peat) and with poplar biochar (P-Bc+peat) in ratios 30:70% v $v^{-1}$: $NH_4$-N (solid) and $NO_3$-N (dashed) in planted (a) and unplanted (b) pots; $K^+$ in planted (c) and unplanted (d) pots (mg $L^{-1}$ substrate).

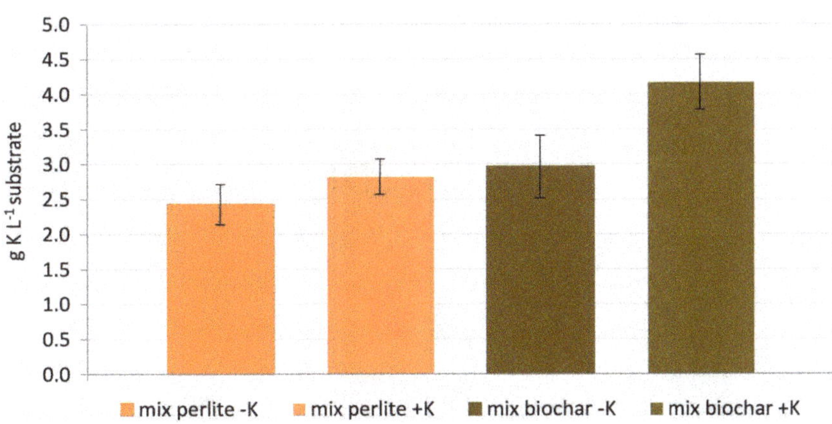


**Figure 3** CAT-extractable potassium in substrates (mix perlite -K; mix perlite +K; mix biochar -K; mix biochar +K) at the end of cultivation (155 days after transplanting) (g $L^{-1}$ substrate). Bars represent standard deviation (n = 3).

The ability of mineral components of growing media to hold and release nutrients has been less documented, except for zeolites (Ming and Mumpton, 1989; Gül et al., 2005), clay (Gelsomino et al., 2010) and tuff (Silber and Raviv, 1996; Silber et al., 1999). There is some evidence that other mineral components in mixtures can affect nutrient release. In a recent study, Geuijen and Verhagen (2021) observed a marked effect of four pumices from different origins (Germany and Iceland) and with different particle sizes, on the composition of nutrient solutions, including a general depletion (from −20% to −70%) of cations (mainly $NH_4^+$ and $K^+$). A study of pumice from a volcanic deposit in Central Italy show that low CEC pumice (13 $cmol_c$ $kg^{-1}$) adsorbs cations from the nutrient solution more than an equal volume of fine-sized peat (0–3 mm, H2-H3), due to the high density of pumice (Fig. 4). The removal of nutrients from the solution were respectively 37% ($NH_4$-N), 77% (K), 66% (Mg) in pumice and 13% ($NH_4$-N), 5% (K), 7% (Mg) in peat. The high density of pumice means it has these effects even mixed with peat in low percentages (10–30% vol). Studies show that about half (52.4%) of $NH_4$-N charged on pumice is water-soluble and easily exchangeable, and 41% is released by distillation with hot NaOH. The retention of $NH_4$-N by pumice offers protection from leaching (Crippa et al., 2019). It may allow gradual nutrient release, supporting long-term plant development. In spite up of 30 times lower amounts of $NH_4$-N detected in the pore water solutions of a peat-pumice mix (70:30% v $v^{-1}$) compared to peat and peat-perlite mix (70:30% v $v^{-1}$), the above-ground parts of ryegrass grown on peat-pumice have similar dry weight and N content (unpublished data).

As mentioned above, the anion exchange capacity of organic and mineral constituents of growing media is much lower than the CEC. Anion exchange capacity mainly involves P ions which have a great affinity for charged

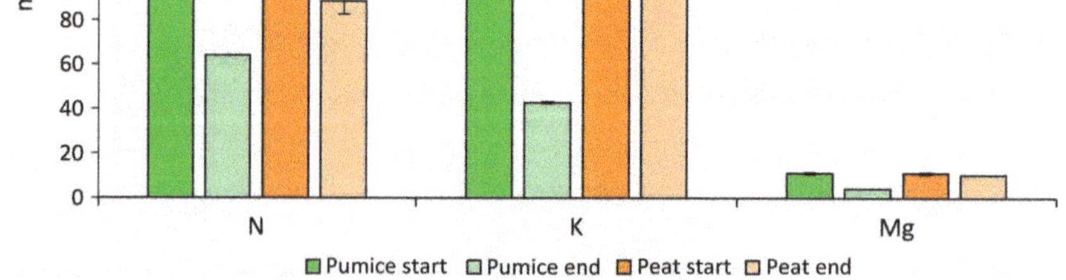


**Figure 4** Nutrient concentrations in solution before (start) and after (end) contact with equal volumes of pumice and peat (mg $L^{-1}$). Bars represent standard deviation (n = 3).

surfaces, as in the case of tuff and calcined clays (Carlile and Bedford, 1988; Silber et al., 1994). This is in contrast to nitrate, which is very weakly affected by adsorption phenomena and therefore extremely soluble. The literature primarily reports studies relating to the interactions of substrate components and phosphorus. P availability in solution is influenced by adsorption and precipitation processes dependent on the pH, ionic strength of the solution and cations present, as discussed, for example, by Silber (2019).

In conclusion, the knowledge of CEC and base saturation of a growing medium can be valuable for appropriate management of plant nutrition, particularly for base dressing and for initial fertigation composition, before an equilibrium is reached between nutrients held by the growing media and the nutrient solution. Briefly, growing media with medium to high CECs need to be differently fertilised than growing media with low CEC. Formulation of the nutrient solution must be modified by increasing or decreasing the number of nutrients, depending on the relative abundance or scarcity of the nutrient present in an exchangeable form in the solid phase, and the base saturation. However, other mechanisms than sorption/desorption can lead to modifications in the nutrient solution, for instance, precipitation of $PO_4^{---}$ with $Fe^{+++}$, $Al^{+++}$ and $Ca^{++}$, $Mg^{++}$ (Sonneveld et al., 1974) depending on pH in the root zone or N mineralisation from compost (Raviv, 2013).

Moreover, the substrate exchange complex does not remain constant over time. It can change due to the development of the root system, the supply of nutrients and the action of microorganisms. These can result in an increase of solid surfaces, ions adsorption on charged surfaces, accumulation of new organic and inorganic compounds derived from degradation of organic materials, excretion of exudates from roots and alteration of the mineral components (Silber, 2019).

## 3 pH and its management

pH is the negative logarithm of the chemical activities of protons ($H^+$): a low pH signifies a high value for protons and a high pH represents a low value for protons. Protons are very unstable and hence very reactive and may have both direct and indirect detrimental effects on biological systems. The pH of solid materials such as growing media depends on the ability of media constituents to give up $H^+$ in pore water as a result of chemical processes, mainly dissolution and cation exchange. Solubility and CEC are critical properties of the materials determining the starting pH of pore water and its maintenance over time.

The pH of growing media components covers a wide range from acidic to alkaline, as shown in Table 1. Sphagnum peat shows a very low average pH from 3.5 to 4.1 (Carlile et al., 2019) due to the abundance of functional groups, mainly carboxylic groups which can be ionised and thus dissociate protons. The origin of peat acidity is related to many factors such as the water quality in the peat bogs where they were formed (e.g. rainwater or mineral-enriched water), degree of humification and composition, for example, mosses or sedges (Puustjarvi and Robertson, 1975; Rippy and Nelson, 2007). Within the peats of the same genus (*Sphagnum*) are species as *S. fuscum* (Schimp.) Klinggr characterised by low pH (3.5) and species such as *S. angustifolium* with higher pH (4–5) (Reddy, 2005). Richness in functional groups enables Sphagnum peat to hold and release protons, resulting in very acidic pH ranges. The high buffering power of peat suggests that a relevant amount of lime must be added to steadily increase its pH (Carlile et al., 2019).

It is well known that peats differ in pH, CEC and base saturation which affects liming. The degree of decomposition of peat may also affect lime requirement in two different ways: through an increase in CEC due to newly formed reactive functional groups and a decrease in particle sizes leading to an increase in bulk density which means a greater quantity of material to be neutralised in the same volume (Puustjarvi and Robertson, 1975). pH alone is not sufficient to establish the quantity of lime required. Peat with lower pH may require less lime than peat with higher pH values because of the higher CEC and/or the higher amount of $H^+$ adsorbed (Bunt, 1988). The knowledge of CEC and base saturation (buffering capacity) and of bulk density makes it possible to calculate lime requirement to reach the ideal medium pH.

Other plant-derived materials are naturally acidic, such as wood fibre, sawdust and bark. However, because they are not subjected to microbial transformations, they do not have a significant buffering power due to the scarcity of acid functional groups. Many barks used as horticultural substrates have higher pH than peat: pine barks are between 4 and 5 (Wright et al., 1999) and spruce barks around 5.5 (Naasz et al., 2009).

Coir is generally moderately acidic (Evans et al., 1996; Abad et al., 2002) even if values up to 8.0 can be found (Zaccheo et al., 2015). The pH of coir is related to the origin of coconuts husks (Konduru et al., 1999), particle size (Noguera et al., 2003) and processing (i.e. retting was found to induce a decrease of 0.5 units of pH of coir, according to a study of Fernando and Amarasinghe, 2017).

For the majority of inorganic components, like vermiculite, perlite, pumice and mineral wool, neutral and slightly basic values of pH and very low CEC are found. Sometimes mineral wool has residues of fine dust resulting from the production process. When hydrated at pH 5.5–5.8, this dust increases pH to about 6–7. A regular irrigation strategy with a solution at pH 5–5.2 allows the pH to normalise in about 1–3 weeks. In the case of sensitive crops, it is advisable to flush in advance before transplanting (Di Pierro, Ageon Ltd., personal comm.).

Compost frequently shows undesirably high pH, due to salts and ammonia and a very high CEC, wholly saturated by bases (Nappi and Barberis, 1993). This is one of the reasons why it is suggested to use compost in mixtures with other components. Other reasons include salinity and high bulk density (Prasad and Carlile, 2008; Raviv, 2013). In order to fulfil EU-Ecolabel requirements, compost used in growing media must be kept within a pH range of 4–7.

pH is a crucial parameter for growing media because it determines whether nutrients and toxic elements are available for root uptake. According to the chemical equilibria of nutrients, the desirable range of pH for most greenhouse crops is between 5.5 and 6.5 (Bos et al., 2002; Sonneveld and Voogt, 2009).

Growing media with too low or high pH need to be pH-adjusted, and timing and way of amendment depend to some extent on the CEC of the materials. In general, with the exception of clay, inorganic materials such as perlite, expanded clay, mineral wool are characterised by a low buffering capacity that allows adjustment 'in situ' at the beginning of cultivation by flushing and by selecting fertilisers, especially N fertilisers, that are able to acidify or to basify media (Sonneveld and Voogt, 2009).

Apart from agronomic management, there are three ways to achieve the ideal pH of a substrate for a particular crop:

i select the component (type, origin, particle size, treatments) with pH quite close to the desired value;
ii choose a blend to achieve the ideal pH; and
iii modify pH of the native component by adding an amendment.

Strategy (i) is limited to those components with almost neutral pH or to specific crops (calcifuges ornamental plants, blueberry etc.). For strategy (ii) it must be kept in mind that the pH of a blend depends in the first instance on components with a higher buffering ability such as peat and compost. Some more recently used components, such as biochar and ashes, have been found to influence the pH of substrates even when added as minor components (Menzies and Aitken, 1996; Nemati et al., 2015).

During the preparation of growing media, it is necessary to adjust the pH of materials with a significant CEC such as bark, compost and particularly peat. The acidity of peat-based substrates is typically adjusted by using acid-neutralising materials such as liming materials. With this process two complementary needs are satisfied: to raise the pH to the desired value and to supply $Ca^{++}$ and $Mg^{++}$ cations as part of nutrients in the circulating solution (Handreck and Black, 1994; Argo and Biernbaum, 1997). Producers and/or growers often adopt routine production recipes (x kg of lime per $m^3$ of peat) that are effective when managing materials from the same batch. Problems can arise with different materials using mixtures of peat and other media constituents. According to Reddy (2005), even when the same amount of lime is added to the same amount of peat, the resulting mix pH is not the same because of different origin of peat and liming materials. Mistakes during the preparation phase may be difficult to correct in the cultivation phase.

Different liming compounds can be utilised to neutralise the acidity of a substrate or its components up to the desired pH value (Argo and Biernbaum, 1997). The most used compounds are hydrated limes ($Ca(OH)_2$ and $Mg(OH)_2$), carbonates (calcite ($CaCO_3$) and dolomite ($CaMg(CO_3)_2$). The choice of the lime depends significantly on the duration of the crop cultivation cycle (Fisher et al., 2006). The time required to reach the target pH (reactivity of the lime), and the residual effect (or buffer capacity, that is, the ability to counteract the pH variations that may occur during crop growth), must be shorter in growing media than in soils (Argo and Biernbaum, 1996; Rippy, 2005). In soilless culture, it is necessary to reach the optimal value for the crops rapidly (in a few hours/days). The residual neutralisation capacity of the unreacted lime needs, in most cases, only to last for a few months, and must face higher pH variations due to the restricted volume in which roots develop.

Many factors influence the reactivity and residual effect of liming materials. The chemical nature of liming material strongly determines the dissolution rate. Hydrated lime reacts faster than carbonates and has no residual effect. Calcitic limestone dissolves faster than dolomitic limestone but also has a shorter residual effect. Particle size also plays a role in neutralisation reactions because smaller particles dissolve faster than larger ones. Other characteristics such as neutralising value, surface area and hardness must also be considered. Huang et al. (2010) investigated how particle size and the amount of liming material affects buffering capacity. They tested two dosages (1.5 and 3.0 g $L^{-1}$) of a dolomitic limestone sieved in different size fractions (850–2000, 250–850, 150–250, 75–150 μm). They concluded that the upper dose and a mixture between relatively fine (<150 μm) and moderately coarse (150–850 μm) fractions ensure both rapid pH adjustment and a long-lasting buffering capacity. The use of finer fractions (<75 μm) and of the coarser fractions (>850 μm) is to be avoided as the former increase excessively the pH without producing residual effect while the latter is practically inert.

As stated earlier, particle size is not the only factor affecting the timing and intensity of the neutralising action of limestones (Rippy et al., 2016). There is a need for more research, especially for dolomite, the most widely used liming material in soilless culture. As early as 1954, Love and Whittaker, in a study on calcitic and dolomitic limestone, highlighted the importance of surface area. They found the specific total surface areas of same-sized but different limestones varied widely and were higher than the geometric surfaces calculated from gross particle dimensions. In research on calcitic and dolomitic Canadian limestones of different origin, Rippy (2005) found that the relationship between particle size and reactivity is stronger for calcites than dolomites. Behaviour is related more to physical characteristics (porosity and hardness) than chemical composition (% of Mg). With the same geometric size (diameter), for example, particles characterised by hard and smoother surfaces have a lower specific surface than particles with an irregular surface and internal porosity and therefore react less intensely with an acidic substrate such as peat. The use of a mixture of calcitic and dolomitic carbonates to neutralise peat acidity could be a good practice for exploiting the more rapid solubilisation of calcitic carbonates and the slower release of dolomitic carbonates as well as supplying both calcium and magnesium (as explained in Fig. 1).

Finally, pH also depends on time and temperatures of storage and moisture levels (Dickinson and Carlile, 1995). Substrates are usually packaged at a humidity level in a range between 45% and 55% which allows only a small part of the added lime to react with the acid constituents. This means the pH measurement carried out after opening the package does not correspond to the value of pH reached after a few days of cultivation under optimal moisture conditions.

In adjusting peat pH, an alternative to traditional liming materials is to mix peat with components to neutralise its acidity. Biochar is a possible partial peat substitute and can act like lime, though not all biochars can fulfil this function. Research has shown that biochars deriving from materials too rich in minerals, such as poultry litter (Steiner and Harttung, 2014) and horticultural crop wastes (Blok et al., 2017), are unsuitable for inclusion in growing media since they may induce excessive salinity. More suitable are biochars obtained from wood residues (softwood or hardwood), which generally have low salinity values and high pH. Steiner and Harttung (2014) tested alkaline biochar (pH 9) prepared from crushed wooden boxes (5–30 mm) in a mixture with Sphagnum peat (0%, 25%, 50%, 75% and 100% volume) and found it to act as a good substitute for lime. Zaccheo et al. (2014) assessed the effectiveness of alkaline biochar (0–10 mm) from pine wood (pH 10.2) in correcting the acidity of white peat (pH 4.2) with varying doses (from 2.5% to 40% v $v^{-1}$) and particle size (10–6 mm, 6–3.3 mm, 3.3–0 mm) of biochar. Doses of 30% and 40% of unsieved biochar were most effective. The finest fraction of biochar at 40% v $v^{-1}$ showed the highest neutralising power, reaching pH stabilisation after 14 days incubation (Fig. 5) (Zaccheo et al., 2014). Prasad et al. (2018) evaluated four commercial biochars as peat substitutes (10%, 25% and 50% vol) and reported an increase of pH in the blend. The neutralisation of peat acidity increases pH and balances the nutrients in the circulating solution, which means the use of biochar must also take into account nutritional aspects (Prasad et al., 2018).

Substrates produced by grinding fresh pine wood have values of pH suitable for plant cultivation without liming (Wright et al., 2008). However, Jackson et al. (2009b) observing sharp acidification after one year of cultivation, due to a very low CEC, suggest that liming could be necessary for long-term

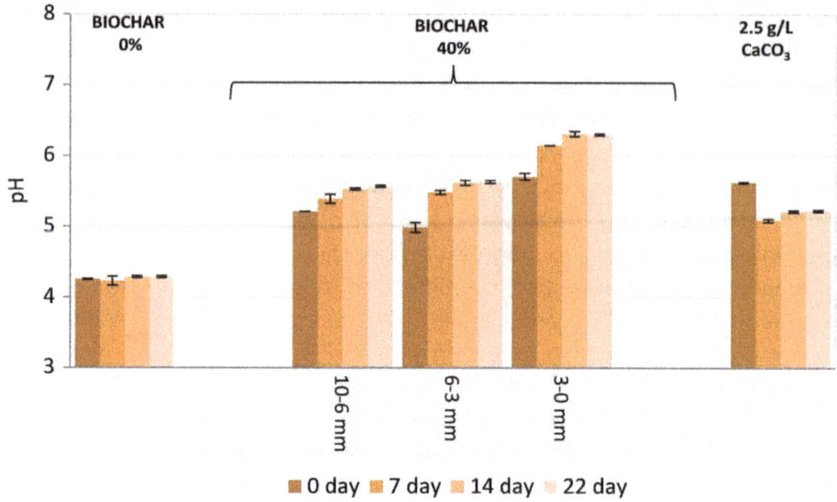


**Figure 5** Liming effect on Sphagnum peat of particle fractions of biochar compared with reagent grade $CaCO_3$ at 2.5 g $L^{-1}$. Bars represent standard deviation (n = 3).

crop production; moreover, when mixed with peatmoss, lime addition is needed (Jackson et al., 2009a).

Barks have higher pH than peat, so lower lime requirements are needed. Altland and Jeong (2016) studied the use of dolomitic lime in the pH adjustment of pine bark substrate for nursery crops, identifying the importance of species cultivated and quality of the water. Detrimental effects from the addition of dolomitic lime included the increased risk of nitrate leaching, lowered availability of micronutrients and the competition of $Ca^{++}$ and $Mg^{++}$ for $K^+$ uptake by plants.

Managing the pH of alkaline substrates is more difficult than for acidic substrates, and is restricted to compost. Acidification of compost-based growing media can be achieved by adding nitric or phosphoric acids to the nutrient solution, before transplanting (Mazuela et al., 2005) or during plant growth and development (Raviv, 2005). Lowering pH can also be achieved by adding acidifying agents such as aluminium sulphate, potassium sulphate, gypsum, ferrous sulphate or sulphur ($S^0$). The latter is most effective, cheaper, requires fewer doses and less time to induce a decrease in pH (Irum et al., 2019). However, anaerobic conditions can induce the production of hydrogen sulphide which is toxic to human health (Velusami et al., 2013). The biological oxidation of elemental $S^0$ to sulphuric acid is carried out by bacteria and fungi, abundant in composted materials, and occurs in 2–4 weeks (Carrión et al., 2008). The sulphur requirement is a function of compost buffering capacity and is related to the carbonate content of the compost (Amberger-Ochsenbauer et al., 2017). It is possible to add $S^0$ and other sulphur-based compounds during composting or the maturation phase, as shown in Fig. 6 (Crippa et al., 2000).

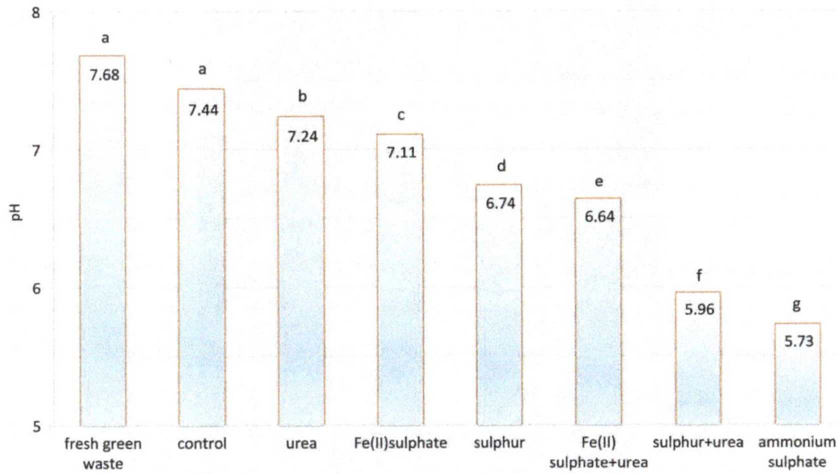


**Figure 6** Effect on compost pH of acidifying materials added to green wastes during composting process. Bars with the same letter do not differ for $p = 0.05$; Tukey b test.

A study by Roig et al. (2004) showed that addition of 0.5% in sulphur on a dry-weight basis during the composting process led to a decrease of about 1 unit of pH in compost from olive mill waste and sheep manure, without increasing EC to levels that could be dangerous to plants. However, Dias et al. (2017) found no differences in nutrient availability and plant growth between compost with and without addition of sulphur to lower pH.

## 4 Electrical conductivity (EC) as a measure of salinity and nutrient potential of growing media

Electric conductivity (EC) provides an indirect way to measure water salinity, based on the ability of a solution to conduct electricity, which in turn depends on the concentrations of ions. Soluble salts also increase the osmotic potential of the water. This is important for the suitability of growing media for soilless cultivation, because of the reduction in plant growth induced by the impairment between external and vacuolar osmotic potential (Eckhard et al., 2012) (Fig. 7). Increasing osmotic potential is frequently adopted as a strategy to improve fruit vegetable quality (Gruda, 2009; Gruda et al., 2018; Silber and Bar-Tal, 2019). Electrical conductivity is measured in deci-Siemens per metre ($dS\ m^{-1}$) or milli-Siemens per centimetre ($mS\ cm^{-1}$).

The comparison of data of EC in different materials is often difficult, because of the different methodologies used in laboratories. The significant difference between the current methods of testing EC is the extraction procedure used, in which a sample of materials and water are mixed in volume ratios. There are three commonly used methods of testing soilless media salinity:

**Figure 7** Wilting effect of excessive salinity in Cyclamen spp.

- 1:1.5 dilution method (Sonneveld et al., 1974).
- 1:5 dilution method (European Committee for Standardization, 2012).
- Saturated media extract, widely used in the United States, in which a paste is made using a sample and water with the liquid portion then separated from the solid portion (U.S. Composting Council, 2001).

All of these methods measure EC, but must be converted to be compared. Frangi et al. (2008) developed an equation to convert values of $EC_{1.5}$ and $EC_5$ of compost-based substrates. A salt risk assessment of substrates, based on $EC_5$, $EC_{1.5}$ and $EC_e$ is reported in Table 3.

Some organic and inorganic component do not significantly increase water salinity, due to their low solubility and/or low basic saturation of CEC (Table 4). This is the case for inorganic materials and for peat, bark and wood fibre. Use of fresh barks may also cause plant toxicity due to the high levels of soluble manganese ions of more than 600 mg $kg^{-1}$ (Solbraa, 1986; Handreck and Black, 1994). Composting of bark usually induces oxidation of manganese from $Mn^{2+}$ to $Mn^{4+}$, that is less soluble and therefore less phytotoxic.

Rock wool and perlite are chemically and biologically inert with no detectable effect on EC of the nutrient solution. Pumice and tuff are inert aluminosilicates with a very low dissolution during cultivation. However, during plant cultivation small grains can be fragmented to promote dissolution of

**Table 3** Evaluation of EC of growing media components according to different extraction ratios

| $EC_5$[a] | $EC_{1.5}$[b] | $EC_e$[a] | Salinity risk |
|---|---|---|---|
| 0–0.11 | <0.7 | 0–0.75 | Very low |
| 0.12–0.35 | 0.7–1.2 | 0.76–2.0 | Low |
| 0.36–0.65 | 1.3–1.8 | 2.0–3.5 | No risk |
| 0.66–0.89 | 1.9–2.7 | 3.5–5.0 | High |
| 0.9–1.1 | 2.8–3.6 | 5.0–6.0 | Severe |
| >1.1 | >3.6 | >6.0 | Extreme |

[a] Handreck and Black (1994).
[b] Bunt (1988).

**Table 4** Electrical conductivity (dS $m^{-1}$) of growing media components and main ions in the water extracts (from Kipp et al., 2000; Mauceri et al., 2019; Sonneveld and Voogt, 2009; Thompson et al., 2001; Verhagen, 1999; Schmilewsky, 2008)

| Component | $EC_5$ | $EC_{1.5}$ | First three most abundant elements (on molar basis) | Harmful elements |
|---|---|---|---|---|
| Peat | 0.1–0.7 | 0.2 | Na, Cl, $NO_3$-N | - |
| Compost | 1.0–16 | 1.5–9 | K, Cl, $NO_3$-N | Heavy metals |
| Coir | - | 0.4–1.4 | Na, Cl, K | - |
| Bark | - | 0.30 | K, Ca, Mn | Mn, B |
| Wood fibre | 0.3 | 0.2 | K, Ca, Mg | Mn, B |
| Perlite | 0.02–0.04 | 0.1 | K, Na, $NO_3$-N | - |
| Pumice | 0.08–0.12 | 0.2 | - | Mn, B |
| Mineral wool | 0.01 | - | - | B |

minor oxides with the release of $Ca^{++}$, $Mg^{++}$ and $Na^+$, especially when combined with root-derived acidification. The process was highlighted by Silber et al. (1999) on tuff, showing a release of sorbed $Ca^{++}$, $Mg^{++}$ and $Al^{+++}$, accomplished by proton consumption.

Sphagnum peat usually contains a very low amount of salts (Puustjärvi, 1977). Low salinity makes peat one of the most versatile organic components of growing media. Fertilisers and liming materials can be added in different amounts and types in order to reach the ideal conditions for the majority of the ornamental and horticultural plants (Gruda et al., 2016).

Salts can be a concern with coir and compost. High levels of salinity in coir may be due to the area of cultivation of coconut plants and to the husk processing method applied to remove the fibre (Abad et al., 2002). Sometimes, raw coconut husks are soaked in saline water before grinding (Evans et al., 1996). Drying techniques can also modify coir salinity. Fernando and Amarasinghe

(2017) found that oven drying results in rupture of cells and releases potassium which increases EC. Soluble salts are also more concentrated in smaller particle-sized coir (Noguera et al., 2003). Compost shows high variability in soluble salts, depending on feedstock type and degree of maturity, but often in higher amounts than plants need (Grebus et al., 1994). Decomposition of labile organic components during storage may also increase the salinity of compost (Carlile, 2004).

Knowledge of the chemical composition of dissolved salts is essential to properly evaluate the contribution of growing media to nutrient solution composition. Common soluble ions and compounds present in growing media components are potassium, calcium, magnesium, sodium, chloride, sulphate and bicarbonate. Smaller quantities of ammonium, nitrate and carbonate are also found (Aendekerk, 2000). They are either essential plant nutrients required in higher amounts like potassium, calcium and nitrogen, or essential nutrients required in smaller amounts such as manganese and chloride, and beneficial elements as sodium. Beneficial elements are not essential for plants but, when supplied in very low amounts, they favour their growth and development (White and Brown, 2010). All elements become harmful when present in concentrations higher than required, as can happen for chloride and sodium (Kronzucker et al., 2013; Geilfus, 2018), especially for sensitive plants or at sensitive stages of growth, as for seedlings and cuttings (Giuffrida et al., 2003; Di Gioia et al., 2018).

The three primary ions responsible for salinity in growing media components (Table 4) differ among materials. Potassium is the most frequent nutrient recorded in the water extracts of various materials. Coir and some barks, especially when composted, may contain significant concentrations of potassium (Abad et al., 2002; Naasz et al. 2009). Nitrogen is found in the two soluble inorganic forms ($NO_3^-$ and $NH_4^+$) mainly in compost and digestate. In contrast, phosphorus, due to the very low solubility of its compounds, is rarely found in water extracts from growing media components. Coir has high variability in soluble P content, depending on origin, age and processing (Evans et al., 1996; Noguera et al., 2000; Abad et al., 2002). Nutrients can accumulate in the solution during cropping (Pardossi et al., 2006; Othman et al., 2019), particularly in high-saline coir and composts.

An agricultural index has been proposed by Crohn (2016) to assess the risk of salinity of composts:

$$\text{Ag Index}\left(g\,g^{-1}\right) = \text{total}\left(N + P_2O_5 + K_2O\right)/\left(Na + Cl\right)$$

Composts with an Ag Index above 10 are considered high quality due to the high ratio of macronutrients to sodium and chloride. Values of 2 or lower are considered poor quality due to the high concentration of Na and Cl or lack of macronutrients in the compost. Combining the Ag Index with EC values provides better evaluation of saline composts (Gondek et al., 2018).

The 'background' salinity of a growing medium is not only due to the main ingredients, but also to liming materials and acidifying compounds, and to starter fertilisers added to the media to support the early growth of plants. Growing media may also increase their salinity during storage, as a result of different processes: biological and mechanical breakdown of the components, solubilisation of residual lime and the release of nutrient from fertilisers (Carlile, 2004; Altland and Jeong, 2016; Jeong and Altland, 2017).

Acidifying compounds may induce an increase in EC (Carrión et al., 2008). Sullivan et al. (2019) found that the EC of compost doubled to pH 5 due to addition of $S^0$. The contribution of fertilisers added to growing media on EC depends on their salt index, which is highest for ammonium nitrate and potassium nitrate, and lowest for calcium phosphate fertilisers. When controlled-release or organic fertilisers are used, the increase in salinity depends on coating thickness, temperature, water content and microbial effects on organic compounds which may occur during storage of growing media (Zaccheo et al., 2013).

Various practices can be used to manage salinity in components of growing media, depending on materials and the origin and type of salts. The salinity of coir is due to highly soluble compounds that can easily be removed by washing. Paramanandham et al. (2013) found sequential washing of different grades of coir to be effective. Noguera et al. (2000) showed that controlled leaching with tap water removed salts in 13 coir samples, even if no differences in plant growth were noticed between washed and unwashed samples.

Managing salinity in high EC compost is a more complex task because of high-water retention ability and CEC. Compost may be leached in order to mitigate salt stress in young plants. Washing releases salts (nutrients and non-nutrients) from compost that can be used as a liquid organic fertiliser for arable crops and as a foliar fertiliser (compost tea) (Pant et al., 2012; Ksheem et al., 2015). A study by Fornes et al. (2010) demonstrated the feasibility of this technique to reduce EC in compost-based growing media, with the leached composts showing a range of physicochemical and chemical characteristics suitable for use as growing media constituents. Wet sieving (Veeken et al., 2005) and zeolite addition (Turan, 2008; Chan et al., 2016) are viable alternative strategies for saline compost, with the former also useful in reducing heavy metal availability (Veeken and Hamelers, 2002).

## 5 Case study: modifications of chemical properties of peat-based growing media during storage

Growing media should be used within a short time but are frequently stored for periods ranging from few days to some months. Even under proper storage conditions, the external temperature and the conditions inside the bags, such as humidity and nutrient content, stimulate biological and chemical processes, particularly in organic-based growing media. As a consequence, chemical properties such as pH, salinity and nutrient status can change, particularly when fertilisers, especially slow and controlled-release fertilisers, are added to the substrate. Controlled-release or slow-release fertilisers are increasingly used for long-term cultivation, for example, woody plants. The combination of hoof and horn is one of the most used organic slow-release fertilisers with long-term release of nitrogen lasting about five months (Handreck and Black, 1994).

A one-year study was performed to evaluate the influence of time and growing media composition on pH, EC and water-soluble nutrients. A peat and a peat-pumice growing media (80:20% v $v^{-1}$) were used with two controlled-release fertilisers (one lasting 8 months coated with a polyurethane-type coating, the other lasting 12 months coated with a thermoplastic resin) and one slow release-fertiliser (hoof and horn, 0–3 mm). The main characteristics of the two growing media are summarised in Table 5.

The treatments on both the substrates were:

- *Control 01*: unfertilised substrate;
- *Control 02*: 1 g $L^{-1}$ of a starter mineral fertiliser (5.2 $NO_3$-N; 8.8 $NH_4$-N; 16 $P_2O_5$; 18 $K_2O$);
- *8M*: 4 g $L^{-1}$ controlled-release fertilisers 8 M longevity (6.6 $NO_3$-N; 8.4 $NH_4$-N; 7 $P_2O_5$; 15 $K_2O$) + 1 g $L^{-1}$ starter mineral fertiliser;
- *12M*: 4 g $L^{-1}$ controlled-release fertilisers 12 M longevity (7 $NO_3$-N; 8 $NH_4$-N; 9 $P_2O_5$; 11 $K_2O$) + g $L^{-1}$ starter mineral fertiliser; and
- *Hoof and horn*: 6 g $L^{-1}$ hoof and horn (14.8 organic N) + 1 g $L^{-1}$ starter mineral fertiliser.

36 bags (1 L) were prepared for each growing medium, brought to a moisture level of 60% and stored at 21°C. At different sampling times, three replicates of each treatment were analysed (destructive sampling) for pH, EC, water-soluble $NO_3^-$, $NH_4^+$ and $K^+$, using a water extraction ratio 1:5 vol/vol. A selection of experimental data is shown in Figs 8 – 12, while the whole study is reported in Cattivello et al. (2013) and Zaccheo et al. (2013).

**Table 5** Main properties of the limed unfertilized growing media

| | pH | EC (dS $m^{-1}$) | Total porosity | Air volume pF1[a] | Water volume pF1[a] |
|---|---|---|---|---|---|
| | | | % v/v | | |
| Peat | 6.37 ± 0.14 | 0.17 ± 0.03 | 93.8 ± 0.24 | 26.9 ± 0.83 | 66.9 ± 0.76 |
| Peat-pumice | 6.91 ± 0.21 | 0.15 ± 0.04 | 85.6 ± 0.18 | 41.2 ± 1.35 | 44.4 ± 1.23 |

[a] pF1 = −1 KPa suction.

Both controlled-release fertilisers induced a strong acidification of growing media, more in peat-pumice than in peat growing media (average pH values were lowered by 2.3 units versus 2.0 units after one-year storage). In contrast hoof and horn increased pH by 0.5 units in the first 15 days on peat-pumice growing media and in 60 days on peat growing media, followed by a significant reduction of average pH (1.8 units of pH after 1-year storage). In both growing media, lower acidification was observed in control 01 and control 02 (0.6 units and 0.9 units of pH after 1-year storage).

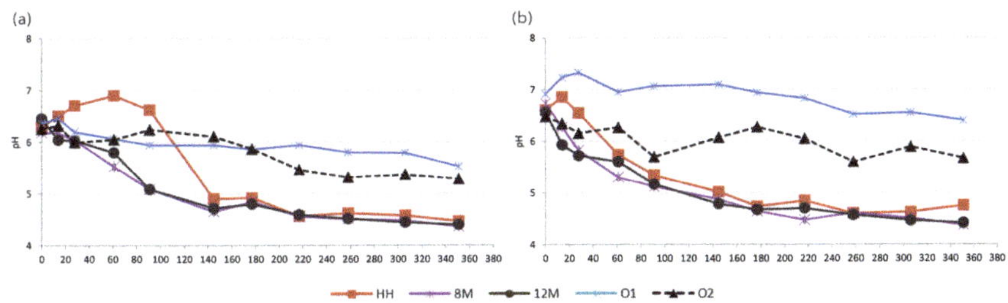


**Figure 8** Time course of pH during storage in peat-based growing media (a) and in peat-pumice growing media (b). HH: Hoof and horn; 8M: CRF 8-month longevity; 12M: CRF 12-month longevity; 01: unfertilised substrate; 02: starter mineral fertiliser.

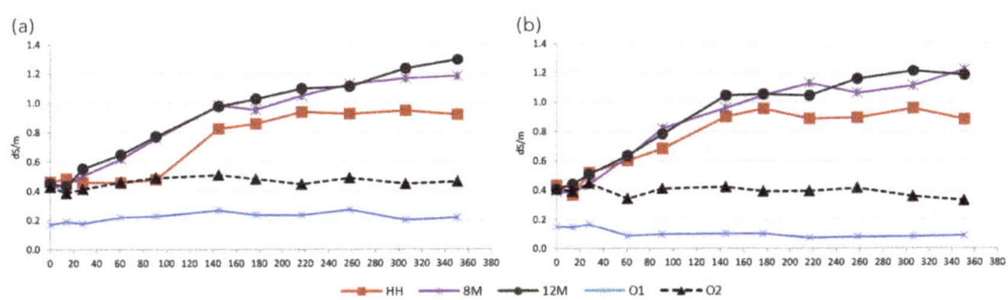


**Figure 9** Time course of EC during storage in peat-based growing media (a) and in peat-pumice growing media (b), dS $m^{-1}$. HH: Hoof and horn; 8M: CRF 8-month longevity; 12M: CRF 12-month longevity; 01: unfertilised substrate; 02: starter mineral fertiliser.

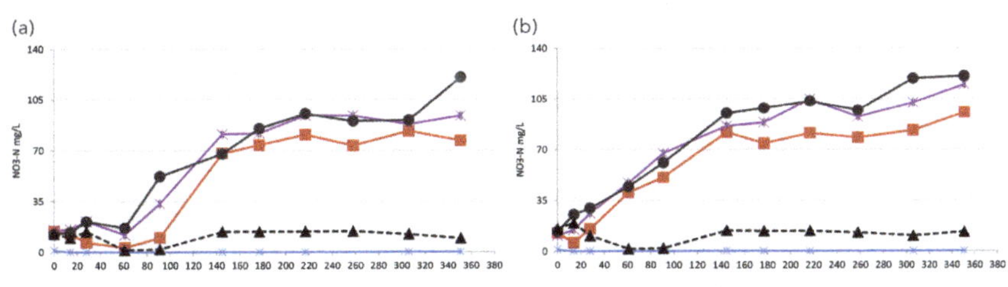


**Figure 10** Time course of $NO_3$-N during storage in peat-based growing media (a) and in peat-pumice growing media (b), mg $L^{-1}$. HH: Hoof and horn; 8M: CRF 8-month longevity; 12M: CRF 12-month longevity; 01: unfertilised substrate; 02: starter mineral fertiliser.

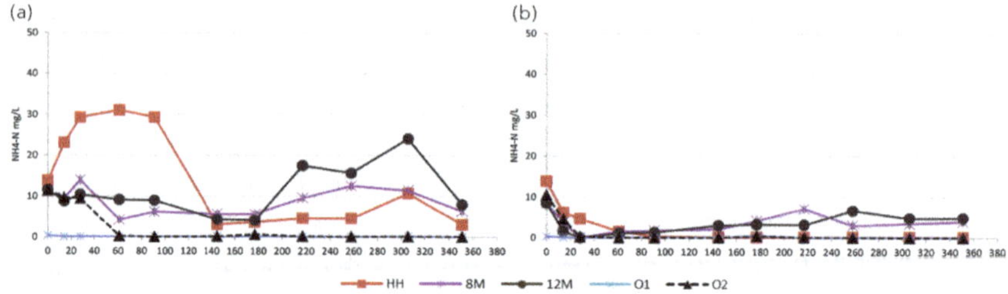


**Figure 11** Time course of $NH_4$-N during storage in peat-based growing media (a) and in peat-pumice growing media (b), mg $L^{-1}$. HH: Hoof and horn; 8M: CRF 8-month longevity; 12M: CRF 12-month longevity; 01: unfertilised substrate; 02: starter mineral fertiliser.

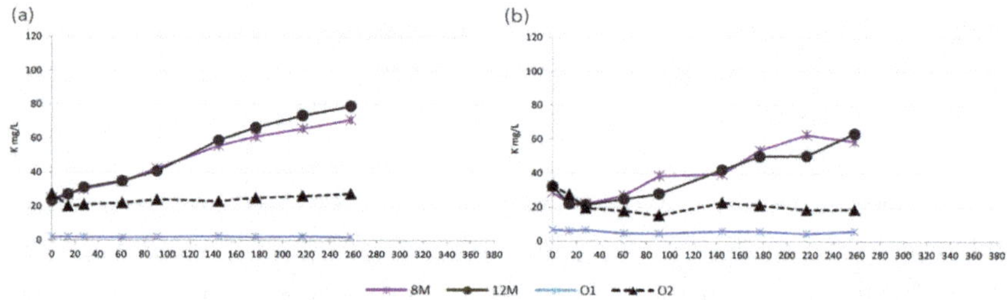


**Figure 12** Time course of K during storage in peat-based growing media (a) and in peat-pumice growing media (b), mg $L^{-1}$. 8M: CRF 8-month longevity; 12M: CRF 12-month longevity; 01: unfertilised substrate; 02: starter mineral fertiliser.

The build up of soluble salts in controlled-release fertiliser treatments led both growing media to exceed 25% of the initial EC just after two months storage. At the end of the incubation, they both reached similar values of EC, irrespective of growing media types. Increase in salinity due to hoof and horn was dependent on growing media types: delayed at 90 days in peat and almost immediately occurring in peat-pumice growing media. In the case of growing media with starter fertilisers (control 02), salinity remained relatively constant over the 12-months storage.

In peat-pumice growing media, nitrate concentration grew steadily from the beginning of storage, due to the release of nitrate from controlled-release fertilisers and to the biological oxidation of ammonium. Ammonium concentration remained lower than in peat growing media, due to nitrification and adsorption of ammonium. In peat-growing media, nitrate increase was delayed up to 60 days. After 220 days storage, nitrate concentrations was similar in both growing media. In the two media with hoof and horn added, nitrate content reached was highest after 150 days. Afterwards nitrification was inhibited, possibly due to the acidity of both growing media at that time. After 1 year of storage, about 40% of the organic nitrogen added from hoof and horn was transformed into inorganic forms.

Both controlled-release fertilisers provided a gradual release of potassium in the peat-growing media over the storage period. In contrast, immobilisation of soluble $K^+$ occurred in the first days of storage in peat-pumice growing media, followed by a gradual release from 60 days. Throughout this time levels were lower than for peat growing media: adsorption onto charged pumice surfaces may have been responsible for this depletion. In control 02, a similar initial $K^+$ depletion is observed in peat-pumice growing media.

The study highlights that several processes take place in growing media during storage: the release of nutrients, biological processes such as ammonification, nitrification, immobilisation of nutrients, and chemical interactions between nutrients and growing media components. The resulting effects on growing media properties are influenced by growing media composition. Well-aerated growing media, containing coarse materials like pumice, enhance all the above-mentioned processes. Nitrogen changes seem to have the strongest impact on pH and EC of growing media. Changes in growing media pH are clearly affected by biological activity, in particular, ammonium release from organic compounds and acidification during oxidation of ammonium. These processes are increased in more aerated conditions (Silber and Bar-Tal, 2008). Changes in pH, in turn, may influence salt solubilisation, together with nutrient release from controlled-release fertilisers and slow-release fertilisers. These changes mean that growing media properties must be checked to ensure they remain suitable for plant cultivation throughout storage, including the effects of adding controlled-release and slow-release fertilisers. It will be important to establish accepted values for EC and pH for products throughout their commercial life.

## 6 Summary and future trends

pH and salinity are essential in determining the suitability of a component for use in growing media. Both have a substantial effect on plant growth. Limitations due to unsuitable pH and EC can be overcome by mixing components with opposite properties, for example, acidifying or alkalinising compounds or performing washing/flushing before use. New components such as biochar have been found to effectively buffer peat acidity, and sulphur-based compounds can be added to compost to reduce its alkalinity. Nevertheless, some novel components are still not yet widely used in the growing media industry because of factors such as operational constraints during manufacturing (i.e. dustiness of biochar), lack of homogeneity of materials (composts from small facilities) or varying availability (waste from food industry, rice hulls etc.).

CEC and base saturation are two properties of growing media that are much less known and managed, but which need to be considered with respect to growing media. CEC and base saturation can be altered in order to improve the ability of a growing media to store nutrients in cationic form. There are several ways to do this: decreasing particle size, increasing pH, selecting humified organic matter and loading materials with a nutrient. The latter strategy can be employed to adsorb volatile compounds (i.e. coir can adsorb ammonia during composting), or to obtain a controlled-release fertiliser (i.e. loaded biochar, loaded zeolite (Eghtedary-Naeini et al., 2016; Ahmad et al., 2018). The strategic development of biofortification of plants with minerals could benefit from a substrate slowly releasing the elements needed by the plant (Sambo et al., 2019).

## 7 Where to look for further information

Many papers deal with all the main chemical characteristics of growing media and their components. Just a few of them are devoted to one aspect of the problem; these are the most relevant reviews on the whole topic:

Physical, chemical and biological properties of growing medium by Lemaire et al. (1989);
Growing Media by Landis (1990);
Growing Media for Ornamental Plants and Turf by Handreck and Black (1994);
International Substrate Manual – Analysis Characteristics Recommendations by Aendekerk (2000);
Plant Nutrition of Greenhouse Crops. Chapter 11 Substrates: Chemical Characteristics and Preparation by Sonneveld and Voogt (2009);
Organic Growing Media: Constituents and Properties by Carlile et al. (2015);
Chemical Characteristics of Soilless Media by Silber (2019);
Hydroponic Technologies in Aquaponics Food Production Systems. Chapter 4 Hydroponic Technologies by Maucieri et al. (2019).

*On the specific issue of compost:*

Production of high-quality composts for horticultural purposes: A mini-review, by Raviv (2005);
The use of composted materials in growing media, by Prasad and Carlile (2008);
Benefits and drawbacks of composted materials as a constituent of peat-based growing media by Prasad and Carlile (2009);
A Review of Soluble Salts in Compost by Composting Council (2018).

*On the specific issue of biochar*

Biochar as a growing media additive by Steiner and Harttung (2014);
Potential Use of Biochar in Growing Media by Nemati et al. (2015).

*On the specific issue of bark*

The dolomitic lime amendment affects pine bark substrate pH, nutrient availability, and plant growth: A review by Altland and Jeong (2016).

## 8 References


Abad, M., Noguera, P., Puchades, R., Maquieira, A. and Noguera, V. (2002). Physico-chemical and chemical properties of some coconut coir dusts for use as a peat substitute for containerised ornamental plants, *Bioresource Technology* 82(3), 241-245. doi: 10.1016/S0960-8524(01)00189-4.

Aendekerk, T. G. L. (2000). *International Substrate Manual: Analysis, Characteristics, Recommendations*. Doetinchen, NL: Elsevier, 94 pp.

Ahmad, M., Usman, A. R. A., Al-Faraj, A. S., Ahmad, M., Sallam, A. and Al-Wabel, M. I. (2018). Phosphorus-loaded biochar changes soil heavy metals availability and uptake potential of maize (Zea mays L.) plants, *Chemosphere* 194, 327-339. doi: 10.1016/j.chemosphere.2017.11.156.

Altland, J. E. and Jeong, K. Y. (2016). Dolomitic lime amendment affects pine bark substrate pH, nutrient availability, and plant growth: a review, *hortTechnology* 26(5), 565-573. doi: 10.21273/HORTTECH03465-16.

Altland, J. E. and Locke, J. C. (2012). Biochar affects macronutrient leaching from a soilless substrate, *HortScience* 47(8), 1136-1140. doi: 10.21273/HORTSCI.47.8.1136.

Altland, J. E., Locke, J. C. and Krause, C. R. (2014). Influence of pine bark particle size and pH on cation exchange capacity, *hortTechnology* 24(5), 554-559. doi: 10.21273/HORTTECH.24.5.554.

Amberger-Ochsenbauer, S., Jauch, M., Schmitz, H.-J. and Meinken, E. (2017) Effect of elemental sulfur and nitrogen form on substrate pH and growth of Calibrachoa in growing media containing compost high in carbonate, *Acta Horticulturae* 1168, 359-364. doi: 10.17660/ActaHortic.2017.1168.46.

André, J. P. (1981). Préparation chimique raisonnée de la tourbe utilisée comme substrat de culture horticole, P.H.M, *Revue Horticole* 214, 33-36.

André, J.-P. (1985). Propriétés chimiques des substrats. In: *Les Cultures Hors De Sol* Blanc, D. (Ed.). Paris: INRA, pp. 127-147.

Argo, W. R. and Biernbaum, J. A. (1996). The effect of lime, irrigation-water source, and water-soluble fertiliser on root-zone pH, electrical conductivity, and macronutrient management of container root media with impatiens, *Journal of the American Society for Horticultural Science* 121(3), 442-452. doi: 10.21273/JASHS.121.3.442.

Argo, W. R. and Biernbaum, J. A. (1997). The effect of root media on root-zone pH, calcium, and magnesium management in containers with impatiens, *Journal of the American Society for Horticultural Science* 122(2), 275-284. doi: 10.21273/JASHS.122.2.275.

Bar-Tal, A., Saha, U. K. and Raviv, M. (2019). Inorganic and synthetic organic components of soilless culture and potting mixtures. In: *Soilless Culture Theory and Practice* Raviv, M., Lieth, J. H. and Bar-Tal, A. (Eds). Academic Press, pp. 259-301.

Bedussi, F., Zaccheo, P. and Crippa, L. (2017). Effects of two nitrogen enriched biochars on barley grown on peat: preliminary observations, *Acta Horticulturae* 1168, 185-192. doi: 10.17660/ActaHortic.2017.1168.24.

Blok, C., Van Der Salm, C., Hofland-Zijlstra, J., Streminska, M., Eveleens, B., Regelink, I., Fryda, L.and Visser, R. (2017). Biochar for horticultural rooting media improvement:

evaluation of Biochar from gasification and slow pyrolysis, *Agronomy* 7(1), 1–23. doi: 10.3390/agronomy7010006.

Bos, E. J. F., Keijzer , R. A. W., Van Schie, W., Verhagen, J. B. G. M. and Zevenhoven, M. (2002). *Potgronden en Subtraten*. Naaldwijk, The Netherlands: Stichting RHP.

Bunt, A. C. (1988). *Media and Mixes for Container Grown Plants*. London UK: Unwin Heyman.

Carlile, W. R. (2004). Changes in organic growing media during storage, *Acta Horticulturae* 648, 153–159.

Carlile, W. R. and Bedford, I. (1988). Plant growth in container media amended with calcinated clay, *Acta Horticulturae* 221, 117–132.

Carlile, W. R., Cattivello, C. and Zaccheo, P. (2015). Organic growing media: constituents and properties, *Vadose Zone Journal* 14(6). doi: 10.2136/vzj2014.09.0125.

Carlile, W. R., Raviv, M. and Prasad, M. (2019). Organic soilless media components. In: *Soilless Culture: Theory and Practice* Raviv, M., Lieth, J. H. and Bar-Tal, A. (Eds). Academic Press, pp. 459–504. doi: 10.1016/B978-0-444-63696-6.00008-6.

Carrión, C., de la Fuente, R. G., Fornes, F., Puchades, R. and Abad, M. (2008). Acidifying composts from vegetable crop wastes to prepare growing media for containerized crops, *Compost Science and Utilization* 16(1), 20–29. doi: 10.1080/1065657X.2008.10702351.

Cattivello, C., Zaccheo, P. and Crippa, L. (2013). Behaviour of hoof and horn during the storage of peat-based and peat/pumice-based growing media, *Acta Horticulturae* 1013, 333–338. doi: 10.17660/ActaHortic.2013.1013.41.

Chan, K. Y. and Xu, Z. H. (2009). Biochar: nutrient properties and their enhancement. In: *Biochar for Environmental Management: Science and Technology* Lehmann, J. and Joseph, S. (Eds). UK: Earthscan Publications, pp. 67–84.

Chan, M. T., Selvam, A. and Wong, J. W. (2016). Reducing nitrogen loss and salinity during "struvite" food waste composting by zeolite amendment, *Bioresource Technology* 200, 838–844. doi: 10.1016/j.biortech.2015.10.093.

Clough, T., Condron, L., Kammann, C. and Müller, C. (2013). A review of biochar and soil nitrogen dynamics, *Agronomy* 3(2), 275–293. doi: 10.3390/agronomy3020275.

Crippa, L., Zaccheo, P. and Bedussi, F. (2019). Influence of pumice on ammonium dynamics in soilless culture, *Acta Horticulturae* 1266, 243–248. doi: 10.17660/ActaHortic.2019.1266.34.

Crippa, L., Zaccheo, P., Corti, C. and Genevini, P. L. (2000). Efficacia di diversi trattamenti di valorizzazione agronomica nel compostaggio di residui lignocellulosici. In: *Proceedings of the XVII Convegno SICA*. Portoferraio, Italy: SICA, pp. 373–378.

Crohn, D. M. (2016). Assessing compost quality for agriculture, *Assessing Compost Quality for Agriculture*, (August). doi: 10.3733/ucanr.8514.

Daniels, W. L. and Wright, L. D. (1988). Cation exchange properties of pine bark growing media as influenced by pH, particle size and cation species, *Journal of the American Society of Horticultural Science* 113, 557–560.

Dias, V., Mechant, E., Hoekstra, B., Perneel, M. and Vandecasteele, B. (2017). Sustainable growing media based on green waste compost and other organic recycled materials: use of elemental sulphur to control pH, *Acta Horticulturae* 1168, 167–174. doi: 10.17660/ActaHortic.2017.1168.22.

Dickinson, K. and Carlile, W. R. (1995). The storage properties of wood-based peat-free growing media, *Acta Horticulturae* 401, 89-96. doi: 10.17660/ActaHortic.1995.401.10.

Di Gioia, F., Rosskopf, E. N., Leonardi, C. and Giuffrida, F. (2018). Effects of application timing of saline irrigation water on broccoliproduction and quality, *Agricultural Water Management* 203, 97–104. doi: 10.1016/j.agwat.2018.01.004.

Domeño, I., Irigoyen, I. and Muro, J. (2010). New wood fibre substrates characterisation and evaluation in hydroponic tomato culture, *European Journal of Horticultural Science* 75(2), 89–94.

Dumont, P. A., Chadwick, D. R. and Robinson, S. (2014).Nutrient removal capacity of wood residues for the Agro-environmental safety of ground and surface waters, *Sustainability, Agri, Food and Environmental Research* 2(2). doi: 10.7770/safer-V2N2-art758.

Eckhard, G., Horst, W. and Neumann, E. (2012). Adaptation of plants to adverse chemical soil conditions. In: *Marschner's Mineral Nutrition of Higher Plants* Marschner, P. (Ed). Academic Press. doi: 10.1016/B978-0-12-3849.

Eghtedary-Naeini, A., Golabadi, M. and Hoodaji, M. (2016). Using enriched zeolite as a slow release iron fertiliser for soilless greenhouse cultivation of cucumber, *Journal of Plant Nutrition*. Taylor & Francis 39(4), 523–530. doi: 10.1080/01904167.2015.1109123.

Enzo, M., Gianquinto, G., Lazzarin, R., Pimpini, F. and Sambo, P. (2001). *Principi tecnico-agronomici della fertirrigazione e del fuori suolo*. Padova Italy: Garbin.

EU Ecolabel. (2020). Regulation (EC) No 66/2010 of the European Parliament and of the Council of 25 November. Available at: https://eur-lex.europa.eu/legal-content/EN/TXT/?uri=CELEX:32010R0066, p. 2009.

European Committee for Standardization. EN 13038 (2012). *Soil Improvers and Growing Media–Determination of Electrical Conductivity*. Milan, Italy: UNI.

Evans, M. R., Konduru, S. and Stamps, R. H. (1996). Source variation in physical and chemical properties of coconut coir dust, *HortScience* 31(6), 965–967. doi: 10.21273/HORTSCI.31.6.965.

Fernando, J. A. K. M. and Amarasinghe, A. D. U. S. (2017). Effects of retting and drying on quality of coir pith and coco discs, *Journal of the National Science Foundation of Sri Lanka* 45(1), 3–12. doi: 10.4038/jnsfsr.v45i1.8032.

Fisher, P. R., Huang, J. and Argo, W. R. (2006). Modeling lime reaction in peat-based substrates, *Acta Horticulturae* 718, 461–468. doi: 10.17660/ActaHortic.2006.718.53.

Fornes, F., Carrión, C., García-de-la-Fuente, R., Puchades, R. and Abad, M. (2010). Leaching composted lignocellulosic wastes to prepare container media: feasibility and environmental concerns, *Journal of Environmental Management* 91(8), 1747–1755. doi: 10.1016/j.jenvman.2010.03.017.

Frangi, P., Castelnuovo, M., Pozzi, A., Valagussa, M., Crippa, L. and Genevini, P. L. (2008) A comparison of methods for the analysis of compost-based growing media, *Acta Horticulturae* 779, 113–119.

Geilfus, C. M. (2018). Chloride: from nutrient to toxicant, *Plant and Cell Physiology* 59(5), 877–886. doi: 10.1093/pcp/pcy071.

Gelsomino, A., Abenavoli, M. R., Princi, G., Attinà, E., Cacco, G. and Sorgonà, A. (2010). Compost from fresh orange waste: a suitable substrate for nursery and field crops?, *Compost Science and Utilization* 18(3), 201–210. doi: 10.1080/1065657X.2010.10736956.

Geuijen, V. H. C. and Verhagen, J. B. G. M. (2021). Nutritional aspects in bioassays according to EN 16086-1, In: *Proceedings of the III International Symposium on Growing Media, Composting and Substrate Analysis, Acta Horticolturae*, in press.

Giuffrida, F., Argento, S., Lipari, V. and Leonardi, C. (2003). Methods for controlling salt accumulation in substrate cultivation, *Acta Horticulturae* 614, 799–803 doi: 10.17660/ActaHortic.2003.614.117.

Gondek, M. D., Thiel, T. and Kleinheink, G. (2018). *Soluble Salts: A Review of Soluble Salts in Compost*. Composting Council Research and Education Foundation. Available at: https://www.compostfoundation.org/Soluble-Salts.

Grebus, M. E., Watson, M. E. and Hoitink, H. A. J. (1994). Biological, chemical and physical properties of composted yard trimmings as indicators of maturity and plant disease suppression, *Compost Science and Utilization* 2(1), 57–71. doi: 10.1080/1065657X.1994.10757919.

Gruda, N. (2009). Do soilless culture systems have an influence on product quality of vegetables?, *Journal of Applied Botany and Food Quality* 82, 141–147. Available at: https://ojs.openagrar.de/index.php/JABFQ/article/view/2093.

Gruda, N., Caron, J., Prasad, M. and Maher, M. J. (2016). Growing media. In: *Encyclopedia of Soil Sciences* Lal, R. (Ed). Boca Raton, FL: CRC Press Press Taylor & Francis Group, pp. 1053–1058.

Gruda, N., Savvas, D., Colla, G. and Rouphael, Y. (2018). Impacts of genetic material and current technologies on product quality of selected greenhouse vegetables–a review, *European Journal of Horticultural Science* 83(5), 319–328 doi: 10.17660/eJHS.2018/83.5.5.

Gül, A., Erŏul, D. and Ongun, A. R. (2005). Comparison of the use of zeolite and perlite as substrate for crisp-head lettuce. *Scientia Horticulturae* 106(4), 464–471.

Handreck, K. H. and Black, N. D. (1994). *Growing Media for Ornamental Plants and Turf*. Sydney, Australia: University of New South Wales Press.

Huang, J., Fisher, P. R., Horner, W. E. and Argo, W. R. (2010). Limestone particle size and residual lime concentration affect pH buffering in container substrates, *Journal of Plant Nutrition* 33(6), 846–858. doi: 10.1080/01904161003654089.

Huber, D., Römheld, V. and Weinmann, M. (2012). Chapter 10–relationship between nutrition, plant diseases and pests. In: *Mineral Nutrition of Higher Plants* Marschner, P. (Ed). San Diego: Academic Press, pp. 283–298. doi: 10.1016/B978-0-12-384905-2.00010-8.

Inbar, Y., Chen, Y. and Hadar, Y. (1989). Solid-state carbon-13 nuclear magnetic resonance and infrared spectroscopy of composted organic matter, *Soil Science Society of America Journal* 53(6), 1695–1701. doi: 10.2136/sssaj1989.03615995005300060014x.

Irum, Ay., Akhtar, S., Iqbal, M. Z. and Nazir, S. (2019). Sulfur-aided acidification of compost ameliorates its negative effects and maximise phosphorous availability in calcareous soils, *International Journal of Botany Studies* 4(3), 26–34.

Jackson, B. E., Wright, R. D. and Gruda, N. (2009a). Container medium pH in a pine tree substrate amended with peat moss and dolomitic limestone affects plant growth, *Hortscience* 44(7), 1983–1987.

Jackson, B. E. J., Wright, R. D. and Seiler, J. R. (2009b). Changes in chemical and physical properties of pine tree substrate and bark during long-term nursery production, *Hortscience* 44(3), 791–799. doi: 10.21273/HORTSCI.44.3.791.

Jeong, K. Y. and Altland, J. E. (2017). Initial substrate moisture content affects chemical properties of bagged substrates containing controlled release fertiliser at two different temperatures, *Hortscience* 52(10), 1429–1434. doi: 10.21273/HORTSCI12216-17.

Kipp, J. A., Wever, G. and De Kreij, C. (2000). *International Substrate Manual*. Doetinchem, The Netherlands: Elsevier International Business Information, 94 pp.

Konduru, S., Evans, M. R. and Stamps, R. H. (1999). Coconut husk and processing effects on chemical and physical properties of coconut coir dust, *HortScience* 34(1), 88–90. doi: 10.21273/HORTSCI.34.1.88.

Kronzucker, H. J., Coskun, D., Schulze, L. M., Wong, J. R. and Britto, D. T. (2013) Sodium as nutrient and toxicant, *Plant and Soil* 369(1–2), 1–23. doi: 10.1007/s11104-013-1801-2.

Ksheem, A. M., Bennett, J. M., Antille, D. L. and Raie, S. R. (2015). Towards a method for optimised extraction of soluble nutrients from fresh and composted chicken manures, *Waste Management* 45, 76–90. doi: 10.1016/j.wasman.2015.02.011.

Landis, T. D. (1990). Chapter 2 Growing media. In: *The Container Tree Nursery Manual* (Vol. II) Landis, T. D. (Ed). Washington, DC: Department Agric. Forest Service, pp. 41–85.

Lemaire, F. (1995). Cultures en pots et conteneurs: principes agronomiques et applications, *Acta Horticulturae* 396, 273–284. doi: 10.17660/ActaHortic.1995.396.33.

Lemaire, F., Dartigues, A. and Rivière, L. M. (1989). *Cultures en pots et conteneurs: principes agronomiques et applications*. Paris: INRA. Available at: http://lib.ugent.be/catalog/rug01:000221554.

Li, B., Chakraborty, S., Sosa, M. F. G., Kusi, N. Y. O. and Weindorf, D. C. (2018). Compost cation exchange capacity via portable X-ray fluorescence (PXRF) spectrometry, *Compost Science and Utilization* 26(4), 271–278. doi: 10.1080/1065657X.2018.1522280.

Love, K. S. and Whittaker, C. W. (1954). Agricultural limestones: surface area and reactivity of typical limestones, *Journal of Agricultural and Food Chemistry* 2(25), 1268–1272. doi: 10.1021/jf60045a002.

Maucieri, C., Nicoletto, C., Os, E., Anseeuw, D., Havermaet, R. V. and Junge, R. (2019). Hydroponic technologies. In: *Aquaponics Food Production Systems* Goddek, S., Joyce, A., Kotzen, B. and Burnell, G. (Eds). Springer International Publishing, pp. 77–110.

Mazuela, P., Salas Sanjuán, MdC. and Urrestarazu, M. (2005). Vegetable waste compost as substrate for melon, *Communications in Soil Science and Plant Analysis* 36(11–12), 1557–1572. doi: 10.1081/CSS-200059054.

Menzies, N. W. and Aitken, R. L. (1996). Evaluation of fly ash as a component of potting substrates, *Scientia Horticulturae* 67(1–2), 87–99. doi: 10.1016/S0304-4238(96)00935-1.

Ming, D. W. and Mumpton, F. A. (1989). Zeolites in soils. In: *Minerals in Soil Environments* Dixon, J. B. and Weed, S. B. (Eds). *SSSA Book Series*. John Wiley & Sons, Ltd, pp. 873–911. https://doi.org/10.2136/sssabookser1.2ed.c18.

Naasz, R., Caron, J., Legault, J. and Pichette, A. (2009). Efficiency factors for bark substrates: biostability, aeration, or phytotoxicity, *Soil Science Society of America Journal* 73(3), 780–791. doi: 10.2136/sssaj2008.0058.

Nappi, P. and Barberis., R. (1993). Compost as a growing medium: chemical, physical and biological aspects, *Acta Horticulturae* 342, 249–256.

Nemati, M. R., Simard, F., Fortin, J. P. and Beaudoin, J. (2015). Potential use of biochar in growing media, *Vadose Zone Journal* 14(6), 1–8. doi: 10.2136/vzj2014.06.0074.

Noguera, P., Abad, M., Noguera, V., Puchades, R. and Maquieira, A. (2000). Coconut coir waste, a new and viable ecologically-friendly peat substitute, *Acta Horticulturae* 517, 279–286.

Noguera, P., Abad, M., Puchades, R., Maquieira, A. and Noguera, V. (2003). Influence of particle size on physical and chemical properties of coconut coir dust as container medium, *Communications in Soil Science and Plant Analysis* 34(3–4), 593–605. doi: 10.1081/CSS-120017842.

Othman, Y., Bataineh, K., Al-Ajlouni, M., Alsmairat, N., Ayad, J., Shiyab, S., Al-Qarallah, B. and St Hilaire, R. (2019). Soilless culture: management of growing substrate, water, nutrient, salinity, microorganism and product quality, *Fresenius Environmental Bulletin* 28(4A), 3249–3260.

Pant, A. P., Radovich, T. J. K., Hue, N. V. and Paull, R. E. (2012). Biochemical properties of compost tea associated with compost quality and effects on pak choi growth, *Scientia Horticulturae* 148, 138–146.

Paramanandham, J., Ronald Ross, P., Vaidehi, J. and Abbiramy, K. S. (2013). Influence of sequential washing on the pH and electric conductivity of graded coir pith, *International Journal of Pure and Applied Zoology* 1, 231–234. Available at: http://www.ijpaz.com.

Pardossi, A., Malorgio, F., Incrocci, L., Carmassi, G., Maggini, R., Massa, D., Tognoni, F. (2006). Simplified models for the water relations of soilless cultures: what they do or suggest for sustainable water use in intensive horticulture, *Acta Horticulturae*, 718, 425–434. doi: 10.17660/ActaHortic.2006.718.49.

Prasad, M. and Carlile, W. R. (2008). Benefits and drawbacks of composted materials as a constituent of peat-based growing media, In: *After Wise Use – The Future of Peatlands, Proceedings of the 13th International Peat Congress: Peat In Horticulture* Farrell, C. and Feehan, J. (Eds), Tullamore, Ireland, pp. 156–159.

Prasad, M. and Carlile, W. R. (2009). Practical experiences and background research on the use of composted materials in growing media for the US market, *Acta Horticulturae* 819, 111–124. doi: 17660/ActaHortic.2009.819.10.

Prasad, M., Chrysargyris, A., McDaniel, N., Kavanagh, A., Gruda, N. S. and Tzortzakis, N. (2020). Plant nutrient availability and pH of biochars and their fractions, with the possible use as a component in a growing media, *Agronomy* 10(1), 1–17. doi: 10.3390/agronomy10010010.

Prasad, M., Tzortzakis, N. and McDaniel, N. (2018). Chemical characterisation of biochar and assessment of the nutrient dynamics by means of preliminary plant growth tests, *Journal of Environmental Management* 216(216), 89–95. doi: 10.1016/j.jenvman.2017.04.020.

Preston, C. M., Axelson, D. E., Lévesque, M., Mathur, S. P., Dinel, H. and Dudley, R. L. (1989). Carbon-13 NMR and chemical characterisation of particle-size separates of peats differing in degree of decomposition, *Organic Geochemistry* 14(4), 393–403. doi: 10.1016/0146-6380(89)90005-3.

Puustjärvi, V. (1977). *Peat and Its Use in Horticulture*. Helsinki: Turveteollisuusliitto.

Puustjarvi, V. and Robertson, R. A. (1975). Physical and chemical properties. In: *Peat in Horticulture* Robinson, D. W. and Lamb, J. G. D. (Eds). London: Academic Press, pp. 23–38.

Raviv, M. (2005). Production of high-quality composts for horticultural purposes: A mini-review, *hortTechnology* 15(1), 52–57. doi: 10.21273/HORTTECH.15.1.0052.

Raviv, M. (2013). Composts in growing media: what's new and what's next?, *Acta Horticulturae* 982, 39–52. doi: 10.17660/ActaHortic.2013.982.3.

Reddy, S. (2005). Harmonizing peat and lime. *the sun Gro'er* 3(3), 1–6. Available at: http://www.sungro.com/issue-33-april-2005-harmonizing-peat-and-lime/.

Rippy, J. (2005). *Factors Affecting pH Establishment and Maintenance in Peat Moss Based Substrates*. PhD Dissertation, North Carolina State University, Raleigh, NC. Available at: https://repository.lib.ncsu.edu/handle/1840.16/5254.

Rippy, J. F. M. and Nelson, P. V. (2007). Cation exchange capacity and base saturation variation among Alberta, Canada, moss peats, *Hortscience* 42(2), 349–352. doi: 10.21273/HORTSCI.42.2.349.

Rippy, J. F. M., Nelson, P. V., Hesterberg, D. L., Niedziela, C. E., Kamprath, E. J. and Bilderback, T. (2016). Importance of limestone specific surface for assessing neutralisation effectiveness in soilless root substrate, *Communications in Soil Science and Plant Analysis* 47(4), 521–526. doi: 10.1080/00103624.2016.1141926.

Robbins, J. A. and Evans, M. (2006). Growing media for container production in a greenhouse or nursery. Part 1 – components and mixes. In: University of Arkansas, Cooperative Extension Service, Greenhouse and Nursery Series, 4 pp. FSA6097-PD-7-11RV. Greenhouse and Nursery Series.

Roig, A., Cayuela, M. L. and Sánchez-Monedero, M. A. (2004). The use of elemental sulphur as organic alternative to control pH during composting of olive mill wastes, *Chemosphere* 57(9), 1099–1105. doi: 10.1016/j.chemosphere.2004.08.024.

Sambo, P., Nicoletto, C., Giro, A., Pii, Y., Valentinuzzi, F., Mimmo, T., Lugli, P., Orzes, G., Mazzetto, F., Astolfi, S., Terzano, R. and Cesco, S. (2019). Hydroponic solutions for soilless production systems: issues and opportunities in a smart agriculture perspective, *Frontiers in Plant Science* 10(July), 923. doi: 10.3389/fpls.2019.00923.

Schmilewski, G. (2008). The role of peat in assuring the quality of growing media. *Mires and Peat* 3(2), 1–8.

Silber, A. (2019). Chemical characteristics of soilless media. In: *Soilless Culture: Theory and Practice Theory and Practice* Raviv, M., Bar-Tal, A and Lieth, J. H. (Eds.). Academic Press, pp. 113–148. https://doi.org/10.1016/C2015-0-01470-8.

Silber, A. and Bar-Tal, A. (2008). Nutrition of substrate-grown plants. In: *Soilless Culture, Theory and Practice* Raviv, M. and Lieth, H. (Eds). Amsterdam: Elsevier, pp. 292–342.

Silber, A. and Bar-Tal, A. D. (2019). Nutrition of substrate-grown plants. In: *Soilless Culture: Theory and Practice* Raviv, M. and Lieth, H. (Eds). Amsterdam: Elsevier Science, p. 712. https://doi.org/10.1016/C2015-0-01470-8.

Silber, A., Bar-Yosef, B. and Chen, Y. (1999). PH-dependent kinetics of tuff dissolution, *Geoderma* 93(1–2), 125–140. doi: 10.1016/S0016-7061(99)00048-8.

Silber, A., Bar-Yosef, B., Singer, A. and Chen, Y. (1994). Mineralogical and chemical composition of three tuffs from northern Israel, *Geoderma* 63(2), 123–144. doi: 10.1016/0016-7061(94)90002-7.

Silber, A. and Raviv, M. (1996). Effects on chemical surface properties of tuff by growing rose plants, *Plant and Soil* 186(2), 353–360. doi: 10.1007/BF02415530.

Solbraa, K. (1986). Bark as growth medium, *Acta Horticulturae* 178, 129–136. doi: 10.17660/ActaHortic.1986.178.17.

Sonneveld, C., Van den Ende, J.and Van Dijk, P. A. (1974). Analysis of growing media by means of 1:11/2 volume extract, *Communications in Soil Science and Plant Analysis* 5(5), 183-202.

Sonneveld, C. and Voogt, W. (2009). Substrates: chemical characteristics and preparation. *Plant Nutrition of Greenhouse Crops*, 227–256. doi: 10.1007/978-90-481-2532-6.

Sposito, G. (1989). *The Chemistry of Soils*. New York: Oxford University Press.

Steiner, C. and Harttung, T. (2014). Biochar as growing media additive and peat substitute, *Solid Earth Discussions* 6(1), 1023–1035. doi: 10.5194/sed-6-1023-2014.

Sullivan, D. M., Costello, R. C. and Andrews, S. B. (2019). A simple titration method to determine elemental sulfur required for compost acidification, *Acta Horticulturae* 1266, 15–20. doi: 10.17660/ActaHortic.2019.1266.3.

Sumner, M. E. and Miller, W. P. (1996). Cation exchange capacity and exchange coefficients. In: *Methods of Soil Analysis* Sparks, D. L. (Ed.), SSSA Book Series 5. Madison, WI: Madison Book Company Society of America, pp. 1201–1230.

Thompson, W., Leege, P., Milner, P. and Watson, M. (2001). *Test Methods for the Examination of Composting and Compost* Smith, J (Ed.). The US Composting Council, US Government Printing Office.

Turan, N. G. (2008). The effects of natural zeolite on salinity level of poultry litter compost, *Bioresource Technology* 99(7), 2097–2101. doi: 10.1016/j.biortech.2007.11.061.

U.S. Composting Council. (2001). Test Methods for the Examination of Composting and Composts (TMECC). Available at: https://compostingcouncil.org/tmecc.

Veeken, A. and Hamelers, B. (2002). Sources of Cd, Cu, Pb and Zn in biowaste, *Science of the Total Environment* 300(1–3), 87–98. doi: 10.1016/S0048-9697(01)01103-2.

Veeken, A. H. M., Blok, W. J., Curci, F., Coenen, G. C. M., Termorshuizen, A . J. and Hamelers, H. V. M. (2005). Improving quality of composted biowaste to enhance disease suppressiveness of compost-amended, peat-based potting mixes, *Soil Biology and Biochemistry* 37(11), 2131–2140. doi: 10.1016/j.soilbio.2005.03.018.

Velusami, B., Curran, T. P. and Grogan, H. M. (2013). Hydrogen sulfide gas emissions during disturbance and removal of stored spent mushroom compost, *Journal of Agricultural Safety and Health* 19(4), 261–275. doi: 10.13031/jash.19.10351.

Verhagen, J. B. G. M. (1999). CEC and the saturation of the adsorption complex of coir dust, *Acta Horticulturae* 481, 151–158.

White, P. J. and Brown, P. H. (2010). Plant nutrition for sustainable development and global health, *Annals of Botany* 105(7), 1073–1080. doi: 10.1093/aob/mcq085.

Wright, A. N., Niemiera, A. X., Harris, J. R. and Wright, R. D. (1999). Preplant lime and micronutrient amendments to pine bark affect growth of seedlings of nine container-grown tree species, *HortScience* 34(4), 669–673. doi: 10.21273/HORTSCI.34.4.669.

Wright, R. D., Jackson, B. E., Browder, J. F. and Latimer, J. G. (2008). Growth of chrysanthemum in a pine tree substrate requires additional fertiliser, *hortTechnology* 18(1), 111–115. doi: 10.21273/HORTTECH.18.1.111.

Yao, Y., Gao, B., Zhang, M., Inyang, M. and Zimmerman, A. R. (2012). Effect of biochar amendment on sorption and leaching of nitrate, ammonium, and phosphate in a sandy soil, *Chemosphere* 89(11), 1467–1471. doi: 10.1016/j.chemosphere.2012.06.002.

Zaccheo, P., Cattivello, C., Minuto, G., Crippa, L. and Orfeo, D. (2015). Indagine conoscitiva di fibra e midollo di cocco presenti sul mercato italiano: aspetti fisici, chimici e microbiologici, *Acta Italus Hortus* 18, 49–53.

Zaccheo, P., Crippa, L. and Bedussi, F. (2019). Biochar can enhance potassium availability during cyclamen cultivation, *Acta Horticulturae* 1266, 35–42. doi: 10.17660/ActaHortic.2019.1266.6.

Zaccheo, P., Crippa, L. and Cattivello, C. (2013). Effect of controlled-release fertilisers on chemical parameters of two growing media during 12 months storage, *Acta Horticulturae* 1013, 327–332. doi: 10.17660/ActaHortic.2013.1013.40.

Zaccheo, P., Crippa, L. and Cattivello, C. (2014). Liming power of different particle fractions of biochar, *Acta Horticulturae* 1034, 363–368. doi: 10.17660/ActaHortic.2014.1034.45.

# Chapter 7

# Understanding and optimising the biological properties of growing media for soilless cultivation

*Francesco Giuffrida and Rosario Paolo Mauro, University of Catania, Italy; and Patrizia Zaccheo, University of Milan, Italy*



## 1 Introduction

For decades biological properties of growing media were considered of minor importance. They have mainly been used to explain modifications in physical and chemical properties due to the degradation of organic matter. A new approach was proposed at the beginning of the 1990s highlighting the importance of biological properties in their own right, including the absence of harmful organisms and weeds as well as beneficial microbial activity in growing media (Carlile and Wilson, 1991).

The biological properties of growing media affect conditioning, storage, use and possible reuse. The latter is important in reducing costs and improving sustainability (Giuffrida et al., 2008; Giuffrida and Consoli, 2016). The use of several new organic materials as growing media components has also highlighted issues

http://dx.doi.org/10.19103/AS.2020.0076.06

related to the microbial degradation of the organic matter and biostability, as well as opportunities relating to microbial composition in the root zone.

The life present or developing in growing media is constituted by a rich and complex array of microorganisms that could have deleterious effects, such as diseases, but also positive effects on plant growth and development. Thanks to recent advances in molecular techniques, a greater understanding of microbial populations will assist the identification and quantification of saprophytic microflora, human and plant pathogens as well as the evaluation of potential biocontrol agents. Beneficial microorganisms in growing media can support plant growth, biological control of insect pests and diseases.

## 2 Biological activity in growing media: weeds, pests and diseases

### *2.1 Weeds*

Weeds are highly competitive for water and nutrients with plants grown in containers. The growth of many ornamental crops can be significantly reduced by a single weed in a restricted rooting environment (Stewart et al., 2018). Weed competition can reduce plant growth in ornamental container production by 24–44% (Whitcomb, 1988). In addition, the market of the ornamental crops appreciates weed-free products (Simpson et al., 2002).

Weed seed contamination of growing media can occur at different points in the life of growing media. The main mineral growing media are produced at very high temperatures and are thus completely weed-free. Organic constituents like pine bark and peat are also generally weed-free (Altland, 2014). Mofidpoor et al. (2009) found just one weed species in 11% of *Sphagnum* bog samples in Manitoba, Canada. In contrast, weeds may be present in poorly or insufficiently composted green and other wastes (Inserra et al., 2006). According to the criteria for the award of the EU Ecolabel for growing media (Commission Decision–EU-2015/2099), final products, except mineral growing media, should not contain more than two units of viable seeds and plant propagules per litre. In important certification standards for growing media in Europe, such as those of the Stichting R.H.P. in The Netherlands and the RAL Quality Assurance Association for Growing Media for Plant Cultivation in Germany, the number of germinating seeds or other viable plant parts in growing media constituents is also strictly limited, for example, to one plant per litre of growing medium (GGS, 2015). To reduce nursery seed contamination by growing media, germination tests should be conducted before planting (Case et al., 2005).

Canadian growing media producers use peat that has been stored in bales for up to 8 or 10 months, whilst European growing media producers stores peat in bales for over 3 months (Mofidpoor et al., 2009). Many other constituents are also stored in the open air before being mixed for growing media production. In case of a long storage period, it is recommended to limit exposure to weed seed being blown in or transmitted via water or animals by covering piles (Case et al., 2005; Stewart et al., 2017).

Wind and water can disseminate weed seed during use of growing media. Because some weeds have very small seeds, they can be distributed by irrigation systems when surface water sources are used. The growth and development of weeds in growing media during the cultivation cycle may result in significant weed infestation when growing media are reused (Fig. 1) or mixed with new materials (Stewart et al., 2017).

Hand removal of weeds from growing media is possible when the weeds are small. However, once weed roots are established in growing media, hand weeding can uproot the crop seedling (Fig. 2). Cultural and chemical control of weeds can be influenced by growing media characteristics. The use of growing media with high air-filled porosity can reduce the germination rate and growth of several weeds (Wada, 2005; Altland, 2014). According to Stewart et al. (2019), the use of the herbicides in growing media to control weeds is challenging because the high organic content of most growing media may inactivate many chemicals, influence their adsorption or cause leaching.

**Figure 1** Weed infestation in reused growing media. Images provided with permission of Salvatore Adamo.

**Figure 2** Seedling container infested by *Polygonum persicaria* weed. Images provided with permission of Costantino Cattivello.

## *2.2 Pests*

Some insects can colonize growing media, causing significant damage in container crop production. The black vine weevil, *Otiorhynchus sulcatus*, and fungus gnats, *Bradysia* spp., are common insect pests of field and greenhouse container-grown plants (Bruck, 2006). Weevil larvae can be found within growing media and plant roots, causing poor plant growth and, with serious infestations, plant death. There are not many studies of the black vine weevil diffusion in growing media. However, there are reports of biocontrol of black vine weevil by the parasitic fungus *Metarhizium anisopliae* in soilless growing media (Bruck, 2005) and the effects of different potting media components on the infectivity of *M. anisopliae* against this insect pest (Bruck, 2006). Fungus gnat adults cause minimal plant damage (Cloyd, 2000) whilst larvae feed on roots, causing wilting, growth reduction and eventual plant death (Broadley et al., 2018). In addition, both larval and adult stages can transmit pathogens, such as *Botrytis*, *Verticillium* and *Pythium*, to healthy plants (Cloyd, 2015).

Growing media may attract fungus gnat adults in different ways because components may affect the population dynamics of the insect differently, including development and reproduction processes (Jagdale et al., 2004). It has been reported that fungus gnat adults are attracted to lay eggs in growing media that provide an abundant level of fungal activity (Gardiner et al., 1990; Olson et al., 2002; Lamb et al., 2013). In general, fungus gnat adults are attracted to moist growing media that can provide a higher level of fungal activity to support eggs and larvae development (Olson et al., 2002). However, Cloyd et al. (2009) hypothesized that the higher attraction of growing media containing composted bark compared to peat may be related to some volatile compounds. Further research is needed to evaluate other growing media and their volatiles to assess which constituents are most attractive to fungus gnat adults. Fungus gnats developed and survived less well in coir than in peat. However, survival was further influenced by growing media texture and moisture (Olson et al., 2002). The most common route for fungus gnat infestation is the use of a growing media previously contaminated in a greenhouse or nursery. Given the importance of preventing insect infestation, specific methods have been developed to quantify eggs (Meers and Cloyd, 2005) and larvae (Zaborski and Cloyd, 2004) in soilless growing media.

### *2.3 Diseases*

Several species of fungi (e.g. *Phytophthora*, *Fusarium*, *Rhizoctonia*, *Pythium*) and rhizogenic agrobacteria (Bosmans et al., 2017) may cause root diseases in container crops. However, the use of growing media has generally reduced the amount of soil-borne diseases and modified the diffusion of different fungal diseases. For instance, *Pythium* root and crown rot are not serious diseases in soil-grown cucumber plants, whereas they are very serious diseases of mineral substrate-grown cucumber and a few other crops (van der Gaag and Wever, 2005).

The higher incidence of *Pythium* diseases in growing media compared to soil may be explained by relatively low microbial activity in some substrates like mineral wool and peat-based substrates, since high microbial activity has often been associated with suppression of *Pythium* and *Rhizoctonia* diseases (Paulitz and Belanger, 2001; van der Gaag and Wever, 2005). However, the conduciveness of some growing media for *Pythium* diseases may also be related to physical characteristics of growing media.

Comparing some inorganic components, van der Gaag and Wever (2005) observed a lower incidence of *Pythium* disease in plants grown in pumice and perlite compared to mineral wool. This is because lower water content plays a fundamental role in the development of root rot caused by *Pythium*. In growing media with high water-holding capacity, such as mineral wool and coir, substrate or container height could be an important tool to modify the water content in the root zone and thus prevent the incidence of soil-borne diseases.

Because it is easier to prevent rather than control root diseases, growing media should be assayed to check they are pathogen-free before purchase. Some evidence suggests that soil-borne diseases may occur in growing media. For example, analysing 38 samples of organic constituents commonly used in the UK (peat, composted green waste, wood-based material, coir and bark),

Waller et al. (2008) found three positive results for pathogenic fungi (*Fusarium oxysporum* f. sp. *melonis* and *Rhizoctonia* spp. in peat and *Rhizoctonia* spp. in a wood-based component). Improper composting procedures may also result in incomplete elimination of pathogens, plant pests and nematode (Noble and Roberts, 2003).

In recent years saprophytic basidiomycetes fungi have begun to appear in growing media. *Peziza* and *Leucocoprinus* spp. can form a coloured and hydrophobic mycelium layer on the surface of some organic materials, such peat as and coir, that reduces their wettability, interfering with irrigation and water uptake (Carlile and Schmilewski, 2010), as well as the market quality of ornamental pot plants (Fig. 3). Withdrawal of fungicides for control of these pathogens may have contributed to their higher incidence (Carlile et al., 2015). When conidia appear on the growing media surface, the mycelium is already extensively spread in the growing media (Verhagen, 2010).

**Figure 3** Saprophytic fungi on an ornamental potting plant. Images provided with permission of Paolo Notaristefano.

Methods for early identification of colonization by these fungi are therefore required. Lohr et al. (2017) suggested using ergosterol as a diagnostic tool to assess the risk of colonization of peat-based and other organic growing media by saprophytic fungi. The development of the biological control of these fungi in growing media is an important research challenge to improve sustainability in soilless culture systems.

Microbial populations of growing media may include human pathogens such as *Legionella longbeachae*, *Salmonella enterica* and *Escherichia coli* O157:H7, posing concerns for human health (Carlile and Hammonds, 2008). These human pathogens, called zoonosis, are increasingly found when animal waste-based composts are used as growing media components. Although composting can be effective in reducing the concentration of pathogens, *E. coli* and *S. enterica* could survive in low concentrations (Shepherd et al., 2007, 2010), probably due to heterogeneity of raw materials and temperature gradients within the compost pile (Agostini et al., 2012). The regrowth of surviving *Salmonella* after composting has been reported in stored composts. Several factors, such as indigenous microflora, moisture content, bio-available nutrients and storage temperature, can have an effect on the incidence of zoonoses (Sidhu et al., 2001; Chen et al., 2018). Sidhu et al. (2001) observed that indigenous microorganisms are the most important factor controlling regrowth of *Salmonella* in composted biosolids, reducing in effectiveness with length of storage. Since contaminated water is also an important source of pathogens, attention should be paid during coco pith extraction and washing to reduce salt content. Some growing media producers reduce the risk of contamination by washing the coir after import from countries of origin.

*E. coli* has been isolated from composted municipal solid wastes from different sources. In most cases the concentration of bacteria was very high, for example, 22 000–1 347 333 CFU $g^{-1}$, and could exclude the use of these composts for the retail trade where manual handling is involved (Dimambro et al., 2007). Soobhany (2018) observed that *E. coli* was undetectable in the vermicompost (generated by *Eudrilus eugeniae*) obtained from the same raw material that produced an infected compost. Extra research is needed to understand the mechanism of pathogenic bacteria control in vermicomposting.

*Legionella longbeachae* have been isolated in composted plant material and composted pine bark used as component in growing media in Australia (Hughes and Steele, 1994) and the United Kingdom, where it was associated with illness among users (Currie et al., 2014). Moreover, three cases of Legionnaires' disease caused by *L. longbeachae* were associated with exposure to potting compost in Scotland (Pravinkumar et al., 2010). The use of purchased compost products was also strongly associated with Legionnaires' disease by Kenagy et al. (2017). The risk of Legionnaires' disease in compost seems to be related to activities that influence its aerosolization and direct contact with face or mouth, presumably leading to inhalation of particles contaminated with *Legionella* bacteria. Advices should be printed on compost bags to avoid contact with the face. The low levels of *Legionella* usually found in composts may explain the low incidence of contracting legionellosis from growing media with composts as components.

Strict limits have been suggested by authorities (e.g. Ecolabel UE) or quality standard (e.g. PAS100) for compost or growing media. Limits for *E. coli* for use in growing media have been set at $10^3$ CFU $g^{-1}$ fresh weight, whilst *Salmonella* spp. has to be absent in 25 g of fresh matter. From a practical point of view, rapid methods to check for human pathogens in growing media components are necessary for growing media manufacturer to avoid purchasing infected components.

In general, growing media are free of plant pathogens and do not require prior sanitization treatment.

According to a survey conducted in the United States and Canada, only 5% of the nurseries routinely treat growing media (Landis, 1989). Treatments are relatively expensive and can double the price of growing media.

Disinfection of growing media is essentially based on the use of heat and chemical fumigants. However, these treatments eliminate not only deleterious microorganisms but also potentially beneficial microbial population, disrupting the equilibrium of the microbiota and putting the root zone at risk of pathogen invasion (Postma, 2009). In addition, these techniques are not seen as environmentally friendly, as reviewed by Panth et al. (2020), who highlight the trend for live growing media.

Pasteurization involves several techniques aimed at increasing the growing media temperature at a thermal threshold and for a minimum time interval sufficient to inactivate most root pathogens and pests. Different heat sources can be used for growing media pasteurization, such as moist heat from steam, aerated steam, hot water, microwave ovens, or dry heat from flame or electric pasteurizers (Landis, 1989). The choice of the best technique is dependent upon operational conditions and economic drivers, but in all cases, it is important to ensure sufficient and uniform heating of growing media in order to reach a sufficient and well-distributed elimination of pathogens/pests. On the other hand, exceeding a certain maximum heat threshold should be avoided in order to preserve the activity of beneficial microorganisms and prevent the formation of toxic compounds (Landis, 1989). Heat treatment of growing media needs to reach a target temperature for a specific minimum period of time that differs according to the specific organism. The lowest temperature level of 50°C for at least 30 min is required to control nematodes, whereas the highest temperature (100°C) is necessary to destroy a heat-resistant plant virus such as tobacco mosaic virus.

Innovative methods for growing media disinfection include the use of non-thermal plasma. Plasmas are neutral ionized gases which have a high proportion of reactive species ($OH^-$, $O_2^-$, $H_2O_2$, or $O_3$) (Moreau et al., 2008) that kill microbes but do not exert residual negative effects on humans or the environment because of the short life span of the reactive species (Akishev et al., 2008). Non-thermal plasma is generated at a low voltage of 1–2 kV, low current density of ~1 mA $cm^{-2}$ and low pressure with an average electron energy of ~1 eV and gas temperatures between 25°C and 50°C. Non-thermal plasma has been used for medical applications and food decontamination (Pankaj et al., 2018) and recently to control fungal pathogen on seeds (Jo et al., 2014). The most interesting results for microbial control refer to bacteria and fungi (Sakudo et al., 2019). To extend the use of plasma to growing media disinfection, further research is necessary to evaluate plasma types, power sources, exposure time and gas species.

## 3 Biological activity in growing media: bulk and rhizosphere microbial communities

In the substrate-plant environment we can identify two important zones: the bulk and the rhizosphere. The bulk is the growing medium without plants and thus without any interaction with the roots, whereas the rhizosphere is the area immediately around the roots. Pang et al. (2017) suggested a separation of the root zone into:

- the rhizosheath, the medium that physically adheres to the root system (less than 1 mm from the root); and
- the rhizosphere, the medium influenced by the root (within 1 cm of the root).

Alsanius and Wohanka (2019) called these two zones the rhizoplane and ectorhizosphere, respectively.

Bulk microbial densities in growing media materials vary. Inorganic materials such as mineral wool, perlite or expanded clay may be considered 'sterile' given exposure to high temperature during production. Other materials such as peat may have a low population of organisms. Peatland conditions (acidity and water saturation) do not favour microbial growth, so untreated peat is usually poorly colonized by microorganisms prior to use, with growth of fungi favoured over growth of bacteria. According to Martínez et al. (2013) colonization is $10^4$ and $10^5$ CFU $mL^{-1}$ for bacteria and fungi, respectively. Coir fibre may have microbial counts two to three orders of magnitude higher than peat, at $10^6$ and $10^6$ CFU $mL^{-1}$ for bacteria and fungi, respectively (Martínez et al., 2013). Colonization is even higher in composted materials, including barks (Dickinson and Carlile, 1995) and other raw materials such as municipal solid waste.

Although the ectorhizosphere acquires microorganisms from the bulk, the microbiome of the rhizosphere (rhizobiome) can be very different from the bulk microbiome. Grunert et al. (2019) found that bacterial community composition in the tomato rhizosphere and bulk growing media were different. In addition, there were significant differences between microbial community structure and activity in the ectorhizosphere and rhizoplane. Once plants are introduced, growing media are colonized with to different degrees by a new range of benign or harmful fungi and bacteria. As an example, the rhizobiome (bacterial and fungal community) in peat-based strawberry cultivation changed in composition and richness over time. In particular, the fungal community composition changed mainly during the first week of plant growth and the bacterial community changed particularly during the first 6 weeks (De Tender et al., 2016). A similar result was found in tomato grown in mineral wool (Calvo-Bado et al., 2006).

The rhizosphere microbial community is the result of a complex exchange of signals between microorganisms and the plant, mainly related to root exudates (Pang et al., 2017). Root exudates may include sugars, amino acids and carboxylic acids characterized by water solubility and a low molecular weight (Alsanius and Wohanka, 2019). Plants can exude up to 21% of the carbon fixed by photosynthesis (Meier et al., 2020) which then feeds the microbial community and influences their activity and diversity (Mendes et al., 2011). Exudate concentration changes depending on the distance from the root and can be influenced by cultivation method (water vs. substrate culture), substrate water content and irrigation management (Alsanius and Wohanka, 2019).

The rhizobiome can be modified not only by the interaction between plant activity and the bulk microbial community but also by physicochemical characteristics of the growing media. Moisture, pH and nutrient contents can modify the abundance of some bacteria (e.g. *Actinobacteridae,* rhizobacteria *and Agrobacterium* spp.) (Grunert et al., 2016). The competition for nutrients between plants and microorganisms in growing media can influence the composition of the microbial community. The use of organic fertilizers in organic growing media could be important because of the saprophytic microbial community that may influence the nutrient status in the rhizosphere (Carlile and Wilson, 1991).

For efficient use of organic nitrogen as fertilizers, it is necessary to control the modification of organic nitrogen into forms absorbed by the plants. Heterotrophic microorganisms in the medium consume organic components for energy and immobilize and mineralize nitrogen (Robertson and Groffman, 2007). The first product of mineralization, ammonium, can be taken up by plants or transformed to nitrite and then to nitrate by chemoautotrophic microorganisms. The observation that the bacterial community involved in nitrogen mineralization and immobilization is different within growing media and its components leads to the conclusion that microbial activity for organic nitrogen utilization (towards a complete nitrification) could be achieved by choosing and mixing different components. However, the interaction between physical and chemical characteristics needs to be considered for future research.

Microbial taxa in the rhizosphere can be stimulated by specific amendments or additives, such as compost, biochar or chitin. In a peat-based media, the addition of chitin significantly increased three genera of fungi (*Morteriella*, *Lecanicillium* and *Pseudogymnoascus*) involved in biocontrol and/or chitin degradation (Debode et al., 2016; De Tender et al., 2019). The increase in plant growth-promoting rhizobacteria (PGPR) and plant growth-promoting fungi (PGPF) in the soil and/or rhizosphere due to the addition of chitin has also been documented (Sarathchandra et al., 1996; Cretoiu et al., 2013). It is well known that a diverse and stable microbial community will help plants resist environmental stresses (Loreau et al., 2001) and root pathogen attack.

The previous discussion shows the microbial life in growing media can represent both an opportunity for soilless cropping systems and potentially an obstacle. Growing media with high microbial activity may suppress both pathogenic microorganisms and plant-beneficial bacteria or fungi inoculated to control plant diseases, improve plant growth or nutrient use efficiency (e.g. mycorrhizal fungi or *Trichoderma*). In general, the survival of inoculated organisms is low; mostly less than 1% survival following incubation (Trifonova et al., 2009). The main reason for low survival is the inability of the new microorganisms to compete with the indigenous microbiota (Van Gerrewey et al., 2020). For this reason, the use of 'sterile' growing media may help in inoculating plant with beneficial microorganisms (Trifonova et al., 2009).

## 4 Biological stability, phytotoxicity and nitrogen immobilization

Organic matter is naturally subjected to degradation by the activity of microorganisms under aerobic conditions. The rate of degradation depends on material properties and on other factors such as water, oxygen and temperature which influence microbial activity. A lower degradation rate corresponds to a higher degree of biological stability.

When unstable organic materials are used as growing media components, organic matter degradation causes growing decay related to nitrogen (N) immobilization, phytotoxicity, changes in cation exchange capacity (CEC), salinity increase, volume reduction (shrinkage) and a decrease in air filled porosity due to particle size reduction (Lemaire, 1995). Organic materials mainly composed of easily degraded polysaccharide components such as hemicelluloses or cellulose and low lignin content are readily degraded.

The growing use of organic growing media makes it particularly important to understand their biostability and phytotoxicity. In general, growing media should maintain their properties over a length of time corresponding to the crop cultivation cycle. Growing media should be biostable enough for bedding plant cultivation but not necessarily for shrub cultivation.

Unstable and/or phytotoxic organic materials could be composted before use as growing media components. During composting, organic material undergoes biological decomposition under controlled conditions and is transformed into a stable humus-like product (Golueke, 1972). Extensive research has been conducted to study the composting process for many organic materials and to develop methods to evaluate the stability, maturity or phytotoxicity of composts (Azim et al., 2018).

Respiration rate has often been considered a reliable method to measure biostability, involving the measurement of $O_2$ consumption or $CO_2$ production. This method shows a high stability of peat-based growing media compared to peat-free growing media. The oxygen uptake rate (OUR) in peat is about 2 mmol $kg^{-1}$ $h^{-1}$ (Veeken et al., 2003; Blok et al., 2017) compared to 10–17 mmol $kg^{-1}$ $h^{-1}$ in some composted organic materials (Veeken et al., 2003). Even measuring the respiration rate as $CO_2$ release, different kinds of peat show lower values of $CO_2$ released (about 20 mg $CO_2$-C $kg^{-1}$ $d^{-1}$) compared to coir fibre (about 82 mg $CO_2$-C $kg^{-1}$ $d^{-1}$) and green waste compost (about 138 mg $CO_2$-C $kg^{-1}$ $d^{-1}$) (Grunert et al., 2016). The higher biostability of peat compared to peat-free media can be explained by high lignin content and initially low microbial population and activity in peat.

Phytotoxicity depends on the chemical composition of growing media. High mineral content and the presence of harmful organic compounds such as phenols, terpenes and organic acids can cause excessive salinity and pH, nutritional disorders and enzymatic or hormonal metabolic alterations (Ortega et al., 1996). The use of fresh or composted organic materials (e.g. forestry products, olive mill, grape marc or manure) as growing media components can cause phytotoxicity problems affecting the growth of plants (Gruda et al., 2009, 2013). Many factors may influence phytotoxin contents. Tree species, age, harvest time, soil type and geographical region are for example factors affecting bark phytotoxicity (Raviv et al., 2002).

To reduce or eliminate phytotoxicity, bark is usually composted or aged (Gruda et al., 2009; Chemetova et al., 2019). Some phytotoxins such as salts and soluble organic compounds can be removed by washing (Olejar et al., 2019). Gruda et al. (2009) reported improvements in germination rate and radicle growth, and thus reduced phytotoxicity, after washing/leaching pine tree bark-based substrates. Chemetova et al. (2018) achieved effective phytotoxicity removal from *Eucalyptus globulus* bark using hydrothermal treatments. Hydrothermal treatment at 40°C for 70 min produces the lowest C mineralization and N-immobilization rates. Several authors have reported that the growth of non-pathogenic fungi (Sampedro et al., 2007) and the activity of some bacteria (Vela-Cano et al., 2019) can decrease phytotoxicity due to the presence of phenolic compounds. More recently, Fornes et al. (2020) observed a reduction of $NH_4^+$ phytotoxicity of composts, probably due to the ability of biochar to retain cations such as $NH_4^+$ in surface anionic charge sites (Laird et al., 2010).

In growing media containing a large amount of degradable carbon, such as hemicelluloses or cellulose, and a small amount of nutrients for microorganisms, microbial activity may remove nutrients from the root-growing media nutrient solution. N-immobilization occurs when organic materials with high C/N ratios are degraded by microorganisms. N-immobilization is influenced by the biochemical composition of the organic matter due to a positive correlation between C mineralization rate and the holocellulose/lignin ratio during the composting process (Francou et al., 2008).

Although nitrogen immobilization occurs in wood fibres, bark, and straw-based materials, important differences can be observed. Spruce bark for example has a higher nitrogen immobilization than pine bark (Prasad and Carlile, 2009) probably due to the much higher concentration of free sugars in spruce bark (Naasz et al., 2009). Vandecasteele et al. (2018) observed higher N-immobilization in organic materials with a high holocellulose/lignin ratio of about 15–20 and proposed a linear regression to predict the risk of N-immobilization based on pH-$H_2O$, C-$H_2O$, hemicellulose (%), lignin (%), holocellulose/lignin and C/N. According to this model, reed and miscanthus straw and flax shives show a higher risk of N-immobilization than peat and wood fibres. Nitrogen immobilization can occur in bagged media that can be stored for a considerable length of time by retailers (Carlile, 2004; Handreck and Black, 2010).

N consumption by microorganisms could lead to plant nutrient deficiencies. Gruda et al. (2000) reported a significant N-immobilization in 100% wood fibres with a consequent reduction in tomato plant growth. Successful cultivation systems based on the use of growing media susceptible to N-immobilization require strategies in N management to promote or maintain desired nutrient amount for plant demands. The main strategy is the addition of N with nutrient solution or solid fertilizer from the beginning of cultivation to counteract microbial N-immobilization (Handreck, 1993; Gruda et al., 2000). In wood-based growing media, treating the substrate before planting (precharging or impregnation) by controlled-released fertilizer or by cheaper urea has been suggested (Niemiera et al., 2014). However, urea can increase pH. The urea precharge rate should therefore be based on the initial pH of the growing media (Niemiera et al., 2014).

Another approach has been proposed by Vandecasteele et al. (2018), based on inhibiting microbial activity in organic material by modifying the pH out of the optimal range or by using biocontrol fungi. However, inoculation of extruded miscanthus fibres with biocontrol fungi (*Trichoderma harzianum* T22, *Metarhizium anisopliae* F52, *Chaetonium globosum* IHEM 3826, *Gliocladium catenulatm* J1446, and *Trichoderma asperellum* T34) did not reduce the risk of N-immobilization, whereas an effective reduction of microbial activity and the N-fixation was obtained by acidification with acetic or citric acid at low values (pH 2.5 and 4.7).

Composting could be used to reduce N-immobilization in hard wood materials by modifying the C pool and lowering the C/N ratio (Prasad, 1997). Modifying the C pool to overcome N-immobilization can be achieved by impregnation of the organic growing media matrix with sulphuric acid in the presence of hot gases; this reduces the decomposable cellulose and, hence, microbial activity and need for nitrogen (Jackson et al., 2009). Stabilisation of the C pool of the growing media can be achieved by adding biochar as growing media constituent (Blok et al., 2017). When produced from nutrient-poor feedstock (e.g. wood), biochar shows poor nutrient content (Gaskin et al., 2008) and an outstanding structural stability, making it highly recalcitrant against microbial decay and so retarding microbial N demand (Steiner and Harttung, 2014).

Growing media for both professional and hobby users could be stored in bales or bags over a long period before use. Although microbial populations are low, stored peat may undergo microbially mediated processes called self-heating and passive heating (Terhoeven-Urselmans et al., 2007) that reduce growing media quality. Self-heating (up to 75°C) may alter the chemical structure of the peat, increase some nutrients and release phytotoxic substances (Wever and Hertogh-Pon, 1993). In contrast to self-heating, passive heating occurs mostly under low oxygen conditions and with temperatures (e.g. 21–52°C) lower than self-heating levels.

A common feature in both processes is the significant increase in $NH_4^+$ and a decrease in $NO_3^-$ content in the growing media, presumably caused by nitrification inhibition at increased temperatures, as observed in pine bark (Walden and Wright, 1995). Self-heating can influence physical, chemical and microbiological characteristics of growing media that may inhibit seed germination and crop growth (Terhoeven-Urselmans et al., 2007; Cattivello, 2009). In a study conducted on peat with different origins and degrees of humification, Terhoeven-Urselmans et al. (2007) concluded that to reduce the negative effects of passive heating, growing media should not be exposed to temperatures higher than 35°C for more than six days during transport and storage.

The possible negative effects of these phenomena on growing media quality imply the need for reliable testing procedures to detect pre-existing heating influencing the usability of growing media. Since self-heating is associated with a transient shift of growing media microbioma toward a thermophilic microbial community, the analysis of phospholipid fatty acid modification in peat and composts has been proposed as a marker of self-heating-driven modifications of growing media (Klamer et al., 2000; Ranneklev and Bååth, 2003). Ranneklev and Bååth (2003) detected the presence of two PFLAs in self-heated peat, indicating the presence of bacteria belonging to the genus *Alicyclobacillus*, whose thermophilic, aerobic, heterotrophic and acidophil growth mirrored the environmental conditions experienced in the heated peat.

## 5 Pathogen and disease suppression and biocontrol

Most of the growing media are initially generally pathogen-free. However, during the growing cycle they can become rapidly infected with potential significant economic losses, especially in closed, recirculating systems (Raviv, 2008). The main cause for growth of root pathogens and pests is poor activity and biological diversity of the microbiota colonizing these substrates. This allows some parasites, even those rarely threatening soil-grown crops, to rapidly colonize and spread into the rhizosphere (van der Gaag and Wever, 2005). This also affects peat moss, the most widely used growing media constituent, because its highly stable matrix does not support microbial activity and thus can promote plant root pathogens (Krause et al., 2001a; Veeken et al., 2005).

Reducing the risk of disease propagation in closed soilless systems focuses on preventing pathogen colonization and dispersal within the crop. However, this is difficult since many root-infecting pathogens can be easily dispersed through the irrigation system, depending on operational conditions (Stewart-Wade, 2011). A key challenge in achieving 'sustainable intensification' in soilless systems is to find growing media with an improved ability to limit the diffusion of root pathogens, pests (substrate 'suppressiveness'), so reducing use of synthetic pesticides (Gruda, 2019).

The increased activity and diversity of the microbiome colonizing organic growing media involves multiple biological and physico-chemical mechanisms responsible for 'pathogen suppression' or 'disease suppression' (Table 1). The first mechanism is the ability of a growing medium to limit the inoculum density of a pathogen. The second mechanism is the capacity of a growing medium to limit disease development in the presence of both the host plant and the pathogen (Saadi et al., 2010). The crucial involvement of the root zone microbiota in conferring suppressiveness in growing media is highlighted by the observation that heat sterilization often eliminates suppressiveness (Boehm and Hoitink, 1992; Wohanka et al., 2010).

Compost is a key alternative organic substrate to peat. The addition of composts to growing media has emerged as one of most interesting ways of tailoring the biological properties and the corresponding rhizosphere

**Table 1** Proposed mechanisms for suppressiveness of composts against root pathogens

| Mechanism of action | Effect | Description | References |
|---|---|---|---|
| Antibiosis | Pathogen suppression | The mechanism refers to the inhibition by the release of antibiotic molecules. Antibiotics may include nitrogen-containing heterocycles (indole derivatives, phenazines, pyrrol-type derivatives and pyo-compounds) or non-nitrogen-containing compounds such as 2,4-diacetylphl-oroglucinol. Despite some antibiotics with broad spectra may be detrimental for beneficial microorganisms, most of them show selectivity against pathogens. Root pathogen inhibition through antibiosis has been reported in several microorganisms, including strains of *Bacillus*, *Pseudomonas* and *Gliocladium virens.* | Lumsden et al. (1992), Dowling et al. (1996), Walker et al. (1998), Boulter et al. (2000) |
| Physical protection of infection sites | Disease suppression | Non-pathogenic microbes inhabiting the rhizoplane operate as a physical barrier preventing the pathogen penetration inside the roots. It has been reported that through this mechanism of action, non-pathogenic strains of *Fusarium oxysporum* limit the occurrence of tomato crown and root rot caused by *Fusarium oxysporum* f. sp. *radicis-lycopersici.* | Pharand et al. (2002) |
| Nutrient competition | Pathogen/ disease suppression | Under conditions of elevated microbial activity, the consumption of nutrients by beneficial microorganisms, mainly carbon and iron, acts to prevent pathogens germination, keeping them in an inactive state. The clamydospores of several *formae spacialis* of pathogenic *Fusarium oxysporum* need an adequate concentration of iron to germinate in the growth substrate. Beneficial microorganism such as *Trichoderma hartianum* or non-pathogenic *Pseudomonas* spp. synthetize siderophores actively chelating the ferric iron for their own nutrition, so preventing pathogen germination. Several actinomycetes were shown to induce autolysis of pathogenic hyphae in response to nutrient starvation. Moreover, extracellular heterolytic enzymes may be produced by a variety of microorganisms in order to reduce the number of competitors for nutrient sources. | Boulter et al. (2000), Maher et al. (2008) |
| Hyperparasitism/ predation | Pathogen suppression | Several beneficial fungi and bacteria are able to hyperparasitize pathogen and pests inhabiting the rhizosphere environment. Hyperparasitism is commonly associated with the production of heterolytic enzymes such as amylases, chitinases, glucanases and proteases degrading the structural matrix of the protective structures of fungi or arthropods, so allowing the parasite penetration inside the targeted microorganism. These BCAs can have either a biotrophic or a necrotrophic life style, depending on their ability to use dead cells or substrate organic matter as alternative nutrient sources. In the second case, BCAs are more favourable to mass production for commercial purposes. The referenced hyperparasites include the bacterium *Bdellovibrio bacteriovorus* and the mycoparasites belonging to the genera *Trichoderma*, *Clonostachys* and *Coniothyrium*. | Köhl et al. (2019) |

| | | | |
|---|---|---|---|
| Futile propagules germination | Pathogen suppression | It is thought that many plant-derived composts suppress soil-borne pathogens by releasing substances mimicking the root exudates of the host plants. This often results in chemical signalling triggering an incorrect identification of the host plant by the pathogen, which is followed by a lowering of the inoculum density if the pathogen does not come in contact with its host. When pathogens having a necrotrophic life phase are concerned, the saprotrophic microbiota can actively compete for nutrient sources, leading to autolysis of pathogen hyphae in response to nutrient starvation. | Boulter et al. (2000), Maher et al. (2008) |
| Induced systemic resistance (ISR) | Disease suppression | Also known as systemic acquired resistance (SAR), it indicates a state of enhanced defensive capacity developed by a plant when appropriately stimulated. Through colonization or stimulation of roots, many BCAs exhibit a range of direct and indirect mechanisms that may prevent or delay infection in the host by pathogens. This type of response is stimulated by microorganisms or their metabolites such as salicylic acid or lipopolysaccharides, inducing an upregulation of plant defense genes such as those encoding for chitinases, jasmonic acid or phenylalanine ammonia lyase. Despite some localized types of response are also known, the resistance response is commonly systemic, so that plants acquire an improved tolerance also against foliar pathogens as in the case of tomato or strawberry *Botrytis cinerea*. | Kavroulakis et al. (2006), Harel et al. (2014), Debode et al. (2018) |
| Reduction/ modification of root chemical signals | Disease suppression | Adsorption or metabolization of chemical signals allowing germination and root colonization by pathogens is another putative mechanism related to disease suppression. Both phenomena reduce the amount of free signal perceivable by root pathogens, so reducing the number of propagules able to germinate and colonize the host plant. Moreover, root colonization by beneficial microorganisms such as arbuscular mycorrhizal fungi, results in a shift in root exudates composition, which sometimes lowers the host attractiveness toward root pathogens. However, conclusive evidences to support these positive mechanisms in compost studies are still unavailable. | Norman and Hooker (2000), Scheffknecht et al. (2006), Akhter et al. (2015) |

microbiome in the growing media, so improving the ecotoxycological profile of horticultural products (Raviv, 2013). Beyond physico-chemical modifications, composted feedstocks show increased biological complexity, the main driver for substrate suppressiveness (Raviv, 2008). The suppressive ability of composts is determined by the diversity of microorganisms colonizing the composts after the 'biological vacuum' caused by high temperatures during the thermophilic phase (López-Mondéjar et al., 2011). For this reason suppressiveness is usually lost if further compost sterilization or pasteurization occurs.

The practice of using compost amendments in growing media to control root pathogens goes back to the pioneering works of Hoitink and coworkers from the 1970s (Hoitink et al., 1975). In an early study, Hoitink and Poole (1976) documented the suppressive effects of composted hardwood bark against *Fusarium oxysporum* f. sp. *chrysanthemi*, the causal agent of chrysanthemum root rot. Since many experiments reporting suppressiveness in a wide array of pathosystem-compost combinations has been described, suggesting that suppression of root pathogens is a global phenomenon related to the nature of the composting process (Noble and Coventry, 2005; Hadar, 2011). Growing media amendment with composts has been proposed as an effective means to control diseases caused by key pathogens such as species of *Fusarium* (Borrero et al., 2004), *Macrophomina phaseolina* (Lodha et al., 2002), *Pythium* (Scheuerell et al., 2005), *Rhizoctonia solani* (Termorshuizen et al., 2006), *Sclerotinia minor* (Pane et al., 2011), *Sclerotium cepivorum* (Coventry et al., 2005) or *Verticillium dahliae* (Malandraki et al., 2008).

Nonetheless, inconsistent compost suppressiveness has been often reported, since suppressiveness depends on many factors including feedstock characteristics, curing and storage of composts, as well as pathosystem (Raviv, 2008; Bonanomi et al., 2010). It has been reported that, in general, composts obtained from plant materials have a higher C/N ratio, immobilize nitrogen and inhibit fungal infection in crops. On the other hand, sludge composts have lower C/N ratio, so their higher N content can favour the attack of phytopathogenic fungi such as *Fusarium* (Hoitink et al., 1997; Cotxarrera et al., 2002). Erhart et al. (1999) reported that bark compost suppressed *Pythium* root rot, whereas grape marc either had no effect or promoted the disease. Vermicomposted animal manure suppressed infection of tomato seedlings caused by *Phytophthora nicotianae*, but was unable to control root and stem rot of cucumber caused by *Fusarium oxysporum* f. sp. *radicis-cucumerinum* (Kannangara et al., 2000; Szczech and Smolinska, 2001).

There have been a number of studies of changes in compost suppressiveness as a function of curing duration, with variable results. Blok et al. (2002) tested pairs of composts before and after curing and found that less cured composts were more suppressive to *Phytophtora cinnamoni*, and usually more suppressive to *Pythium ultimum*, but no relation was found between compost age and *Rhizoctonia solani* suppression. In biosolids compost made of sewage sludge and yard waste (1:1 v/v), Danon et al. (2007) found that prolonged curing was associated with lower suppressiveness toward *Sclerotium rolfsii* sclerotia, reflecting a decrease of basal respiration, pH and dissolved organic carbon (DOC) and $NH_4^+$ concentration as well as an increase in $NO_3^-$ concentration and specific UV absorbance values. This implies that excessively stabilized compost is unable so support sufficient microbial activity, with subsequent loss of their suppression activity.

Since suppressiveness is closely linked to microbial abundance, activity and composition, there are also concerns about the effects that prolonged storage periods may have on compost disease suppression. Since compost products may not be sold immediately, the effect of storage on different suppression mechanisms is of crucial importance. Widmer et al. (1998) showed that, when composted municipal waste was stored for more than 3 months, its suppressive capacity against *Phytophthora nicotianae* was reduced, possibly because of the disappearance of the hyperparasitic *Acremonium* spp., a Hypocreaceae fungus able to coil around and penetrate the *P. nicotianae* hyphae. However, Saadi et al. (2010) found that suppressiveness against *Fusarium oxysporum* f. sp. *melonis* (FOM) of mature compost prepared from tomato plants and separated cow manure was maintained or even improved after one year of storage, irrespective of the storage conditions they tested. This long-lasting suppressiveness was attributed to the high stability of mature compost, which generated low-nutrient conditions unable to sustain the saprophytic phase of FOM. Raviv et al. (2005) found that, under similar conditions, propagule population of different strains of *Fusarium* almost could also be eliminated. The intrinsic variability in performance of composts, given the biological processes involved in their formation, makes it difficult to ensure uniform and predictable suppressiveness, seriously hindering their wider use since growers require consistency and reliability.

The addition of biological control agents (BCAs) to composts has proved to be a valid way to strengthen the suppressive characteristics of growing media. During the heat peak of the composting process, enrichment of composts with specific BCAs increases the ability of compost to suppress certain pathogens (Postma et al., 2003). These microorganisms have important advantages over synthetic pesticides, as they have a lower environmental impact, are effective against pesticide-resistant pests, potentially reduce the onset of resistance and can be used in organic farming systems, where use of synthetic fungicides is restricted (Brimner and Boland, 2003; Castaño et al., 2013).

Several BCAs have been tested in soilless systems, such as non-pathogenic *Fusarium* species against *Fusarium oxysporum* f. sp. *basilici* on basil (Fravel and Larkin, 2002), *Gliocladium virens* against *Rhizoctonia solani* and *Pythium ultimum* of zinnia, cotton and cabbage, or *Muscodor albus* against *Rhizoctonia* damping-off of broccoli, (Lumsden and Locke, 1989). Potting mixtures containing dark peat, light peat, or composted pine bark possess inconsistent suppressiveness against *Rhizoctonia*, but the inoculation of these media with *Trichoderma hamatum* 382 and *Chryseobacterium gleum* 299 significantly enhanced disease suppression (Krause et al., 2001b). Similar results were obtained by Ryckeboer (2001), who introduced *T. hamatum* 382 and *C. gleum* 299 to vegetable, fruit, and garden waste compost and to green waste compost, where they were used as an amendment in potting soil to improve suppressiveness towards *Pythium ultimum* and *R. solani*.

The addition of the fungal antagonists *Verticillium biguttatum* and a non-pathogenic isolate of *Fusarium oxysporum* to different plant-derived composts has proven to significantly reduce the disease index in the pathosystems *R. solani*/sugar beet and *F. oxysporum* f. sp. *dianthi*/carnation (Postma et al., 2003). When inoculated with selected microorganisms, composts enhance BCA efficacy against the root pathogens and pests, for example by increasing their persistence in the substrate mixture in the absence of the parasitized organism.

Noble et al. (2018) demonstrated that spent mushroom compost of *Agaricus bisporus* added to a peat/wood fibre matrix (60:40 v/v) improved the efficacy and persistence of the entomopathogenic fungi *Metarhizium brunneum* and *Beauveria bassiana* against the black vine weevil (*Otiorhynchus sulcatus*) in greenhouse potted strawberry plants, possibly providing an alternative nutrient source to insect-derived chitin. This implies the need for an in-depth characterization of the ecophysiology of many BCAs, in order to ensure more consistent performance. The addition of inoculated composts produces extensive modifications of biotic and abiotic substrate characteristics, so contributing to the deployment of a complex array of physico-chemical mechanisms involved in pathogen suppressiveness.

It has been demonstrated that, under high pH conditions in the growth substrate, natural competitors such as *Trichoderma hartianum* or *Pseudomonas* spp. are stimulated to produce siderophore to cope with iron (Fe) unavailability, so increasing competition for this mineral (Elad and Baker, 1985; Blaya et al., 2013). This increases pathogen suppressiveness toward several *formae speciales* of *Fusarium oxysporum*, since the germination capability of their clamydospores is tightly linked to Fe availability in the growth substrate (Srinivasan et al., 1992; Segarra et al., 2010). It has been demonstrated that the addition of BCAs such as *T. hartianum* affect microbial community composition in the growth substrate, modifying availability of carbon, chitin and other oligomers in the substrate, so increasing rhizosphere microbiome competition for energy sources and thus substrate suppressiveness (Debode et al., 2018).

Another reason for the growing popularity of some BCAs is their ability to promote plant growth, probably as a consequence of the hormone substances they synthetize which affect plant physiology, as well as their ability to stimulate plant defence systems by activating plant-induced systemic resistance (Bahramisharif and Rose, 2019; Scotti et al., 2020). This resistance represents the enhanced defensive state of the plant upon appropriate stimulation by one of more beneficial root microorganism and has been demonstrated in crops such as tomato (Harel et al., 2014) and strawberry (Debode et al., 2018). Debode et al. (2018) found that root colonization by *T. hartianum* T22 was associated with an increased expression of genes-encoding chitinases (FaChi2-1, FaChi2-2) and phenylalanine ammonia lyase and to enhanced post-harvest disease suppressiveness against *Botrytis cinerea*.

## 6 Methods to measure biological properties: stability, N-immobilization and phytotoxicity

### 6.1 Biological stability

Many methods have been proposed to measure the biostability of organic growing media based, for example, on the loss in organic matter (Lemaire et al., 1980), on the measure of water-soluble organic matter (Van der Sloot et al., 2003), on forms of carbon (Lineres and Djakovitch, 1993; Hue and Liu, 1995) and on self-heating (Dewar flask). The last method is used in many European countries for compost biostability but has the disadvantage in that it mostly distinguishes only very mature compost from very immature compost (NiChualain and Prasad, 2009).

More recently, organic growing media biostability has often been assessed by respirometric tests that usually involve either measurement of $O_2$ consumption or $CO_2$ evolution. Determination of aerobic biological activity by measuring the oxygen uptake rate (OUR) could be achieved using the Oxitop system. The method consists of 7 days incubation at 30°C. Samples, suspended in nutritive solution (pH 6.5), are stirred in closed vessels (1 L) equipped with $CO_2$ traps (KOH, sodalime) and lids fitted with a pressure transducer (OxiTop® C, WTW). The pressure drops in the headspace in proportion to the amount of oxygen used by decomposers and is recorded twice an hour by the OxiTop device. Respiration rate is measured in the range between 48 h and 120 h and is expressed as following: OUR = mmol $O_2 \cdot kg^{-1}$ OM$\cdot$hour$^{-1}$. Veeken et al. (2003) proposed some values, based on OxiTop® measurements, for the stability of compost from biowaste and green waste as follows: Very stable <5, Stable 5–15, Unstable 15–30, Very unstable >30. Table 2 reports some values of OUR measured for several organic materials used as growing media components.

Measurement of self-heating of composts is performed in a 1.5 L Dewar vessel, at a moisture content adjusted according to the results of a fist test (EN 16086-1: 2012) at a room temperature of 22°C for a maximum of 10 days.

**Table 2** Respiration rates (oxygen uptake rate–OUR) of different organic components

| Organic materials | OUR (mmol $O_2$·$kg^{-1}$ OM $h^{-1}$) |
|---|---|
| Peat | 1.59 ± 0.08 |
| Coir[a] | 1.73 ± 0.53 |
| Almond shells | 1.76 ± 0.00 |
| Wood fibre | 4.10 ± 1.34 |
| Compost | 15.3 ± 2.05 |
| Biowaste compost[b] | 17.0 ± 7.8 |
| Green waste compost[b] | 10.1 ± 1.68 |
| Digestate 1 | 6.31 ± 1.02 |
| Digestate 2 | 9.19 ± 3.20 |

[a] 14 different coir.
[b] After Veeken et al. (2003).

Maximum temperature is an indicator of the state of aerobic biological activity, and it is usually reached after two to five days.

Among indirect measurements of stability water soluble carbon content has been suggested as a simple and reliable indicator. Pigoli et al. (2021) found a significant non-linear correlation between OUR and soluble carbon on 26 samples from different growing media ingredients (compost at different maturity stages, peat, coir, bark etc.), suggesting that the soluble C method can be used to screen organic components of growing media. Although a single method for measuring components biostability might be insufficient to predict its behaviour throughout the cultivation, this is a cheap, easy to use and quick method that gives reproducible results independent from the materials. NiChualain and Prasad (2009) compared three methods (Dewar, OxiTop® and SolvitaTM) to determine the stability of composted materials and concluded that of OUR was effective in determining the stability of compost for use in growing media. The European Regulation (2019/1009) on EU-fertilizing products for evaluating compost stability accepts both the OUR method (EN 16087-1: 2019) and the self-heating test (EN 16087-2: 2012); the former method can also be applied on digestates.

## 6.2 N-immobilization

The most frequently used and accepted method for determining N-immobilization in growing media is the nitrogen drawdown index proposed by Handreck (1993) that involves incubating media (at 22°C) with a $KNO_3$ solution containing 75 or 150 mg $L^{-1}$ N and determining the amount of N left after 4 days. The nitrate nitrogen ($NO_3$-N) levels in the substrate solution are determined at the start and then again after 4 days (the incubation period). The nitrogen drawdown index is based on the ratio between the $NO_3$-N content in the substrate solution on Day 4 and that measured on Day 0. The index can range between 1.0 and 0.0 with a value of 1.0 representing no N loss during the incubation and an index value of 0.0 indicating complete N loss. Sharman and Whitehouse (1993) suggest saturating solutions with concentrations of 150, 200 or 300 mg $L^{-1}$ N be used in N-immobilization tests on materials with high C/N ratios.

### 6.3 Phytotoxicity

Plant growth bioassays evaluate both the ability of the physical environment of growing media to allow root development and the phytotoxicity of some organic and inorganic compounds. Two European Standards meet these objectives: EN 16086-1: 2012 Soil improvers and growing media–Determination of plant response–Part 1: Pot growth test with Chinese cabbage and EN 16086-2: 2012 Soil improvers and growing media–Determination of plant response–Part 2: Petri dish test using cress.

The pot growth test is performed using Chinese cabbage though spring barley, lettuce or other plants may be used. It can be conducted on the solid sample or diluted with sphagnum peat or on a water extract. Three pots/ sample are put in a climate cabinet and seed germination is checked after 5 days. The total weight and number of plants in each pot are determined as soon as the fifth true leave is clearly visible on at least 50% of the plants of the control sample in fertilized sphagnum peat.

The Petri dish test using cress (EN 16086-2: 2012) measures germination and early root elongation of cress seeds (*Lepidium sativum* L.), kept for 3 days in contact with moistened materials (contact method) or with a filter paper on top of perlite thoroughly wetted with water extracts of materials (extract method). Controls consist of limed sphagnum peat (pH 5.5–6.5) and pure water. After 3 days germination percentage and root length are measured in samples and the control and results expressed as Index of Length (IL %) and Munoo-Lisa Vitality Index (MLV %) that take into account both root elongation and germination percentage.

A root elongation test (ISO 11269-1, 2012), originally proposed to determine effects of contaminated soils or contaminated samples on root elongation of plants has been used by Zaccheo et al. (2009) to test the physical properties of growing media and for estimating the impact of cultivation on the structural stability of growing media.

## 7 Case study: the use of green compost in the production of container nursery stock

There has been an increasing interest recently in new organic components for growing media soilless cultivation both for cost and environmental reasons. Agricultural, industrial and urban activities produce a large amount of organic residue that can be recycled. Ornamental nurseries in regions such as Southern Italy need high volumes of organic growing media such as peat. However, 55% of the cost of peat sourced from Lithuania for growers in Southern Italy is the cost of shipment, suggesting the need for cheaper, locally sourced materials.

For these reasons, some experiments were conducted in East Sicily (an important area for nursery production) to produce green compost from pruning wastes as substitute for peat as a growing medium in gazania pot cultivation. Pruning wastes of trees (mainly *Schinus molle*, *Schinus terebinthifolius*, *Olea europaea*, *Cinnamomum camphora*, *Ficus microcarpa*, *Ficus australis*, *Ceratonia siliqua*) and shrubs (*Laurus nobilis*, *Metrosideros excelsa*) were ground into fragments between 2 and 15 cm and then mixed to ensure homogenization of the composting pile. Pile moisture was set at 55–60%, monitored weekly and piles were watered when necessary to ensure microbial activity.

In experiment I, two aeration rates were used during the bio-oxidation phase: 0.45 L $kg^{-1}$ dry matter (DM) $min^{-1}$ for pile 1 and 0.30 L $kg^{-1}$ DM $min^{-1}$ for pile 2. During the maturation phase, piles were turned every two weeks. In experiment II the pile was aerated at a rate of 0.30 L $kg^{-1}$ DM $min^{-1}$ during the bio-oxidative phase. During the maturation phase, the pile was divided into three sub-piles: untreated pile; pile aerated with a rate of 0.20 L $kg^{-1}$ DM $min^{-1}$ and pile turned every 15 days.

Air was blown for 45 min every 60 min through pipes placed at the base of the pile with an intermittent pump. The piles were turned approximately every two weeks in order to improve the homogeneity of the composted material. The bio-oxidation phase finished when temperature was stable at 45 ± 2°C for two weeks. The end of the maturation phase and of the whole process was determined when pile temperature was similar to the environmental temperature and after the evaluation of maturity and stability parameters of compost: the carbon to nitrogen ratio (C/N) was below 20, the degree of humification was above 50% and germination index after 72 h was above 80%.

The results of experiment I showed that the highest aeration rate (0.45 L $kg^{-1}$ DM $min^{-1}$) during the bio-oxidation phase reduced the length of both phases and subsequently the whole composting process (170 days in pile 1 vs. 226 days in pile 2), whereas the application of a moderate aeration (0.30 L $kg^{-1}$ DM $min^{-1}$) promoted the biodegradation process, improving the maturity and stability of compost (Table 3). Nevertheless, the humification parameters and germination indexes recorded at the end of the process indicate a good maturation and stability degree using both aeration rates. From the experiment II it may be concluded that forced aeration static pile, natural ventilation static pile and mechanical turning windrow applied during the maturation phase produced a stable and mature compost (Table 4), but in a shorter period using

**Table 3** Effects of different aeration rates applied during the bio-oxidation phase on compost stability and maturity

| Compost characteristics | P1 | P2 | Significance[a] |
|---|---|---|---|
| Carbon to nitrogen ratio | 17.7 | 12.8 | ** |
| Organic matter (g $kg^{-1}$) | 684 a | 552 b | *** |
| Cellulose to lignin ratio | 0.38 a | 0.16 b | ** |
| Humification index | 0.33 b | 0.42 a | * |
| Humification rate (%) | 31.9 b | 38.1 a | * |
| Humification degree (%) | 75.3 a | 70.6 b | ** |
| Germination index 48 h (%)[b] | 83.3 b | 180 a | *** |
| Germination index 72 h (%)[b] | 116 b | | * |

[a] NS = not significance. *, **, *** represent significance at P ≤ 0.05, 0.01, and 0.001. Values followed by different letters differ according SNK test P = 0.05.
[b] Indicate germination at 48 and 72 h compared to the distilled water (control).

**Table 4** Effects of different aeration methods applied during the maturation phase on stability and maturity of the compost

| Compost characteristics | PT (turned) | PA (aerated) | PU (untreated) | Significance[a] |
|---|---|---|---|---|
| C/N | 12.7 b | 14.2 a | 14.8 a | * |
| $N\text{-}NH_4^+/N\text{-}NO_3^-$ | 0.13 b | 0.26 a | 0.16 b | * |
| Organic matter (g $kg^{-1}$ DW) | 575 b | 620 a | 615 a | *** |
| Cellulose content (g $kg^{-1}$ DW) | 53.8 b | 50.2 b | 79.2 a | ** |
| Humification index | 0.42 | 0.47 | 0.42 | NS |
| Humification rate (%) | 38.1 a | 34.2 b | 35.3 ab | * |
| Humification degree (%) | 70.6 | 68.2 | 70.3 | NS |
| Germination index 48 h (%)[b] | 230 a | 87 b | 206 a | *** |
| Germination index 72 h (%)[b] | 138 | 125 | 133 | NS |

[a] NS = not significance; *, **, *** represent significance at P ≤ 0.05, 0.01, and 0.001. Values followed by different letters differ according SNK test P = 0.05.
[b] Indicate germination at 48 and 72 h compared to the distilled water (control).

forced aeration (170 days in pile aerated and 226 days in pile turned and pile untreated).

In experiment III, the pile 1 compost of experiment II was used as alternative to peat as component of the usually used growing media for gazania cultivation as follows:

- T1 (control): 60% peat, 20% pumice and expanded clay, 10% agricultural soil, 10% mature manure;
- T2: 20% compost, 40% peat, 20% pumice and expanded clay, 10% agricultural soil, 10% mature manure;
- T3: 40% compost, 20% peat, 20% pumice and expanded clay, 10% agricultural soil, 10% mature manure; and
- T4: 60% compost, 20% pumice and expanded clay, 10% agricultural soil, 10% mature manure.

pH of the substrates was measured on saturated extract 1:5 (substrate:water v:v) and a correction of substrate pH was performed when values were out of the optimal range (5.3–6.5).

The correction of pH was carried out using a 3 g $L^{-1}$ solution of calcium carbonate ($CaCO_3$) for T1, and two sulphuric acid solutions ($H_2SO_4$) (2 and 6 mL $L^{-1}$) for T3 and T4, respectively. Volumes added for each solution were related to the water retention of each substrate. After correction, the substrates were washed to leach the calcium carbonate and sulphuric acid in excess. The basal dressing of the substrates was carried out with different amounts of fertilizer to ensure similar nutrient content within treatments. Irrigation was performed by a drip system according to the hydrological characteristics of each substrate. The agronomic result from the experiment III demonstrated the suitability of compost from pruning waste for the partial or total replacement of peat in substrates for *Gazania rigens* cultivation.

The use of different percentages of compost strongly influenced chemical and physical characteristics of the substrates. Therefore, for a correct agronomic evaluation of compost-based growing media, it is essential to adjust substrate management to a high pH, salinity and air-filled porosity that usually characterize compost from green waste. Taking into consideration these factors, compost from pruning waste can be useful to promote peat-free substrates and sustainable disposal of organic wastes, ensuring good production results in ornamental potted plant production. However, the use of a material as growing media constituent in the nursery requires consistent characteristics. Because pruning wastes change in the nursery according to the season, further studies are necessary to study and set up the composting conditions to obtain a homogeneous compost through the year. Finally, it is important to consider suppressiveness and biocontrol.

## 8 Summary and future trends

Soilless culture systems are growing strongly (Savvas and Gruda, 2018). As a result, the demand for growing media requires looking for new and cheaper material, compared to the traditional growing media mainly used for potted plants production, that could be reused for many cultivation cycles. The use of new and renewable organic materials as growing media has required the development of knowledge of its biological characteristics. The biological evaluation of growing media has changed from preference for low microbial activity to a growing media rich in microorganisms creating resilience to root disease attack. As early as the 1990s Carlile and Wilson (1991) noticed the potential for microbial manipulation of growing media. Schmilewski (2000) has suggested the concept of a third generation of growing media ('intelligent growing media') to be developed through to 2020. After 30 years, many new insights have been achieved in the microbiology of the root zone in soilless systems thanks to the rapid development of DNA-based methods for the microbial genome characterization of different growing media. This new knowledge has highlighted the very complex relationship within the microbial community in the root zone-growing media-plant system.

A fruitful area of research might be the use of biocontrol agents against the main pests and diseases of root system. Some experiments have been conducted on the characterization and modification of the microbiome in the root system mainly in peat-based growing media, but a huge amount of information is still necessary for biocontrol at the commercial level. There is a gap in knowledge on the:

- the efficacy of BCAs in different growing media considering the interaction with indigenous microflora (e.g. wood fibre and coir);
- efficacy of biocontrol agents against the key root pathogens and pests and on the interaction between BCAs necessary to control different diseases (BCA consortium);
- effects of BCAs on pH mediated nutrient availability; and
- duration of the biocontrol effects.

The demand of next generation growing media has been partially satisfied by the European Regulation (2019/1009) that allow the commercialization of growing media inoculated with microorganism for improved fertilization. The use of biocontrol agents is not yet included in the EU regulation.

# 9 Where to look for further information

## *9.1 Further reading*

Root zone microbiology of soilless cropping systems, by Alsanius and Wohanka (2019).

Organic soilless media components by Carlile et al. (2019).

The status of biological control of plant diseases in soilless cultivation by Postma (2010).

## 9.2 Key conferences/project/research groups

Project Interreg 2 Seas Mers Zeeën Horti-BlueC (www.horti-bluec.eu).

Growing substrates (https://www.ishs.org/growing-substrates) and Substrate Analysis (https://www.ishs.org/substrate-analysis) ISHS workgroups.

International Symposium on Growing Media, Soilless Cultivation, and Compost Utilization in Horticulture (ISHS).

International Symposium on Growing Media, Composting and Substrate Analysis (ISHS).

# 10 References

Agostini, F., Sundberg, C. and Navia, R. 2012. Is biodegradable waste a porous environment? A review. *Waste Manag. Res.* 30(10), 1001–1015. https://doi.org/10.1177/0734242X12452444.

Akhter, A., Hage-Ahmed, K., Soja, G. and Steinkellner, S. 2015. Compost and biochar alter mycorrhization, tomato root exudation, and development of *Fusarium oxysporum* f. sp. *lycopersici*. *Front. Plant Sci.* 6, 529. https://doi.org/10.3389/fpls.2015.00529.

Akishev, Y., Grushin, M., Karalnik, V., Trushkin, N., Kholodenko, V., Chugunov, V., Kobzev, E., Zhirkova, N., Irkhina, I. and Kireev, G. 2008. Atmospheric-pressure, nonthermal plasma sterilization of microorganisms in liquids and on surfaces. *Pure Appl. Chem.* 80(9), 1953–1969.

Alsanius, B. W. and Wohanka, W. 2019. Root zone microbiology of soilless cropping systems. In: Raviv, M., Lieth, J. H. and Bar-Tal, A. (Eds). *Soilless Culture. Theory and Practice* (2nd edn.). Academic Press, London, pp. 149–194.

Altland, J. 2014. Weed control in herbaceous perennial container production. *GPN* 12-16. https://doi.org/10.16526/j.cnki.11-4762/tp.2014.11.051.

Azim, K., Soudi, B., Boukhari, S., Perissol, C., Roussos, S. and Thami Alami, I. 2018. Composting parameters and compost quality: a literature review. *Org. Agr.* 8(2), 141–158 https://doi.org/10.1007/s13165-017-0180-z.

Bahramisharif, A. and Rose, L. E. 2019. Efficacy of biological agents and compost on growth and resistance of tomatoes to late blight. *Planta* 249(3), 799–813. https://doi.org/10.1007/s00425-018-3035-2.

Blaya, J., López-Mondéjar, R., Lloret, E., Pascual, J. A. and Ros, M. 2013. Changes induced by *Trichoderma harzianum* in suppressive compost controlling *Fusarium* wilt. *Pestic. Biochem. Physiol.* 107(1), 112–119. https://doi.org/10.1016/j.pestbp.2013.06.001.

Blok, C., Van Der Salm, C., Hofland-Zijlstra, J., Streminska, M., Eveleens, B., Regelink, I., Fryda, L. and Visser, R. 2017. Biochar for horticultural rooting media improvement: evaluation of biochar from gasification and slow pyrolysis. *Agronomy* 7(1). https://doi.org/10.3390/agronomy7010006.

Blok, W. J., Coenen, G. C. M., Pijl, A. S. and Termorshuizen, A. J. 2002. Disease suppression and microbial communities in potting mixes amended with composted biowastes. In: Michel, F. J., Rynk, R. F. and Hoitink, H. A. J. (Eds). *Proceedings of the International Symposium Composting and Compost Utilization*. JG Press, Columbus, pp. 630–644.

Boehm, M. J. and Hoitink, H. A. J. 1992. Sustenance of microbial activity in potting mixes and its impact on severity of pythium root rot of poinsettia. *Phytopathology* 82(3), 259–264.

Bonanomi, G., Antignani, V., Capodilupo, M. and Scala, F. 2010. Identifying the characteristics of organic soil amendments that suppress soilborne plant diseases. *Soil Biol. Biochem.* 42(2), 136–144. https://doi.org/10.1016/j.soilbio.2009.10.012.

Borrero, C., Trillas, M. I., Ordovás, J., Tello, J. C. and Avilés, M. 2004. Predictive factors for the suppression of fusarium wilt of tomato in plant growth media. *Phytopathology* 94(10), 1094–1101. https://doi.org/10.1094/PHYTO.2004.94.10.1094.

Bosmans, L., De Bruijn, I., Gerards, S., Moerkens, R., Van Looveren, L., Wittemans, L., Van Calenberge, B., Paeleman, A., Van Kerckhove, S., De Mot, R., Rozenski, J., Rediers, H., Raaijmakers, J. M. and Lievens, B. 2017. Potential for biocontrol of hairy root disease by a *Paenibacillus* clade. *Front. Microbiol.* 8, 447. https://doi.org/10.3389/fmicb.2017.00447.

Boulter, J. I., Boland, G. J. and Trevors, J. T. 2000. Compost: a study of the development process and end-product potential for suppression of turfgrass disease. *World J. Microbiol. Biotechnol.* 16(2), 115–134. https://doi.org/10.1023/A:1008901420646.

Brimner, T. A. and Boland, G. J. 2003. A review of the non-target effects of fungi used to biologically control plant diseases. *Agric. Ecosyst. Environ.* 100(1), 3–16 https://doi.org/10.1016/S0167-8809(03)00200-7.

Broadley, A., Kauschke, E. and Mohrig, W. 2018. Black fungus gnats (Diptera: Sciaridae) found in association with cultivated plants and mushrooms in Australia, with notes on cosmopolitan pest species and biosecurity interceptions. *Zootaxa* 4415(2), 201–242. https://doi.org/10.11646/zootaxa.4415.2.1.

Bruck, D. J. 2005. Ecology of Metarhizium anisopliae in soilless potting media and the rhizosphere: implications for pest management. *Biol. Control* 32(1), 155–163. https://doi.org/10.1016/j.biocontrol.2004.09.003.

Bruck, D. J. 2006. Effect of potting media components on the infectivity of *Metarhizium anisopliae* against the black vine weevil (Coleoptera: Curculionidae). *J. Environ. Hortic.* 24(2), 91–94.

Calvo-Bado, L. A., Petch, G., Parsons, N. R., Morgan, J. A. W., Pettitt, T. R. and Whipps, J. M. 2006. Microbial community responses associated with the development of oomycete plant pathogens on tomato roots in soilless growing systems. *J. Appl. Microbiol.* 100(6), 1194–1207. https://doi.org/10.1111/j.1365-2672.2006.02883.x.

Carlile, W. R. 2004. Growing media and the environment lobby in the UK 1997–2001. *Acta Hortic.* 644, 107–113. https://doi.org/10.17660/ActaHortic.2004.644.12.

Carlile, W. R., Cattivello, C. and Zaccheo, P. 2015. Organic growing media: constituents and properties. *Vadose Zone J.* 14(6), vzj2014.09.0125. https://doi.org/10.2136/vzj2014.09.0125.

Carlile, W. R. and Hammonds, S. J. 2008. Micro-organisms of human health importance in growing media. *Acta Hortic.* 779, 67–74. https://doi.org/10.17660/ActaHortic.2008.779.5.

Carlile, W. R., Raviv, M. and Prasad, M. 2019. Organic soilless media components. In: Raviv, M., Lieth, J. H. and Bar-Tal, A. (Eds). *Soilless Culture: Theory and Practice* (2nd edn.). Academic Press, London pp. 303–378.

Carlile, W. R. and Schmilewski, G. 2010. Life in growing media–the good, the bad and the ugly. Proceedings of the Integrability Peat Sociaal Peat Hortícolas–Life Growing. Media, 7–14.

Carlile, W. R. and Wilson, D. P. 1991. Microbial activity in growing media–a brief review. *Acta Hortic.* 294, 197–206. https://doi.org/10.17660/ActaHortic.1991.294.21.

Case, L. T., Mathers, H. M. and Senesac, A. F. 2005. A review of weed control practices in container nurseries. *Horttechnology* 15(3), 535–545. https://doi.org/10.21273/HORTTECH.15.3.0535.

Castaño, R., Borrero, C., Trillas, M. I. and Avilés, M. 2013. Selection of biological control agents against tomato fusarium wilt and evaluation in greenhouse conditions of two selected agents in three growing media. *BioControl* 58(1), 105–116. https://doi.org/10.1007/s10526-012-9465-z.

Cattivello, C. 2009. Autoriscaldamento. In: Zaccheo, P. and Cattivello, C. (Eds). *I substrati di coltivazione*. Edagricole Edizioni Agricole Agrisole de Il Sole 24 Ore Business Media s.r.l., Bologna, pp. 321–329.

Chemetova, C., Braga, S., Fabiao, A., Gominho, J. and Ribeiro, H. 2019. The potential of aged *Acacia melanoxylon* bark as a source for alternative substrate. *Acta Hortic.* 1266, 389–394.

Chemetova, C., Fabiao, A., Gominho, J. and Ribeiro, H. 2018. Range analysis of *Eucalyptus globulus* bark low-temperature hydrothermal treatment to produce a new component for growing media industry. *Waste Manag.* 79, 1–7.

Chen, Z., Kim, J. and Jiang, X. 2018. Survival of *Escherichia coli* O157:H7 and *Salmonella enterica* in animal waste-based composts as influenced by compost type, storage condition and inoculum level. *J. Appl. Microbiol.* 124(5), 1311–1323. https://doi.org/10.1111/jam.13719.

Cloyd, R. A. 2015. Ecology of fungus gnats (*Bradysia* spp.) in greenhouse production systems associated with disease-interactions and alternative management strategies. *Insects* 6(2), 325–332. https://doi.org/10.3390/insects6020325.

Cloyd, R. A. 2000. Fungus gnat and shore fly management strategies: panel discussion. In: Florists, S. A. (Ed). *Proceedings of the 16th Conference on Insect and Disease Management in Ornamentals*, Alexandria, VA, pp. 57–59.

Cloyd, R. A., Marley, K. A., Larson, R. A. and Arieli, B. 2009. Attractiveness of parboiled rice hulls to the fungus gnat, *Bradysia* sp. nr *coprophila* (Diptera: Sciaridae), adult relative to standard growing medium components. *Hortscience* 44(5), 1366–1369. https://doi.org/10.21273/HORTSCI.44.5.1366.

Cotxarrera, L., Trillas-Gay, M. I., Steinberg, C. and Alabouvette, C. 2002. Use of sewage sludge compost and *Trichoderma asperellum* isolates to suppress Fusarium wilt of tomato. *Soil Biol. Biochem.* 34(4), 467–476. https://doi.org/10.1016/S0038-0717(01)00205-X.

Coventry, E., Noble, R., Mead, A. and Whipps, J. M. 2005. Suppression of allium white rot (*Sclerotium cepivorum*) in different soils using vegetable wastes. *Eur. J. Plant Pathol.* 111(2), 101–112. https://doi.org/10.1007/s10658-004-1420-0.

Cretoiu, M. S., Korthals, G. W., Visser, J. H. M. and Van Elsas, J. D. 2013. Chitin amendment increases soil suppressiveness toward plant pathogens and modulates the actinobacterial and oxalobacteraceal communities in an experimental agricultural field. *Appl. Environ. Microbiol.* 79(17), 5291–5301. https://doi.org/10.1128/AEM.01361-13.

Currie, S. L., Beattie, T. K., Knapp, C. W. and Lindsay, D. S. J. 2014. *Legionella* spp. in UK composts-a potential public health issue? *Clin. Microbiol. Infect.* 20(4), O224–O229. https://doi.org/10.1111/1469-0691.12381.

Danon, M., Zmora-Nahum, S., Chen, Y. and Hadar, Y. 2007. Prolonged compost curing reduces suppression of *Sclerotium rolfsii*. *Soil Biol. Biochem.* 39(8), 1936–1946. https://doi.org/10.1016/j.soilbio.2007.02.012.

Debode, J., De Tender, C., Cremelie, P., Lee, A. S., Kyndt, T., Muylle, H., De Swaef, T. and Vandecasteele, B. 2018. Trichoderma-inoculated miscanthus straw can replace peat in strawberry cultivation, with beneficial effects on disease control. *Front. Plant Sci.* 9, 213. https://doi.org/10.3389/fpls.2018.00213.

Debode, J., De Tender, C., Soltaninejad, S., Van Malderghem, C., Haegeman, A., Van der Linden, I., Cottyn, B., Heyndrickx, M. and Maes, M. 2016. Chitin mixed in potting soil alters lettuce growth, the survival of zoonotic bacteria on the leaves and associated rhizosphere microbiology. *Front. Microbiol.* 7, 565. https://doi.org/10.3389/fmicb.2016.00565.

De Tender, C. A., Debode, J., Vandecasteele, B., D'Hose, T., Cremelie, P., Haegeman, A., Ruttink, T., Dawyndt, P. and Maes, M. 2016. Biological, physicochemical and plant health responses in lettuce and strawberry in soil or peat amended with biochar. *Appl. Soil Ecol.* 107, 1–12. https://doi.org/10.1016/j.apsoil.2016.05.001.

De Tender, C., Mesuere, B., Van der Jeugt, F., Haegeman, A., Ruttink, T., Vandecasteele, B., Dawyndt, P., Debode, J. and Kuramae, E. E. 2019. Peat substrate amended with chitin modulates the N-cycle, siderophore and chitinase responses in the lettuce rhizobiome. *Sci. Rep.* 9(1), 9890. https://doi.org/10.1038/s41598-019-46106-x.

Dickinson, K. and Carlile, W. R. 1995. The storage properties of woodbased peat-free growing media. *Acta Hortic.* 401, 89–96.

Dimambro, M. E., Lillywhite, R. D., Rahn, C. R. 2007. The physical, chemical and microbial characteristics of biodegradable municipal waste derived composts. *Compost Sci. Util.* 15(4), 243–252. https://doi.org/10.1080/1065657X.2007.10702340.

Dowling, D. N., Sexton, R., Fenton, A., Delany, I., Fedi, S., McHugh, B., Callanan, M., Moenne-Loccoz, Y. and O'Gara, F. 1996. Iron regulation in plant-associated *pseudomonas fluorescens* M114: implications of biological control. In: Nakazwa, T., Furukawa, K., Haas, D. and Silver, S. (Eds). *Molecular Biology of Pseudomonads*. ASM Press, Washington, DC, pp. 502–511.

Elad, Y. and Baker, R. 1985. The role of competition for iron and carbon in suppression of chlamydospore germination of *Fusarium* spp. by *Pseudomonas* spp. *Phytopathology* 75(9), 1053–1059. https://doi.org/10.1094/Phyto-75-1053.

EN 16086-1: 2012. Soil improvers and growing media–Determination of plant response–Part 1. Pot Growth Test. Chin. Cabbage 1–25.

EN 16086-2: 2012. Soil improvers and growing media–Determination of plant response–Part 2. Petri Dish Test. Using Cress 1–18.

EN 16087-1: 2019. Soil Improvers and growing media–Determination of the aerobic biological activity–Part 1. Oxygen Uptake Rate.

EN 16087-2: 2012. Soil Improvers and growing media–Determination of the aerobic biological activity–Part 2. Self Heating Test for Compost.

Erhart, E., Burian, K., Hartl, W. and Stich, K. 1999. Suppression of *Pythium ultimum* by biowaste composts in relation to compost microbial biomass, activity and content of phenolic compounds. *J. Phytopathol.* 147(5), 299–305. https://doi.org/10.1111/j.1439-0434.1999.tb03834.x.

European Parliament. 2019. Regulation 2019/1009 laying down rules on the making available on the market of EU fertilising products. Available at: https://eur-lex.europa.eu/legal-content/EN/TXT/?uri=CELEX%3A32019R1009.

Fornes, F., Liu-Xu, L., Lidón, A., Sánchez-García, M., Cayuela, M. L., Sánchez-Monedero, M. A. and Belda, R. M. 2020. Biochar improves the properties of poultry manure compost as growing media for sosemary production. *Agronomy* 10(2), 261.

Francou, C., Linères, M., Derenne, S., Villio-Poitrenaud, M. L. Le and Houot, S. 2008. Influence of green waste, biowaste and paper-cardboard initial ratios on organic matter transformations during composting. *Bioresour. Technol.* 99(18), 8926–8934. https://doi.org/10.1016/j.biortech.2008.04.071.

Fravel, D. R. and Larkin, R. P. 2002. Reduction of fusarium wilt of hydroponically grown basil by *Fusarium oxysporum* strain CS-20. *Crop Prot.* 21(7), 539–543. Https://doi.org/10.1016/S0261-2194(01)00143-0.

Gardiner, R. B., Jarvis, W. R. and Shipp, J. L. 1990. Ingestion of *Pythium* spp. by larvae of the fungus gnat *Bradysia impatiens* (Diptera: Sciaridae). *Ann. Appl. Biol.* 116(2), 205–212. https://doi.org/10.1111/j.1744-7348.1990.tb06600.x.

Gaskin, J. W., Steiner, C., Harris, K., Das, K. C. and Bibens, B. 2008. Effect of low-temperature pyrolysis conditions on biochar for agricultural use. *Trans. ASABE* 51(6), 2061–2069. https://doi.org/10.13031/2013.25409.

GGS. 2015. RAL German Institute for Quality Assurance and characterisation registered society, Sankt Augustin, Germany. *RAL Gütesicherung GZ 250 Substrate für Pflanzen (RAL Quality Assurance GZ 250 Growing Media for Plants).* RAL Deutsches Institut für Gütesicherung und Kennzeichnung e.V, p. 56.

Giuffrida, F. and Consoli, S. 2016. Reusing perlite substrates in soilless cultivation: analysis of particle size, hydraulic properties, and solarization effects. *J. Irrig. Drain. Eng.* 142(2). https://doi.org/10.1061/(ASCE)IR.1943-4774.0000968.

Giuffrida, F., Leonardi, C. and Marfà, O. 2008. Substrate reuse in tomato soilless cultivation. *Acta Hortic.* 801, 1577–1582. https://doi.org/10.17660/ActaHortic.2008.801.195.

Golueke, C. G. 1972. *Composting: A Study of the Process and Its Principles*. Rodale Press, Emmaus, PA.

Gruda, N. 2019. Increasing sustainability of growing media constituents and stand-alone substrates in soilless culture systems. *Agronomy* 9(6), 1–24. https://doi.org/10.3390/agronomy9060298.

Gruda, N., Qaryouti, M. M. and Leonardi, C. 2013. Growing media. In: Baudoin, W., Nono-Womdim, R., Lutaladio, N. B., Hodder, A., Castilla, N., Leonardi, C., De Pascale, S., Qaryouti, M. (Eds). *Good Agricultural Practices for Greenhouse Vegetable Crops–Principles for Mediterranean Climate Areas*. Food and Agriculture Organization, Rome, pp. 271–301.

Gruda, N., Rau, B. J. and Wright, R. D. 2009. Laboratory bioassay and greenhouse evaluation of a pine three substrate used as container substrate. *Europ.J.Hort.Sci.* 74(2), 73–78.

Gruda, N., Von Tucher, S. and Schnitzler, W. H. 2000. N-immobilisierung in Holzfasersubstraten bei der Anzucht von Tomatenjungpflanzen (*Lycopersicon lycopersicum* (L.) Karst. ex Farw.). *J. Appl. Bot.* 74, 32–37.

Grunert, O., Hernandez-Sanabria, E., Vilchez-Vargas, R., Jauregui, R., Pieper, D. H., Perneel, M., Van Labeke, M. C., Reheul, D. and Boon, N. 2016. Mineral and organic growing media have distinct community structure, stability and functionality in soilless culture systems. *Sci. Rep.* 6, 18837. https://doi.org/10.1038/srep18837.

Grunert, O., Robles-Aguilar, A. A., Hernandez-Sanabria, E., Schrey, S. D., Reheul, D., Van Labeke, M. C., Vlaeminck, S. E., Vandekerckhove, T. G. L., Mysara, M., Monsieurs, P., Temperton, V. M., Boon, N. and Jablonowski, N. D. 2019. Tomato plants rather than fertilizers drive microbial community structure in horticultural growing media. *Sci. Rep.* 9(1), 9561. https://doi.org/10.1038/s41598-019-45290-0.

Hadar, Y. 2011. Suppressive compost: when plant pathology met microbial ecology. *Phytoparasitica* 39(4), 311–314. https://doi.org/10.1007/s12600-011-0177-1.

Handreck, K. A. 1993. Use of the Nitrogen Drawdown Index to predict fertilizer nitrogen requirements in soilless potting media. *Commun. Soil Sci. Plant Anal.* 24(17–18), 2137–2151. https://doi.org/10.1080/00103629309368943.

Handreck, K. and Black, N. 2010. *Growing Media for Ornamental Plants and Turf* (4th edn.). University of New South Wales Press, Sydney, NSW, Australia.

Harel, Y. M., Haile Mehari, Z. H., Rav-David, D. and Elad, Y. 2014. Systemic resistance to gray mold induced in tomato by benzothiadiazole and *Trichoderma harzianum* T39. *Phytopathology* 104(2), 150–157. HTTPS://doi.org/10.1094/PHYTO-02-13-0043-R.

Hoitink, H. A. J. and Poole, H. A. 1976. Composted bark medium for control of soil-borne plant pathogens. *Am. Nurserym* 144, 88–89.

Hoitink, H. A. J., Schmitthenner, A. F. and Herr, L. J. 1975. Composted bark for control of root rot in ornamentals. *Ohio Rep. Res. Dev.* 2, 25–26.

Hoitink, H. A. J., Stone, A. G., and Han, D. Y. 1997. Supresiòn de enfermedades de plantas mediante compost. *Agron. Costarric* 21, 25–33.

Hue, N. V. and Liu, J. 1995. Predicting compost stability. *Compost Sci. Util.* 3(2), 8–15.

Hughes, M. S. and Steele, T. W. 1994. Occurrence and distribution of *Legionella* species in composted plant materials. *Appl. Environ. Microbiol.* 60(6), 2003–2005.

Inserra, R. N., Ozores-Hampton, M., Schubert, T. S., Stanley, J. D., Brodie, M. W. and O'Bannon, J. H. 2006. Guidelines for compost sanitation. *Annu. Proc. Soil Crop Sci. Soc. Fla* 65, 31–37.

ISO 11269-1. 2012. Soil Quality–Determination of the Effects of Pollutants on Soil Flora Method for the Measurement of Inhibition of Root Growth.

Jackson, B. E., Wright, R. D. and Alley, M. M. 2009. Comparison of fertilizer nitrogen, nitrogen immobilization, carbon dioxide efflux, and leaching in peat-lite, pine, and pine tree substrates. *Hortscience* 44(3), 781–790.

Jagdale, G. B., Casey, M. L., Grewal, P. S. and Lindquist, R. K. 2004. Application rate and timing, potting medium, and host plant effects on the efficacy of *Steinernema feltiae* against the fungus gnat, *Bradysia coprophila*, in floriculture. *Biol. Control* 29(2), 296–305. https://doi.org/10.1016/S1049-9644(03)00164-6.

Jo, Y. K., Cho, J., Tsai, T. C., Staack, D., Kang, M. H., Roh, J. H., Shin, D. B., Cromwell, W. and Gross, D. 2014. A non-thermal plasma seed treatment method for management of a seedborne fungal pathogen on rice seed. *Crop Sci.* 54(2), 796–803. https://doi.org/10.2135/cropsci2013.05.0331.

Kannangara, T., Utkhede, R. S., Paul, J. W. and Punja, Z. K. 2000. Effects of mesophilic and thermophilic composts on suppression of Fusarium root and stem rot of greenhouse cucumber. *Can. J. Microbiol.* 46(11), 1021–1028. https://doi.org/10.1139/w00-082.

Kavroulakis, N., Papadopoulou, K. K., Ntougias, S., Zervakis, G. I. and Ehaliotis, C. 2006. Cytological and other aspects of pathogenesis-related gene expression in tomato plants grown on a suppressive compost. *Ann. Bot.* 98(3), 555–564. https://doi.org/10.1093/aob/mcl149.

Kenagy, E., Priest, P. C., Cameron, C. M., Smith, D., Scott, P., Cho, V., Mitchell, P. and Murdoch, D. R. 2017. Risk factors for *Legionella longbeachae* Legionnaires' disease, New Zealand. *Emerg. Infect. Dis.* 23(7), 1148–1154. https://doi.org/10.3201/eid2307.161429.

Klamer, M., Eiland, F., Lind, A.-M., Leth, M., Iversen, J. J. L., Söchting, U. and Bååth, E. 2000. Changes in chemical composition and microbial biomass during composting of straw and pig slurry. In: Warman, P. R. and Taylor, B. R. (Eds). *Proceedings of the International Composting Symposium (ICS'99)*. Catholic Biblical Association of America Press, New York, NY, pp. 322–334.

Köhl, J., Kolnaar, R. and Ravensberg, W. J. 2019. Mode of action of microbial biological control agents against plant diseases: relevance beyond efficacy. *Front. Plant Sci.* 10, 845. https://doi.org/10.3389/fpls.2019.00845.

Krause, M., Hoitink, H. and Han, D. 2001a. Spectrum and mechanisms of plant disease control with composts. In: Stoffella, P. J., Kahn, B. A. (Eds). *Compost Utilization in Horticultural Cropping Systems*. CRC Press, pp. 263–273. https://doi.org/10.1201/9781420026221.sec3.

Krause, M. S., Madden, L. V. and Hoitink, H. A. J. 2001b. Effect of potting mix microbial carrying capacity on biological control of Rhizoctonia damping-off of radish and Rhizoctonia crown and root rot of poinsettia. *Phytopathology* 91(11), 1116–1123. https://doi.org/10.1094/PHYTO.2001.91.11.1116.

Laird, D., Fleming, P., Wang, B., Horton, R. and Karlen, D. 2010. Biochar impact on nutrient leaching from a Midwestern agricultural soil. *Geoderma* 158(3–4), 436–442.

Lamb, E., Eshenaur, B., Mattson, N. S. and Sanderson, J. P. 2013. Practical suggestions for managing fungus gnat in the greenhouse. Report. Cornell University, New York State IPM Program, USA.

Landis, T. D. 1989. Disease and pest management. In: Landis, T. D., Tinus, R. W., McDonald, S. E. and Barnett, J. P. (Eds). *The Container Tree Nursery Manual* (Vol. 5). United States Department of Agriculture, Forest Service, Washington, DC, pp. 1–99.

Lemaire, F. 1995. Physical, chemical and biological properties of growing medium. *Acta Hortic.* 396, 273–284. https://doi.org/10.17660/ActaHortic.1995.396.33.

Lemaire, F., Dartigues, A. and Riviere, L. M. 1980. Properties of substrates with ground pine bark. *Acta Hortic.* 99, 67–80.

Lineres, M. and Djakovitch, J. L. 1993. Caracterisation de la stabilité biologique des apports organiques par par l'analyse biochimique. In: Decroux, S. and Ignazi, J.C. (Eds). *Matière organique et agriculture. Actes des 4ème journées d'analyse de terre GEMAS et 5ème forum de la fertilisation raisonnée*. GEMAS, Blois, pp. 159–168.

Lodha, S., Sharma, S. K. and Aggarwal, R. K. 2002. Inactivation of *Macrophomina phaseolina* propagules during composting and effect of composts on dry root rot severity and on seed yield of clusterbean. *Eur. J. Plant Pathol.* 108(3), 253–261. https://doi.org/10.1023/A:1015103315068.

Lohr, D., Woeck, C. and Meinken, E. 2017. Use of ergosterol as an indicator for colonization of peat-based growing media by saprophytic fungi. *Europ.J.Hortic.Sci.* 82(1), 3–11. https://doi.org/10.17660/eJHS.2017/82.1.1.

López-Mondéjar, R., Ros, M. and Pascual, J. A. 2011. Added-value of *Trichoderma* amended compost as biopesticide organic substrates: alternative to traditional organic substrates. *Acta Hortic.* 898, 189–196. https://doi.org/10.17660/ActaHortic.2011.898.23.

Loreau, M., Naeem, S., Inchausti, P., Bengtsson, J., Grime, J. P., Hector, A., Hooper, D. U., Huston, M. A., Raffaelli, D., Schmid, B., Tilman, D. and Wardle, D. A. 2001. Biodiversity and ecosystem functioning: current knowledge and future challenges. *Science* 294(5543), 804–808. https://doi.org/10.1126/science.1064088.

Lumsden, R. D. and Locke, J. C. 1989. Biological control of damping-off caused by *Pythium ultimum* and *Rhizoctonia solani* with *Gliocladium virens* in soilless mix. *Phytopathology* 79(3), 361. https://doi.org/10.1094/Phyto-79-361.

Lumsden, R. D., Locke, J. C., Adkins, S. T., Walter, J. F. and Ridout, C. J. 1992. Isolation and localization of the antibiotic gliotoxin produced by *Gliocladium virens* from alginate prill in soil and soilless media. *Phytopathology* 82(2), 230–235. https://doi.org/10.1094/Phyto-82-230.

Maher, M., Prasad, M. and Raviv, M. 2008. Organic soilless media components, in: soilless Culture. *Theor. Pract.* 459–504. Https://doi.org/10.1016/B978-044452975-6.50013-7.

Malandraki, I., Tjamos, S. E., Pantelides, I. S. and Paplomatas, E. J. 2008. Thermal inactivation of compost suppressiveness implicates possible biological factors in disease management. *Biol. Control* 44(2), 180–187. https://doi.org/10.1016/j.biocontrol.2007.10.006.

Martínez, F., Castillo, S., Borrero, C., Pérez, S., Palencia, P. and Avilés, M. 2013. Effect of different soilless growing systems on the biological properties of growth media in strawberry. *Sci. Hortic.* 150, 59–64. https://doi.org/10.1016/j.scienta.2012.10.016.

Meers, T. L. and Cloyd, R. A. 2005. Egg-laying preference of female fungus gnat *Bradysia* sp. nr. coprophila (Diptera: Sciaridae) on three different soilless substrates. *J. Econ. Entomol.* 98(6), 1937–1942. https://doi.org/10.1093/jee/98.6.1937.

Meier, I. C., Tückmantel, T., Heitkötter, J., Müller, K., Preusser, S., Wrobel, T. J., Kandeler, E., Marschner, B. and Leuschner, C. 2020. Root exudation of mature beech forests across a nutrient availability gradient: the role of root morphology and fungal activity. *New Phytol.* 226(2), 583–594. https://doi.org/10.1111/nph.16389.

Mendes, R., Kruijt, M., De Bruijn, I., Dekkers, E., Van Der Voort, M., Schneider, J. H. M., Piceno, Y. M., DeSantis, T. Z., Andersen, G. L., Bakker, P. A. H. M. and Raaijmakers, J. M. 2011. Deciphering the rhizosphere microbiome for disease-suppressive bacteria. *Science* 332(6033), 1097–1100. https://doi.org/10.1126/science.1203980.

Mofidpoor, M., Krzic, M. and Principe, L. 2009. Effects of peat source and length of storage time in bales on selected properties of Sphagnum peat. *Can. J. Soil Sci.* 89(5), 635–644. https://doi.org/10.4141/CJSS08051.

Moreau, M., Orange, N. and Feuilloley, M. G. J. 2008. Non-thermal plasma technologies: new tools for bio-decontamination. *Biotechnol. Adv.* 26(6), 610–617 https://doi.org/10.1016/j.biotechadv.2008.08.001.

Naasz, R., Caron, J., Legault, J. and Pichette, A. 2009. Efficiency factors for bark substrates: biostability, aeration, or phytotoxicity. *Soil Sci. Soc. Am. J.* 73(3), 780–791.

NiChualain, D. and Prasad, M. 2009. Evaluation of three methods for determination of stability of composted material destined for use as a component of growing media. *Acta Hortic.* 819, 303–310. https://doi.org/10.17660/ActaHortic.2009.819.35.

Niemiera, A. X., Taylor, L. L. and Shreckhise, J. H. 2014. Urea hydrolysis in pine tree substrate is affected by urea and lime rates. *horts* 49(11), 1437–1443. https://doi.org/10.21273/HORTSCI.49.11.1437.

Noble, R. and Coventry, E. 2005. Suppression of soil-borne plant diseases with composts: a review. *Biocontrol Sci. Technol.* 15(1), 3–20. https://doi.org/10.1080/09583150400015904.

Noble, R., Dobrovin-Pennington, A., Fitzgerald, J., Dew, K., Wilson, C., Ross, K. and Perkins, C. 2018. Improving biocontrol of black vine weevil (*Otiorhynchus sulcatus*) with entomopathogenic fungi in growing media by incorporating spent mushroom compost. *BioControl* 63(5), 697–706. https://doi.org/10.1007/s10526-018-9877-5.

Noble, R. and Roberts, S. J. 2003. *A Review of the Literature on Eradication of Plant Pathogens and Nematodes during Composting, Disease Suppression and Detection of Plant Pathogens in Compost*. The Waste and Resources Action Programme. The Old Academy, Oxon, UK.

Norman, J. R. and Hooker, J. E. 2000. Sporulation of *Phytophthora fragariae* shows greater stimulation by exudates of non-mycorrhizal than by mycorrhizal strawberry roots. *Mycol. Res.* 104(9), 1069–1073. Https://doi.org/10.1017/S0953756299002191.

Olejar, K. J., Vandermeer, C., Fedrizzi, B. and Kilmartin, P. A. 2019. A horticultural medium established from the rapid removal of phytotoxins from winery grape marc. *Horticulturae* 5(4), 69. https://doi.org/10.3390/horticulturae5040069.

Olson, D. L., Oetting, R. D. and Van Iersel, M. W. 2002. Effect of soilless potting media and water management on development of fungus gnats (Diptera: Sciaridae) and plant growth. *Hortscience* 37(6), 919–923. https://doi.org/10.21273/HORTSCI.37.6.919.

Ortega, M. C., Moreno, M. T., Ordovas, J. and Aguado, M. T. 1996. Behavior of different horticultural species in phytotoxicity bioassays of bark substrates. *Sci. Hort.* 66(1–2), 125–132.

Pane, C., Spaccini, R., Piccolo, A., Scala, F. and Bonanomi, G. 2011. Compost amendments enhance peat suppressiveness to *Pythium ultimum*, *Rhizoctonia solani* and *Sclerotinia minor*. *Biol. Control* 56(2), 115–124. https://doi.org/10.1016/j.biocontrol.2010.10.002.

Pang, J., Ryan, M. H., Siddique, K. H. M. and Simpson, R. J. 2017. Unwrapping the rhizosheath. *Plant Soil* 418(1-2), 129–139. https://doi.org/10.1007/s11104-017-3358-y.

Pankaj, S. K., Wan, Z. and Keener, K. M. 2018. Effects of cold plasma on food quality: a review. *Foods* 7(1). https://doi.org/10.3390/foods7010004.

Panth, M., Hassler, S. C. and Baysal-Gurel, F. 2020. Methods for management of soilborne diseases in crop production. *Agriculture* 10(1). https://doi.org/10.3390/agriculture10010016.

Paulitz, T. C. and Belanger, R. R. 2001. Biological control in greenhouse systems. *Annu. Rev. Phytopathol.* 39(1), 103–133.

Pharand, B., Carisse, O. and Benhamou, N. 2002. Cytological aspects of compost-mediated induced resistance against *Fusarium* crown and root rot in tomato. *Phytopathology* 92(4), 424–438. HTTPS://doi.org/10.1094/PHYTO.2002.92.4.424.

Pigoli, A., Crippa, L. and Zaccheo, P. 2021. Relationship between dissolved organic carbon and oxygen uptake rate in growing media components. *Acta Hortic*, In press.

Postma, J. 2009. The status of biological control of plant diseases in soilless cultivation. In: Gullino, M. L. (Ed), *Recent Developments in Management of Plant Diseases*. Springer Dordrecht, pp. 133–146.

Postma, J. 2010. The status of biological control of plant diseases in soilless cultivation. In: Gisi, U., Chet, I. and Gullino, M. (Eds). *Recent Developments in Management of Plant Diseases*. Plant Pathology in the 21st Century (Contributions to the 9th International Congress), vol 1. Springer, Dordrecht. https://doi.org/10.1007/978-1-4020-8804-9_11.

Postma, J., Montanari, M. and Van Den Boogert, P. H. J. F. 2003. Microbial enrichment to enhance the disease suppressive activity of compost. *Eur. J. Soil Biol.* 39(3), 157–163. https://doi.org/10.1016/S1164-5563(03)00031-1.

Prasad, M. 1997. Nitrogen fixation of various material from a number of European countries by three nitrogen fixation tests. *Acta Hortic.* 450, 353–362. https://doi.org/10.17660/ActaHortic.1997.450.43.

Prasad, M. and Carlile, W. R. 2009. Practical experiences and background research on the use of composted materials in growing media for the UK market. *Acta Hortic.* 819, 111–124.

Pravinkumar, S. J., Edwards, G., Lindsay, D., Redmond, S., Stirling, J., House, R., Kerr, J., Anderson, E., Breen, D., Blatchford, O., McDonald, E. and Brown, A. 2010. A cluster of Legionnaires' disease caused by *Legionella longbeachae* linked to potting compost in Scotland, 2008–2009. *Eurosurveillance* 15, 1–3. https://doi.org/10.2807/ese.15.08.19496-en.

Ranneklev, S. B. and Bååth, E. 2003. Use of phospholipid fatty acids to detect previous self-heating events in stored peat. *Appl. Environ. Microbiol.* 69(6), 3532–3539. https://doi.org/10.1128/AEM.69.6.3532-3539.2003.

Raviv, M. 2008. The use of compost in growing media as suppressive agent against soil-borne diseases. *Acta Hortic.* 779, 39–50. https://doi.org/10.17660/ActaHortic.2008.779.2.

Raviv, M. 2013. Composts in growing media: what's new and what's next? *Acta Hortic.* 982, 39–52. https://doi.org/10.17660/ActaHortic.2013.982.3.

Raviv, M., Oka, Y., Katan, J., Hadar, Y., Yogev, A., Medina, Sh., Krasnovsky, A. and Ziadna, H. 2005. High-nitrogen compost as a medium for organic container- grown crops. *Bioresour. Technol.* 96(4), 419–427.

Raviv, M., Wallach, R., Silber, A. and Bar-Tal, A. 2002. Substrates and their analysis. In Savvas, D. and Passam, H. (Eds). Hydroponic Production of Vegetables and Ornamentals, Embryo Publications, Athens, pp. 25–101.

Robertson, G. and Groffman, P. 2007. Nitrogen transformations. *Soil Microbiol. Ecol. Biochem.* 3, 341–364.

Ryckeboer, J. 2001. *Biowaste and Yard Waste Composts: Microbiological and Hygienic Aspects–Suppressiveness to Plant Diseases.* Katholieke Universiteit Leuven, Belgium.

Saadi, I., Laor, Y., Medina, S., Krassnovsky, A. and Raviv, M. 2010. Compost suppressiveness against *Fusarium oxysporum* was not reduced after one-year storage under various moisture and temperature conditions. *Soil Biol. Biochem.* 42(4), 626–634. https://doi.org/10.1016/j.soilbio.2009.12.016.

Sakudo, A., Yagyu, Y. and Onodera, T. 2019. Disinfection and sterilization using plasma technology: fundamentals and future perspectives for biological applications. *Int. J. Mol. Sci.* 20(20). https://doi.org/10.3390/ijms20205216.

Sampedro, I., D'Annibale, A., Ocampo, J. A., Stazi, S. R. and Garcia Romera, I. 2007. Solid-state cultures of *Fusarium oxysporum* transform aromatic components of olive-mill dry residue and reduce its phytotoxicity. *Bioresour. Technol.* 98(18), 3547–3554.

Sarathchandra, S. U., Watson, R. N., Cox, N. R., Di Menna, M. E., Brown, J. A., Burch, G. and Neville, F. J. 1996. Effects of chitin amendment of soil on microorganisms, nematodes, and growth of white clover (*Trifolium repens* L.) and perennial ryegrass (*Lolium perenne* L.). *Biol. Fertil. Soils* 22(3), 221–226. https://doi.org/10.1007/BF00382516.

Savvas, D. and Gruda, N. 2018. Application of soilless culture technologies in the modern greenhouse industry–a review. *Europ. J. Hortic. Sci.* 83(5), 280–293. https://doi.org/10.17660/eJHS.2018/83.5.2.

Scheffknecht, S., Mammerler, R., Steinkellner, S. and Vierheilig, H. 2006. Root exudates of mycorrhizal tomato plants exhibit a different effect on microconidia germination of *Fusarium oxysporum* f. sp. *lycopersici* than root exudates from non-mycorrhizal tomato plants. *Mycorrhiza* 16(5), 365–370. https://doi.org/10.1007/s00572-006-0048-7.

Scheuerell, S. J., Sullivan, D. M. and Mahaffee, W. F. 2005. Suppression of seedling damping-off caused by *Pythium ultimum*, *P. irregulare*, and *Rhizoctonia solani* in container media amended with a diverse range of Pacific Northwest compost sources. *Phytopathology* 95(3), 306–315. https://doi.org/10.1094/PHYTO-95-0306.

Schmilewski, G. 2000. Sustainable horticulture with peat–a German case study. Proceedings of the 11th Interface Peat Congress, 6–12 August 2000, Québec. Canadian, 27–30.

Scotti, R., Mitchell, A. L., Pane, C., Finn, R. D. and Zaccardelli, M. 2020. Microbiota characterization of agricultural green waste-based suppressive composts using omics and classic approaches. *Agriculture* 10(3). https://doi.org/10.3390/agriculture10030061.

Segarra, G., Casanova, E., Avilés, M. and Trillas, I. 2010. *Trichoderma asperellum* strain T34 controls Fusarium wilt disease in tomato plants in soilless culture through competition for iron. *Microb. Ecol.* 59(1), 141–149. https://doi.org/10.1007/s00248-009-9545-5.

Sharman, K. V. and Whitehouse, M. 1993. Nitrogen drawdown index as a predictor of nitrogen requirements for nephrolepis in sawdust media. *Sci. Hortic.* 54(1), 23–33. https://doi.org/10.1016/0304-4238(93)90080-A.

Shepherd, M. W., Liang, P., Jiang, X., Doyle, M. P. and Erickson, M. C. 2007. Fate of *Escherichia coli* O157:H7 during on-farm dairy manure-based composting. *J. Food Prot.* 70(12), 2708–2716. https://doi.org/10.4315/0362-028X-70.12.2708.

Shepherd, M. W., Singh, R., Kim, J. and Jiang, X. 2010. Effect of heat-shock treatment on the survival of *Escherichia coli* O157:H7 and *Salmonella enterica* Typhimurium in dairy manure co-composted with vegetable wastes under field conditions. *Bioresour. Technol.* 101(14), 5407–5413. https://doi.org/10.1016/j.biortech.2010.01.147.

Sidhu, J., Gibbs, R. A., Ho, G. E. and Unkovich, I. 2001. The role of indigenous microorganisms in suppression of *Salmonella* regrowth in composted biosolids. *Water Res.* 35(4), 913–920. https://doi.org/10.1016/S0043-1354(00)00352-3.

Simpson, C. V., Gillian, C. H., Altland, J. E., Wehtje, G. R. and Sibley, J. L. 2002. Post-emergence oxalis control in container-grown crops. In Proceedings of the Southern Nursery ASSN. Res. Conference, pp. 376–379.

Soobhany, N. 2018. Preliminary evaluation of pathogenic bacteria loading on organic municipal solid waste compost and vermicompost. *J. Environ. Manag.* 206, 763–767. https://doi.org/10.1016/j.jenvman.2017.11.029.

Srinivasan, U., Staines, H. J. and Bruce, A. 1992. Influence of media type on antagonistic modes of *Trichoderma* spp. against wood decay basidiomycetes. *Mater. Org.* 27, 301–321.

Steiner, C. and Harttung, T. 2014. Biochar as a growing media additive and peat substitute. *Solid Earth* 5(2), 995–999. https://doi.org/10.5194/se-5-995-2014.

Stewart, C. J., Christopher Marble, S. C., Jackson, B. E., Pearson, B. J. and Wilson, P. C. 2018. Effects of three fertilization methods on weed growth and herbicide performance in soilless nursery substrates. *J. Environ. Hortic.* 36(4), 133–139.

Stewart, C. J., Marble, S. C., Jackson, B., Pearson, B. J., Wilson, P. C. and Lauer, D. K. 2019. Influence of pine bark substrate age on performance and leaching of nursery preemergence herbicides. *Hortscience* 54(5), 896–902. https://doi.org/10.21273/HORTSCI13748-18.

Stewart, C. J., Marble, S. C., Pearson, B. J. and Wilson, P. C. 2017. Impact of container nursery production practices on weed growth and herbicide performance. *Hortscience* 52(11), 1593–1600. https://doi.org/10.21273/HORTSCI12241-17.

Stewart-Wade, S. M. 2011. Plant pathogens in recycled irrigation water in commercial plant nurseries and greenhouses: their detection and management. *Irrig. Sci.* 29(4), 267–297. https://doi.org/10.1007/s00271-011-0285-1.

Szczech, M. and Smolinska, U. 2001. Comparison of suppressiveness of vermicomposts produced from animal manures and sewage sludge against *Phytophthora nicotianae* Breda de Haan var. *nicotianae*. *J. Phytopathol.* 149(2), 77–82. https://doi.org/10.1046/j.1439-0434.2001.00586.x.

Terhoeven-Urselmans, T., Bruns, C., Schmilewski, G. and Ludwig, B. 2007. Effects of passive heating and storage on the quality of hand-bagged and pre-packed growing media. *Sci. Hortic.* 115(1), 82–90. https://doi.org/10.1016/j.scienta.2007.07.007.

Termorshuizen, A. J., van Rijn, E., van der Gaag, D. J., Alabouvette, C., Chen, Y., Lagerlöf, J., Malandrakis, A. A., Paplomatas, E. J., Rämert, B., Ryckeboer, J., Steinberg, C. and Zmora-Nahum, S. 2006. Suppressiveness of 18 composts against 7 pathosystems: variability in pathogen response. *Soil Biol. Biochem.* 38(8), 2461–2477. https://doi.org/10.1016/j.soilbio.2006.03.002.

Trifonova, R., Postma, J., Schilder, M. T. and van Elsas, J. D. 2009. Microbial enrichment of a novel growing substrate and its effect on plant growth. *Microb. Ecol.* 58(3), 632–641. https://doi.org/10.1007/s00248-009-9518-8.

Vandecasteele, B., Muylle, H., De Windt, I., Van Acker, J., Ameloot, N., Moreaux, K., Coucke, P. and Debode, J. 2018. Plant fibers for renewable growing media: potential of defibration, acidification or inoculation with biocontrol fungi to reduce the N drawdown and plant pathogens. *J. Clean. Prod.* 203, 1143–1154. https://doi.org/10.1016/j.jclepro.2018.08.167.

van der Gaag, D.J. and Wever, G. 2005. Conduciveness of different soilless growing media to pythium root and crown rot of cucumber under near-commercial conditions. *Eur. J. Plant Pathol.* 112(1), 31–41. https://doi.org/10.1007/s10658-005-1049-7.

Van der Sloot, H. A., Comans, R. N. J. and Bakker, F. P. 2003. Role of Stable Organic Matter, Degradable Organic Matter, Combustible Organic Matter and Dissolved Organic Matter in Waste Characterisation. Report. ECN, Petten, The Netherlands.

Van Gerrewey, T., Ameloot, N., Navarrete, O., Vandecruys, M., Perneel, M., Boon, N. and Geelen, D. 2020. Microbial activity in peat-reduced plant growing media: identifying influential growing medium constituents and physicochemical properties using fractional factorial design of experiments. *J. Clean. Prod.* 256. https://doi.org/10.1016/j.jclepro.2020.120323.

Veeken, A. H. M., Blok, W. J., Curci, F., Coenen, G. C. M., Termorshuizen, A. J. and Hamelers, H. V. M. 2005. Improving quality of composted biowaste to enhance disease suppressiveness of compost-amended, peat-based potting mixes. *Soil Biol. Biochem.* 37(11), 2131–2140. https://doi.org/10.1016/j.soilbio.2005.03.018.

Veeken, A. H. M., de Wilde, V., Hamelers, H. V. M., Moolenaar, S. W. and Postma, R. 2003. OxiTop® measuring system for standardized determination of the respiration rate and N-mineralization rate of organic matter in waste material, compost and soil. Report. Wageningen University, Agrotechnology and Food Science, Wageningen, The Netherlands.

Vela-Cano, M., Cervera-Mata, A., Purswani, J., Pozo, C., Delgado, G. and González-López, J. 2019. Bacterial community structure of two Mediterranean agricultural soils amended with spent coffee grounds. *Appl. Soil Ecol.* 137, 12–20.

Verhagen, J. B. G. M. 2010. Can we get a grip on saprophytic fungi in growing media?. In: International Peat Conference Peat in Horticulture–Life in Growing Media, pp. 25–28.

Wada, S. 2005. Nursery container weeds response to modification of substrate pH, substrate particle size and applied nitrogen form. M. Sc. Thesis. Oregon State University, p. 114.

Walden, R. F. and Wright, R. D. 1995. Interactions of high temperature and exposure time influence nitrification in a pine bark medium. *Hortscience* 30(5), 1026–1028. https://doi.org/10.21273/HORTSCI.30.5.1026.

Walker, R., Powell, A. A. and Seddon, B. 1998. Bacillus isolates from the spermosphere of peas and dwarf French beans with antifungal activity against *Botrytis cinerea* and *Pythium species*. *J. Appl. Microbiol.* 84(5), 791–801. https://doi.org/10.1046/j.1365-2672.1998.00411.x.

Waller, P. L., Thornton, C. R., Farley, D. and Groenhof, A. 2008. Pathogens and other fungi in growing media constituents. *Acta Hortic.* 779, 361–366. https://doi.org/10.17660/ActaHortic.2008.779.45.

Wever, G. and Hertogh-Pon, M. H. 1993. Effects of self-heating on biological, chemical and physical characteristics of peat. *Acta Hortic.* 342, 15–24.

Whitcomb, C. E. 1988. *Plant Production in Containers* (2nd edn.). Lacebark Publications, USA.

Widmer, T. L., Graham, J. H. and Mitchell, D. J. 1998. Composted municipal waste reduces infection of citrus seedlings by *Phytophthora nicotianae*. *Plant Dis.* 82(6), 683–688. https://doi.org/10.1094/PDIS.1998.82.6.683.

Wohanka, W., Molitor, H.-D. and Bruckner, U. 2010. Microbiological and disease suppressive properties of peat from different sources. In: International Peat Symposium Peat in Horticulture-Life in Growing Media. International Peat Society, Amsterdam, pp. 35–43.

Zaborski, E. R. and Cloyd, R. A. 2004. Method for quantitative sampling of fungus gnat larvae in soilless growing media. *J. Econ. Entomol.* 97(2), 678–686. https://doi.org/10.1093/jee/97.2.678.

Zaccheo, P., Crippa, L. and Orfeo, D. 2009. Application of ISO 11269-1 root elongation bioassay for testing the physical properties of growing media. *Acta Hortic.* 819, 427–434. https://doi.org/10.17660/ActaHortic.2009.819.52.

# Part 2

## Technologies

# Chapter 8

# Advances in liquid- and solid-medium soilless culture systems

*Yüksel Tüzel, Ege University, Turkey; and Astrit Balliu, Agricultural University of Tirana, Albania*



## 1 Introduction

Due to increasing problems in soil-based crop production such as loss of arable land (Dockril, 2015), soil degradation (Arsenault, 2014) and, particularly, the impacts of climate change and soil-borne pathogens (Mihajlovic et al., 2017), soilless culture systems (SCS) represent a good opportunity for the horticultural production sector (Sambo et al., 2019). Globally the hydroponics market is projected to reach an annual growth rate of 6.8% for the period 2020–2025 and to increase to €38.9 billion by 2024 (Mordor Intelligence, 2020). Currently, soilless culture is one of the fastest growing sectors in horticulture, used in both simple greenhouses and in more advanced controlled-environment conditions. In this chapter, the most common liquid- and solid-medium systems are described, together with improvements in traditional systems as well as innovative advanced systems. Since production depends on the optimization of water/nutrient supply, the main factors related to water and nutrient optimization are also discussed.

In soilless culture, plants are grown either directly in a nutrient solution (known as water or liquid culture or hydroponics) or in a growing medium with substrate (known as a solid-medium or solid systems) where nutrient elements

http://dx.doi.org/10.19103/AS.2020.0076.08

and water are provided with a nutrient solution. SCS are further categorized as open or closed systems.

The decision to use the appropriate soilless system is affected by the size of the production area (commercial or small scale), the crop itself, resource availability, installation cost and technical skill requirements. During the last two decades, intensive high-tech systems are able to achieve a high yield and quality as well as efficient resource-use has been achieved year-round, particularly for leafy vegetables and medicinal and aromatic plants. However simple operating systems with low investment remain a useful alternative for small-scale production.

## 2 Liquid systems

In liquid systems, which are also known as water culture or hydroponic systems, plants are grown in either a static aerated nutrient solution or in a flowing mist of nutrient solution. In these systems, root aeration, light conditions and plant support affect the success of production. Among liquid systems, floating systems (including deep flow techniques in which plants are placed onto floating materials), the nutrient film technique (NFT) and the aeroponic systems are used in commercial greenhouse production (Savvas and Gruda, 2018). Liquid systems can be categorized as the following types:

- Deep water culture (DWC) system.
- Nutrient film technique (NFT).
- Floating system.
- Deep flow technique (DFT).
- Aeroponics.
- Dry hydroponics.

These systems are described in the sections that follow.

### *2.1 Deep water culture (DWC) system*

The DWC system involves the growing of plants in tanks or reservoirs holding the nutrient solution. A net or a cloth is stretched over the top of the tank. Above it, a thin layer of 1–3 cm of sand, sawdust, straw or peat moss is placed to support the plant. Floating styrofoam to suspend the plant pots from a lid can also be used to hang the roots in the nutrient solution.

DWC systems were first described by Gericke in the *American Journal of Botany* in 1929 (Edwards, 2016). However, the system was not much used commercially due to poor aeration in the solution. The system has been improved by providing aeration and dissolved oxygen (i.e. via pumps) (see Fig. 1) or using recirculating DWC systems in which many containers are connected to a central reservoir.

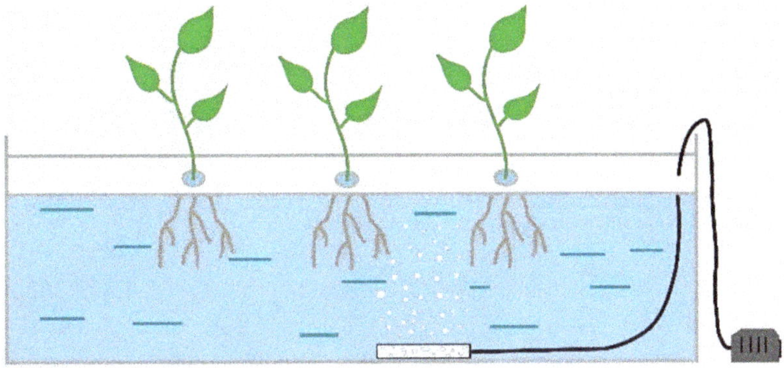

**Figure 1** Deep water culture (drawn by Tunc Durdu).

## 2.2 Nutrient film technique (NFT)

NFT is characterized by the flow of a very shallow nutrient solution stream circulated in troughs, channel or gullies. Plant roots lie in the channels placed with a slope of 1.2–3% (see Fig. 2). The depth of the solution is usually less than 1 cm. The channels are generally made of plastic materials such a polyethylene sheets, polyvinylchloride, polyester with a rectangle or triangle shaped cross-section. The width ranges between 10 cm and 30 cm and depends on the crop and length of the production. Larger plants and long-term crops require wider channels. The length of the channels is 12–20 m. The nutrient solution is introduced from the elevated end of the channel at 1–2 L per minute (see Figs 3 and 4).

The NFT system does not require a substrate and the amount of required nutrient solution is low. Therefore, environmental impact is minimized due to a lower amount of disposal and low water and fertilizer use. However, temperature changes in the nutrient solution, such as high temperature in spring season,for example, in the Mediterranean region, can affect production. Moreover, the system requires a higher degree of management since there is no buffer capacity (Burrage, 1999) and the risk of spread of soil-borne pathogens is high.

**Figure 2** Slope provided by the greenhouse floor (left and middle) and catchment tanks (right) (Photos: Yüksel Tüzel, private collections).

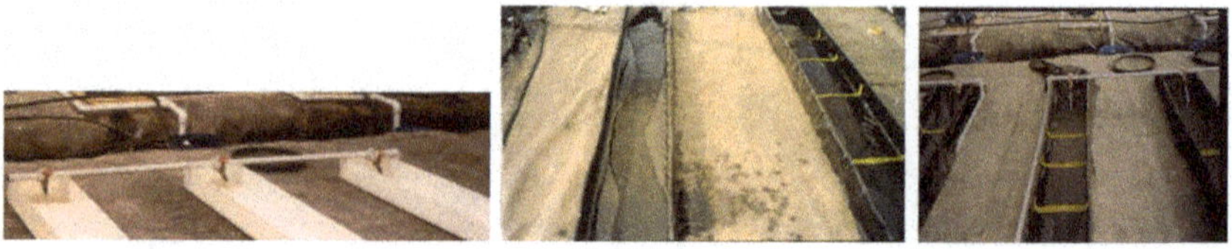

**Figure 3** Polyester channels (left) and flow of nutrient solution (middle and right) (Photos: Yüksel Tüzel, private collections).

**Figure 4** Tomato plants (left) and the roots (right) (Photos: Yüksel Tüzel, private collections).

In a NFT system the nutrient solution is circulated continuously or intermittently. Intermittent flow is practised mainly to control growth and improve oxygenation and at the same time to save electricity and to reduce wear of the pumps (El-Behairy et al., 1991; Economakis, 1993). Comparison of NFT with conventional soil systems in different conditions has been well documented (Abou-Hadid et al., 1999; El-Shinawy et al., 1999; Pardossi et al., 2001; El-Behairy et al., 2001; Valenzano et al., 2008). Compared with closed substrate culture, NFT increased yield in a range between 8% and 22%. However, the increase of solution temperature increases the risk of infections from some pathogens (*Pythium* spp.), which could limit production (Gul et al., 2001). Intermittent flow has been shown to increase early yield but had no effect on final yield (El-Behairy et al., 1991; Economakis, 1993). The effects of the nutrient solution circulation regime on tomato yield changes according to the production season. Intermittent flow resulted in a marked increase compared to continuous flow in terms of early yield in autumn, but there was no significant difference between continuous and intermittent circulation regimes in respect to both early and total yields in spring (Gul et al., 2001).

## 2.3 Floating system

A floating system is a form of water culture in which plants float above the surface of the solution on a lightweight material (i.e. polystyrene or styrofoam) (see Fig. 5). This system is used for lettuce, other leafy crops, baby-leaf vegetable crops and medicinal and aromatic plants. The nutrient solution is contained in tanks with a depth of 15–30 cm (see Fig. 6). Air-compressors or air-pumps are generally used to generate bubbles which affects how much oxygen is dissolved (Butcher, 2016) (see Fig. 7). Different aeration devices, such as venturi oxygenators or hydrogen peroxide application, are also used to ensure an oxygen level of 6 ppm.

**Figure 5** Styrofoam plant supports (left) and lettuce production in floating systems (middle and right) (Photos: Yüksel Tüzel, private collections).

**Figure 6** Leafy vegetables in floating systems in different solution depths and types of tanks/trays (Photos: Yüksel Tüzel, private collections).

**Figure 7** Air bubble generation in the tanks (left) and leafy vegetables transferred to tanks (middle) and harvest time (right) (Photos: Yüksel Tüzel, private collections).

## 2.4 Deep flow technique (DFT)

DFT is a variant of the NFT. The basic procedure is the same, but instead of a thin nutrient solution film, the system keeps the plant roots in the nutrient solution with a depth of 5–15 cm. Higher volume of nutrient solution simplifies control and buffers the temperature. Plants are inserted to the polystyrene rafts/trays placed on the nutrient solution (see Fig. 8). There are differences between NFT and DFT according to the crop and growing season. For instance, DFT system produced higher biomass for dandelion due to more space available for root growth (Gill, 2015) while the choice of DFT or NFT system did not cause significant changes in growing basil (dos Santos et al., 2019).

**Figure 8** Deep flow technique (drawn by Tunc Durdu).

### 2.5 Aeroponics

Aeroponics is a system in which the roots of the plants are kept continuously or discontinuously under fine drops/mist of nutrient solution (see Fig. 9). Plants are grown in holes on the panels of a polystyrene material placed horizontally or on a slope with a square or triangle section (A frame) (see Fig. 10). The nutrient solution is applied with a misting system turning on for only a few seconds in every 2–3 min (Schwarz, 1995).

**Figure 9** Aeroponic system at Epcot (USA) (Photos: Yüksel Tüzel, private collections).

**Figure 10** A-frame aeroponic system (Photos: Yüksel Tüzel, private collections).

### 2.6 Dry hydroponics

A dry hydroponics system is designed as a closed system to grow short-cycle crops such as different types of lettuce, herbs and flowers. The system is efficient in terms of water, fertilizer and energy consumption and labour and space use; however, it requires mechanization. The crops are grown in crop holders placed on floats placed into ponds filled with nutrient solution (see Fig. 11). The system provides an ideal microclimate for the above ground part of the crop while

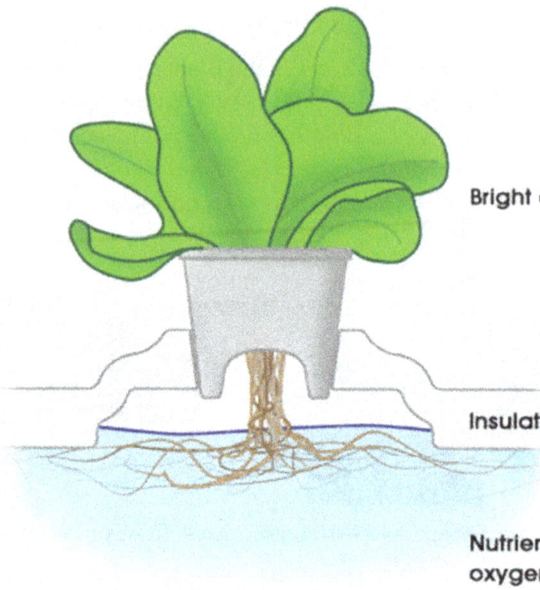


**Figure 11** Dry hydroponics system (Dry Hydroponics, 2020) (Source: Maurice van der Knaap).

nutrient solution and oxygen are always available for the below ground part of the plant. One of the advantages of the system is the ability to adjust crop spacing multiple times during the cultivation period without disrupting growth, resulting in an increase in production volume without necessitating an increase in production space (Dry Hydroponics, 2020).

## 3 Solid systems

In such systems, a solid substrate provides physical support, aeration, a supply of water and nutrients to the plants. Several inorganic, organic or synthetic materials are used alone or as a part of a mixture. The nutrient solution is delivered directly to the plant roots. Those systems may be either open or closed and are named according to the substrate used:

- gravel culture;
- sand culture; and
- rockwool culture.

They are also named according to where the substrate is placed:

- bag culture;

- container culture;
- trough culture.

These are discussed in the sections that follow.

### 3.1 Gravel culture

Gravel culture is a system where plants are grown in troughs filled with gravel placed in soil or above the ground as raised beds built from timber, concrete or plastic (see Fig. 12). It was the most widely used system from the 1940s to the 1960s. In almost all systems sub-irrigation system is used by pumping water into the troughs until flooding and then letting the solution drain out for a while. In this case, half of the total gravel particles should be about 10–15 mm in diameter (Resh, 2002). If drip irrigation is used, particle size should vary between 3 mm and 10 mm. The particles should not be of calcareous material to prevent changes in pH (Resh, 2002; Savvas et al., 2013). The drained solution is collected and reused. The growing cycle in such a closed system typically lasts for 2–6 weeks (Resh, 2002).

**Figure 12** Tomato (left), eggplant (middle) and pepper (right) growing in raised gravel beds (Photos: Astrit Balliu, private collections).

**Figure 13** Enarenado (left) (Source: Maria T. Lao), cucumber (middle) (Source: Miguel de Cara) and tomato production (right) (Photo: Astrit Balliu, private collections).

'Enarenado' is based on digging out the top of the original soil and replacing it with a layer of compacted clay (25–30 cm), followed by a layer of 2–3 cm fermented manure or organic material (composted crop residues). A layer of 10 cm of washed beach sand or coarse grit finally sits on the bed (Cantliffe and Vansickle, 2001) (see Fig. 13).

### 3.2 Bag culture

In bag culture a substrate is filled into plastic bags that are placed on channels or panels (see Figs 14 and 15). Granulated materials such as perlite, coir, pumice, tuff or a mixture of these materials (see Fig. 14) or slabs such as rock wool (see Fig. 16) or coir are used in bag culture (see Fig. 17). Each bag has holes for plants and one dripper is used for each one. Two small slits are made to the base for drainage and leaching (see Fig. 15). The bags or slabs are 90–100 cm long and 15–30 cm wide and 7–10 cm deep. The number of plants per bag changes depending on the crop, substrate volume per plant and length of the growing season. Wider

**Figure 14** Tomato grown in perlite bags in raised channels in Spain (Photo: Yüksel Tüzel, private collections).

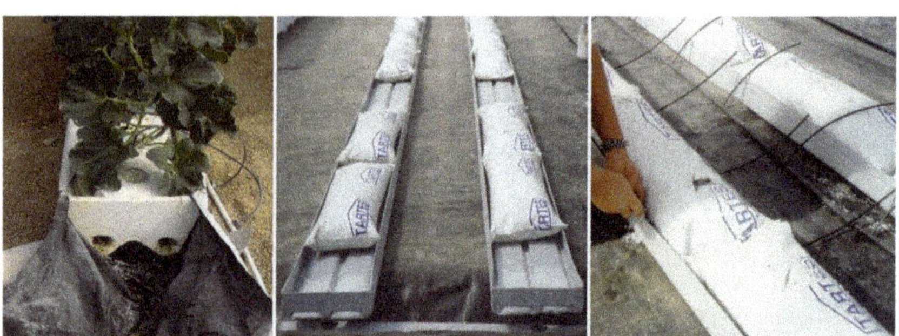

**Figure 15** Melon grown in perlite in polystyrene channels (left) (Photo: Yüksel Tüzel, private collections), panels on the floor for bag culture (middle) and slits for drainage (right) (Source: Ayse Gül).

**Figure 16** Rockwool (left) and coir slabs (right) for tomato production (Photos: Yüksel Tüzel, private collections).

**Figure 17** Coir slabs for tomato (left) (Photo: Yüksel Tüzel, private collections) and strawberry (right) production (Photo: Astrit Balliu, private collections).

bags are used for cucumbers (i.e. 30 cm-wide), while narrower (15–20 cm) slabs with 3–5 plants are sufficient for tomatoes. However, vertical bags with open tops can be used for growing single plants (Maloupa, 2002) (see Fig. 18).

### *3.3 Container culture*

Different types of plastic or polystyrene containers, such as pots, cases and buckets, are used (see Fig. 19). The number of plants in each one depends on the container size and the crop. Substrate volume per plant is 5–8 L. Higher container height with higher substrate depth (more than 20 cm) has better drainage and enhances aeration. Pots can be lined up vertically (see Fig. 20).

### *3.4 Trough culture*

In trough culture plants are grown in beds or plastic troughs on the greenhouse floor (see Figs 21 and 22). A slope (0.5%) is required. The depth of troughs change varies from 10 cm to 35 cm (Savvas et al., 2013). The drained solution should be collected through drainage pipes placed on the bottom.

**Figure 18** Vertical plastic bags for a single plant growth (Photo: Astrit Balliu, private collections).

**Figure 19** Different types of containers (Photos top left and right: Yüksel Tüzel, private collections) (Photos bottom left and right: Astrit Balliu, private collections).

**Figure 20** Vertically lined pots (Photos: Yüksel Tüzel, private collections).

**Figure 21** Beds prepared in the greenhouse (Source: Ayşe Gül).

**Figure 22** Plastic troughs used for ornamental and vegetable crops (Photos left and middle: Yüksel Tüzel, private collections) (Photo right: Astrit Balliu, private collections).

# 4 Aquaponics, bioponics and vertical farming

## 4.1 Aquaponics

Aquaponics is the combination of aquaculture (fish farming) and hydroponics. Nutrient-rich effluent from the fish tanks is used by the plants (see Fig. 23). Naturally occurring nitrifying bacteria convert fish wastes into nutrients available for plants. Most of the required nutrient elements are supplied by the aquaculture effluent. Only a few nutrients (for instance Ca, K and Fe) may need supplementation (Rakocy et al., 2004).

**Figure 23** Aquaponics (left) (drawn by Tunc Durdu) and lettuce production in floating aquaponics system (right) (Source: Rasit Zeki Eltez).

**Figure 24** Grass crap *(Ctenopharyngodonidellavalenciennes)* (left bottom) and lettuce (left top) and tomato (right) production with NFT system in aquaponics (Source: Rasit Zeki Eltez).

There are three main aquaponics techniques widely in use worldwide (Thorarinsdottir et al., 2015; see Fig. 24):

- trough culture;
- floating rafts; and
- DWC and NFT.

Various edible species (i.e. leafy vegetables, herbs, cucumber, tomato, okra) have been cultivated successfully (Danner et al., 2019). A recent study comparing NFT, using standard commercial nutrient solution, and an NFT aquaponic system, using fish waste from grass carp showed that the performance of lettuce and herbs, namely dill, rocket, coriander and parsley were equal to or better than the standard system (Lennard and Ward, 2019).

### 4.2 Bioponics

Bioponics is a hybrid of hydroponics and organic horticulture. The use of soil is one of the core characteristics of organic production (i.e. EU Council Regulation No 834/2007). However the US National Standard Board has considered including hydroponics (not aeroponics) in 'organic' labelling and defined bioponics as 'a contained and controlled growing system in which plants in growing media derive nutrients from natural animal, plant and mineral substances that are released by the biological activity of microorganisms and delivered in water' (NOSB, 2016). Bioponics utilizes flood and drain or ebb and flow techniques using growing media troughs, NFT and DWC (Ahmad, 2018). Fang and Chung (2018) have used bioponics for lettuce production in a plant factory with artificial lighting and found promising results, though it is still controversial to consider these products as organic.

Plants are grown with organic fertilizers resulting in the colonization of beneficial organisms on and around the roots, which contributes in biofiltration and improves nutrient uptake. More options are available for materials to use as organic fertilizer compared with aquaponics but water and fertilizer requirements are higher for the adjustment of appropriate nutrient concentrations. The system may be subject to losses or reductions in yield due to the water chemistry, lack of oxygen, diseases, temperature, EC and pH level of the solution and risk of system failure (Texier, 2016; Ahmad, 2018; Bioponica, 2020).

### 4.3 Vertical farming

Vertical farming involves stacked growing systems using SCS at different scales either in the greenhouses or controlled environment conditions, often in urban settings, enabling rapid, preciselycontrolled production and eliminating the need for transport (Sarkar and Majumder, 2015; Kalantari et al., 2018). Vertical farming enables the use of vertical and horizontal spaces more efficiently resulting in higher yield. Additionally, resource-use efficiency, in particular water-use efficiency, is increased (Rajan et al., 2019).

Recently Beacham et al. (2019) divided vertical farming systems into two categories:

- systems with multiple layers of traditional horizontal growing systems; and
- systems growing plants on a vertical surface.

The first category involves multiple layers of horizontal platforms stacked on top of each other in order to utilize potentially unused space either in greenhouses or controlled environment conditions (i.e. plant factories) (see Fig. 25).

Mostly leafy vegetables and microgreens are grown (Skygreens, 2019; Rajan et al., 2019; 'http://growing-underground.com' accessed on 13 March

**Figure 25** Vertical farming in controlled environment conditions (Source: Farminova).

**Figure 26** Cylindrical vertical growth units (Photos: Yüksel Tüzel, private collections).

2020). If each level of multiple layers is located on different floors, the system is called a 'multi-floor tower' system. Production of plants on balconies in domestic tower blocks can be seen as a basic form of vertical farming (Beacham et al., 2019). The second category are the vertical growth surfaces such as living wall systems comprising modular systems attached to the walls for plant growth or green roofs (Köhler, 2008) or cylindrical growth units (Beacham et al., 2019) (see Fig. 26).

## 5 Open- and closed-nutrition systems

In open systems, the nutrient solution is supplied as in conventional soil culture and the drained solution runoff is discarded. The amount of drain is between 5 and 50% of the supply (Nederhoff and Stanghellini, 2010). The waste water and nutrients are a potential environmental hazard. However, the maintenance of the nutrient solution is easier and infection risk is reduced.

In closed systems surplus solution is recovered, replenished and reused. Closed systems are more 'environmental-friendly' since these systems increase water and nutrient-use efficiency and decrease their impact on the environment (Rouphael et al., 2016). Closed systems have been found to achieve an average water and nutrient saving of 22% and 35%, respectively, in cucumber (Gul et al., 1999; Tuzel et al., 1999) in comparison to open-loop systems. Closed-loop systems for tomatoes achieved a nutrient saving of 32–34% (Tuzel et al., 2000), and 48% and 15%, respectively, in systems for roses and chrysanthemums (van Os, 1999).

Closed systems need close monitoring and proper maintenance of nutrient solution due to ion accumulation at the root zone (Savvas and Gizas, 2002; Lykas et al., 2006). Closed systems can result in higher ion concentration in the water causing an unbalanced ratio of nutrients and higher EC in the solution (Savvas et al., 2007). To overcome such a problem nutrient solution can be bled or dumped. Bleeding is a continuous discharge of the nutrient solution by 10% while dumping involves emptying the mixing tank and refilling it again. These systems are called 'semi-closed' systems with lower discharges compared to open systems. In fully closed systems, the nutrient solution is not bled or dumped (Nederhoff and Stanghellini, 2010).

## 6 Optimizing water and nutrient supply

The optimization of water and nutrient supply is a critical issue that determines both the performance of soilless grown plants and the efficacy of soilless production systems. Several factors should be considered to ensure the right amount of water and nutrients are delivered to the plant and effectively absorbed by them.

## 6.1 Growing medium and solution volume per plant

The growth of plants in soilless culture is commonly undertaken in small volume containers. In response the plants increase root density to compensate for the limited root zone volume. Although root overproduction in a limited space is consistent with the effects of a larger soil volume available to the plants (Hess and De Kroon, 2007), the increased root density involves greater oxygen and nutrient consumption per unit volume of the root zone. Root restriction can also reduce the nutrient uptake of plants. Dubik et al. (1990) have reported that the concentrations of N, P, Ca, Mg and Cu in the leaves of plants grown in small-volume containers were 23–48% less than those in large-volume containers.

However, when the growing system enables the maintenance of a constant concentration of each nutrient at the root surface, as is the case of soilless production systems, small container size is no longer a limiting factor. The frequent fertigation improves the uptake of nutrients, especially of the less mobile elements such as P and K, through continuous replenishment of nutrients in the depletion zone at the vicinity of root interface and enhances the transport of dissolved nutrients by mass flow (Silber et al., 2003a,b). As an example, a significantly higher level of phosphorus content in the leaves of plants irrigated at 6 and 8 times/day, compared to those plants irrigated 4 times/day, has been reported in chrysanthemums (Taweesak et al., 2014).

One might expect that growing medium volume affects the crop yield. However, the results are contradictory. Taweesak et al. (2014) reported that the total leaf area, number of internodes, leaf length and leaf width of chrysanthemums grown in 140 $cm^3$ substrate were significantly higher than those grown in 73 $cm^3$ substrate. Similarly, Hess and De Kroon (2007) reported that sweet potato plants grown in small-sized pots (1.6 L) decreased the fresh weight of tuberous roots compared with plants grown in 3.0 L and 4.5 L pots.

However, Pires et al. (2011) found that the substrate volume did not affect the number of fruits and bunches or the total yield of tomatoes grown in pots (5, 7.5, and 10 L $plant^{-1}$) filled with coconut fibre substrate. Similarly, studying the effect of nutrient solution volume on the yield and quality of spinach grown in floating system, Lenzi et al. (2011) found that reduction of the volume of the solution to 60 L $m^{-2}$ of cultivated area (versus 252 and 126 L $m^{-2}$) induced a decrease in nitrate accumulation without a negative effect on yield.

A high fertigation frequency can induce a significant increase in plant yield even under restricted container volume. With the same container volume, Tuzel et al. (2001) found that total yield obtained from tomato plants fertigated once a day (6.4 kg $plant^{-1}$) was lower than that obtained from the plants irrigated twice (6.9 kg $plant^{-1}$) and four times (7.1 kg $plant^{-1}$) daily. This is in accordance with studies by Taweesak et al. (2014) and Pires et al. (2011).

The depth of the growing container is also important because it determines the free-draining water content of pots and thereby the water potential as well as oxygen availability in the pots (Passioura, 2006). A ratio of container depth to its diameter of 1–4 is recommended as the optimum (Dominguez-Lerena et al., 2006). For plants growing in liquid culture, commercial settings rely mostly on buffering through a large volume of liquid in the system. As such, DFT, with its large amount of water, has less rapid changes than aeroponics, which is, in turn, more stable than NFT (Raviv et al., 2008).

## 6.2 Root zone temperature

Root growth is significantly impacted by changes in soil temperature. The temperature affects not only the growth rate of single root tips but also the total root system architecture, especially the formation and orientation of lateral roots (Nagel et al., 2009). In addition to morphological changes, root function and metabolisms can be influenced by such changes (McMichael and Burke, 1998).

One of the most remarkable effects induced by low root temperature is the alteration of water (Kafkafi, 2008) and nutrient uptake (Adebooye et al., 2010; Kawasaki et al., 2014). The magnitude of the effect depends on the crop's physiological stage (Kafkafi, 2008), the growing season and cultivar (He et al., 2019). Decreased root-sourced water supply also leads to reduced leaf growth (Poiré et al., 2010) and stomatal conductance (Dodd et al., 2000), which negatively affect the overall assimilation capability of a plant (Wang et al., 2016).

High root zone temperature also affects the functionality of the root system. Heat effects on roots and nutrient relations are often long-lasting, with incomplete recovery in severely stressed plants even after several days of post-heat recovery (Giri et al., 2017). In an experiment aimed to study the effects of high solution temperature (35°C) versus optimum temperature (22°C) on water and nutrient uptake by roots of tomato plants cultivated in a NFT system, Falah et al. (2010) found that the effects of high temperature were significant over both short and long periods. In the short term, high solution temperature activated water and nutrient uptake through a decrease in water viscosity and affected membranes transport. On the other hand, the long-term effects of high solution temperature caused growth depression and browning in roots, and these resulted in depressed water and nutrient uptake rates. In a similar experiment with strawberry plants using a deep flow technique Sakamoto et al. (2016) found that high root zone temperature treatment (30°C) decreased oxygen consumption and cell viability of the roots, resulting in the withering of most of the plants after 2 months of treatment.

Either cooling of the root zone under high air temperature conditions, or heating of the root zone under low air temperature conditions, can alleviate the adverse effects of the non-optimum air temperature. Kawasaki et al. (2014) reported that root zone heating of tomato plants enhances root growth under low air temperatures. The authors found that the root zone heating influences root growth first, and this subsequently promotes shoot growth. In addition, temperature treatments encourage the development of the internal xylem's structure near the root tip (Kawasaki et al., 2013, 2014). This, in turn, increases both xylem exudation and root respiration, which improves nutrient transport to the shoot, thereby increasing shoot growth (Kawasaki et al., 2013, 2014).

The temperature at the root zone influences growth, chemical composition and yield of many plants (Sakamoto and Suzuki, 2015a,b; Wang et al., 2016). Reduction of marketable yield per plant (7.9–20.9%) was observed in two different cocktail tomato cultivars in response to root cooling (He et al., 2019). In contrast strawberry plants treated with a low root zone temperature increased the biomass of reproductive organs, such as inflorescences and fruits, and the proportion of fruit biomass to total plant biomass (Sakamoto et al., 2016). The increase of root zone temperature from 13°C to 19°C had a significant positive effect on the growth of cucumber seedlings (Wang et al., 2016). Tomato fruit yield was higher using heating treatment due to increased individual fruit dry weight (Kawasaki et al., 2014). However, high root zone temperatures reduced shoot and root growth and water content in carrots grown in a DFT hydroponic system, whereas total phenolic compounds and soluble-solid content increased (Sakamoto and Suzuki, 2015b).

## *6.3 Oxygen enrichment*

The uptake of most essential nutrients depends on processes which are closely related to the metabolic activity of the root system. These processes require energy produced by root cell respiration corresponding to the degradation of photosynthates in the presence of oxygen (Morard et al., 2000). The available oxygen is mainly determined by the layout of the hydroponic system and the physical properties of the substrate, while oxygen diffusion rates into water depend directly on volumetric air content, the partial oxygen pressure and temperature (Schroeder and Lieth, 2004).

Even well-aerated substrate crops may suffer oxygen deficiency as the result of factors such as high root density, high root respiration rates and high substrate temperature. Hydroponically grown plants may also suffer from oxygen deficiency, especially in the nutrient film technique where oxygen concentration can deplete heavily during the daytime (Acuña et al., 2008). Oxygen deficiency can affect water and nutrient uptake. For instance, root asphyxia of a tomato plant caused a 20–30% decrease of water uptake after 48 h, while the active uptake of nitrate, potassium and phosphate were rapidly reduced (Morard et al., 2000). Aeration of the nutrient solution promoted plant growth; leaf K, P and Mg and water uptake (Alvarado-Camarillo et al., 2020).

To keep plants healthy and fully efficient, oxygen concentration should not fall below a plant-specific critical value (Suhl et al., 2019). According to Morard and Silvestre (1996), the root respiration depends on plant species and can differ from 1.44 μmol $O_2$ $h^{-1}$ $g^{-1}$ fresh matter (FM) to 7.8 μmol $O_2$ $h^{-1}$ $g^{-1}$ FM. However, conflicting results have been found regarding the use of oxygen-enriched methods on substrate-grown greenhouse crops (Bonachela et al., 2010). No effects of oxygen enrichment were found on the yield of pepper and cucumber plants grown in cedar sawdust and perlite (Ehret et al., 2010). The study demonstrated that oxygen enrichment of porous substrates under typical hydroponic conditions is difficult and, possibly because of this, effects on yield are inconsistent. Acuña et al. (2008) found that the final yield of marketable and first-class melon fruits was significantly higher for the oxygen-enriched rock wool. Similarly, Öztekin and Tüzel (2020) reported increases in yield of lettuce plants grown in aerated nutrient solution.

An explanation for the contradictory results regarding oxygenation is provided by Bonachela et al. (2010). The authors found a 7% increase in total and marketable yield, associated with a higher fruit number, for oxygen-enriched melon grown on rockwool slabs, whereas no significant differences were found for the melon grown in perlite grow-bags. This study concluded that oxygen enrichment should be applied to rockwool substrates and crop periods when a high oxygen demand coincides with low oxygen availability. Moreover, Conesa et al. (2015) add that the effect of aeration also depends on the growing cycle and climatic conditions.

A number of techniques have been developed to facilitate oxygen enrichment in SCS. Ozone and $K_2O_2$ have been used to increase the $O_2$ concentration in the nutrient solutions. Oxyfertigation techniques have been developed to improve rhizosphere oxygen availability. These involve the supply of oxygen in the irrigation water at oversaturated concentrations, using simple fertigation equipment (Marfà et al., 2005) or through a circular air-stone (Öztekin and Tüzel, 2020). In addition, direct aeration treatments into the root zone have recently been studied (Lee et al., 2014). Interestingly, excessive aeration inhibited root respiration, nutrients, bioactivity, and water uptake and resulted in the reduction of plant growth and fruit yield.

## *6.4 Planting density*

Plant density is one of the most important aspects of a production system, since it greatly affects leaf area and canopy photosynthesis of a crop. A number of experiments have demonstrated the benefits of increasing plant density on the yield of greenhouse crops (Jovicich et al., 1999; Rodriguez et al., 2007; Maboko and Du Plooy, 2013a). Although wide within-row plant spacing increases per plant yield, it decreases production per unit area in a greenhouse (Jovicich et al., 1999). The main factors responsible for the increase infruit yield with narrow spacing are greater crop biomass and increased availability of assimilates for distribution to the fruits (Papadopoulos and Pararajasingham, 1997).

By studying the effect of plant population density (2, 3 and 4 plant·$m^{-2}$) and shoot pruning (1, 2 and 4 main stems) of greenhouse-grown sweet pepper, Jovicich et al. (1999) found that the marketable yield per $m^2$ increased linearly with plant density. A greenhouse experiment conducted to identify the most suitable plant density (2, 4, 6, 8 and 10 plants per $m^2$) for growing cucumber in a recirculating hydroponic system found that an increase in plant density from 2 to 6 plants per $m^2$ significantly increased the yield; a declining trend in yield and fruit number was observed at more than 6 plants per $m^2$ (Pant et al., 2005). Similarly, in an experiment to determine the effect of muskmelon (*Cucumis melo* L.) plant density (1.7, 2.5, 3.3, and 4.1 plants $m^{-2}$) on yield and fruit quality found that the marketable yield was increased linearly from 11.0 to 20.0 kg $m^{-2}$ in the autumn/fall and from 21.9 to 48.3 kg $m^{-2}$ in the spring. Mean fruit size was unaffected by plant density during the autumn/fall but reduced linearly during the spring from 1.8 kg at 1.7 plants $m^{-2}$ to 1.5 kg at 4.1 plants$^{-2}$ (Rodriguez et al., 2007).

Increased yields due to increased planting density are also reported in leafy vegetables. Maboko and Du Plooy (2009) found that plant population significantly affected plant height, fresh and dry leaf mass, leaf area and leaf number $m^{-2}$, with four cultivars of lettuce (*Lactuca sativa* L.) with the highest yield at a density of 50 plants $m^{-2}$. Similarly, a plant density of 40 plants $m^{-2}$ significantly improved growth and yield of basil during the summer/fall season, although a density between 20 and 25 plants $m^{-2}$ was more cost-effective during the spring/summer season (Maboko and Du Plooy, 2013a).

Light conditions also affect the response to differing plant densities. According to Peet and Welles (2005), the number of plant stems or side shoots allowed to develop in greenhouse-grown tomato should be based on light intensity. This ensures not only high yield but also optimal quality, including taste. Logendra et al. (2001) reports that under the higher light conditions of spring (17.2 mol $m^{-2}$ $d^{-1}$), although fruit yield per plant and fruit weight of tomato plant grown in rockwool blocks in an ebb-and-flood hydroponics system were greater for the lower density treatments, no significant effect was found when calculated on a unit area basis. Meantime, under the lower light conditions in winter (7.0 mol $m^{-2}$ $d^{-1}$), spacing affected neither yield per plant nor yield per unit area (Logendra et al., 2001). Similarly, Stadler (2010) reports that the marketable yield of sweet pepper was increased with the increase of artificial light intensity. At the lowest light intensity, the accumulated marketable yield was not influenced by stem density, but with higher light intensity the positive effect of a higher stem density became evident; the marketable yield was significantly higher with 9 stems $m^{-2}$ than with 6 stems $m^{-2}$. The yield increase was attributed to more fruits, whereas the average fruit weight was not influenced (Stadler, 2010).

Planting density is also strongly related to plant architecture. Tomato is a typical example of this relationship. Commonly it is grown with a single stem, and in general, under a plant density varying from 2.5 to 3.5 plants $m^{-2}$. In Mediterranean countries, the growing point of the tomato plant is removed after 6–8 trusses are set, and usually, two growing seasons per year are conducted. With the high-wire system used in northern Europe and the United States, topping during the growing season is practised infrequently and plant stems continue to grow from December of one year to November of the following year (Peet and Welles, 2005). Cultivation of tomato with a single or two bunches has also been studied (Logendra et al., 2001; Maboko and Du Plooy, 2013b; Cardoso et al., 2018). This allows high planting densities since the plants are pruned with reduced height. The crop cycle slows down and, therefore, reduces the risk of pest and disease attacks (Cardoso et al., 018). Comparing four planting densities (11.1, 8.3, 6.6 or 5.5 plants $m^{-2}$) with one or two bunches per plant Cardoso et al. (2018) have found that the total yield was influenced by the interaction planting density × number of bunches per plant. The treatment that allowed the highest yield was 11.1 plants $m^{-2}$ with two bunches, with 22.61 kg $m^{-2}$, in a crop cycle of 134 days (Cardoso et al., 2018).

The response of nine tomato cultivars with early decapitation of growing points, after the second inflorescence, at different plant densities (10, 16, 20 and 25 plants $m^{-2}$) was investigated in a closed hydroponic system under a 40% white shade-net structure. It was found that plant densities of 20 or 25 plants $m^{-2}$produced significantly higher marketable and total yield compared to other variants (Maboko and Du Plooy, 2013b). The discrepancy between these two experiments regarding optimum plant density may be explained by the differences in light intensity between the two experimental sites. Logendra et al. (2001) have reported that planting tomato in a density of 5.5, 7.4 and 9.2 plants $m^{-2}$ affected fruit yield per plant, but not the fruit yield per unit area. Given the higher costs for materials and labour associated with higher plant densities, a two-cluster crop at 5.5 plants $m^{-2}$ with two leaves above the cluster was found to be best (Logendra et al., 2001).

## 7 Case study: rocket production in liquid and solid medium systems

Herbs are used for food flavouring and garnishing, for example, to improve colour, fragrance and aroma as well as for their nutraceutical properties (FAO, 2011). Recently, there has been an increasing interest in the health benefits of such plants (Güney, 2019). The global market value of the herbal industry was US$60 billion in 2000 and increased to US$105 billion in 2017. The projection of the World Bank is around 7.6% annual growth and it is expected to reach US$5 trillion by 2050 (Hafizudin et al., 2019). According to the World Health Organization (WHO), approximately 80% of the world population relies on herbal sources for their health (Christaki et al., 2012).

Rocket (*Eruca sativa*), a member of Brassica family, is one of the vegetables with nutraceutical compounds such as glucosinolates and flavonoids. The health effects of rocket (including the seeds) have been well documented as including antioxidant, antithrombotic, antifungal, antibacterial, anticancer and antimicrobial properties (Jilani et al., 2015; Punetha et al., 2018; Gugliandolo et al., 2018). As a result, over the last two decades, rocket has become an attractive alternative crop (Hall et al., 2012). In Turkey as well as in some other Mediterranean countries, leafy vegetables, such as lettuce, parsley, rocket, are also grown in unheated greenhouses. The growing area has increased 13-fold in Turkey during the last two decades and almost half of the production comes from protected cultivation.

The increasing interest of consumers in ready-to-eat product with attractive nutraceutical and organoleptical properties such as rocket have encouraged testing of different growing systems since the end of 1990s. Soilless culture systems, that is, floating trays, were compared with soil culture using criteria such as plant density (Nicola et al., 2005; Fontana and Nicola, 2009), oxygen level in the system (Ferrante et al., 2003) and oxygen source (Carrasco et al., 2011), differences in cell size and substrate (Jakse et al., 2013), nutrient solution (full dose vs half dose) (Duyar and Kilic, 2016; Öztekin et al., 2019), salinity (Bonasia et al., 2017; Urlic et al., 2017) or light regimes (Mattson and Harwood, 2012; Matysiak and Kowalski, 2019).

In our study a floating system was used to compare two depths of nutrient solution (20 cm vs. 30 cm) and four nutrient concentrations. Complete solution (N-150, P-50, K-150, Ca-150, Mg-50, Fe-5, Mn-0.5, Zn-0.05, B-0.5, Cu-0.03, Mo-0.02 ppm) prepared according to Resh (1991) was the control and others were adjusted by multiplication of macroelements with a coefficient of 1.25, 075 and 0.5. Styrofoam pots with a volume of 100 L and 150 L were placed in the tanks at the beginning of February. No significant difference was found in yield while the nitrate content increased with increasing macronutrient concentrations and the vitamin C content increased with decreasing nutrient concentration in the solution (Table 1) (Tuzel et al., 2020). Those results are consistent with a previous experiment in which a complete solution was compared with a half dose (macroelements reduced by 50%) and later without nutrients (Öztekin et al., 2019).

In another study conducted in substrate, rocket plants were grown in peat, perlite or peat:perlite (1:1 v:v) during the hot season (May or July).

**Table 1** Effects of nutrient concentration and solution depth on rocket yield, nitrate and vitamin C content

| Dose | Depth (cm) | Yield (1. harvest) | Yield (2. harvest) | Yield (3. harvest) | Total yield (kg m$^{-2}$) | $NO_3$-N (mg kg$^{-1}$) | $NO_3$ content (mg kg$^{-1}$) | Vit. C content (mg 100 g$^{-1}$) |
|---|---|---|---|---|---|---|---|---|
| 0.50 | | 1.33 | 1.46 | 1.18 | 3.97 | 597.11 b | 2644.35 b | 75.63 a |
| 0.75 | | 1.41 | 1.48 | 1.12 | 4.02 | 657.52 ab | 2911.9 ab | 73.32 ab |
| 1.00 | | 1.52 | 1.38 | 1.61 | 4.52 | 666.20 ab | 2950.33 ab | 62.49 ab |
| 1.25 | | 1.59 | 1.55 | 1.58 | 4.72 | 725.29 a | 3212.02 a | 60.91 b |
| | | **P=0.0957** | **P=0.9496** | **P=0.1017** | **P=0.0626** | **P=0.0074** | **P=0.0074** | **P=0.0116** |
| | 20 | 1.42 | 1.36 | 1.34 | 4.12 | 641.53 | 2841.08 | 65.60 a |
| | 30 | 1.51 | 1.57 | 1.41 | 4.49 | 681.53 | 3018.22 | 70.58 a |
| | | **P=0.2272** | **P=0.3245** | **P=0.6933** | **P=0.1084** | **P=0.0910** | **P=0.0910** | **P=0.0116** |
| 0.50 | 20 | 1.29 | 1.21 | 1.27 | 3.78 | 593.55 | 2628.61 | 72.65 |
| | 30 | 1.37 | 1.71 | 1.08 | 4.17 | 600.66 | 2660.09 | 78.61 |
| 0.75 | 20 | 1.39 | 1.74 | 1.14 | 4.26 | 651.07 | 2883.35 | 73.22 |
| | 30 | 1.43 | 1.22 | 1.11 | 3.77 | 663.97 | 2940.44 | 73.41 |
| 1.00 | 20 | 1.48 | 1.00 | 1.33 | 3.82 | 643.97 | 2851.87 | 56.33 |
| | 30 | 1.56 | 1.75 | 1.90 | 5.21 | 688.43 | 3048.78 | 68.65 |
| 1.25 | 20 | 1.51 | 1.50 | 1.63 | 4.63 | 677.52 | 3000.47 | 60.19 |
| | 30 | 1.67 | 1.60 | 1.54 | 4.81 | 773.06 | 3423.56 | 61.63 |
| | | **P=0.4573** | **P=0.9579** | **P=0.0067** | **P=0.0154** | **P=0.0277** | **P=0.0277** | **P=0.0823** |

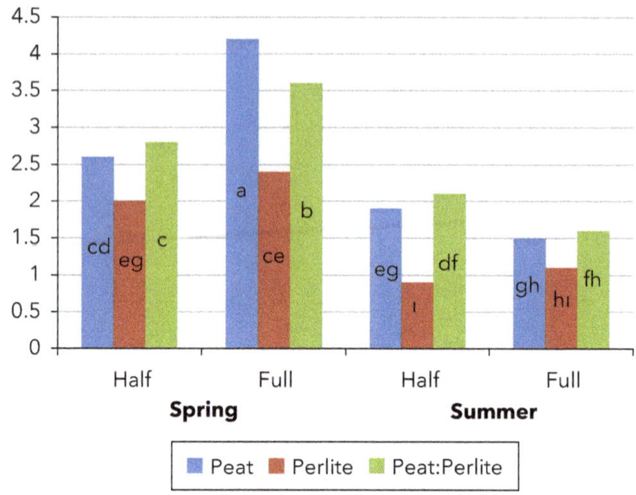


**Figure 27** Effects of substrates, growing season and nutrient solution concentration on rocket yield (kg $m^{-2}$) (Sahin, 2020).

Two concentrations (complete solution based on Resh (1991) vs. half dose of macroelements) of nutrient solution were compared. The yield was significantly affected by growing season, substrate and nutrient concentration. The highest yield was obtained in peat with the use of a complete nutrient solution. The yield was lower in summer months and the performance of a complete solution was better at high temperatures (Fig. 27) (Sahin, 2020).

In addition to commercial production in floating systems and/or pot culture, growing in chambers/cabinets as microgreen has also attracted interest. Microgreens are the seedlings of plants harvested and consumed within 10–20 days after emergence (Lee et al., 2004). Rocket is one of the potential plants used as a microgreen. Bulgari et al. (2017) studied microgreens with a density of 48.5, 242, and 45 g $m^{-2}$ for basil, Swiss chard and rocket, respectively to achieve a crop density of 21 700 plants per $m^2$. After germination, trays were transferred into polyethylene tanks containing half-strength Hoagland's nutrient solution. The nitrate content was lower than the baby leaf or adult vegetables. However, some nutraceutical components and sugars were decreased. Further research is needed to improve the growing conditions.

## 8 Future trends

The use of SCS is a potential solution to meet food demand particularly in urban areas. There has been significant research on the development of innovative sustainable systems, water and nutrients optimization and the improvement of the performance of soilless grown plants. Among other issues, root zone temperature control and manipulation, effective oxygenation and proper plant spacing, specific to different soilless production systems need further research and innovative solutions to practical implementation.

The effects of root temperature are not well documented and the interrelationships among root and shoot temperature influence on growth are still unidentified (Lakhiar et al., 2018). Further investigation is required to develop a deeper understanding of plant root-shoot communication and proper strategies for environmental control under non-optimal or stressful temperatures (Wang et al., 2016). In addition, research is needed for the development of control techniques suitable for year-round greenhouse production and energy conservation (Kawasaki et al., 2014). A differential thermal regulation of shoots and roots would be an effective strategy to increase plant growth (Sakamoto et al., 2016). Root zone temperature effects on growth characteristics, photosynthesis-related parameters and crop components associated with human health (Sakamoto and Suzuki, 2015b) should also be investigated.

A lot of research has been undertaken into the best alternatives to maximize light interception by specific soilless grown crops. One option is to design greenhouses which ensure maximum solar radiation transmission into the greenhouse and using greenhouse covering materials which allow maximum possible light transmission (Roberts, 1998). The use of rotatable stacked arrays of plants in vertical farming and the use of LED lightening (Benke and Tomkins, 2017) are new alternatives to increase light availability to the plants. The other alternative is to devise cropping systems which maximize natural radiation interception and its efficient use by greenhouse crops. There is a need to define the right plant density/shoot density and training methods of specific crops according to the soilless production methods used and lightening conditions. Given the phenotypic differences, the plant spacing should be established for each cultivar of a certain crop (Lakhiar et al., 2018). Of a specific interest will be investigating how density affects the phenotypic characteristics of individual plants (Poorter et al., 2016). In addition, there is a need for research in plant genetics for yield optimization, extending the range of crop types, and fine-tuning for optimal response to controlled variables such as wavelength of LED lighting, temperature, humidity and $CO_2$ levels (Benke and Tomkins, 2017) in specific soilless production systems.

## 9 Where to look for further information

- The following book chapter is a good introduction for anyone new to the subject. (Savvas, D., Gianquinto, G., Tuzel, Y. and Gruda, N. (2013). 'Soilless Culture' in W. Baudoin, R. Nono-Womdim, N. Lutaladio, A. Hodder, N. Castilla, C. Leonardi, S. De Pascale, M. Qaryouti (Eds), *Good*

*Agricultural Practices for Greenhouse Vegetable Crops. Principles for Mediterranean Areas*, FAO Plant Production and Protection Paper 217, pp. 303–354).

- The authors identify that the following articles could shape the subject (Sambo, P., Nicoletto, C., Giro, A., Pii, Y., Valentinuzzi, F., Mimmo, T., Lugli, P., Orzes, G., Mazzetto, F., Astolfi, S., Terzano, R. and Cesco, S. (2019). Hydroponic Solutions for Soilless Production Systems: Issues and Opportunities in a Smart Agriculture Perspective. *Front. Plant Sci.* 10. https://doi.org/10.3389/fpls.2019.00923.) (Savvas, D. and Gruda, N. (2018). Application of Soilless Culture Technologies in the Modern Greenhouse Industry–A Review. *Eur. J. Hortic. Sci.* 83(5), 280–293. https://doi.org/10.17660/eJHS.2018/83.5.2).
- Key research in this area to keep up to date with trends can be found at the following websites (https://www.wur.nl/en/Research-Results/Projects-and-programmes/Euphoros.htm) (http://www.adaptiveagrotech.com/) (https://www.commercial-hydroponic-farming.com/) (https://www.greenhousegrower.com/) (http://blogs.cornell.edu/cornellcea/) (http://www.ceaglobal.org/).
- International Society for Horticultural Science (ISHS) Division of Protected Cultivation and Soilless Culture and its relevant working group (Hydroponics and Aquaponics) is open to all interested researchers, educators, students and horticultural industry professionals with specific interest in the fields of hydroponics and aquaponics. Calendar of Symposia on soilless culture can be found at: https://www.ishs.org/cal.

## 10 References

Abou-Hadid, A. F., Medany, M. A. and Benjamien, I. S. S. (1999). Water requirements for improved cantaloupe cultivation, *Acta Horticulturae* 491, 295–300.

Acuña, R., Gil, I., Bonachela, S. and Magán, J. J. (2008). Oxyfertigation of a greenhouse melon crop grown in rockwool slabs in a Mediterranean area, *Acta Horticulturae* 779, 447–454. https://doi.org/10.17660/ActaHortic.2008.779.56.

Adebooye, O. C., Schmitz-Eiberger, M., Lankes, C. and Noga, G. J. (2010). Inhibitory effects of sub-optimal root zone temperature on leaf bioactive components, photosystem II (PS II) and minerals uptake in *Trichosanthes cucumerina* L. Cucurbitaceae, *Acta Physiologiae Plantarum* 32(1), 67–73. https://doi.org/10.1007/s11738-009-0379-z.

Ahmad, P. (2018). The nomadic farm soilless culture, MSc Thesis, Eindhoven University of Technology, 101.

Alvarado-Camarillo, D., Valdez-Aguilar, L. A., González-Fuentes, J. A., Rascón-Alvarado, E. and Peña-Ramos, F. M. (2020). Response of hydroponic lettuce to aeration, nitrate and potassium in the nutrient solution, *Acta Agriculturae Scandinavica, Section B–Soil and Plant Science* 0, 1–8. https://doi.org/10.1080/09064710.2020.1730430.

Arsenault, C. (2014). Only 60 years of farming left if soil degradation continues, *Scientific American*, Available at:https://www.scientificamerican.com/article/only-60-years-of-farming-left-if-soil-degradation-continues/# (Accessed 11 February 2020).

Beacham, A. M., Vickers, L. H. and Monaghan, J. M. (2019). Vertical farming: a summary of approaches to growing skywards, *The Journal of Horticultural Science and Biotechnology* 94(3), 277–283. https://doi.org/10.1080/14620316.2019.1574214.

Benke, K. and Tomkins, B. (2017). Future food-production systems: vertical farming and controlled-environment agriculture, *Sustainability: Science, Practice and Policy* 13(1), 13–26. https://doi.org/10.1080/15487733.2017.1394054.

Bioponica (2020). Deep water culture in a 40'biogarden trough, Available at: www.bioponica.net (Accessed 13 April 2020).

Bonachela, S., Acuña, R. A., Magan, J. J. and Malfa, O. (2010). Contenido de oxígeno y respuesta de cultivos hortícolas en sustrato e invernadero al enriquecimiento de oxígeno de la solución nutritiva, *Spanish Journal of Agricultural Research* 8(4),1231–1241. https://doi.org/10.5424/sjar/2010084-1203.

Bonasia, A., Lazzizera, C., Elia, A.and Conversa, G. (2017). Nutritional, biophysical and physiological characteristics of wild rocket genotypes as affected by soilless cultivation system, salinity level of nutrient solution and growing period, *Frontiers in Plant Science* 8, 300. https://doi.org/10.3389/fpls.2017.00300.

Bulgari, R., Baldi, A., Ferrante, A. and Lenzi, A. (2017). Yield and quality of basil, Swiss chard, and rocket microgreens grown in a hydroponic system, *New Zealand Journal of Crop and Horticultural Science* 45(2), 119–129. https://doi.org/10.1080/01140671.2016.1259642.

Burrage, S. W. (1999). The Nutrient Film Technique (NFT) for crop production in the mediterranean region, *Acta Horticulturae* 486, 301–306.

Butcher, J. D. (2016). A comparative study of oxygenation techniques in the hydroponic cultivation of *Pelargonium tomentosum*, MSc Thesis, Cape Peninsula University of Technology, 109 p.

Cantliffe, D. and Vansickle, J. J. (2001). Competitiveness of the Spanish and Dutch greenhouse industries with the Florida fresh vegetable industry, *Proceedings of the Florida State Horticultural Society* 114, 121–135.

Cardoso, F. B., Martinez, H. E. P., da Silva, D. J. Hd, Milagres, CdC. and Barbosa, J. G. (2018). Yield and quality of tomato grown in a hydroponic system, with different planting densities and number of bunches per plant, *Pesquisa Agropecuária Tropical* 48(4), 340–349. https://doi.org/10.1590/1983-40632018v4852611.

Carrasco, G., Gajardo, J. M., Álvaro, J. E. and Urrestarazu, M. (2011). Rocket production (*Eruca sativa* Mill.) in a floating system using peracetic acid as oxygen source compared with substrate culture, *Journal of Plant Nutrition* 34(9), 1397–1401. https://doi.org/10.1080/01904167.2011.581045.

Christaki, E., Bonos, E., Giannenas, I. and Florou-Paneri, P. (2012). Aromatic plants as a source of bioactive compounds, *Agriculture* 2(3), 228–243. https://doi.org/10.3390/agriculture2030228.

Conesa, E., Fernandez, J. A., Ninirola, D. and Egea-Gilabert, C. (2015). Nutrient solution aeration and growing cycles affect quality and yield of fresh-cut baby leaf red lettuce, *Agricultural and Food Science* 24(4), 313–322.

Danner, R. I., Mankasingh, U., Anamthawat-Jonsson, K. and Thorarinsdottir, R. I. (2019). Designingaquaponic production systems towards integration into greenhouse farming, *Water* 11(10), 2123. https://doi.org/10.3390/w11102123.

Dockril, P. (2015). The World has lost a third of its farmable land in the last 40 years, *Science Alert, Environment*, Available at: https://www.sciencealert.com/the-world-has-lost-a-third-of-its-farmable-land-in-the-last-40-years (Accessed 11 February 2020).

Dodd, I. C., He, J., Turnbull, C. G. N., Lee, S. K. and Critchley, C. (2000). The influence of supra-optimal root-zone temperatures on growth and stomatal conductance in *Capsicum annuum* L., *Journal of Experimental Botany* 51(343), 239–248. https://doi.org/10.1093/jexbot/51.343.239.

Dominguez-Lerena, S., Herrero Sierra, N., Carrasco Manzano, I., Ocaña Bueno, L., Peñuelas Rubira, J. L. and Mexal, J. G. (2006). Container characteristics influence on *Pinus pinea* seedling development in the nursery and field, *Forest Ecology and Management* 221(1–3),63–71. https://doi.org/10.1016/j.foreco.2005.08.031.

dos Santos, J. F., Filho, M. A. C., Cruz, T. M., Soares, T. M. and Cruz, A. M. L. (2019). Growth, water consumption and basil production in the hydroponic system under salinity, *Revista Ceres* 66(1), 46–53.

Dry Hydroponics (2020). The dry hydroponic system. Available at: www.dryhydroponics.nl (Accessed 13 April 2020).

Dubik, S. P., Krizek, D. T. and Stimart, D. P. (1990). Influence of root zone restriction on mineral element concentration, water potential, chlorophyll concentration, and partitioning of assimilate in spreading euonymus (*E. kiautschovica* Loes, *Journal of Plant Nutrition* 13(6), 677–699, https://doi.org/10.1080/01904169009364109.

Duyar, H. and Kilic, C. C. (2016). A research on production of rocket and parsley in floating system, *Journal of Agricultural Science* 8(7), 54–60.

Economakis, C. D. (1993). The influence of solution heating and intermittent solution circulation on the production of flowers and fruits of tomatoes in nutrient film culture. Proceedings of the of the 8th Int. Cong. on Soilless Culture, 131–143.

Edwards, J. (2016). History of hydroponics, part II: the birth of hydroponics, *Garden Culture Magazine, US-CA Issue* 10, 64–69.

Ehret, D. L., Edwards, D., Helmer, T., Lin, W., Jones, G., Dorais, M. and Papadopoulos, A. P. (2010). Effects of oxygen-enriched nutrient solution on greenhouse cucumber and pepper production, *Scientia Horticulturae* 125(4), 602–607. https://doi.org/10.1016/j.scienta.2010.05.009.

El-Behairy, U. A., Abou-Hadid, A. F., El-Beltagy, A. S. and Burrage, S. W. (1991). Intermittent circulation for earlier tomato yield under nutrient film technique (NFT), *Acta Horticulturae* 287, 267–272.

El-Behairy, U. A., Abou-Hadid, A. F., Medany, M. A. and Awad, M. M. (2001). The effect of different cultivars, orientation and soilless culture systems on production and quality of strawberry, *Acta Horticulturae* 548, 59–64.

El-Shinawy, M. Z., Abd-Elmoniem, E. M. and Abou-Hadid, A. F. (1999). The use of organic manure for lettuce plants grown under NFT conditions, *Acta Horticulturae* 491, 315–318.

Falah, M. A. F., Wajima, T., Yasutake, D., Sago, Y. and Kitano, M. (2010). Responses of root uptake to high temperature of tomato plants (*Lycopersicon esculentum* Mill.) in soilless culture, *Agriculture and Technology* 6, 543–558.

Fang, W. and Chung, H. (2018). Bioponics for lettuce production in a plant factory with artificial lighting, *Acta Horticulturae* 1227, 593–598.

FAO (2011). *Spices and Herbs for Home and Fresh Market* (Matthews, M. and Jack, M., Eds.). Diversification Booklet, Number 20, 65 p.

Ferrante, A., Incrocci, L., Maggini, R., Tognoni, F. and Serra, G. (2003). Preharvest and postharvest strategies for reducing nitrate content in rocket (*Eruca sativa*), *Acta Horticulturae* 628, 153–159. Https://doi.org/10.17660/ActaHortic.2003.628.18.

Fontana, E. and Nicola, S. (2009). Traditional and soilless culture systems to produce corn salad (*Valerianella olitoria* L.) and rocket (*Eruca sativa* Mill.) with low nitrate content, *Journal of Food, Agriculture and Environment* 7(2), 405–410.

Gill, R. (2015). Nutrient management for growing dandelion (*Taraxacum officinale L.*) in nutrient film and deep flow hydroponics, Univ. of Arkansas, Theses and Dissertations 1505, Available at: http://scholarworks.uark.edu/etd/1505 (Accessed 26 February 2020).

Giri, A., Heckathorn, S., Mishra, S. and Krause, C. (2017). Heat stress decreases levels of nutrient-uptake and -assimilation proteins in tomato roots, *Plants* 6(1), 443–448. https://doi.org/10.3390/plants6010006.

Gugliandolo, A., Giacoppo, S., Ficicchia, M., Aliquo, A., Bramanti, P. and Mazzon, E. (2018). *Eruca sativa* seed extract: a novel natural product able to counteract neuroinflammation, *Molecular Medicine Reports* 17(5), 6235–6244.

Gul, A., Tuzel, I. H., Tuncay, O., Eltez, R. Z. and Zencirkıran, E. (1999). Soilless culture of cucumber in glasshouses: I. A comparison of open and closed systems on growth, yield and quality, *Acta Horticulturae* 491, 389–393.

Gul, A., Tuzel, I. H., Tuzel, Y. and Eltez, R. Z. (2001). Effect of continuous and ıntermittent solution circulation on tomato plants grown inNFT, *Acta Horticulturae* 554, 205–212. https://doi.org/10.17660/actahortic.2001.554.21.

Güney, O. I. (2019). Consumption attributes and preferences on medicinal and aromatic plants: a consumer segmentation analysis, *Ciência Rural* 49(5), 1–10.

Hafizudin, M., Roslina, A., Nor AmnaA'liah, M. N. and Nik Rahimah, N. O. (2019). Transformation of herbal industry in Malaysia, *FFTC Agricultural Policy Platform*, Available at: http://ap.fftc.agnet.org (Accessed 14.03.2020).

Hall, M., Jobling, J. J. and Rogers, G. S. (2012). Some perspectives on rocket as a vegetable crop: a review. Vegetable Crop Bulletin, *Vegetable Crops Research Bulletin* 76(1), 21–41.https://doi.org/10.2478/v10032-012-0002-5.

He, F., Thiele, B., Watt, M., Kraska, T., Ulbrich, A. and Kuhn, A. J. (2019). Effects of root cooling on plant growth and fruit quality of cocktail tomato during two consecutive seasons, *Journal of Food Quality* 2019, 1–15. https://doi.org/10.1155/2019/3598172.

Hess, L. and De Kroon, H. (2007). Effects of rooting volume and nutrient availability as an alternative explanation for root self/non-self discrimination, *Journal of Ecology* 95(2), 241–251. https://doi.org/10.1111/j.1365-2745.2006.01204.x.

Jakse, M. H., Hacin, C. and MarsicKacjan, N. (2013). Production of rocket (*Eruca sativa* Mill.) on plug trays and on a floating system in relation to reduced nitrate content, *Acta Agriculturae Slovenica* 101(1), 59–68.

Jilani, M. I., Ali, A., Rehman, R., Sadique, S. and Nisar, S. (2015). Health benefits of arugula: a review, *International Journal of Chemical and Biochemical Sciences* 8, 65–70.

Jovicich, E., Cantliffe, D. J. and Hochmuth, G. J. (1999). Plant density and shoot pruning on yield and quality of a summer greenhouse sweet pepper crop in Northcentral Florida, KD Batal 28th Natl. Proceedings of the Amer. Soc. Plast., Talahassee, Fla. Agr. Plast. Congress 19–22 May, ASP. State, Coll., 184–190.

Kafkafi, U. (2008). Functions of the root system. In: Raviv, M. and Lieth, J. H. (Eds), *Soilless Culture: Theory and Practice*, Elsevier, London, 13–41.

Kalantari, F., Tahir, O. M., Joni, R. A. and Fatemi, E. (2018). Opportunities and challenges in sustainability of vertical farming: a review, *Journal of Landscape Ecology* 11(1), 35–60.

Kawasaki, Y., Matsuo, S., Kanayama, Y. and Kanahama, K. (2014). Effect of root-zone heating on root growth and activity, nutrient uptake, and fruit yield of tomato at low air temperatures, *Journal of the Japanese Society for Horticultural Science* 83(4), 295–301. https://doi.org/10.2503/jjshs1.MI-001.

Kawasaki, Y., Matsuo, S., Suzuki, K., Kanayama, Y. and Kanahama, K. (2013). Root-zone cooling at high air temperatures enhances physiological activities and internal structures of roots in young tomato plants, *Journal of the Japanese Society for Horticultural Science* 82(4), 322–327. https://doi.org/10.2503/jjshs1.82.322.

Köhler, M. (2008). Green facades–a view back and some visions, *Urban Ecosystems Urban* 11(4), 423–436. https://doi.org/10.1007/s11252-008-0063-x.

Lakhiar, I. A., Gao, J., Syed, T. N., Chandio, F. A. and Buttar, N. A.(2018). Modern plant cultivation technologies in agriculture under controlled environment: a review on aeroponics, *Journal of Plant Interactions* 13(1), 338–352. https://doi.org/10.1080/17429145.2018.1472308.

Lee, J. S., Pill, W. G., Cobb, B. B. and Olszewski, M. (2004). Seed treatments to advance greenhouse establishment of beet and chard microgreens, *Journal of Horticultural Science and Biotechnology* 79(4),565–570.

Lee, J. W., Lee, B. S., Kang, J. G., Bae, J. H., Ku, Y. G., Gorinstein, S. and Lee, J. H. (2014). Effect of root zone aeration on the growth and bioactivity of cucumber plants cultured in perlite substrate, *Biologia* 69(5), 610–617. https://doi.org/10.2478/s11756-014-0360-1.

Lennard, W. and Ward, J. (2019). A comparison of plant growth rates between an NFT hydroponic system and an NFT aquaponic system, *Horticulturae* 5(2), 2–16. https://doi.org/10.3390/horticulturae5020027.

Lenzi, A., Baldi, A. and Tesi, R. (2011). Growing spinach in a floating system with different volumes of aerated or non aerated nutrient solution, *Advances in Horticultural Science* 25, 21–25. https://doi.org/10.13128/ahs-12780.

Logendra, L. S., Gianfagna, T. J., Specca, D. R. and Janes, H. W. (2001). Greenhouse tomato limited cluster production systems: crop management practices affect yield, *HortScience: A Publication of the American Society for Horticultural Science* 36(5), 893–896. https://doi.org/10.21273/hortsci.36.5.893.

Lykas, C., Katsoulas, N., Giaglaras, P. and Kittas, C. (2006). Electrical conductivity and pHprediction in a recirculated nutrient solution of a greenhouse soilless rose crop, *Journal of Plant Nutrition* 29(9), 1585–1599.

Maboko, M. M. and Du Plooy, C. P. (2009). Effect of plant spacing on growth and yield of lettuce (*Lactuca sativa* L.) in a soilless production system, *South African Journal of Plant and Soil* 26(3), 195–198. https://doi.org/10.1080/02571862.2009.10639954.

Maboko, M. M. and Du Plooy, C. P. (2013a). High-plant density planting of basil (*Ocimum basilicum*) during summer/fall growth season improves yield in a closed hydroponic system, *Acta Agriculturae Scandinavica, Section B–Soil and Plant Science* 63(8), 748–752. https://doi.org/10.1080/09064710.2013.861921.

Maboko, M. M. and Du Plooy, C. P. (2013b). High-density planting of tomato cultivar's with early decapitation of growing point increased yield in a closed hydroponic system, *Acta Agriculturae Scandinavica, Section B–Soil and Plant Science* 63(8), 676–682. https://doi.org/10.1080/09064710.2013.851276.

Maloupa, E. (2002). Hydroponic systems. In: Savvas, D. and Passam, H. (Eds), *Hydroponic Production of Vegetables and Ornamentals*, Embryo Publications Athens, Greece, 143–175.

Marfà, O., Cáceres, R. and Guri, S. (2005). Oxyfertigation: a new technique for soilless culture under Mediterraneanconditions, *Acta Horticulturae* 697, 65–72. https://doi.org/10.17660/ActaHortic.2005.697.6.

Mattson, N. S. and Harwood, E. D. (2012). Effect of light regimen on yield and flavonoid content of warehouse grown aeroponic *Eruca sativa*, *Acta Horticulturae* 956, 417–422. https://doi.org/10.17660/actahortic.2012.956.49.

Matysiak, B. and Kowalski, A. (2019). White, blue and red LED lighting on growth, morphology and accumulation of flavonoid compounds in leafy greens, *Zemdirbyste-Agriculture* 106(3), 281–286.

McMichael, B. L. and Burke, J. J. (1998). Soil temperature and root growth, *HortScience* 33(6), 947–951.https://doi.org/10.21273/HORTSCI.33.6.947.

Mihajlovic, M., Rekanovic, E., Hrustic, J., Grahovac, M. and Tanokovic, B. (2017). Methods for management of soilborne plant pathogens, *Pesticides and Phytomedicine* 32(1), 9–24. https://doi.org/10.2298/PIF1701009M.

Morard, P., Lacoste, L. and Silvestre, J. (2000). Effect of oxygen deficiency on uptake of water and mineral nutrients by tomato plants in soilless culture, *Journal of Plant Nutrition* 23(8), 1063–1078. https://doi.org/10.1080/01904160009382082.

Morard, P. and Silvestre, J. (1996). Plant injury due to oxygen deficiency in the root environment of soilless culture: a review, *Plant and Soil* 184(2), 243–254. https://doi.org/10.1007/BF00010453.

Mordor Intelligence (2020). Global hydroponics market (2019–2024), Available at: www.mordorintelligence.com (Accessed 11 February 2020).

Nagel, K. A., Kastenholz, B., Jahnke, S., van Dusschoten, D., Aach, T., Mühlich, M., Truhn, D., Scharr, H., Terjung, S., Walter, A. and Schurr, U. (2009). Temperature responses of roots : impact on growth, root system architecture and implications for phenotyping, *Functional Plant Biology : FPB* 36(11), 947–959.

National Organic Standards Board (NOSB) (2016). Hydroponic and aquaponics task force report, Available at: https://www.ams.usda.gov/sites/default/files/media/2016%20Hydroponic%20Task%20Force%20Report.PDF (Accessed 11 February 2020).

Nederhoff, E. and Stanghellini, C. (2010). Water use efficiency of tomatoes- in greenhouses and hydroponics, *Practical Hydroponics and Greenhouses* 115, 53–59.

Nicola, S., Hoeberechts, J. and Fontana, E. (2005). Comparison between traditional and soilless culture systems to produce rocket (*Eruca sativa*) with low nitrate content, *Acta Horticulturae* 697, 549–555. https://doi.org/10.17660/ActaHortic.2005.697.72.

Öztekin, G. B. and Tüzel, Y. (2020). Effects of oxyfertigation and plant growth promoting rhizobacteria on greenhouse lettuce grown in perlite, *Acta Scientiarum Polonorum Hortorum Cultus* 19(1), 97–105. https://doi.org/10.24326/asphc.2020.1.9.

Öztekin, G. B., Uludağ, T., Tüzel, Y. and Tepecik, M. (2019). Farklı Besin Solüsyonlarının Yüzen Su Kültüründe Rokanın Verim ve Kalite Özelliklerine Etkileri, *Turkish Journal of Agriculture–Food Science and Technology* 7(2), 258–265.

Pant, T., Bhatt, R. P., Bhoj, A. S. and Kumar, N. (2005). Exploring suitable plant density for hydroponically grown cucumber (*Cucumis sativus* L.) in greenhouse conditions, *Journal of Applied Horticulture* 7(1), 58–60.

Papadopoulos, A. P. and Pararajasingham, S. (1997). The influence of plant spacing on light interception and use in greenhouse tomato (*Lycopersicon esculentum* Mill.):

a review, *Scientia Horticulturae* 69(1-2), 1-29. https://doi.org/10.1016/S0304-4238(96)00983-1.

Pardossi, A., Malorgio, F., Incrocci, L., Tognoni, F., Campiotti, C. A. (2001). Recirculating nutrient solution culture of melon (*Cucumis melo* L.): physiological and cultural aspects. *Acta Horticulturae* 554, 213-219.

Passioura, J. B. (2006). The peril of pot experiments, *Functional Plant Biology: FPB* 33(12), 1075-1079. https://doi.org/10.1071/fp06223.

Peet, M. M. and Welles, G. (2005). Greenhouse tomato production, In: Heuvelink, E. (Ed.), *Tomatoes*, CABI Publishing, Wallingford, UK, 257-304.

Pires, R. C. de M., Furlani, P. R., Ribeiro, R. V., Bodine Junior, D., Sakai, E., Lourenção, A. L. and Neto, A. T. (2011). Irrigation frequency and substrate volume effects in the growth and yield of tomato plants under greenhouse conditions, *Scientific Agriculture* 68, 400-405. https://doi.org/10.1590/S0103-90162011000400002.

Poiré, R., Schneider, H., Thorpe, M. R., Kuhn, A. J., Schurr, U. and Walter, A. (2010). Root cooling strongly affects diel leaf growth dynamics, water and carbohydrate relations in *Ricinus communis, Plant, Cell and Environment* 33(3), 408-417. https://doi.org/10.1111/j.1365-3040.2009.02090.x.

Poorter, H., Fiorani, F., Pieruschka, R., Wojciechowski, T., van der Putten, W. H., Kleyer, M., Schurr, U. and Postma, J. (2016). Pampered inside, pestered outside? Differences and similarities between plants growing in controlled conditions and in the field, *New Phytologist* 212(4), 838-855. https://doi.org/10.1111/nph.14243.

Punetha, H., Kumar, S., Mudila, H. and Prakash, O. (2018). Chapter 5. Brassica meal as source of health protecting neuraceutical and its antioxidative properties. In: Verma, K. S., Singh, S. and Singh, H. (Eds), *Nutraceuticals and Innovative Food Production for Healthy Living and Preventive Care*, IGI Global, Hershey, PA: Medical Information Science Reference,108-132.

Rajan, P., Lada, R. R. and MacDonald, M. T. (2019). Advancement in indoor vertical farming for microgreen production, *American Journal of Plant Sciences* 10(8),1397-1408. https://doi.org/10.4236/ajps.2019.108100.

Rakocy, J., Shultz, R. C., Bailey, D. S. and Thoman, E. S. (2004). Aquaponic production of tilapia and Basil: comparing a batch and staggered cropping system, *Acta Horticulturae* 648, 63-69.

Raviv, M., Lieth, J. H., Bar-Tal, A. and Silber, A. (2008). Growing plants in soilless culture: operational conclusions. In:Raviv, M. and Lieth, J. H. (Eds), *Soilless Culture: Theory and Practice*, Elsevier, London, 545-573.

Resh, H. M. (1991). *Hydroponic Food Production: A Definitive Guidebook of Soilless Food-Growing Methods* (4th edn). Woodbridge Press Publishing Co, U.S., 462 p.

Resh, H. M. (2002). *Hydroponic Food Production: A Definitive Guidebook of Soilless Food-Growing Methods* (6th edn.), Newconcept Press Inc., Mahwah, NJ, 557 pp.

Roberts, W. J. (1998). *Glazing Materials, Structural Design, and Other Factors Affecting Light Transmission in Greenhouses*, CCEA Rutgers Univ., Cook College, 1-8.

Rodriguez, J. C., Shaw, N. L. and Cantliffe, D. J. (2007). Influence of plant density on yield and fruit quality of greenhouse-grown galia muskmelons, *HortTechnology* 17(4), 580-585. https://doi.org/10.21273/HORTTECH.17.4.580.

Rouphael, Y., Raimondi, G., Caputo, R. and De Pascale, S. (2016). Fertigation strategies for improving water use efficiency and limiting nutrient loss in soilless Hippeastrum production, *HortScience* 51(6), 684-689.

Sahin, E. (2020). Production opportunities of rocket and purslane in growing media and determination of postharvest life, MSc Thesis, Ege Univ. Graduate School of Natural and Applied Science, 87.

Sakamoto, M. and Suzuki, T. (2015a). Effect of root-zone temperature on growth and quality of hydroponically grown red leaf lettuce (*Lactuca sativa* L. cv. Red Wave), *American Journal of Plant Sciences* 06(14), 2350–2360. https://doi.org/10.4236/ajps.2015.614238.

Sakamoto, M. and Suzuki, T. (2015b). Elevated root-zone temperature modulates growth and quality of hydroponically grown carrots, *Agricultural Sciences* 06(8), 749–757. https://doi.org/10.4236/as.2015.68072.

Sakamoto, M., Uenishi, M., Miyamoto, K. and Suzuki, T. (2016). Effect of root-zone temperature on the growth and fruit quality of hydroponically grown strawberry plants, *Journal of Agricultural Science* 8(5), 122. https://doi.org/10.5539/jas.v8n5p122.

Sambo, P., Nicoletto, C., Giro, A., Pii, Y., Valentinuzzi, F., Mimmo, T., Lugli, P., Orzes, G., Mazzetto, F., Astolfi, S., Terzano, R. and Cesco, S. (2019). Hydroponic solutions for soilless production systems: issues and opportunities in a smart agriculture perspective, *Frontiers in Plant Science* 10, 923. https://doi.org/10.3389/fpls.2019.00923.

Sarkar, A. and Majumder, M. (2015). Opportunities and challenges in sustainability of vertical eco-farming: a review, *Journal of Advanced Agricultural Technologies* 2(2), 98–105.

Savvas, D., Gianquinto, G., Tuzel, Y. and Gruda, N. (2013). Chapter 12. Soilless culture.In: Baudoin, W., Nono-Womdim, R., Lutaladio, N., Hodder, A., Castilla, N., Leonardi, C., De Pascale, S. and Qaryouti, M. (Eds), *Good Agricultural Practices for Greenhouse Vegetable Crops, Principles for Mediterranean Areas, Plant Production and Protection Paper 217*, Food and Agriculture Organization of the United Nations (FAO), Rome, 303–354.

Savvas, D. and Gizas, G.(2002). Response of hydroponically grown gerbera to nutrient solution recycling and different nutrient cation ratios, *Scientia Horticulturae* 96(1–4), 267–280.

Savvas, D. and Gruda, N. (2018). Application of soilless culture technologies in the modern greenhouse industry – a review, *European Journal of Horticultural Science* 83(5), 280–293. https://doi.org/10.17660/eJHS.2018/83.5.2.

Savvas, D., Mantzos, N., Barouchas, P. E., Tsirogiannis, I. L., Olympios, C. and Passam, H. C. (2007). Modelling salt accumulation by a bean crop grown in a closed hydroponic system in relation to water uptake, *Scientia Horticulturae* 111(4), 311–318. https://doi.org/10.1016/j.scienta.2006.10.033.

Schroeder, F. G. and Lieth, J. H. (2004). Gas composition and oxygen supply in the root environment of substrates in closed hydroponic systems, *Acta Horticulturae* 644, 299–305. https://doi.org/10.17660/ActaHortic.2004.644.40.

Schwarz, M. (1995). *Soilless Culture Management, Advanced Series in Agricultural Sciences 24*, Springer-Verlag, Berlin, Heidelberg, New York, 197.

Silber, A., Xu, G. and Wallach, R. (2003a). High irrigation frequency: the effect on plant growth and on uptake of water and nutrients, *Acta Horticulturae* 627, 89–96. https://doi.org/10.17660/ActaHortic.2003.627.10.

Silber, A., Xu, G., Levkovitch, I., Soriano, S., Bilu, A. and Wallach, R. (2003b). High fertigation frequency: the effects on uptake of nutrients, water and plant growth, *Plant and Soil* 253(2), 467–477.

Skygreens (2019). Skygreen vegetables, Available at: https://www.skygreens.com/skygreens-vegetables/ (Accessed 14 February 2020).

Stadler, C. (2010). Effects of plant density, interlighting, light intensity and light quality on growth, yield and quality of greenhouse sweet pepper, Final Report, RitLbhĺ30, 65.

Suhl, J., Oppedijk, B., Baganz, D., Kloas, W., Schmidt, U. and van Duijn, B. (2019). Oxygen consumption in recirculating nutrient film technique in aquaponics, *Scientia Horticulturae* 255, 281–291. https://doi.org/10.1016/j.scienta.2019.05.033.

Taweesak, V., Lee Abdullah, T., Hassan, S. A., Kamarulzaman, N. H. andWan Yusoff, W. A. (2014). Growth and flowering responses of cut chrysanthemum grown under restricted root volume to irrigation frequency, *The Scientific World Journal* 2014, 254867. https://doi.org/10.1155/2014/254867.

Texier, W. (2016). Bioponics a strange hybird, Available at: https://www.maximumyield.com/bioponics-a-strange-hybrid/2/1396 (Accessed 13 April 2020).

Thorarinsdottir, R., Kledal, P. R., Skar, S. L. G., Sustaeta, F., Ragnarsdottir, K. V., Pantanella, E., van de Ven, R. and Shultz, R. C. (2015). *Aquaponics Guidelines*, R. I. Thorarinsdottir (Eds.), Haskolaprent, Reykjavik, Iceland, 58.

Tuzel, I. H., Irget, M. E., Gul, A., Tuncay, Ö. and Eltez, R. Z. (1999). Soilless culture of cucumber in glasshouses: II a comparison of open and closed systems on water and nutrient consumptions, *Acta Horticulturae* 491, 395–400.

Tuzel, I. H., Tuzel, Y., Gul, A., Meric, K. M. and Eltez, R. Z. (2000). Comparison of open and closed systems on yield, water and nutrient consumption and their environmental impact, *Acta Horticulturae* 554, 221–228.

Tuzel, I. H., Tüzel, Y., Gül, A., Altunlu, H. and Eltez, R. Z. (2001). Effect of different irrigation schedules, substrates and substrate volumes on fruit quality and yield of greenhouse tomato, *Acta Horticulturae* 548, 285–292.

Tuzel, Y., Oztekin, G. B., Gul, A. and Tuzel, I. H. (2020). Production of aromatic herbs in water culture, Report of BAP Project 2016-ZRF-046, 56.

Urlic, B., Dumicic, G., Romic, M. and Ban, S. G. (2017). The effect of N and NaCl on growth, yield, and nitrate content of salad rocket (*Eruca sativa* Mill.), *Journal of Plant Nutrition* 40(18), 2611–2618.

Valenzano, V., Parente, A., Serio, F. and Santamaria, P. (2008). Effect of growing system and cultivar on yield and water-use efficiency of greenhouse-grown tomato, *Journal of Horticultural Science and Biotechnology* 83(1), 71–75.

van Os, E. A. (1999). Closed soilless growing systems: a sustainable solution for Dutch greenhouse horticulture, *Water Science and Technology* 39(5), 105–112.

Wang, X., Zhang, W., Miao, Y. and Gao, L. (2016). Root-zone warming differently benefits mature and newly unfolded leaves of *Cucumis sativus* L. seedlings under sub-optimal temperature stress, *PLOS ONE* 11(5), e0155298. https://doi.org/10.1371/journal.pone.0155298.

# Chapter 9

# Advances in irrigation/fertigation techniques in greenhouse soilless culture systems (SCS)

*Georgios Nikolaou and Damianos Neocleous, Ministry of Agriculture, Rural Development and Environment, Cyprus; and Evangelini Kitta and Nikolaos Katsoulas, University of Thessaly, Greece*



## 1 Introduction

Modern greenhouse production systems are being upgraded to high-tech agricultural systems aiming to increase plant production and labor productivity (Shamshiri et al., 2018). Research has shifted from hardware-software interfacing and instrumentation control to computer-generated interactions between greenhouse climatic parameters and plant growth (Shamshiri et al., 2018). More traditional databases, application programs and interfaces are being replaced by information systems that can autonomously execute decisions locally or cooperate to implement specific tasks, for example, Cooperative Information Systems (CIS) (King and Sigrimis, 2001). An enabler for precision agriculture (PA) (smart farming) system is the adaptation of Internet of Things (IoT) applications, which improves efficient use of resources, such as, for example,

http://dx.doi.org/10.19103/AS.2020.0076.10

water, fertilizers, energy, to reduce production costs and improve yields (Harun et al., 2019).

Soilless culture systems (SCS) are increasingly making use of computational intelligence, that is, the ability of a computer-based system to learn from experimental observation for automatic, independent monitoring and control (Domingues et al., 2012; Savvas and Gruda, 2018). SCS are recognized globally for their ability to support efficient and intensive plant production using environmentally friendly technology (Gruda, 2009; Barrett et al., 2016). Closed hydroponic systems, which involve nutrient solution recycling, have been implemented to take into account environmental concerns regarding fertilizer leaching and substrate disposal (Putra and Yuliando, 2015). However, despite their advantages, SCS have not achieved their potential, for example, in developing countries. A key problem is the high initial investment needed (Souza et al., 2019), the continuous need for monitoring and the technical skill required by growers.

Irrigation and fertigation are two critical inputs which enable farmers to control plant development, fruit yield and quality (Abidin and Ibrahim, 2015; Bar-Yosef, 1999). Automated hydroponic fertigation units are essential in supplying water and nutrients to crops at the required amount and frequency (Magán et al., 2004). However, any attempt to transfer a certain irrigation approach to a different climatic environment or geographical location is bound to face particular challenges. That has to do with the effect of the prevailing greenhouse environment on crop transpiration rate, the different types of SCS, the unique physical and hydraulic properties of individual substrates, crop types and growth stages, which are constantly changing. Therefore, irrigation control techniques have to be matched to local growing conditions.

Traditionally, irrigation scheduling was based either on decisions by growers or irrigation applications at fixed time intervals (i.e. hourly irrigation), changing the irrigation dose with plant size and development. That leads to excessive or insufficient irrigation and either to water and fertilizer losses, or crop water stress conditions (Lizarraga et al., 2003). To address this problem, extensive research has been conducted over the years investigating the effects of climatic parameters on crop water uptake (e.g. transpiration rate), in an attempt to match the crops' irrigation needs with daily transpiration fluctuations. This has led to the development of several transpiration models (Katsoulas and Stanghellini, 2019) based on the Penman-Monteith approach. However, model validation under different climatic conditions does not always successfully fit calibration data. Plant-monitoring systems have also been developed in an attempt to optimize the efficiency of irrigation by identifying water stress threshold values. In the context of the 'fourth agricultural revolution', sensing, smart and sustainable technologies have been developed to improve SCS using emerging information and communication technologies (ICTs) (Rose and

Chilvers, 2018; Miranda et al., 2019). The objective of this chapter is to describe the most common SCS irrigation and fertigation methods and identify new directions in the application of the IoT and phyto-sensing technology.

## 2 Types of irrigation and fertigation system in SCS

Ferti(irri)gation has become the most common practice in SCS for supplying nutrients along with water (i.e. nutrient solution) simultaneously to the crop (Pérez-Castro et al., 2017; Garcia-Caparros et al., 2017). Nutrient solutions are prepared at 100 or 200 times higher than the average concentration supplied to plants (Savvas et al., 2017; Van Os et al., 2008). Two (Tank A and Tank B) or even up to eight stock nutrient tanks and one tank for acid (Tank C) supply nutrients to a smaller mixing tank (Van Os et al., 2008), adding irrigation water in order to achieve a predefined electrical conductivity (EC) and pH value in the outgoing fresh nutrient solution based on crops' nutritional needs (Savvas and Gruda, 2018). The whole preparation process of the fresh nutrient solution is automatically controlled through hydroponic fertigation head units; in some cases mixing of the stock nutrient solution is done directly in the irrigation mainline (Savvas and Gruda, 2018) (Fig. 1).

In open SCSs, where the drainage water is discarded, irrigation is commonly based on a at a fixed irrigation dose, according to the substrate/container type and shape, and a changeable frequency according to the crop evapotranspiration rate. A surplus amount of 25–50% of the nutrient solution in each irrigation is recommended to overcome any inconsistencies in irrigation distribution and to prevent gradual salt accumulation in the root zone (Kläring, 2001). Overhead systems, which apply the nutrient solution using sprinklers, are mainly used in nurseries due to excessive water losses and the need for frequent foliage wetting (Goodwin et al., 2003).

Closed SCSs enable the reuse of drainage water without discarding irrigation solution to the surroundings (Fig. 2). The benefit of recycling the nutrient solution without adversely affecting yields presupposes that it is replenished with appropriate amounts of nutrients before reuse. In particular, the uptake of water and nutrients by the crop has significant variations, mainly linked to the type and stage of plant growth and the cultivation and environmental conditions. Plants' needs for water and nutrients depend on the plant itself and its growing environment and roughly remain the same regardless of whether the plants are cultivated in open or closed hydroponic systems. However, in the closed recycling systems, some nutrients and non-nutrients are accumulated excessively in the root zone, causing an increase of salts above the salinity threshold value (Carmassi et al., 2007). This is a result of the presence of ballast ions in the primary water such as $Na^+$ and $Cl^-$ when poor quality irrigation water is used, as is the case in many Mediterranean countries,

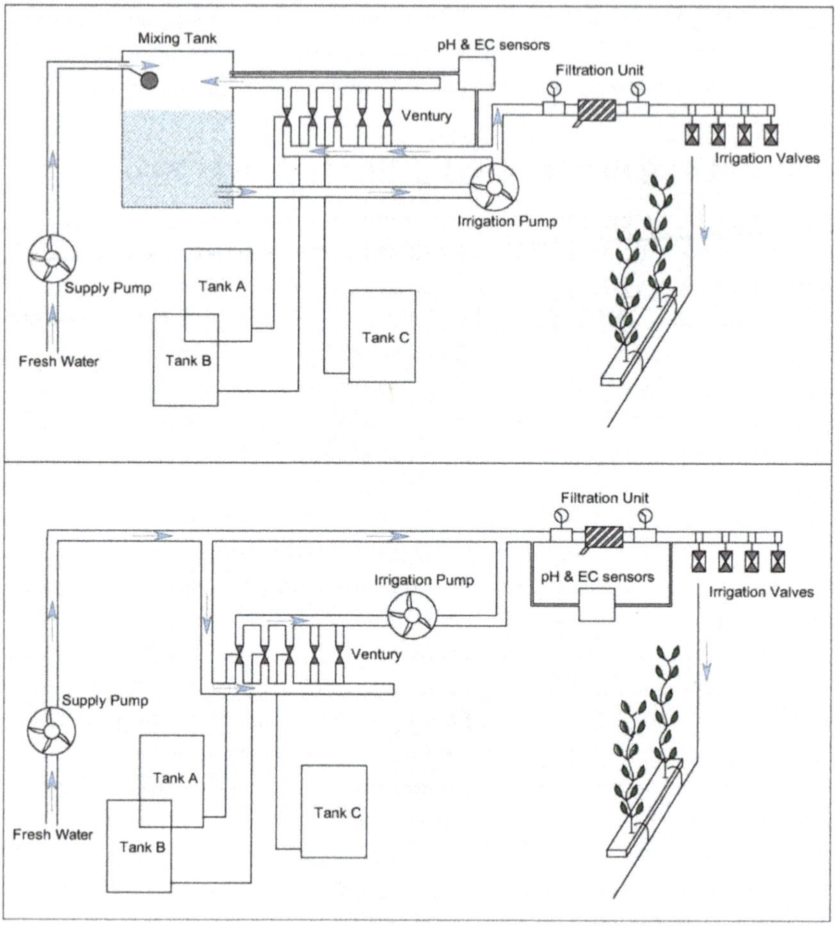


**Figure 1** Schematic diagram of the preparation of nutrient solution with the use of a hydroponic fertigation head (i.e. a system of pumps and sensors) using a mixing tank (upper part) and supplying the concentrated nutrients solution directly to the irrigation mainline (lower part).

which constrains the potential of complete nutrient recycling throughout the cropping period (Katsoulas and Voogt, 2014). In fact, apart from EC measurements, there is a difficulty in ensuring individual real-time corrections to each nutrient in response to actual crop nutritional demands (Cloutier, 1997; Bouchaaba et al, 2015). This is why the nutrient solution in recirculation systems may be periodically discarded during crop growth or at least partially replaced (Carmassi et al., 2007). Despite this, fertilizers and water losses are respectively 15–65% and 15–35% lower in closed compared to open systems (Bres, 2009).

The use of a semi-closed SCS is recommended when excess quantities of specific ions like $Na^+$, $Cl^-$ so as $Ca^{2+}$, $Mg^{2+}$ and $SO_4^{2-}$ are present in the

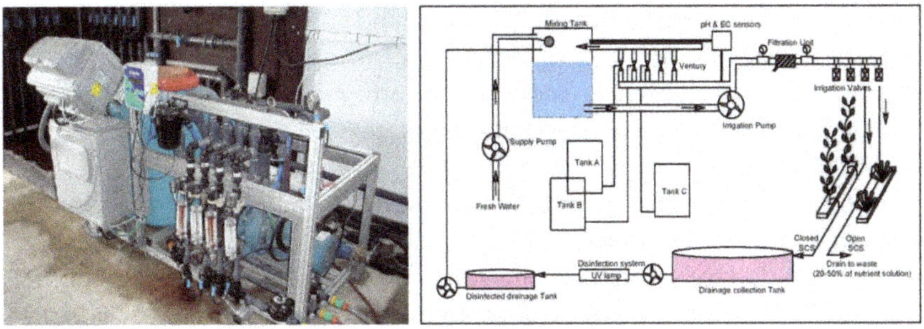


**Figure 2** Fertigation head unit with a mixing tank and a schematic diagram of an open and a closed SCS.

water. In that case the nutrient solution is partially or totally replaced when a predetermined criterion (e.g. salinity) is met. Katsoulas et al. (2015) reported that a semi-closed SCS efficiently controlled salt accumulation in the root zone when primary water is of low quality. In this content, mathematical and empirical models simulating specific ion accumulation in the root-zone environment have also been developed (Carmassi et al., 2005; Massa et al., 2010; Varlagas et al., 2010; Katsoulas et al., 2015; Neocleous and Savvas, 2017).

Micro-irrigation based SCS are probably the most commonly used system since they are a highly efficient delivery method. Drippers, trickle devices or emitters are operated using low inlet pressure heads usually 100–150 kPa and with small discharge turbulent flow rates such as 2, 4 and 8 L $h^{-1}$ (Benami and Ofen, 1983). In trickle irrigation it is very important that the design of the irrigation system is correctly sized in order to deliver water at the desired pressure and flow rate (Lieth and Oki, 2008). The use of pressure-compensated drippers is recommended to increase uniform application of water within each lateral device. Drippers may connect to laterals directly on-line or with the use of a micro-tube where the crop is elevated above ground level. Laterals with fixed interval labyrinth-type in-line drippers, or the use of a drip tape, are alternative systems.

Sub-irrigated SCS, for example, capillary mats, capillary wicks or ebb and flow systems, use the capillary action of a medium to deliver the nutrient solution from a water source that is below the growing bed (Semananda et al., 2018). Capillary irrigation allows the use of lower nutrient solution concentrations because there is a minimum amount of leaching. The main constraints are the accumulation of salts in the top layer of the substrate and the problem of leaching (Ferrarezi et al., 2015).

## 3 Factors affecting irrigation decisions

Water scarcity, climate change and other environmental issues put pressure on agricultural producers to improve efficiency (Gruda, 2019). This includes efficient irrigation management (Loubna and Bouchaib, 2018; Cahn and Johnson, 2019). The timing and the dose of irrigation scheduling can affect plant performance, optimize economic return, increase water use efficiency (WUE), maximize yield and stimulate vegetative or generative growth (Saha et al., 2008; Andrew Lee, 2010).

In optimizing WUE and improving fruit quality characteristics, a number of studies have investigated the application of partial root-zone drying and regulated deficit irrigation in substrate-based SCS (Sun et al., 2013; Gholamhoseini et al., 2018; Hooshmand, et al., 2019). However, special attention needs to be given to substrate water-holding capacity (WHC) and the restricted root area of crops in a SCS in order to avoid plant water deficiency.

Irrigation scheduling has to ensure that nutrient solution supply matches plant water needs. Growers follow various irrigation regimes according to the crop plant size, climatic conditions, for example, light, temperature, humidity, as well as substrate or hydroponic system (Nikolaou et al., 2017c). Similarly, target nutrient concentrations vary among different crops, and nutrient uptake is also affected by external factors, particularly weather conditions (Sonneveld, 2002; Adams, 2002). Accordingly, different types of SCSs require different irrigation scheduling approaches (Fig. 3).

Irrigation scheduling requires modification according to the microclimate inside the greenhouse. The daily transpiration of a tomato plant in Northern Europe is estimated to be between 3 mm and 5 mm, increasing in the Mediterranean region to 6–9 mm (Van Os et al., 2008). Growers have to make decisions on how much water and when to apply it; however, the optimal irrigation schedule must be matched to the production system as well. Different methods for controlling the greenhouse environment could lead to differences in the transpiration rate even in the same greenhouse (Nikolaou et al., 2019). Ferentinos et al. (2017) reported significant spatial variability in air temperature and humidity differences in a commercial greenhouse of up to 3.3°C and 9%, respectively. A strong genotype effect on the transpiration efficiency was also observed under greenhouse conditions (Bogeat-Triboulot et al., 2019).

The frequency and duration of each irrigation event are also important. For example, increasing the irrigation frequency of cucumber crops grown in rockwool from 6 events to 16 (same daily amount) increased the transpiration rate to 24% (Nikolaou et al., 2017a). Meric et al. (2011) have also shown that the rate of transpiration of tomato increases at higher frequency interval treatments (i.e. starting irrigation even at 1 MJ $m^{-2}$ rather than 4 MJ $m^{-2}$). The benefits of higher irrigation frequency on the vegetative growth and fruit yield of tomato plants grown in different volumes of coconut coir were also demonstrated by Pires et al. (2011).

The significant role of the substrate type in affecting watering has also been shown in many cases (Kim et al., 2015; Gruda, 2019). In an open SCS,

**Figure 3** Different SCSs stressing the necessity of different irrigation scheduling approaches.

Mavrogiannopoulos (2015) found a great potential for saving fertilizer and water by calculating the irrigation amount based on the WHC of the substrate, using the equation:

$$Q = \frac{Y \cdot Ww \cdot N}{(1 - dr)} \quad (1)$$

where $Q$ is the irrigation amount (L slub$^{-1}$); $Y$ is the WHC of the substrate (L slub$^{-1}$); $Ww$ is the percentage of the WHC, that is, easily available water (EAW; %), $N$ referred to a critical percentage of the EAW that when reached irrigation should start; $dr$ is the drainage amount (%).

To calculate WHC during different crop growth stages, the substrate should be weighed dry, then filled up with water for up to 24 h to complete the wetting process, drained up to 12 h, and reweighed. The difference in weight between the two measurements represents the substrate WHC. Ww values (the percentage of the WHC that is easily available water (EAW)) are estimated to be 24% for perlite, 58% for coir and about 85% for rockwool of the WHC. Lower percentages are used when water with higher electrical conductivity is used. The percentage of drainage is proportionally increased from 10 to 30% when the salts in the irrigation water are increased. In hydroponic systems without a substrate (e.g. nutrient film technique (NFT), the deep flow technique or aeroponics), Blok et al. (2017) concluded that the maximum plant uptakes for nutrients, oxygen and water are not always met by the irrigation rate. Additionally, in those systems, in order to reduce the operational cost of pumping and disinfection, flow rates must be kept to a minimum. For example, in lettuce crops grown in the nutrient film technique (NFT) system, Al-Tawaha et al. (2018) found that water and nutrients uptake were positively affected when the flow rate was 20 L $h^{-1}$ rather than 30 L $h^{-1}$.

Water quality analysis (e.g. mineral ions, pH, salinity and alkalinity) is crucial in the proper preparation of the nutrient solution and final nutrient solution composition (Savvas, 2002). Different water sources of different quality may be used depending on the country and region. The most common sources are rainwater, surface and groundwater, municipal tap water and (in some cases) treated wastewater (Van Os et al., 2016). Taking into account the limited amount of good-quality freshwater resources, seawater use has also been reported in a SCS (Atzori et al. 2019). Generally, the quality of source water is evaluated based on the type of the soilless system. For open systems, for several crops such as, for example, rose, gerbera, tomato and cucumber, the EC, $Na^+$, and $Cl^-$ should be lower than 1.0 dS $m^{-1}$, 3.0 and 2.8–3.0 mmol $L^{-1}$, respectively while in a system with less drainage those values should be lower than 0.5 dS $m^{-1}$, 1.5 and 1.5 mmol $L^{-1}$ respectively, and even lower values are expected for closed hydroponic crop systems (de Kreij et al., 2003; Schröder and Lieth, 2002). High salinity in water means that a dilute nutrient solution is needed to reach the target EC value of the outgoing irrigation solution and a more rigorous flushing of the system may be required. The salinity of the nutrient solution in a SCS is quantified based on EC measurements, which correspond to the total amount of dissolved ions in the solution (Steidle Neto et al., 2014).

In automatically controlling nutrient and water supply, Kläring (2001) developed an irrigation technique based on a feed-back system taking into account the specific ion concentration in the drainage solution to predict nutrient and water uptake of different crops. Kim et al. (2013) and Cho et al. (2018) developed an ion monitoring system based on ion-selective electrodes that automatically calibrate sensors and measure the concentrations of individual ions ($NO_3^-$, $K^+$ and $Ca^{2+}$) in hydroponic solutions. Domingues et al. (2012) showed that, for a lettuce NFT cultivation system, specific macronutrients and micronutrients (symbols such as nitrogen, potassium, calcium and copper) could be monitored by the use of selective electrodes.

## 4 Common irrigation management systems

As indicated by Bar-Yosef (2008), problems in transforming a model into an effective decision support system have to do mainly with engineering issues, the lack of validation under commercial conditions and negative feedback from growers, extension personnel and scientists. In general irrigation supply in SCS can be managed using the following methods listed below, from the simplest to the more sophisticated:

1 Monitoring greenhouse climate and vegetative parameters in order to calculate transpiration rates (Ondrašek et al., 2007).
2 Adding amounts of nutrients and water (quantity concept) estimated to be required by the crop (Kläring, 2001).
3 Controlling the substrate or drainage concentrations of water and nutrients (Kläring, 2001; Neocleous and Savvas, 2019).
4 Controlling the water content of plant tissues, based on the 'speaking plant approach (SPA)' approach (Udink ten Cate et al., 1978; Hashimoto, 1989).

These methods can be combined.

Automatic irrigation controller units may rely on (Nikolaou et al., 2019; Romero et al., 2012):

- a feed-forward (open loop) support system, and/or
- a feed-back (closed loop) support system.

The feed-forward process measures changes in a certain variable before the process needs changing (Nandong, 2015). A feed-back strategy can take action only after a problem occurs (Singh et al., 2015). For these reasons, feed-forward and feed-back irrigation support systems are often used simultaneously.

Traditionally, irrigation scheduling is based on decisions by the grower (e.g. look and feel method) where plants are inspected daily to identify symptoms of wilting or by lifting substrates in order to judge the weight loss from water uptake (Lieth and Oki, 2008). This method is still commonly used in organic substrates which are characterized by high WHC, needing only occasional irrigation, and when capillary irrigation applies in closed SCSs (Semananda et al., 2018).

However, on substrates with high aeration and little WHC, such as rockwool or wood fiber, irrigation is needed several times a day. In that case, growers use a fixed irrigation dose, changing the irrigation frequency with plant size (e.g. time clock scheduling-irrigation intervals on an hourly basis). However, this method of irrigation results in supplying excess water and fertilizers. In addition, it does not address transpiration fluctuations within a day, increasing the danger of substrates drying out in pockets (Lizarraga et al., 2003; Lieth and Oki, 2008).

To match water supply to crop evapotranspiration rate, which is largely driven by the intensity of the light (Shelford et al., 2004), many scientists have suggested the 'solar radiation method'. According to this method, the frequency of irrigation is based on measurements of internal or external greenhouse solar radiation. When a specific amount of accumulated solar radiation has been intercepted, irrigation is triggered (Meric et al., 2011; Shin et al., 2014; Shelford et al., 2004; Katsoulas et al., 2006). Accumulated radiation serves to start irrigation (Katsoulas et al., 2006; Tsirogiannis et al., 2010; Nikolaou et al., 2017c). It is common, in days with high radiation, to irrigate up to six times per hour especially during midday as transpiration rates reach their highest values (Nikolaou et al., 2017a). Although the solar radiation method is widely used in greenhouses, this method does not take into account other climatic factors which also affects evapotranspiration such as the greenhouse vapor-pressure deficit and when the crops should be irrigated during the night (Seginer et al., 1990; Shelford et al., 2004; Medrano et al., 2005). Furthermore, irrigation may be delayed during periods with low radiation on cloudy days. For these reasons, researchers have highlighted the need to combine the 'solar radiation method' with another irrigation method, for example, time clock scheduling, to avoid drying out of the substrate and crop stress (Lizarraga et al., 2003). Similarly, other methods of irrigation approaches based on monitoring a single variable (e.g., drainage percentage or substrate electrical conductivity) are not able to estimate the exact amount of water used by the crop (Shin et al., 2014).

Evapotranspiration models have been developed based on measuring the daily fluctuation of crop evapotranspiration at relative short time intervals (i.e. every few minutes). These models take into account internal climatic and crop variables, based on the originally proposed 'big leaf' Penman-Monteith model (Katsoulas and Stanghellini, 2019). Nevertheless, in order to increase the accuracy of evapotranspiration models, there is a need for additional recalibration for climatic conditions (Yan et al., 2018). The direct estimation of evapotranspiration through the use of lysimeters is the simplest and most accurate way to determine the water uptake by a crop (Ehret et al., 2001), though their commercial application is limited due to their high cost. Similarly, the use of evaporation pans (e.g. class A pan or reduced evaporation pans) or atmometers to estimate evapotranspiration rate within greenhouses is still a matter of debate (Fernandes et al., 2003).

A common feed-back irrigation monitoring control system for SCS is based on measurements of leachate volume at a particular time in a day over a few days in order to re-adjust a fixed-time irrigation program (Sigrimis et al., 2001). Variations of the drainage amount within a day can cause an increase in root-zone electrical conductivity (Li et al., 2002). That is the reason why Lizarraga et al. (2003) have argued that it is not realistic to measure drainage once a day but ideally after each irrigation even. In order to avoid plant stress, it is also important to monitor the volumetric water content or matric potential and electrical conductivity of the substrate. Several kinds of substrate moisture sensors have been developed for this purpose, including:

- Dielectric sensors: these measure substrate moisture and electrical conductivity. Sensors have several needles which are embedded in the substrate and create an oscillating wave that changes according to the dielectric properties of the material in order to measure the dielectric permittivity of the surrounding medium. They need calibration according to the substrate type.
- Tensiometers: these measure the water availability via matric potential. They consist of a plastic or a glass tube with a porous ceramic cup, filled with water and buried in the soil/substrate. The measuring range of a typical device is usually between −10 and −200 kPa.
- Electrical resistance sensors; these measure water tension, between −10 and −20 kPa. As the tension changes with water content, the resistance changes as well, and the monitor correlates resistance to kPa.

They can be used separately or in combination.

Sigrimis et al. (2001) have developed a feed-forward irrigation support system for hydroponics based on a simple evapotranspiration model, combined with a feed-back system based on a drain-measuring device. Jose Steidle Neto et al. (2014) successfully developed an automated control system for fertigation control in soilless crops based on the Penman-Monteith equation to estimate evapotranspiration, and real-time monitoring of the electrical conductivity of the drainage nutrient solution in order to maintain the preset EC values ($3 \pm 0.8$ dS $m^{-1}$).

## 5 Plant-based sensing and monitoring systems

Plant-based monitoring can be successfully applied to irrigation scheduling (Fernández, 2017). This is because crop water status is directly related to the available water content in the substrate (Alvino and Marino, 2017). However, biotic and abiotic stress factors also affect crop water status, limiting the use of a phyto-sensing irrigation approach (Katsoulas et al., 2016). Different kinds of sensors can be used either in direct contact with the plant (e.g., measuring leaf temperature, stem micro-variation, sap-flow, and leaf thickness) or remotely for example, by using radiometric sensors.

Katsoulas et al. (2016) and Jones (2004) have described different types of remote and plant-based sensors based, such as crop reflectance. On the whole, plant-sensing technologies have been mainly applied in soil-based systems rather than in SCS. This is because crop water status in crops such as vegetables and ornamentals grown in soilless conditions is constantly changing due to fast crop growth, rates of crop defoliation and so on. The ideal plant sensors for SCS need to run continuously and automatically to ensure accurate measurement (Fernández, 2017). Increasing the number of contact sensors within a greenhouse increases measurement reliability.

Plant-based irrigation scheduling also requires the definition of reference or threshold plant stress values to determine when irrigation is needed (Jones, 2004). Leaf temperature is considered as one of the most promising plant-sensing parameters in implementing an automated irrigation scheduling system. Nikolaou et al. (2017b) validated a simple equation for the estimation of crop evapotranspiration, based on reformulating the Pennman-Monteith equation which, in turn, was based on single leaf temperature measurements in soilless cucumber crop cultivation. In addition, Morales et al. (2014) suggested that infrared thermography can be used for identifying water stress symptoms. Other plant-based sensors include:

- Stem micro-variation sensors: these are used for monitoring micro-variations in stem diameter in a micron range. The diurnal variations mostly represent fluctuations of water content in plants. Two diameter-based indices are commonly used for evaluating plant water status, daily contraction amplitude and trend of daily maxima. The stem micro-variation sensor makes it possible to analyze the effect of irrigation rate and other environmental factors on water balance and growth in plants.
- Leaf temperature sensors: these are used for measuring the absolute temperature of a leaf with almost negligible disturbance by natural leaf temperature. The sensor consists of a probe with a high precision glass-encapsulated thermistor about a millimeter in diameter.
- Sap-flow sensors: these are used for monitoring relative variations in sap flow rate in a leaf petiole or small shoot. The sensor's probe is a hollow collapsible heat-insulating cylinder. A spring-loaded heater and a pair of bead thermistors are located inside the cylinder. A signal conditioner powers the heater and conditions the output signal.
- Leaf thickness sensors: these are used for measuring variations in leaf thickness. The output shows the plant response to environmental changes

(e.g. irrigation, lighting, temperature changes) and is also useful for plant stress detection (e.g. drought, diseases).

- Radiometric sensors: these are used for measuring the reflectance of electromagnetic radiation from crop surfaces. They can measure crop water stress-induced changes in the daily patterns of canopy temperature and canopy radiation reflectance.

These sensors are often combined.

## 6 Case study: effects of greenhouse microclimate on irrigation scheduling and the use of contact plant-sensing technology in SCS

Four experiments were carried out with a substrate-based soilless cucumber crop in the autumn-winter and spring cropping seasons. The system used was an open SCS. The aim was to evaluate the effects of irrigation scheduling, irrigation amount and frequency upon growth, as affected by different greenhouse cooling systems in Eastern Mediterranean climatic conditions. The efficiency and the precision of using wireless phyto-sensing technology was also investigated. Relative changes in plant water status and growth are affected by greenhouse microclimate and irrigation scheduling. These were estimated with stem micro-variations and leaf temperature sensors together with substrate measurements (i.e. substrate temperature) (Fig. 4).

It was found that, while the frequency of irrigation may not affect cucumber production or growth rate, it significantly affects the crop transpiration rate and the percentage of nutrients lost. The study found that, as the frequency of irrigation increased from 6 events to 12 and 16 per day, crop water uptake increased, resulting in a lower percentage of drainage and thus less nutrient and water outflow into the environment. The high irrigation frequency treatment (0.19 mm) minimized the drainage amount by 24%, in comparison to a lower irrigation frequency (0.45 mm). The total volume of irrigation applied in each treatment for a 10-week period was 180 mm. Lower WUE values and higher greenhouse emissions were recorded in both autumn-winter and spring cropping periods. Irrespectively of the cooling system tested in the fan-wetted evaporative system comparing with the force ventilation system, experiments show that an irrigation dose of 0.24 mm maintained a drainage fraction of 20–25% rather than 35–40% (irrigation dose of 0.32 mm), ensuring better water use.

A simplified Penman-Monteith model equation for predicting short-term transpiration was also calibrated and validated through the use of contact plant-sensing technology (e.g. stem micro-variation and leaf temperature sensors). It was clearly demonstrated that substrate temperature, leaf temperature and stem micro-variation measurement could be significant predictors of

**Figure 4** (a) Mounting a lysimeter in a greenhouse structure, (b) plants grown in a lysimeter, (c) a stem micro-variation sensor, (d) a pyranometer located above cucumber crop, (e) a device for drainage measurement and monitoring the substrate (EC, pH, temperature and % drain water), (f) a wireless substrate monitoring sensor.

the transpiration rate in SCS with a high-frequency irrigation schedule. Leaf temperature measurements with a single leaf temperature sensor had a higher correlation coefficient with transpiration, suggesting irrigation scheduling based on transpiration values may be feasible for real-time monitoring. A simplified Penman-Monteith evapotranspiration model, modified with input parameters of leaf temperature measurements was used as follows:

$$\lambda T = A \cdot \left(1 - \exp\left(-K \cdot LAI\right)\right)\left(70.694T - 1376.69\right) + B \cdot LAI\left(0.192T - 3.156\right) \quad (2)$$

where T is the mean crop transpiration rate (kg $m^{-2}$ $s^{-1}$), K is the light extinction coefficient, LAI represents the calculated leaf area index ($m^2$ leaf $m^{-2}$ ground) and T is the leaf temperature (°C).

The model based on leaf temperature measurement worked well, even under different environmental conditions than those calibrated and therefore could be applied in decision support systems in Mediterranean greenhouses.

The leaf-air temperature differential estimated in different parts of the greenhouse (using different cooling methods) indicated that plants were not facing any water stress conditions as the leaf temperatures were consistently lower than the temperature of the air. However, within the same greenhouse environment, leaf temperature analysis showed that cucumber plants watered using a low-irrigation frequency schedule (0.45 mm) did face water stress conditions (Fig. 5). During the midday period, the irrigation interval time was greater than one hour, resulting in an inability of irrigation supply to meet increasing rates of transpiration demand, resulting in leaf stomata closure and an increase of the leaf temperature.

Estimates of crop growth rate and relative water loss from stem micro-variation measurements show good accuracy within different greenhouse environments and irrigation-scheduling regimes. Measurements show that, under

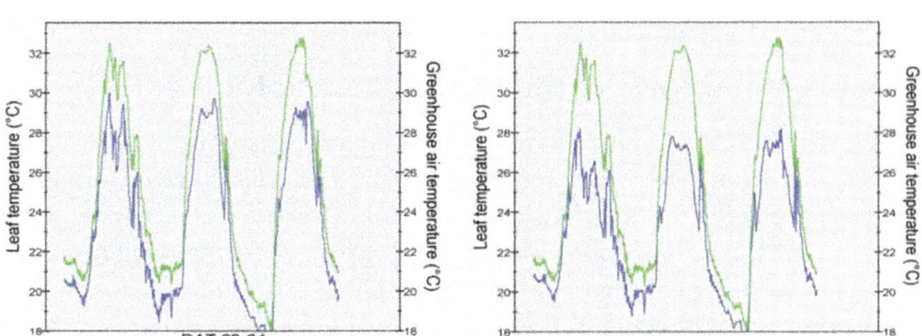


**Figure 5** Greenhouse air (green line) and leaf temperature (blue line; °C) variations during a three-day representative period for an irrigation treatment applying 12 irrigation events daily (left side) and with 6 irrigation events daily (right side).

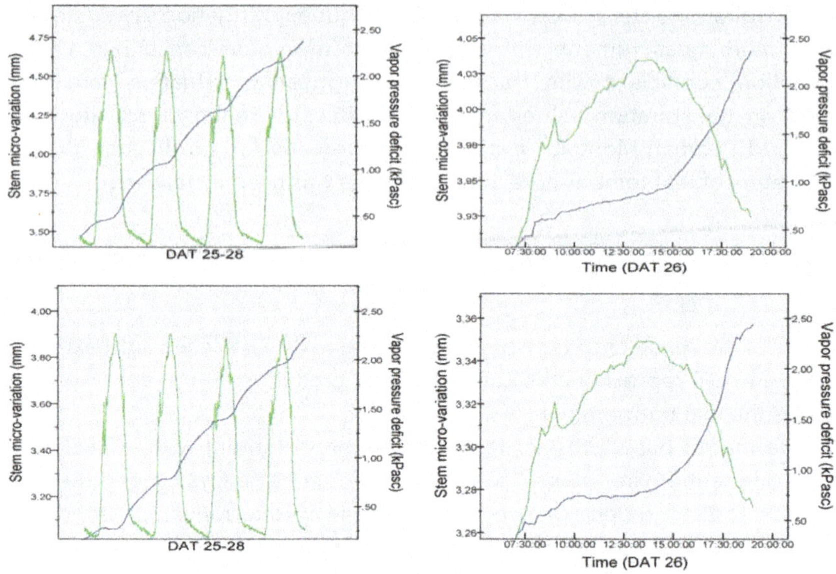


**Figure 6** Stem micro-variation changes (mm) (blue line), and vapor pressure deficit (kPa) (green line), during a four-day measurement period for an irrigation treatment applying 12 irrigation events daily (upper left side) and with 6 irrigation events daily (bottom left, right side) and stem micro-variation changes during a day.

high irrigation frequency treatments, significantly less-explicit trunk depression occurred (Fig. 6). Similarly, for a representative 4-day period, the general growth rate of a plant (i.e. daily diameter increment) showed a significant increase as the irrigation events were divided into shorter intervals. The estimated stem growth rate was 1.154 mm in the treatment with a lower interval rate in comparison with 0.705 mm in the treatment applying a higher interval rate. Regarding temperature changes in the substrate, as expected, higher irrigation frequency resulted in lower substrate temperature values with mean values of 1°C.

## 7 The use of internet of things (IoT) technologies

In conventional agricultural systems, automation of irrigation includes systems such as microcontrollers, programmable logic controllers and crop/sensor interfaces, programmed to control specific tasks which are characterized by a sequential evolution (Rodriguez-Ortega et al., 2016). However, plant responses to their environment are characterized by nonlinearity, prompting the need for open integrated architecture systems involving cooperation between different devices (Morimoto and Hashimoto, 1998; King and Sigrimis, 2001).

The internet of things (IoT) offers the opportunity to improve PA in terms of productivity, food security and sustainability in agriculture (Kamilaris et al.,

2017; Zamora-Izquierdo et al., 2019). The IoT bridges the gaps between the cyber and physical worlds through the integration of sensor-based information, data communication, data cloud or fog-computing acquisition and intelligent processing in real time (Goap et al., 2018; Mehra et al., 2018; Zamora-Izquierdo et al., 2019). Even though the term 'IoT' was first mentioned by a British visionary Kevin Ashton back in 1999, IoT and platforms for agricultural systems are still at an early stage of development (Madakam et al., 2015; Tzounis et al., 2017; Karim et al., 2017).

Terms such as the IoT and cyber-physical systems describe hardware systems (platforms) that gather data from sensors, communicate with actuators and cooperate together to cover automation needs (Zamora-Izquierdo et al., 2019). At the same time cloud-, fog- and edge-computing software application store are able to analyze collected data ('big data analysis') in real time (Tzounis et al., 2017). According to Khanna and Kaur (2019), the most common IoT technologies include:

- Radio frequency identification,
- Wireless sensor networks,
- Bluetooth,
- Near-field communication,
- ZigBee, and
- Wireless fidelity.

At a low level of IoT development, as cited by Shi et al. (2019), data were simply processed and presented in a spreadsheet program such as Excel. With the development of cloud computing, decisions are implemented autonomously through artificial intelligence, developed from machine-to-machine interaction to replicate human intelligence (King and Sigrimis, 2001; Mehra et al., 2018). The most common artificial intelligence systems according to Kirtan et al. (2019) and Yong et al. (2018) are:

- Artificial neural networks,
- Fuzzy logic and neuro-fuzzy logic,
- Expert systems, and
- Bayesian networks.

Common to all these systems is the central processing unit (CPU) responsible for primary functions and using basic components such as microcontrollers and motherboards such as (Lakhiar et al., 2018):

- Arduino: a microcontroller motherboard, with easy to use hardware and software, which is able to read analog input signals from different sensors. Two versions are Arduino Uno and Arduino Yun.

- Raspberry Pi: a small single-board computer, with no storage on board, which can run multiple programs at a time. It can be easily connected to the internet.

In addition, searchable encryption protocols and public-key encryption with keyword search protocols have been developed in order to ensure the security of the data in the cloud (Yong et al., 2018). A recent concept is the internet of underground things, which consists of different kinds of sensors (e.g., soil moisture and salinity sensors) which are partly or completely embedded in the soil (Vuran et al., 2018).

There have been a number of studies on the use of IoT technologies in SCSs. Jiménez-Carvajal et al. (2017) developed a weighting lysimetric system for the direct estimation of evapotranspiration for potted crops. Instead of Arduino Uno which was commonly used by other researches (Sihombing et al., 2018; Rahman et al., 2018) the authors used an Arduino Yun platform programmed to sample data from load cells and water meters (water balance method) every 10 s and averaged every minute. The system was also capable of sending data wirelessly to a server in the cloud, so data could later be queried from any device with internet access.

Software to automate irrigation of potted crops based on weighting lysimeters was also developed by Vera-Repulloa et al. (2015). Each lysimetric unit consisted of a metallic structure to hold the pots and four load cells to measure changes in weight and drainage water. A communication network (Modbus/RS-485) received data from load cells and processed them to control irrigation valves in real time. The software allowed for data acquisition and logging as well as irrigation scheduling based on real-time water balance estimates obtained from the lysimeters. Helmer et al. (2005) described the 'CroppAssist system' to continuously monitor a greenhouse vine crop using 11 physiological and irrigation parameters measured simultaneously, based on real-time data received with pairs of load cells. Steidle Neto et al. (2014) showed that the frequency of irrigation cycles in tomato crops could be managed through the 'Hidro-Control' algorithm, estimating plant transpiration rates by monitoring the electrical conductivity of leachate under preset limits.

Sihombing et al. (2018) developed an automated architecture system to control the level and temperature of a nutrient solution based on different sensors using a 'logic if else' approach. The system was used for small-scale closed growing of spinach, connected to an Arduino Uno microcontroller with communication via low-cost wireless fidelity (Wi-Fi; ESP8266) connected to an Android smart phone. In line, Palande et al. (2018) developed the 'Titan Smartronic' system for fully automating a hydroponic unit, eliminating any human interaction after transplanting the plants. The system used several sensors (e.g. EC and pH monitoring and a water temperature sensor) supported by a Raspberry Pi, Arduinos and open-source software for maintaining the required threshold values. The system incorporates an IoT network allowing monitoring and control from Android or Ios devices.

Mehra et al. (2018) used deep neural networks to develop a system using several parameters (e.g. pH, nutrition level, ambient temperature, humidity and lighting) collected in the cloud in real time to manage tomato plants grown in a hydroponic tank. The prototype was developed using Arduino, Raspberry Pi3 and tensor flow measurement.

Zamora-Izquierdo et al. (2019) developed a multi-tier PA management platform based on edge and cloud computing. The IoT system was implemented in tomato hydroponic units in real greenhouse conditions. The IoT application achieved water savings of more than 30–80% in some nutrients. Similarly, Katsoulas et al. (2017) developed a web-based irrigation scheduling algorithm in order to optimize the irrigation interval rate and the amount of nutrient solution in each irrigation event in SCS. To determine irrigation frequency the crop transpiration rate was estimated from greenhouse climatic data (e.g. air temperature and relative humidity) and solar radiation values. The amount of irrigation was estimated by taking into account substrate water content and water balance. Use of the algorithm resulted in a 100% increase in water and fertilizer use efficiency, in two commercial hydroponic units, as compared to common irrigation practices.

An automatic irrigation control system for soilless culture of lettuce was also proposed by Liu and Xu (2018). Irrigation scheduling was based on software (e.g., modeling lettuce evapotranspiration and substrate wetting pattern) and a ZigBee wireless network. Irrigation water-use efficiency of the intelligent irrigation system was found to be 68% and 98% higher than manual control irrigation in the spring and summer, respectively. Shin and Son (2015) developed an irrigation control system for paprika plants grown on rockwool media. The system consisted of pairs of load cells to estimate the amount of irrigation (input) and drainage rate (output) based on a transpiration rate over a 60-s interval. Data was transmitted throughout ZigBee wireless technology. In addition, irrigation strategies were developed based on substrate EC and drainage pH values. The irrigation control system improved both water-use efficiency and crop productivity.

In order to optimize fertilizer use in hydroponics fertigation systems, Bueno-Delgado et al. (2016) developed an android application called 'Ecofert'. The quantities and types of proposed fertilizers for preparation of stock solutions were estimated according to crop type and specific nutrient needs, the available commercialize fertilizer type and cost. 'Ecofert' used linear programming and a cloud-based database of common fertilizers which was updated daily. Pérez-Castro et al. (2017) also used a linear programming algorithm developed for android mobile devices (named 'cFertigUAL') to determine optimal fertigation parameters for a particular crop.

## 8 Conclusion and future trends

This chapter has reviewed a range of common SCS irrigation and fertigation methods and discussed new applications of IoT and phyto-sensing technology. The majority of irrigation methods used in the past implemented a static feed-back or a feed-forward irrigation decision support system. Increasing irrigation intervals is also increasing the operating costs of irrigation. Smart technologies have been developed to improve SCS using advanced sensors, communication and big data technologies. This is important because, in high technology greenhouses, where optimum climatic conditions can be achieved, crop production only becomes profitable when productivity is optimized.

In optimizing production it is essential to understand, on the one hand, the effects of greenhouse climatic conditions on crop growth and, on the other, the catalytic effect of transpiration on the greenhouse microclimate. That is why it is necessary to establish a network of sensors to assess plant physiological conditions in real time. Wireless sensor technologies offer huge opportunities for research and development as well as for commercial applications. This is a consequence of increasingly smaller wireless phyto-sensing devices and advances in the technology of radio frequency and digital circuits.

By monitoring plants, rather than their surrounding environment, it is easier to understand the plant's physiological status, allowing for improved irrigation decisions. Early detection of changes in plant status has proved to be very helpful for taking cost-effective, well-timed action to prevent undesirable sudden shifts in optimum ranges for healthy crop growth. This is in contrast to traditional visual diagnosis which typically spotted symptoms after damage occurred (e.g. symptoms of plant wilting or nutrient imbalances) which, in many cases, were only partially reversible.

A possible future trend could include monitoring of plant water stress indices on a real-time basis through the use of remote sensors, utilizing non-destructive technology such as the use of reflectance indices. There is a need for the development of a commercial irrigation controller unit in order to model and monitor soil-plant-atmosphere conditions utilizing artificial intelligence. In addition, the implementation of mathematical and empirical models, in combination with decision support systems, will be a useful tool for the better management of nutrient solutions in SCS crops.

## 9 Where to look for further information

For commercialized substrate sensors you can visit:

- Watermark® by Irrometer (https://www.irrometer.com/sensors.html).
- Delta-T device Ltd (UK) (https://www.delta-t.co.uk/).

Companies which are specialized in the use of phyto-sensing:

- Phytec (https://www.phytech.com/).
- Supplant (https://supplant.me/).

## 10 Acknowledgment

The work is carried out in the frame of the PRECIMED project that is funded by the General Secretariat for Research and Technology of the Ministry of Development and Investments of Greece under the PRIMA Programme. PRIMA is an Art.185 initiative supported and co-funded under Horizon 2020, the European Union's Programme for Research and Innovation (Project application number: 155331/I4/19.09.18).

## 11 References

Abidin, S. A. H. Z. and Ibrahim, S. N. 2015. Web-based monitoring of an automated fertigation system: an IoT application. In: IEEE 12th Malaysia international conference Commun MICC Kuching, Malaysia (vol. 2015), pp. 1-5.

Adams, P. 2002. Nutritional control in hydroponics. In: Savvas, D. and Passam, H. C. (Eds), *Hydroponic Production of Vegetables and Ornamentals*. Embryo Publications, Athens, Greece, pp. 213-246.

Al-Tawaha, A. R., Al-Karaki, G., Al-Tawaha, A. R., Sirajuddin, S. N., Makhadmeh, I., Wahab, P. E. M., Youssef, R. A., Sultan, W. A. and Massadeh, A. 2018. Effect of water flow rate on quantity and quality of lettuce (Lactuca sativa L.) in nutrient film technique (NFT) under hydroponics conditions. *Bulg. J. Agric. Sci.* 24(5):793-800.

Alvino, A. and Marino, S. 2017. Remote sensing for irrigation of horticultural crops. *Horticulturae* 3(2):40.

Atzori, G., Mancuso, S. and Masi, E. 2019. Seawater potential use in soilless culture: a review. *Sci. Hortic.* 249:199-207.

Barrett, G. E., Alexander, P. D., Robinson, J. S. and Bragg, N. C. 2016. Achieving environmentally sustainable growing media for soilless plant cultivation systems–a review. *Sci. Hortic.* 212:220-234.

Bar-Yosef, B. 1999. Advances in fertigation. *Adv. Agron.* 65(C):1-77.

Bar-Yosef, B. 2008. Fertigation management and crops response to solution recycling in semi-closed greenhouses. In: Raviv, M. and Lieth, J. H. (Eds). *Soilless Culture: Theory and Practice* (1st edn.). Elsevier, London, pp. 343-419.

Benami, A. and Ofen, A. 1983. Trickle irrigation. In: *Irrigation Engineering; Sprinkler, Trickle, Surface Irrigation*. Irrigation Engineering Scientific Publications, Haifa, Israel, pp. 147-190.

Blok, C., Jackson, B. E., Guo, X., de Visser, P. H. B. and Marcelis, L. F. M. 2017. Maximum plant uptakes for water, nutrients, and oxygen are not always met by irrigation rate and distribution in water-based cultivation systems. *Front. Plant Sci.* 8:1-15.

Bogeat-Triboulot, M. B., Buré, C., Gerardin, T., Chustea, P. A., Le Thiec, D., Hummel, I., Durand, M., Wildhagenb, H., Douthec, C., Molins, A., Galmésc, J., Smith, H. K., Flexas, J., Polle, A., Taylor, G. and Brendel, O. 2019. Additive effects of high growth rate and low transpiration rate drive differences in whole plant transpiration efficiency among black poplar genotypes. *Environ. Exp. Bot.* 166:2-11.

Bouchaaba, Z., Santamaria, P., Choukr-Allah, R., Lamaddalena, N. and Montesano, F. F. 2015. Open-cycle drip vs closed-cycle subirrigation: effects on growth and yield of greenhouse soilless green bean. *Sci. Hortic.* 182:77-85.

Bres, W. 2009. Estimation of nutrient losses from open fertigation systems to soil during horticultural plant cultivation. *Pol. J. Environ. Stud.* 18(3):341-345.

Bueno-Delgado, M. V., Molina-Martínez, J. M., Correoso-Campillo, R. and Pavón-Mariño, P. 2016. Ecofert: an android application for the optimization of fertilizer cost in fertigation. *Comput. Electron. Agric.* 121:32-42.

Cahn, M. D. and Johnson, L. 2019. New approaches to irrigation scheduling of vegetables. *Horticulturae* 3(2):28.

Carmassi, G., Incrocci, L., Maggini, R., Malorgio, F., Tognoni, F. and Pardossi, A. 2005. Modeling salinity build-up in recirculating nutrient solution culture. *J. Plant Nutr.* 28(3):431-445.

Carmassi, G., Incrocci, L., Maggini, R., Malorgio, F., Tognoni, F. and Pardossi, A. 2007. An aggregated model for water requirements of greenhouse tomato grown in closed rockwool culture with saline water. *Agr. Wat. Manag.* 88(1-3):73-82.

Cho, W. J., Kim, H. J., Jung, D. H., Kim, D. W., Ahn, T. I. and Son, J. E. 2018. On-site ion monitoring system for precision hydroponic nutrient management. *Comput. Electron. Agric.* 146:51-58.

Cloutier 1997. Evaluation of sensor technologies for automated control of nutrient solutions in life support systems using higher plants. In: Proceedings of the Sixth European Symposium on Space Environmental Control Systems, Noordwijk, The Netherlands (vol. 21) (3), pp. 295-316.

de Kreij, C., Voogt, W. and Bass, R. 2003. Nutrient solutions and water quality for soilless cultures. Applied plant research–Divion glasshouse ISSN 1385-3015 horticulture Naaldwijk Office P.O. Box 8:2670.

Domingues, D. S., Takahashi, H. W., Camara, C. A. P. and Nixdorf, S. L. 2012. Automated system developed to control pH and concentration of nutrient solution evaluated in hydroponic lettuce production. *Comput. Electron. Agric.* 84:53-61.

Ehret, D. L., Lau, A., Bittman, S., Lin, W. and Shelford, T. 2001. Automated monitoring of greenhouse crops. *Agronomie* 21(4):403-414.

Ferentinos, K. P., Katsoulas, N., Tzounis, A., Bartzanas, T. and Kittas, C. 2017. Wireless sensor networks for greenhouse climate and plant condition assessment. *Biosyst. Eng.* 153:70-81.

Fernandes, C., Corá, J. E. and de Araújo, J. A. Cd 2003. Reference evapotranspiration estimation inside greenhouses. *Sci. Agric. (Piracicaba Braz.)* 60(3):591-594.

Fernández, J. 2017. Plant-based methods for irrigation scheduling of woody crops. *Horticulturae* 3(2):35.

Ferrarezi, R. S., Weaver, G. M., van Iersel, M. W. and Testezlaf, R. 2015. Subirrigation: historical overview, challenges, and future prospects. *Horttechnology* 25(3):262-276.

Garcia-Caparros, P., Contreras, J. I., Baeza, R., Segura, M. L. and Lao, M. T. 2017. Integral management of irrigation water in intensive horticultural systems of Almería. *Sustainability* 9(12):2271.

Gholamhoseini, M., Habibzadeh, F., Atae, R., Hemmati, P. and Ebrahimian, E. 2018. Zeolite and hydrogel improve yield of greenhouse cucumber in soil-less medium under water limitation. *Rhizosphere* 6:7-10.

Goap, A., Sharma, D., Shukla, A. K. and Rama Krishna, C. 2018. An IoT based smart irrigation management system using Machine learning and open source technologies. *Comput. Electron. Agric.* 155:41–49.

Goodwin, P. B., Murphy, M., Melville, P. and Yiasoumi, W. 2003. Efficiency of water and nutrient use in containerised plants irrigated by overhead, drip or capillary irrigation. *Aust. J. Exp. Agric.* 43(2):189–194.

Gruda, N. 2009. Do soilless culture systems have an influence on product quality of vegetables? *J. Appl. Bot. Food Qual.* 82(2):141-147.

Gruda, N. S. 2019. Increasing sustainability of growing media constituents and stand-alone substrates in soilless culture systems. *Agronomy* 9(6):298.

Harun, A. N., Mohamed, N., Ahmad, R., Rahim, A. R. A. and Ani, N. N. 2019. Improved Internet of Things (IoT) monitoring system for growth optimization of Brassica chinensis. *Comput. Electron. Agric.* 164:104836.

Hashimoto, Y. 1989. Recent strategies of optimal growth regulation by the speaking plant concept. *Acta Hortic.* 260(260):115–122.

Helmer, T., Ehret, D. L. and Bittman, S. 2005. Crop assist, an automated system for direct measurement of greenhouse tomato growth and water use. *Comput. Electron. Agric.* 48(3):198–215.

Hooshmand, M., Albaji, M., Boroomand Nasab, S. and Alam zadeh Ansari, N. 2019. The effect of deficit irrigation on yield and yield components of greenhouse tomato (*Solanum lycopersicum*) in hydroponic culture in Ahvaz region, Iran. *Sci. Hortic.* 254:84–90.

Jiménez-Carvajal, C., Ruiz-Peñalver, L., Vera-Repullo, J. A., Jiménez-Buendía, M., Antolino-Merino, A. and Molina-Martínez, J. M. 2017. Weighing lysimetric system for the determination of the water balance during irrigation in potted plants. *Agric. Water Manag.* 183:78–85.

Jones, H. G. 2004. Irrigation scheduling: advantages and pitfalls of plant-based methods. *J. Exp. Bot.* 55(407):2427–2436.

Kamilaris, A., Kartakoullis, A. and Prenafeta-Boldú, F. X. 2017. A review on the practice of big data analysis in agriculture. *Comput. Electron. Agric.* 143:23–37.

Karim, F., Karim, F. and Frihida, A. 2017. Monitoring system using web of things in precision agriculture. *Procedia Comput. Sci.* 110:402–409.

Katsoulas, N., Kittas, C., Dimokas, G. and Lykas, Ch. 2006. Effect of irrigation frequency on rose flower production and quality. *Biosyst. Eng.* 93(2):237–244.

Katsoulas, N. and Stanghellini, C. 2019. Modelling crop transpiration in greenhouses : different models for different applications. *Agronomy* 9(7):392.

Katsoulas, N. and Voogt, W. 2014. Recent trends in salinity control for soilless growing systems management. *Acta Hortic.* 1034(1034):433–442.

Katsoulas, N., Bartzanas, T. and Kittas, C. 2017. Online professional irrigation scheduling system for greenhouse crops. *Acta Hortic.* 1154(1154):221–228.

Katsoulas, N., Elvanidi, A., Ferentinos, K. P., Kacira, M., Bartzanas, T. and Kittas, C. 2016. Crop reflectance monitoring as a tool for water stress detection in greenhouses: a review. *Biosyst. Eng.* 151:374–398.

Katsoulas, N., Savvas, D., Kitta, E., Bartzanas, T. and Kittas, C. 2015. Extension and evaluation of a model for automatic drainage solution management in tomato crops grown in semi-closed hydroponic systems. *Comput. Electron. Agric.* 113:61–71.

Khanna, A. and Kaur, S. 2019. Evolution of Internet of Things (IoT) and its significant impact in the field of Precision Agriculture. *Comput. Electron. Agric.* 157:218–231.

Kim, H. J., Kim, W. K., Roh, M. Y., Kang, C. I., Park, J. M. and Sudduth, K. A. 2013. Automated sensing of hydroponic macronutrients using a computer-controlled system with an array of ion-selective electrodes. *Comput. Electron. Agric.* 93:46–54.

Kim, M., Kim, S., Kim, Y., Choi, Y. and Seo, M. 2015. Infrared estimation of canopy temperature as crop water stress indicator. *Korean J. Soil Sci. Fert.* 48(5):499–504.

King, R. E. and Sigrimis, N. 2001. Editorial. Computational intelligence in crop production. *Comput. Electron. Agric.* 31(1):1–3.

Kirtan, J., Aalap, D., Poojan, P. I. and Manan, S. 2019. A comprehensive review on automation in agriculture using artificial intelligence. *Artif. Intell. Agr.* 2:1–12. ISSN 2589-7217, https://doi.org/10.1016/j.aiia.2019.05.004.

Kläring, H. P. 2001. Strategies to control water and nutrient supplies to greenhouse crops. A review. *Agronomie* 21(4):311–321.

Lakhiar, I. A., Jianmin, G., Syed, T. N., Chandio, F. A., Buttar, N. A. and Qureshi, W. A. 2018. Monitoring and control systems in agriculture using intelligent sensor techniques: a review of the aeroponic system. *J. Sens.* 2018:1–18.

Lee, A. 2010. Water and EC management. *Pract. Hydroponics Greenhouses* 48–54.

Li, Y. L., Stanghellini, C. and Challa, H. 2002. Response of tomato plants to a step-change in root-zone salinity under two different transpiration regimes. *Sci. Hortic.* 93(3–4):267–279.

Lieth, J. H. and Oki, L. R. 2008. Irrigation in soilless production systems. In: Raviv, M. and Lieth, J. H. (Eds). *Soilless Culture: Theory and Practice* (1st edn.). Elsevier, London, pp. 117–155.

Liu, Z. and Xu, Q. 2018. An automatic irrigation control system for soilless culture of lettuce. *Water* 10(11):1–9.

Lizarraga, A., Boesveld, H., Huibers, F. and Robles, C. 2003. Evaluating irrigation scheduling of hydroponic tomato in Navarra, Spain. *Irrig. and Drain.* 52(2): 177–188.

Loubna, H. and Bouchaib, N. 2018. Integration of irrigation system with wireless sensor networks: prototype and conception of intelligent irrigation system. In: Proceedings of the World Congress on Engineering and Computer Science 2018 Vol I WCECS, San Francisco, USA, pp. 23–25.

Madakam, S., Ramaswamy, R. and Tripathi, S. 2015. Internet of things (IoT): a literature review. *J. Comput. Commun.* 03(5):164–173.

Magán, J. J., Moreno, N. I., Meca, D. and Cánovas, F. 2004. Response to salinity of a tomato crop in Mediterranean climate conditions. *Acta Hortic.* 644(644):479–484.

Massa, D., Incrocci, L., Maggini, R., Carmassi, G., Campiotti, C. A. and Pardossi, A. 2010. Strategies to decrease water drainage and nitrate emission from soilless cultures of greenhouse tomato. *Agric. Water Manag.* 97(7):971–980.

Mavrogiannopoulos, G. N. 2015. Irrigation dose according to substrate characteristics, in hydroponic systems. *Open Agric.* 1(1):1–6.

Medrano, E., Lorenzo, P., Sánchez-Guerrero, M. C. and Montero, J. I. 2005. Evaluation and modeling of greenhouse cucumber-crop transpiration under high and low radiation conditions. *Scientia Horticul.*, 105:163–175.

Mehra, M., Saxena, S., Sankaranarayanan, S., Tom, R. J. and Veeramanikandan, M. 2018. IoT based hydroponics system using deep neural networks. *Comput. Electron. Agric.* 155:473–486.

Meric, M. K., Tuzel, I. H., Tuzel, Y. and Oztekin, G. B. 2011. Effects of nutrition systems and irrigation programs on tomato in soilless culture. *Agric. Water Manag.* 99(1):19–25.

Miranda, J., Ponce, P., Molina, A. and Wright, P. 2019. Sensing, smart and sustainable technologies for Agri-Food 4.0. *Comput. Ind.* 108:21–36.

Morales, I., Álvaro, J. E. and Urrestarazu, M. 2014. Contribution of thermal imaging to fertigation in soilless culture. *J. Therm. Anal. Calorim.* 116:1033–1039.

Morimoto, T. and Hashimoto, Y. 1998. An intelligent control technique based on fuzzy controls, neural networks and genetic algorithms for greenhouse automation. *IFAC Artif. Intell. Agriculture* 31(5):61–66.

Nandong, J. 2015. A unified design for feedback-feedforward control system to improve regulatory control performance. *Int. J. Control Autom. Syst.* 13(1):1–16.

Neocleous, D. and Savvas, D. 2017. Modelling $Ca^{2+}$ accumulation in soilless zucchini crops: physiological and agronomical responses. *Agric. Water Manag.* 203:197–206.

Neocleous, D. and Savvas, D. 2019. The effects of phosphorus supply limitation on photosynthesis, biomass production, nutritional quality, and mineral nutrition in lettuce grown in a recirculating nutrient solution. *Sci. Hortic.* 252:379–387.

Nikolaou, G., Neocleous, D., Katsoulas, N. and Kittas, C. 2017a. Effect of irrigation frequency on growth and production of a cucumber crop under soilless culture. *Emirates J. Food Agric.* 29(11):863–871.

Nikolaou, G., Neocleous, D., Katsoulas, N. and Kittas, C. 2017b. Modelling transpiration of soilless greenhouse cucumber and its relationship with leaf temperature in a Mediterranean climate. *Emirates J. Food Agric.* 29(12):911–920.

Nikolaou, G., Neocleous, D., Katsoulas, N. and Kittas, C. 2017c. Irrigation management techniques used in soilless cultivation. In: Webster, D. J. (Ed), *Advances in Hydroponic Research* (vol. 2017). Nova Science Publishers, Inc., New York, NY. pp. 1–33.

Nikolaou, G., Neocleous, D., Katsoulas, N. and Kittas, C. 2019. Irrigation of greenhouse crops. *Horticulturae* 5(1):7.

Ondrašek, G., Romić, D., Borošić, J., Romić, M. and Stricević, I. 2007. Comparison of transpiration models in tomato soilless culture. *Agric. Conspect. Sci.* 72(2):113–116.

Palande, V., Zaheer, A. and George, K. 2018. Fully automated hydroponic system for indoor plant growth. *Procedia Comput. Sci.* 129:482–488.

Pérez-Castro, A., Sánchez-Molina, J. A., Castilla, M., Sánchez-Moreno, J., Moreno-Úbedac, J. C. and Magán, J. J. 2017. FertigUAL : a fertigation management app for greenhouse vegetable crops. *Agric. Water Manag.* 183:186–193.

Pires, R. C. M., Furlani, P. R., Ribeiro, R. V., Bodine Junior, D., Sakai, E., Lourenção, A. L. and Torre Neto, A. 2011. Irrigation frequency and substrate volume effects in the growth and yield of tomato plants under greenhouse conditions. *Sci. Agric. (Piracicaba Braz.)* 68(4):400–405.

Putra, P. A. and Yuliando, H. 2015. Soilless culture system to support water use efficiency and product quality: a review. *Agric. Agric. Sci. Procedia* 3:283–288.

Rahman, M. K. I. A., Buyamin, S., Abidin, M. S. Z. and Mokji, M. M. 2018. Development of automatic mixing process for fertigation system in rock melon cultivation. *IJECE* 8(3):1913–1919.

Rodriguez-Ortega, W. M., Martinez, V., Rivero, R. M., Camara-Zapata, J. M., Mestre, T. and Garcia-Sanchez., F. 2016. Use of a smart irrigation system to study the effects of irrigation management on the agronomic and physiological responses of tomato plants grown under different temperatures regimes. *Agric. Water Manag.* 183:158–168.

Romero, R., Muriel, J. L., García, I. and Muñoz de la Peña, D. 2012. Research on automatic irrigation control: state of the art and recent results. *Agric. Water Manag.* 114:59–66.

Rose, D. C. and Chilvers, J. 2018. Agriculture 4.0: broadening responsible innovation in an era of smart farming. *Front. Sustain. Food Syst.* 2:0–7.

Saha, U. K., Papadopoulos, A. P., Hao, X. and Khosla, S. 2008. Irrigation strategies for greenhouse tomato production on rockwool. *Hortscience* 43(2):484–493.

Savvas, D. 2002. Nutrient solution recycling. In: Savvas, D. and Passam, H. C. (Eds), *Hydroponic Production of Vegetables and Ornamentals*. Embryo Publications, Athens, Greece, pp. 299–344.

Savvas, D. and Gruda, N. 2018. Application of soilless culture technologies in the modern greenhouse industry–a review. *Europ. J. Hortic. Sci.* 83(5):280–293.

Savvas, D., Ntatsi, G. and Barouchas, P. E. 2017. Soil conservation, soil fertility, and plant nutrition management. In: Baudoin, W., Nersisyan, A., Shamilov, A., Hodder, A., Gutierrez, D., de Pascale, S., Nicola, S., Gruda, N., Urban, L. and Tanny, J. (Eds), *Good Agricultural Practices for Greenhouse Vegetable Production in the South East European Countries for Greenhouse Vegetable; Principles for Sustainable Intensification of Smallholder Farms; Plant Production and Protection Paper 230*. Food and Agriculture Organization, Rome, Italy, pp. 53–79.

Schröder, F.-G. and Lieth, J. H. 2002. Irrigation control in hydroponics. In: Savvas, D. and Passam, H. C. (Eds), *Hydroponic Production of Vegetables and Ornamentals*. Embryo Publications, Athens, Greece, pp. 211–263.

Seginer, I., Kantz, D., Levav, N. and Peiper, U. M. 1990. Night-time transpiration in greenhouses. *Sci. Hortic.* 41(3):265–276.

Semananda, N. P. K., Ward, J. D. and Myers, B. R. 2018. A semi-systematic review of capillary irrigation: the benefits, limitations, and opportunities. *Horticulturae* 4(3):23.

Shamshiri, R. R., Kalantari, F., Ting, K. C., Hammed, I. A., Weltziec, C., Ahmad, D. and Shad, Z. M. 2018. Advances in greenhouse automation and controlled environment agriculture: a transition to plant factories and urban agriculture. *Int. J. Agric. Biol. Eng.* 11(1):1–22.

Shelford, T. J., Lau, A. K., Ehret, D. L. and Chieng, S. T. 2004. Comparison of a new plant-based irrigation control method with light-based irrigation control for greenhouse tomato production. *Can. Biosyst. Eng.* 46:1.1–1.6.

Shi, X., An, X., Zhao, Q., Liu, H., Xia, L., Sun, X. and Guo, Y. 2019. State-of-the-art internet of things in protected agriculture. *Sensors* 19(8):1833.

Shin, J. H., Park, J. S. and Son, J. E. 2014. Estimating the actual transpiration rate with compensated levels of accumulated radiation for the efficient irrigation of soilless cultures of paprika plants. *Agric. Water Manag.* 135:9–18.

Shin, J. H. and Son, J. E. 2015. Development of a real-time irrigation control system considering transpiration, substrate electrical conductivity, and drainage rate of nutrient solutions in soilless culture of paprika (Capsicum annuum L.). *Europ.J.Hort. Sci.* 80(6):271–279.

Sigrimis, N., Arvanitis, K. G., Pasgianos, G. D. and Ferentinos, K. 2001. Hydroponics water management using adaptive scheduling with an on-line optimiser. *Comput. Electron. Agric.* 31(1):31–46.

Sihombing, P., Karina, N. A., Tarigan, J. T. and Syarif, M. I. 2018. Automated hydroponics nutrition plant systems using Arduino uno microcontroller based on android. *J. Phys.: Conf. Ser.* 978(1):012014.

Singh, R., Muzzio, F. J., Ierapetritou, M. and Ramachandran, R. 2015. A combined feed-forward/feed-back control system for a QbD-based continuous tablet. *J. Manuf. Process* 3:339-356.

Sonneveld, C. 2002. Composition of nutrient solution. In: Savvas, D. and Passam, H. C. (Eds), *Hydroponic Production of Vegetables and Ornamentals*. Embryo Publications, Athens, Greece, pp. 180-207.

Souza, S. V., Gimenes, R. M. T. and Binotto, E. 2019. Economic viability for deploying hydroponic system in emerging countries: a differentiated risk adjustment proposal. *Land Use Policy* 83:357-369.

Steidle Neto, A. J., Zolnier, S. and de Carvalho Lopes, D. 2014. Development and evaluation of an automated system for fertigation control in soilless tomato production. *Comput. Electron. Agric.* 103:17-25.

Sun, Y., Feng, H. and Liu, F. 2013. Comparative effect of partial root-zone drying and deficit irrigation on incidence of blossom-end rot in tomato under varied calcium rates. *J. Exp. Bot.* 64(7):2107-2116.

Tsirogiannis, I., Katsoulas, N. and Kittas, C. 2010. Effect of irrigation scheduling on gerbera flower yield and quality. *horts* 45(2):265-270.

Tzounis, A., Katsoulas, N., Bartzanas, T. and Kittas, C. 2017. Internet of Things in agriculture, recent advances and future challenges. *Biosyst. Eng.* 164:31-48.

Udink ten Cate, A. J., Bot, G. P. A. and van Dixhoorn, J. J. 1978. Computer control of greenhouse climates. *Acta Hortic.* 87(87):265-272.

Van Os, E., Blok, C., Voogt, W. and Waked, L. 2016. *Water Quality and Salinity Aspects in Hydroponic Cultivation*, Wageningen University, Wageningen, The Netherlands, pp. 1-7.

Van Os, E., Gieling, T. H. and Lieth, J. H. 2008. Technical equipment in soilless production systems. In: Raviv, M. & Lieth, J. H. (Ed.). *Soilless Culture: Theory and Practice* (1st edn.). Elsevier, London, pp. 157-204.

Varlagas, H., Savvas, D., Mouzakis, G., Liotsos, C., Karapanos, I. and Sigrimis, N. 2010. Modelling uptake of Na+ and Cl− by tomato in closed-cycle cultivation systems as influenced by irrigation water salinity. *Agric. Water Manag.* 97(9):1242-1250.

Vera-Repulloa, J. A., Ruiz-Penalverb, L., Jiménez-Buendíaa, M., Rosillob, J. J. and Molina-Martínez, J. M. 2015. Software for the automatic control of irrigation using weighing-drainage lysimeters. *Agric. Water Manag.* 151:4-12.

Vuran, M. C., Salam, A., Wong, R. and Irmak, S. 2018. Internet of underground things in precision agriculture: architecture and technology aspects. *Ad Hoc Netw.* 81:160-173.

Yan, H., Zhang, C., Gerrits, M. C., Acquaha, S. J., Zhang, H., Wu, H., Zhao, B., Huang, S. and Fu, H. 2018. Parametrization of aerodynamic and canopy resistances for modeling evapotranspiration of greenhouse cucumber. *Agric. For. Meteorol.* 262:370-378.

Yong, W., Shuaishuai, L., Li, L., Minzan, L., Ming, L., Arvanitis, K. G., Georgieva, C. and Sigrimis, S. 2018. Smart sensors from ground to cloud and web intelligence. *IFAC Papers OnLine* 17:31-38.

Zamora-Izquierdo, M. A., Santa, J., Martínez, J. A., Martínez, V. and Skarmeta, A. F. 2019. Smart farming IoT platform based on edge and cloud computing. *Biosyst. Eng.* 177:4-17.

# Chapter 10

# Advances in nutrient management modelling and nutrient concentration prediction for soilless culture systems

*Jung Eek Son, Tae In Ahn and Taewon Moon, Seoul National University, Korea*



## 1 Introduction

Closed-loop nutrient-management techniques are essential for sustainable soilless culture systems (SCS) (Gruda, 2019). In closed-loop SCS, ion concentration and ionic balance are important factors for stable management of nutrient solutions (Son and Takakura, 1989). In order to manage the nutrient solutions, direct measurements of the ion concentrations of drained and reused nutrient solutions are required but are not yet commercially feasible (Gieling et al., 2005; Gutierrez et al., 2008). Electrical conductivity (EC), which reflects the apparent concentration of the whole ions, is used as a practical alternative measurement (Son, 1998). The EC of a solution is linearly related to the total equivalents of ions in the solution (Griffin and Jurinak, 1973), and so nutrient solution concentrations can potentially be controlled by measuring EC.

EC measurement of nutrient solutions needs to be precise to optimise nutrient reuse and plant growth while minimising waste (Ahn and Son, 2019). However, attempts to use EC to measure and manage plant nutrition in closed-loop SCS in practice have had mixed results (Savvas and Gizas, 2002). While

http://dx.doi.org/10.19103/AS.2020.0076.11

some studies have shown positive long-term effects on plant growth (Raviv et al., 1998; Ehret et al., 2005), other studies have shown negative effects (Ehret et al., 2005; Zekki et al., 1996). Other studies have shown that EC variation in closed-loop SCS is similar to the variation in open-loop systems, suggesting limited success in control (Hao and Papadopoulos, 2002; Massa et al., 2011; Signore et al., 2016).

Given the range of variables involved, most studies of EC to measure nutrient solution composition have been conducted using controlled experiments within a particular set of parameters; however, it has been difficult to apply these approaches successfully to the more complex environment represented by real production conditions (Zekki et al., 1996; Hao and Papadopoulos, 2002; Incrocci et al., 2006; Rouphael et al., 2016; Signore et al., 2016). The problems associated with the accuracy of using EC are presumed to be inevitable, as nutrient uptake concentration is affected by a wide range of environmental factors (Signore et al., 2016; van Noordwijk, 1990). As an example, in closed-loop SCS the root zone is a complex environment since plants uptake ions selectively. In closed-loop SCS, drainage is also restricted, further complicating changes in the root zone. This can result in potential nutrient imbalances (Olsen, 1950; Neocleous and Savvas, 2016; Savvas and Gizas, 2002; Ko et al., 2013). Accumulation of specific ions can increase deviations in nutrient solutions and reduce crop yields.

Deep learning is one of a range of machine-learning methods based on artificial neural networks. The technique has increasingly been used to draw meaningful conclusions from complex nonlinear data in sectors such as agriculture (Wachinger et al., 2018; Heffernan et al., 2017; Howard et al., 2017). Deep learning has shown higher accuracy than other techniques, making it possible, for example, to analyse complex root-zone conditions in SCS (Fischer and Krauss, 2018; Meng et al., 2017).

## 2 Analysing the relationship between ion activity and electrical conductivity (EC) measurement

Improved measurement and control require addressing a range of problems, including the calculation of ion activity and ionic contribution to EC readings, as well as taking into account changes in ion concentration and balance. These are discussed in this section.

### *2.1 Calculation of ion activity and ionic contribution to EC reading*

The ratio at which each ion contributes to nutrient solution EC is determined by the equivalence and mobility of the ions. It is assumed that, at low nutrient concentration conditions such as SCS, the ratio of the equivalent ionic conductivity is the same as the ratio of the infinite ionic equivalent conductivity (Eq. 1):

$$EC_i = \frac{|z_i|T_i\lambda_i}{\sum|z|T\lambda} \cong \frac{|z_i|T_i\lambda_\infty}{\sum|z|T\lambda_\infty} \quad (1)$$

where $EC_i$ is EC contribution ratio of an ion, $T_i$ is ion activity, $\lambda_i$ is ionic equivalent conductivity, $\lambda_\infty$ is infinite ionic equivalent conductivity, $i$ is an ion.

For a more detailed analysis, the ionic activity except for the ion pair is used as the concentration contributing to the EC, and the ionic activity is calculated using the numerical analysis in the chemical equilibrium state in the solution (Adams, 1977). The radius and the infinite ionic equivalent conductance of the ions used are shown in Table 1.

The chemical equilibrium state is obtained by substituting $C_i$ for the initial value of $A_i$ to obtain the ionic strength ($I$) of the nutrient solution (Eq. 2), and then calculating the activity coefficient $r_i$ from the equation of Debye-Hückel (Eq. 3). The activity ($T_i$) of each ion is the product of the molarity of the ion and the activity coefficient (Eq. 4). The ion pairs are calculated using the activity and the equilibrium constant ($K_{ij}$) of each ion, and then $A_i$ is obtained (Eq. 5). This process is repeated until equilibrium is reached, assuming that $Cl^-$ and $NO_3^-$ do not form cations and ion pairs and that $PO_4^{3-}$ exists in the form of $H_2PO_4^-$.

$$I = \frac{1}{2}\Sigma\left(A_i \cdot z_i^2\right) \quad (2)$$

$$-\log r_i = \frac{Az_i^2\sqrt{I}}{1 + B\overline{a_i}\sqrt{I}\sqrt{I}} \quad (3)$$

$$T_i = r_i A_i \quad (4)$$

$$A_i = C_i - \Sigma\frac{T_i T_j}{K_i} \quad (5)$$

where $z_i$ is ion charge, $I$ is ionic strength, $A$ and $B$ are 0.509 and 0.329 (1atm, 25°C), respectively, $\overline{a_i}$ is effective ion radius (Å), $A_i$ is the molarity of the ion in the chemical equilibrium state, $r$ is activity coefficient, $T_i$ is ion activity (mole/L), $C_i$ is

**Table 1** Effective ion radius and limit equivalent ionic conductivity

| | $K^+$ | $Ca^{2+}$ | $Mg^{2+}$ | $NH_4^+$ | $NO_3^-$ | $SO_4^{2-}$ | $H_2PO_4^-$ | $Na^+$ | $Cl^-$ |
|---|---|---|---|---|---|---|---|---|---|
| Ion radius (Å) | 3.0 | 6.0 | 8.0 | 2.5 | 3.0 | 4,0 | 4.5 | 4.5 | 3.0 |
| Limit equivalent ionic conductivity ($\Omega\cdot cm^2\cdot mol^{-1}$) | 73.5 | 59.5 | 53.5 | 73.55 | 71.46 | 80.02 | 36.0 | 50.1 | 76.4 |

estimated molarity, $K_{ij}$ is equilibrium constant, *i* is ion, *j* means each ion having an opposite polarity to *i*.

### 2.2 Changes in ion concentration and balance

Because a closed-loop SCS reuses its drainage, the variation in nutrient concentration can negatively affect plant growth as the reuse period becomes longer (Kläring, 2001; Moon et al., 2018; van Noordwijk, 1990; Zekki et al., 1996). Equivalent concentrations of individual ions in the mixing tanks of each treatment vary with time. Research suggests ion concentration in the mixing tank does not change according to the replacement cycle. This could mean that other factors besides the replacement cycle may have an effect on changes in concentration. The drainage ratio affects the EC of the drainage (Corwin et al., 2007; Ahn et al., 2010), and reuse of this drainage will affect the concentration of the nutrient solution in the mixed tank and the concentration of each ion. In particular, the volume and concentration of drainage in an EC-based closed-loop SCS is an important factor in determining the rate of dilution of raw water and the rate of input of new nutrients (Savvas and Manos, 1999). The fact that there is no significant difference in growth or yield in closed-loop SCS in the short term indicates that a period of time is required until the imbalance of nutrients is observed as a significant effect (Raviv et al., 1998; Ehret et al., 2005).

Ion balances change over time in SCS, particularly in closed-loop SCS. An example is the change in ratio of $K^+$:$Ca^{2+}$:$Mg^{2+}$ in the cation and $NO_3^-$:$PO_4^{3-}$:$SO_4^{2-}$ in a SCS growing sweet pepper (Fig. 1). Although the absorption of nutrients is different depending on species, Marti and Mills (1991) reported that the absorption of $K^+$ was relatively high in sweet pepper growth, while the absorption of $Mg^{2+}$ was relatively low in $Ca^{2+}$. In this case, changes in the ratio of cations to that of $K^+$ and $Ca^{2+}$ are mainly due to the nutrient absorption characteristics of the plant.

### 2.3 Modelling changes in ionic contribution to EC measurement

The activity coefficients of monovalent ions such as $K^+$, $NO_3^-$ and $H_2PO_4^-$ and divalent ions such as $Ca^{2+}$, $Mg^{2+}$ and $SO_4^{2-}$ in the nutrient solution are typically maintained between 0.8 and 0.9 and between 0.5 and 0.6, respectively. The difference in the activity coefficients of valency of 1 and 2 is due to the ionic strength of each ion and the effective ionic radius of each ion. The divalent ion has a large effective ionic radius (Table 1). Therefore, when the divalent ions of $Ca^{2+}$, $Mg^{2+}$, and $SO_4^{2-}$ remain in the nutrient solution, the ionic strength is increased and the activity is decreased (Son and Takakura, 1989). Models for estimating the ion activity coefficient by various methods have been developed (Pazuki and Rohhani, 2006).

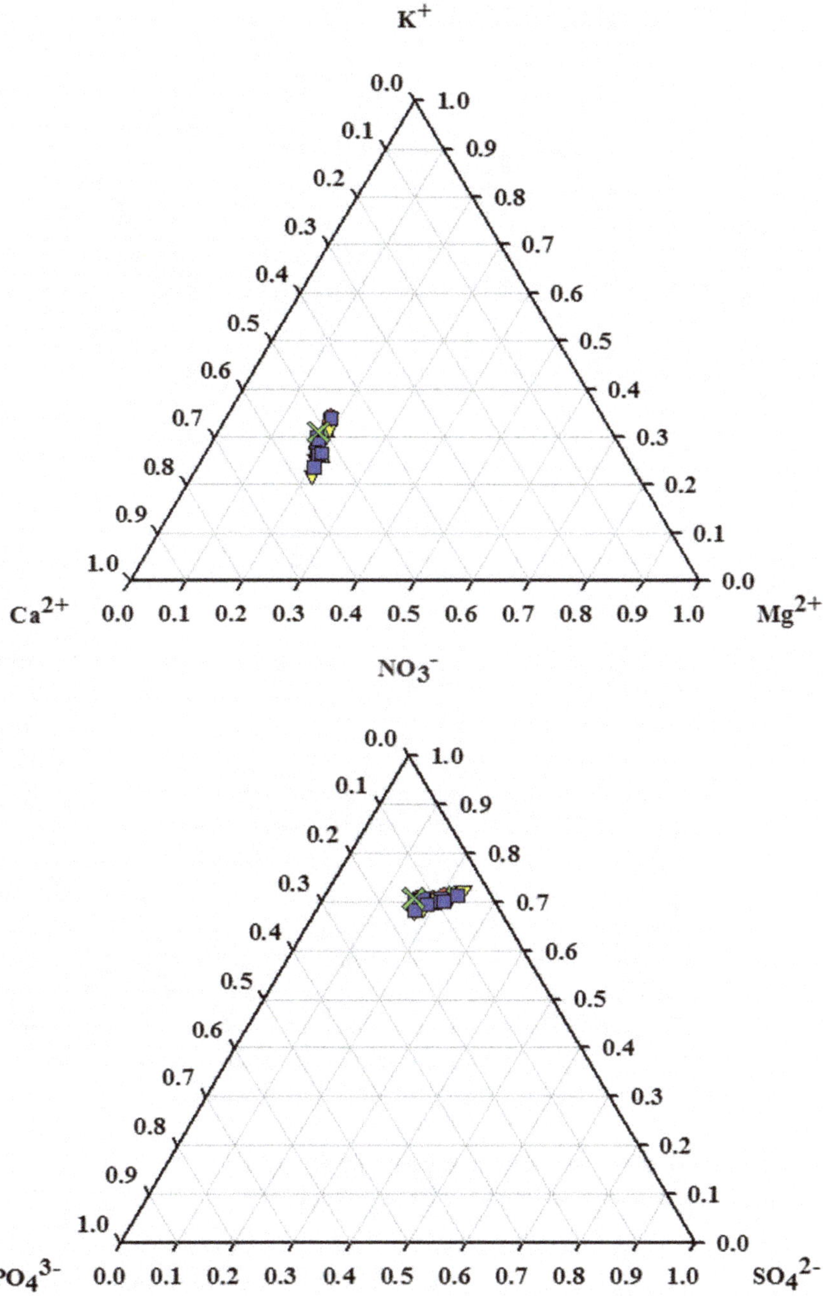


**Figure 1** Change in cation (top) and anion (bottom) balances of the nutrient solution in EC-based closed-loop soilless culture of sweet pepper at different renewal intervals (**×** initial ratio, ● 1- week, ▲ 2- week, ▼ 3- week, and ■ 4- week renewal intervals) (Ahn et al., 2010).

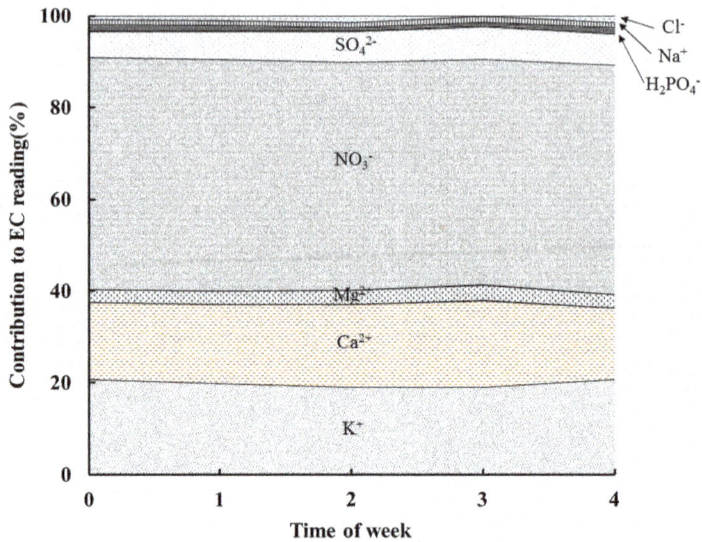


**Figure 2** Change in individual ionic contributions to EC reading in the nutrient solution (Ahn et al., 2010).

In one example, the ratio of each ion contributed to EC in the nutrient solution was highest in $K^+$ and $NO_3^-$, followed by $Ca^{2+}$, $SO_4^{2-}$, $Mg^{2+}$ (Fig. 2). This is because the ratio of $K^+$, $NO_3^-$, and $Ca^{2+}$ in the initial nutrient solution is high. In the case of $Ca^{2+}$, however, because the bivalent ion has extremely low infinite equivalent ionic conductivity and the activity coefficient, the EC contribution rate is usually low. Using EC control of nutrient supply of certain components, if the absorption of the monovalent ions and the persistence of the divalent ions continue, the low monovalent ions have a high EC contribution, and the high divalent ions have a low EC contribution (Son, 1998; Son and Okuya, 1991). Therefore, irregularity of the ion component, accompanied by reduction of monovalent ions and an increase of divalent ions, can be accelerated by the nutrient solution supplied irrespective of the change of the ion component in the nutrient solution. These variables need to be taken account of if EC is to provide an accurate measure of ion concentration and balance.

## 3 Nutrient management modelling in open and closed-loop soilless culture systems

Nutrients in SCS are managed primarily with an open-loop nutrient supply (Bouchaaba et al., 2015; Olympios, 1999). Open-loop SCS are more comfortable to implement, but resource losses are inevitable. Moreover, due to the intensive use of fertilisers, the threat posed to aquatic environments by repeated discharges has required regulation by national governments

(Beerling et al., 2014; Raviv and Lieth, 2008; van Noordwijk, 1990). Closed-loop SCS avoid this problem.

In both closed-loop and open-loop SCS, the EC of the nutrient solution in the mixing tank is adjusted to a target value before the solution is applied to the plant (Putra and Yuliando, 2015; Savvas and Manos, 1999; Zekki et al., 1996). However, unlike an open-loop system, the mixing ratio of tap water to stock solution in a closed-loop system is adjusted by considering the change in nutrient concentration due to the inflow of drainage (Savvas and Manos, 1999). Alternatively, in simple systems, a premixed standard nutrient solution of a certain EC is supplied based on the difference between initial and current water levels in the circulation tank, which simultaneously performs drainage collection and nutrient-solution feeding (Massa et al., 2011; Signore et al., 2016). In an EC-based closed-loop SCS, a supply of stock solution or standard nutrient solution and tap water is intended to replenish nutrients and water consumed in the system (Savvas and Manos, 1999). For a single system in which the plants are grown directly in a nutrient solution container, the nutrients and water consumed due to the absorption of plants in the system can be estimated almost exactly (Anpo et al., 2018). However, errors may occur in systems in which the root zone and nutrient supply are separate from drainage collection. Both are widely used in commercial farming conditions.

An appropriate model is needed to take account of these variables (Eggert, 2005). SCS and plant growth models are constructed by referring to the nutrient transport model in a substrate condition (Ahn and Son, 2019; Raviv and Lieth, 2008; Silberbush and Ben-Asher, 2001; Snape et al., 1995). These models simulate nutrient changes in a SCS with an automated nutrient-mixing system (Fig. 3). This section discusses a number of variables to take account of in developing a suitable model.

### *3.1 Water and nutrient transport in a substrate*

According to standard practices for automated irrigation of a SCS, the mixing process for a nutrient solution is initiated in the mixing tank, and the nutrient solution is supplied to the substrate after mixing. The target nutrients for the simulation were selected as macronutrient cations ($K^+$, $Ca^{2+}$, and $Mg^{2+}$).

$$\frac{dV_n}{dt} = Q_{n-1} - Q_n - T_n \tag{6}$$

The volume of water in a substrate layer ($V_n$, L) is calculated depending on the flow rate of the water from the former ($Q_{n-1}$, L min$^{-1}$) and to the next layer ($Q_n$, L min$^{-1}$) and the evapotranspiration rate ($T_n$, L min$^{-1}$). The flow rate of the water to the first substrate layer ($V_1$) is the irrigation flow rate ($Q_0$). $Q_n$ is the difference between the flow rate of the water from the former layer and the

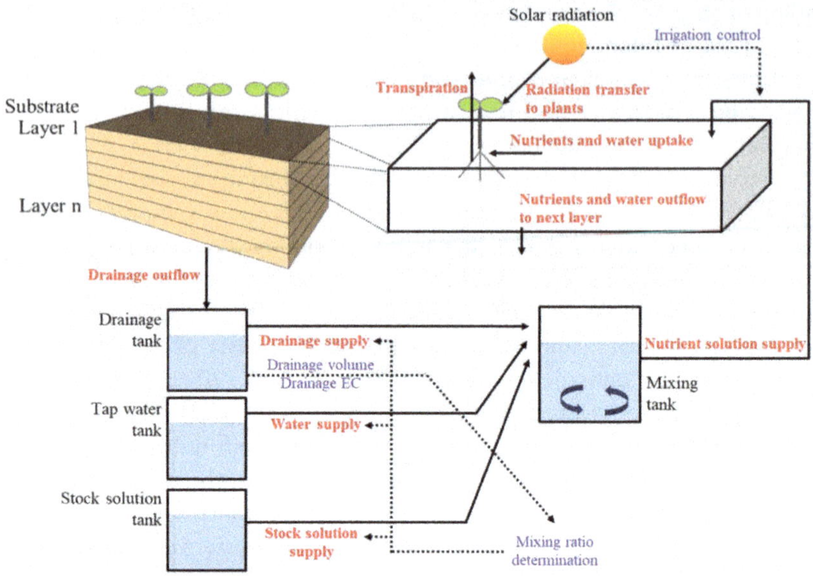


**Figure 3** Schematic description of a SCS in a simulated condition. Solid lines indicate water and nutrient flow, and dotted lines indicate data flow for nutrient solution mixing and irrigation control (Ahn and Son, 2019).

evapotranspiration rate ($Q_{n-1} - T_n$) or the difference between the irrigation rate and the evapotranspiration rate in the first substrate layer ($Q_0 - T_1$) (Shin et al., 2014). The field capacity ($F$, dimensionless) and available water ($W_{DAW}$, dimensionless), respectively, restrict $Q_n$ and $T_n$. The flow for $Q_n$ occurs only when $V_n > S_{sub,n}F$, and $T_n$ flows only when $V_n > S_{sub,n}W_{DAW}$. $S_{sub,n}$ (L) is the volume of the substrate layer $n$.

The flow of nutrients in the medium is generated by the flow rate of water:

$$V_n \frac{dC_n^I}{dt} = Q_{n-1} \cdot C_{n-1}^I - Q_n \cdot C_n^I - P_{RSA} \cdot J_n^I \tag{7}$$

$C$ is the molar concentration of nutrient (mM), superscript $I$ is the type of ions ($K^+$, $Ca^{2+}$, $Mg^{2+}$), $J_n^I$ is the uptake rate of nutrients (mmol $m^{-2}$ $min^{-1}$), and $P_{RSA}$ is the specific root surface area ($m^2$), which is described as the root length density and specific root surface area.

### 3.2 Plant variables and growth parameters

Models need to take into account evapotranspiration and nutrient uptake rates as plant variables in the substrate. Changes in evapotranspiration and nutrient uptake rates occur with plant growth. The relationship between solar radiation and evapotranspiration is adjusted by the leaf area index (Baille et al., 1994). The parameters related to the nutrient uptake rate are derived from the characteristics of the plant ion transporters and are modelled as increasing with growth of the root surface area (Bassirirad, 2000). Variables to take account of include:

- Leaf area index;
- Evapotranspiration;
- Root length density; and
- Nutrient uptake.

These are discussed below.

The Boltzmann sigmoid equation was used to apply changes in leaf area index to the evapotranspiration rate:

$$P_{LAI,t} = \frac{a_{LAI}}{\left[1+e^{\frac{x_0-t}{b_{LAI}}}\right]} \quad (8)$$

where $a_{LAI}$, $b_{LAI}$, and $x_0$ are constants, and $t$ is time.

The evapotranspiration rate is modelled using the simplified Penman-Monteith equation by Baille et al. (1994).

$$T_n = a_T\left[1-e^{-k_{LAI}P_{LAI}}\right]\frac{R}{\lambda}+b_T \quad (9)$$

$T_n$ (L min$^{-1}$, numbers are converted from kg min$^{-1}$) is calculated depending on the radiation for a minute ($R$, MJ m$^{-2}$ min$^{-1}$), the latent heat of vaporisation ($\lambda$, MJ kg$^{-1}$), the light extinction coefficient ($k_{LAI}$), and the leaf area index ($P_{LAI}$). $a_T$ (dimension-less) and $b_T$ (kg m$^{-2}$ min$^{-1}$) are regression parameters.

Root length density is used to calculate specific root surface area and modelled using a logistic function of time (Barber, 1995; Silberbush and Ben-Asher, 2001):

$$P_{len,t} = \frac{RLD_{\max}}{1+K_1e^{-k_1t}} \quad (10)$$

$$P_{RSA,t} = 2\pi r_0 P_{len,t} \quad (11)$$

where $RLD_{\max}$ (m m$^{-3}$) is the maximal root length density, and $K_1$ and $k_1$ are coefficients. $r_0$ is the mean root radius (m).

Root length density is set to start at the top layer of the substrate and be sequentially assigned to the subsequent layer as the value increased. The allocation of root length density for each layer was calculated by dividing $RLD_{max}$ by the total number of layers.

The nutrient uptake rate of the plant in the substrate was simulated as a function of Michaelis–Menten:

$$J_n^l = \frac{J_{\max}^l \left(C_n^l - C_{\min}^l\right)}{K_m^l + \left(C_n^l - C_{\min}^l\right)} \tag{12}$$

where $J_{\max}^l$ is the maximum absorption rate of nutrient $l$, $K_m^l$ is the Michaelis-Menten constant, and $C_{\min}^l$ is the minimal concentration at which $J_n^l = 0$.

### 3.3 Mixing of nutrient solutions

The conventional mixing process for the stock solution, tap water, and drainage under automated closed-loop SCS is performed in the mixing tank (Putra and Yuliando, 2015; Savvas and Manos, 1999). When the system receives an irrigation command, the entire volume of collected drainage is diluted with tap water within the range of the irrigation volume, and the stock solution is added to the target EC. In the case of the open-loop SCS, the collected drainage is not used for the preparation of the nutrient solution supply.

However, because drainage is included in the automated mixing process in closed-loop SCS, the model to allow for mixing stock solution, drainage, and water (Savvas and Manos, 1999). The nutrient solution mixing process occurs intermittently according to the irrigation interval, and the basic equation for conventional nutrient replenishment can be summarised based on the dilution equation:

$$V_T C_T = V_D C_D + V_W C_W + V_S C_S \tag{13}$$

$$V_w = V_T - V_D - V_S \tag{14}$$

$$V_S = \frac{C_T V_T - C_w V_T + C_W V_D - C_D V_D}{C_S - C_W} \tag{15}$$

where $V_T$ (L) is the target irrigation volume per event, $C_T$ (mEq $L^{-1}$) is the target total equivalent concentration, $V_D$ is the drainage volume, $C_D$ (mEq $L^{-1}$) is the total equivalent concentration in drainage, $V_W$ (L) is the amount of tap water input to the mixing tank, $V_S$ (L) is the amount of stock solution input to the mixing tank, $C_w$ (mEq $L^{-1}$) is the total equivalent concentration in tap water, and $C_S$ (mEq $L^{-1}$) is the total equivalent concentration of the stock solution.

Equation 13 can be summarised as Equation 15 by substituting Equation 14 for $V_W$. Equation 15 is calculating the amount of stock solution input based on the total equivalent concentration. In this simulation, the total equivalent concentration is assumed as EC based on the linear relationship between EC and the total equivalent concentration of nutrient solution presented by Savvas and Manos (1999).

The amount of stock solution input to the mixing tank is calculated through this process, and when the irrigation control command is generated during the simulation, the mixing process begins based on the volume of drainage stored in the drainage tank at that moment. If the calculated value of the Equation 15 is less than zero, dilution using tap water could not be adjusted to the target concentration within the range of irrigation amount. In this case, the amount of tap water required for diluting the drainage to target total equivalent concentration ($C_T$) is calculated, and then the ratio between the drainage and calculated tap water was multiplied by $V_T$. In the open-loop SCS, the volume of the drainage in the Equation 13 is set to zero. When the doses of $V_D$, $V_W$, and $V_S$ is determined through the above-mentioned calculation, a flow rate is generated until the corresponding amount is transferred to the mixing tank according to $Q_{drg}$, $Q_{wrt}$, and $Q_{stk}$, respectively. In the simulation, irrigation is controlled by an integral radiation method, which is conventionally used in automated irrigation control (Shin and Son, 2016).

### 3.4 Model refinement

From a whole-system perspective, water consumed by evapotranspiration can be eventually detected through changes in the stored volume of nutrient solution. It is therefore technically advantageous to maintain a certain water level in the system. Furthermore, even though a cultivation system is divided into sub-systems such as root zone, drainage tank, and mixing tank, the approximate concentration of total nutrients in the system can be fed back via EC measurements. This means that, if stock solution input ($V_S$ in Eq. 15) is adjusted according to EC measurement, total nutrient concentration in a closed-loop SCS can be maintained at an average steady-state.

However, almost all closed-loop nutrient management research to date has been based on the problem definition and interpretation domain of an open-loop SCS. This means that the system for determining the influx of nutrients and water into the system has been designed in the same way as the open-loop. The effects of nutrient and water replenishment in a closed-loop system divided into sub-systems such as root zone, drainage tank, and mixing tank (a simple system) cannot be adequately interpreted the same way they would be in a cultivation system with only a single root-zone container. Starting from a single system that controls only the liquid container comprising the root zone, the problem is defined as follows:

$$V_0C_0 = V_1C_1 + V_UC_U \tag{16}$$

where $V_0$ is the initial volume of nutrient solution in the plant root container, $C_0$ is the initial total nutrient concentration in the plant root container, $V_1$ is the current volume of nutrient solution, $C_1$ is the current concentration of total nutrient, $V_U$ is the amount of water absorbed by the plant, and $C_U$ is average total nutrient uptake concentration.

However, when a closed-loop SCS is divided into a substrate, drainage, and mixing tank, $V_0$ and $C_0$ in Equation 16 become the irrigation volume in the mixing tank and the target total nutrient concentration, respectively. That is, replenishment of nutrients and water is not calculated for all nutrients in the system but only for those in the mixing tank. To theoretically remove the errors in calculation associated with nutrients and water replenishment, the system-wide nutrients and water need to be considered as follows:

$$V_0C_0 = V_{drg}C_{drg} + V_{sub}C_{Sub} + V_{mix}C_{mix} + V_UC_U \tag{17}$$

where $V_0$ is the initial volume of water in the system; $C_0$ is the initial total concentration of the system; $V_{drg}$, $V_{sub}$, and $V_{mix}$ are the volumes of water stored in the drainage tank, substrate, and mixing tank, respectively; and $C_{drg}$, $C_{sub}$, and $C_{mix}$ are the total nutrient concentrations in the drainage tank, substrate, and mixing tank.

Summarising the equation with respect to the amount of nutrient uptake ($V_UC_U$), it is possible to calculate the number of total nutrients to be added to maintain an average steady-state in the system.

To calculate the amount of nutrient replenishment in the actual system, Equation 17 needs to be rewritten, with the amount of nutrient uptake ($V_UC_U$) substituted for the input volume of stock solution ($V_{stk}$) for $V_U$ and stock solution concentration for $C_U$. $V_{stk}$ can then be expressed as follows:

$$V_{stk} = \frac{C_0V_0 - C_{drg}V_{drg} - C_{mix}V_{mix} - C_{sub}V_{sub}}{C_{stk}} \tag{18}$$

The variables in Equation 18 are available for setting control parameters in the simulation. For total nutrient concentration, the average steady-state error is observed in the system using a conventional nutrient replenishment method, and while in the system using Equation 18 for nutrient replenishment by stock solution, the total nutrient concentration shows repeated fluctuations near the initial concentration (Ahn and Son, 2019). The amount of total nutrients in the system also increase in the conventional nutrient replenishment system; in the Equation 18 applied system, it stayed near the initial value (Ahn and Son, 2019).

Precise measurements for the variables in Equation 18 in a real cultivation system have technical limitations. In particular, the amounts of total nutrients $C_{sub}$ and $V_{sub}$ in the substrate are difficult to estimate. In a SCS, the field capacity ($F$) of a substrate corresponds to the parameters of the system, and the volume of water cannot exceed the volume of the root zone multiplied by field capacity ($F$). The EC of the drainage ($C_{drg}$) can be indicative of a change in concentration of substrate. Considering this, the Equation 18 can be modified as follows:

$$V_{stk} = \frac{C_0 V_0 - C_{drg} V_{drg} - C_{mix} V_{mix} - C_{drg} F}{C_{stk}} \tag{19}$$

where $C_{drg}$ and $F$ are substituted for $C_{sub}$ and $V_{sub}$, respectively.

However, in this case, total ion concentration fluctuates around the initial concentration (Fig. 4).

For these analyses to be implemented, the Equation 19 must be reviewed from the perspective of control engineering. In Equation 19, all, but $C_{drg}$, can be viewed as parameters. That is, the process of calculating the difference between $C_0 V_0$ and the product of the parameters and $C_{drg}$ is performed in every mixing process. From this point of view, Equation 19 is structurally similar to proportional control:

$$u(t) = K_p e(t) \tag{20}$$

where $u$ is controller output, $t$ is time, $K_p$ is the proportional gain, and $e$ is instantaneous process error at $t$ (Golnaraghi and Kuo, 2010).

In the existing problem definition, the EC variation in the closed-loop SCS is derived from the dynamic change of the uptake concentration; thus, there are restrictions on active control and interpretation. However, a series of analysis steps leading to Equation 20 makes it possible to convert EC control

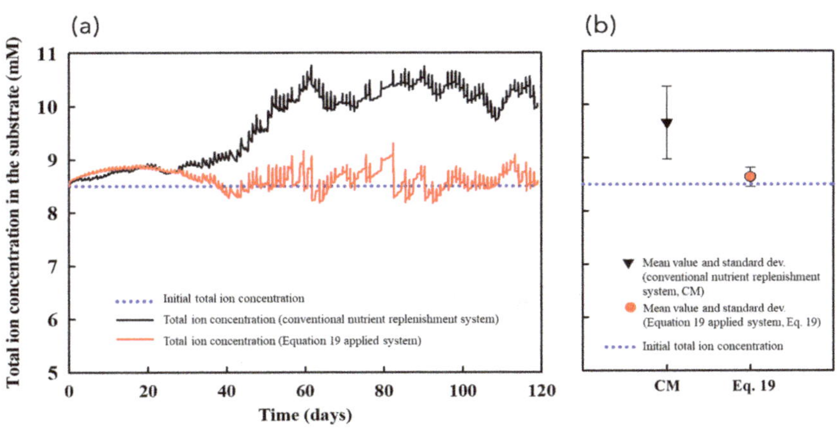


**Figure 4** (a) Changes in total ion concentration in the substrate according to the nutrient-replenishment method (b) and mean and standard deviation of total ion concentration in the substrate according to the nutrient-replenishment method (Ahn and Son, 2019).

in the closed-loop SCS to the problem of proper gain search through arbitrary adjustment of system parameters.

### 3.5 Modelling in practice

Figure 5 shows the experimental results of the EC control of the rock wool substrate and drainage using Equation 19 in a closed-loop SCS. The EC of substrate and drainage in the Equation 19 applied system was maintained near the initial value of the system. In a closed-loop SCS, the increase in EC can lead to the discharge of drainage (Massa et al., 2011; Savvas and Manos, 1999; Signore et al., 2016). The Equation 19 applied system ensured stability in stored drainage volume in the drainage tank compared to the conventional replenishment method. The average level of stored drainage volume in the system applied with the conventional replenishment method was higher than in the Equation 19 applied system, and the range of variation was wider (Ahn and Son, 2019).

The mixing ratio of drainage, water, and stock solution in the conventional nutrient solution mixing process depends on the target EC for the irrigation solution. However, this could generate significant fluctuations in the stored volume of drainage. No increasing or decreasing trend in EC or stored drainage volume can be inferred over the entire experimental period in the closed-loop system. In the conventional replenishment method, the EC of the rock wool substrate was higher, and a gradual increase was observed. A high EC value in a closed-loop SCS where concentration control of the recycled nutrient solution by an operator is carried out can lead to an increase in the volume of stored drainage solution and subsequently to discharge of drainage when it exceeds system capacity (Massa et al., 2011; Signore et al., 2016). This can be a factor in system instability.

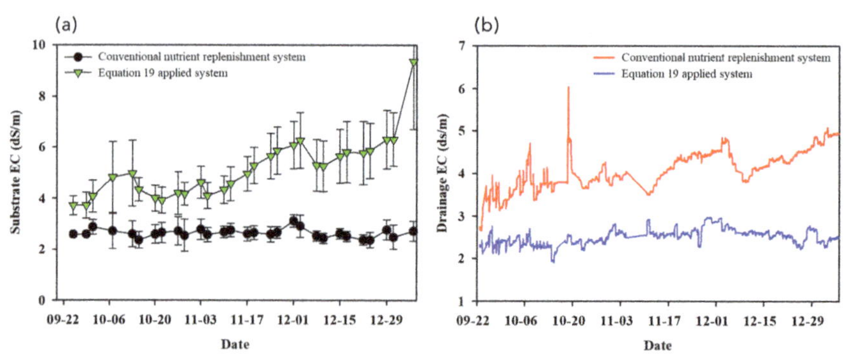


**Figure 5** (a) Comparison of electrical conductivity (mean ± SD) in the rock wool substrate and the drainage (b) of the closed-loop SCS during the experimental period between conventional and Equation (19)-applied nutrient-replenishment methods (Ahn and Son, 2019).

However, not only the control of the total nutrient concentration, the behaviour of concentration changes in individual nutrients according to the EC control needs to be considered. In the conventional replenishment method, increasing $Ca^{2+}$ and $Mg^{2+}$ and decreasing $K^+$ are observed (Fig. 6). In the Equation 19 applied system, $Ca^{2+}$ and $Mg^{2+}$ concentrations are stable at a level relatively close to the initial value, but $K^+$ values show a rapid decline and then fluctuated at a low concentration. For the conventional replenishment method, variations in nutrient concentrations similar to those reported in previous studies were observed (Hao and Papadopoulos, 2002; Signore et al., 2016; Zekki et al., 1996). Previous research on closed-loop SCS has determined that nutrient variations are a result of dynamic changes in nutrient uptake concentrations, and following those changes is challenging. (Carmassi et al., 2007; Savvas et al., 2007; Signore et al., 2016; van Noordwijk, 1990). Figure 6a–c, however, indicates that a more deterministic change occurred in the system when nutrient replenishment was synchronised with total nutrient uptake through the Equation 19 applied system.

The cumulative standard deviations of the Equation 19 applied system maintain a lower level than those of the conventional replenishment method during the entire experimental period, and gradually decreasing tendencies are observed in $K^+$ and $Mg^{2+}$ for the Equation 19 applied system (Fig. 6d–f). Considering the nutrient variations of the Equation 19 applied system itself, there may be a limit in defining it as steady-state in the strict sense. However, in the actual cultivation conditions in this experiment, the input of nutrients and water into the root zone by irrigation occurs intermittently, and the variation

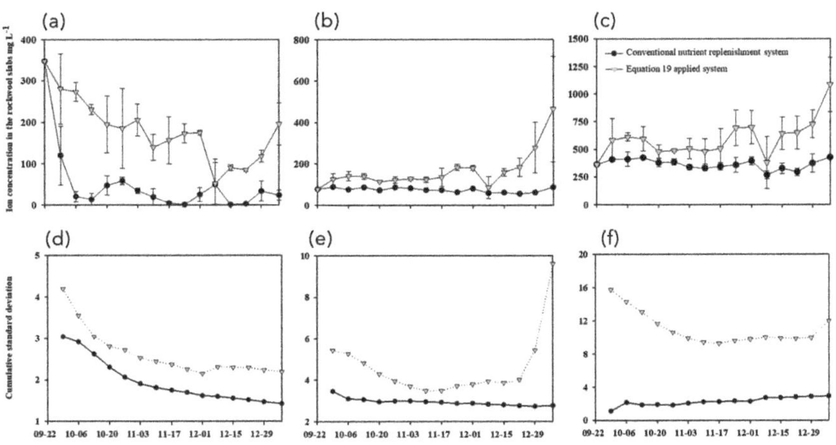


**Figure 6** Changes in nutrient concentrations (mean ± SD) of $K^+$ (a), $Mg^{2+}$ (b), and $Ca^{2+}$ (c) and changes in cumulative standard deviation of nutrient concentrations of $K^+$ (d), $Mg^{2+}$ (e), and $Ca^{2+}$ (f) in the rock wool substrates under the conventional and Equation (19)-applied nutrient-replenishment methods, respectively (Ahn and Son, 2019).

in the section where no input occurs cannot be controlled until the next input event. Furthermore, the frequency of changes in such input can affect system fluctuations (Savvas and Gruda, 2018; Ta et al., 2012; Xu et al., 2004), and the Equation 19 applied system is also affected. When these constraints are considered, it can be assumed that the Equation 19 applied system entered an average steady-state that fluctuates within a certain range. The nutrient concentration control in the SCS can, therefore, be seen as shifting the fluctuation range of the average steady-state to the target range through a compositional change in the stock nutrient solution.

## 4 Prediction of electrical conductivity (EC) and macronutrient ion concentrations using deep learning algorithms in closed-loop soilless culture systems

As a deep learning methodology, recurrent neural networks (RNNs) are used to analyse time-series data, such as voice recognition, video recognition, and natural language processing. Long short-term memory (LSTM) is a representative structure of RNN algorithms (Hochreiter and Schmidhuber, 1997). LSTM has shown state-of-the-art performance and higher accuracy than previous algorithms in many applications. The core of the LSTM algorithm is a cell with several gates (Fig. 7). Symbols *h* and $\sigma$ represent a hyperbolic tangent function and a sigmoid function, respectively. LSTM not only multiplies but also

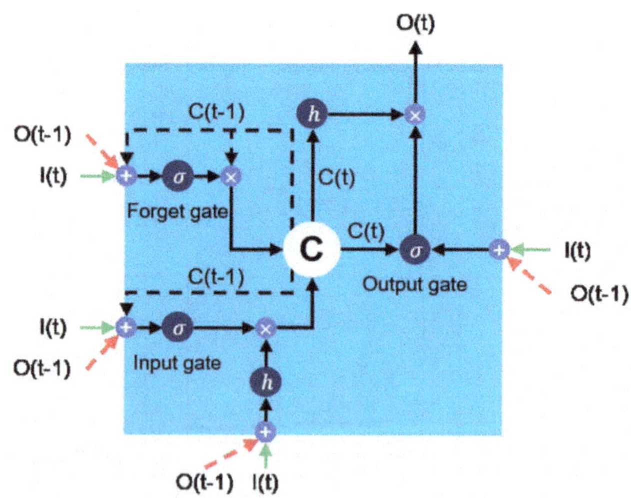


**Figure 7** A structure of long short-term memory (LSTM). I, inputs; O, outputs; C, cell states; *h*, *tanh* activation function; σ, sigmoidal activation function; *t* and *t-1*, current and previous times, respectively (Moon et al., 2018).

adds sequence information, which makes LSTM deliver previous information within a long sequence. The root-zone environment is also influenced by the accumulation of time-series factors such as the greenhouse environment, water supply, water drainage, and plant growth. Therefore, EC and macronutrient ion concentrations in closed-loop SCS can be predicted with LSTM using greenhouse environmental data and plant growth data.

As an example, a LSTM of a single layer with 64 nodes with a test accuracy with $R^2 = 0.72$ was the best model with the highest accuracy. Inputs included: EC of a substrate, moisture content of the substrate, EC of nutrient solutions in the drainage tank, the volume of nutrient solutions in the drainage tank, cumulative drainage volume per day, EC of nutrient solutions in the mixing tank, the volume of nutrient solutions in the mixing tank, mixing volume of drainage, mixing volume of water, mixing volume of stock solution, cumulative irrigation volume per day, preset radiation integral for irrigation control, the target volume of nutrient solutions per irrigation event per dripper, $CO_2$ concentration, light intensity, temperature, relative humidity, growth stage, plant height, and the number of nodes. Based on the flexibility of the deep learning, inputs can be differed by users.

In general, deep learning algorithms can be more accurate when tested with large volumes of data to generalise all possible conditions (Lopez et al., 2001). Other agricultural studies using deep learning have been conducted with big data sets with long collecting periods to cover variables such as seasonal influences (Trejo-Perea et al., 2009; Wang et al., 2017). Virtual conditions via simulation can also be helpful for training the neural network (Beltramo et al., 2016). Moreover, if similar data is collected, the trained LSTM can be applied to other periods or other plants using transfer learning (Gao et al., 2014; Shin and Son, 2016).

The trained LSTM was able to predict changes in EC with little deviation between predicted and measured values (Fig. 8). The difference between accuracies resulted from generalising the entire range of data. Predicting EC changes during the day is difficult because the plants affected water and nutrients via transpiration. Underestimated or overestimated predictions between 6 am and 12 am could be the result of increasing transpiration. In addition, transpiration significantly affects the uptake of nutrients, which is related to the change in root-zone EC and varies with the growth stage (Van Noordwijk, 1990; Baille et al., 1994; Le Bot et al., 1998). Therefore, a variation of the root-zone EC can be larger depending on the growth stage, even when the drainage rate is controlled (Massa et al., 2011; Shin and Son, 2016).

The trained LSTM can be applied to control nutrient solutions in closed-loop SCS based on the prediction of future EC. The algorithm could make planned management of nutrient solutions possible, reducing resource wastes. Deep learning algorithms can be more accurate with additional data, and thus other environmental factors or plant growth data can improve model robustness.

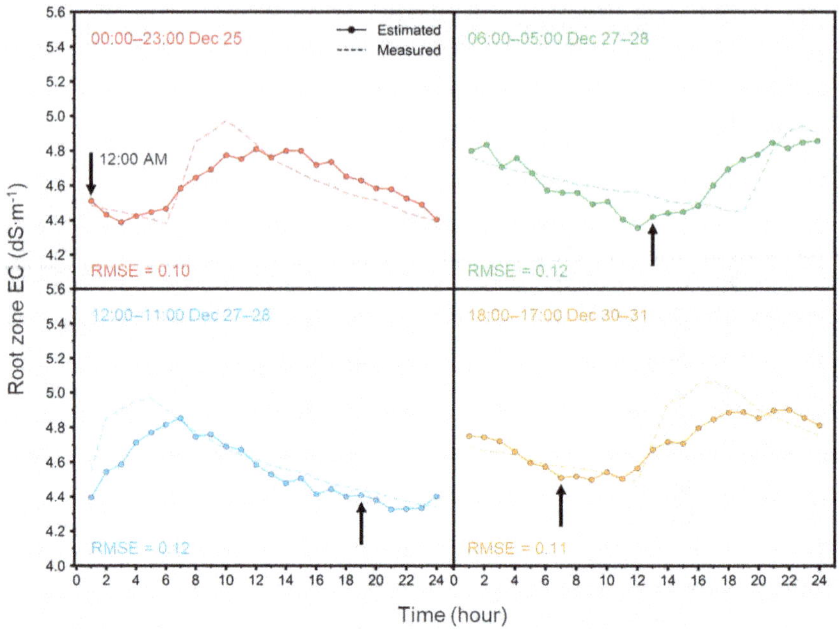


**Figure 8** Chronological comparisons of predicted root-zone electrical conductivity (EC) via trained LSTM and measured data from 0:00 to 23:00 on December 25, from 6:00 on December 26 to 5:00 on December 27, from 12:00 on December 27 to 11:00 on December 28, and from 18:00 on December 29 to 17:00 on December 30. Arrows represent the point of 00:00 (Moon et al., 2018).

## *4.1 Applying deep learning in practice*

Ion uptake in plants differs depending on the individual ions and varies with the growth stage. In a case study, the LSTM was appropriately trained for six different ions. After model training, accuracy $R^2$ values ranged between 0.51 and 0.78 for six macro ions. In an open-loop SCS, only the concentration of water and stock solution should be considered when replenishing the nutrient solution (Pardossi et al., 2006; Massa et al., 2010). However, restricted drainage in closed-loop SCS resulted in wider variation in ion concentrations, which made prediction of ion concentrations difficult. Despite the wider variation in ion concentrations, the accuracies were still high enough to say the model was relatively robust.

Predicting the concentration of $Ca^{2+}$ showed the highest accuracy, while the concentration of $K^+$ had the lowest accuracy. The accuracies of the ion concentrations varied per ion, which may be typical for nutrient solutions as previous studies have shown that variability in ion concentrations are dependent on each ion (Jung et al., 2015; Cho et al., 2018). Interpolation of ion concentrations was smooth at the beginning of the experiment, but significant variations occurred later (Fig. 9). The deviance of both the ions and actual measured ion concentrations was large (Fig. 9a), but the ratio of the ions changed in a reasonable range (Fig. 9b).

One characteristic of a closed-loop condition is frequent variation in ion concentrations towards the later stages of cultivation (Ko et al., 2013). The model will have higher accuracy if the interval and time step are adjusted. However, the intervals cannot improve accuracy in cases when the time step was longer than one week. Similarly, the structure of the LSTM cannot significantly affect accuracy. The dimension of the input data processed at one time is usually much larger than the dimension of the data used for this study when the LSTM is used (Brunner et al., 2018; Cho et al., 2014). Because the dimension of the

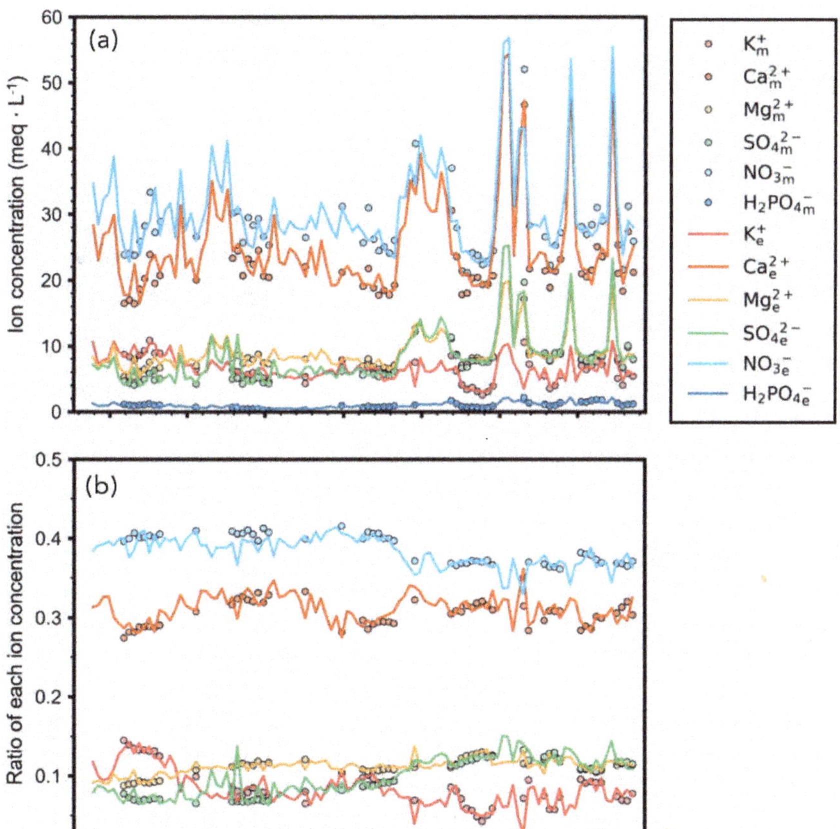


**Figure 9** Plant height, diameter, number of nodes, SPAD of upper leaves, and SPAD of lower leaves measured from January 12 to April 26, 2018. Average values of five plants from each line were used to train the LSTM (Moon et al., 2019).

data in this case is about 20, the relationship between the ion concentrations and the plant environment may be found once a sufficient time step and interval are determined. Likewise, the number of layers cannot affect accuracy when the data are simple enough that a multilayer is unnecessary. In addition, the LSTM is a model-free method, and thus the application and modification of the model are relatively easy. Because most of the greenhouse environmental data is time-series data, it can be used to estimate plant growth if the same model is trained using different data. The trained LSTM can also be applied to different domains or conditions using relatively small amounts of data if transfer learning is used (Glorot et al., 2011), that is, the trained LSTM could be applied to estimate ion concentrations under different conditions such as other greenhouses, plant factories, or crops.

## 5 Conclusion and future trends

In closed-loop SCS, ion concentration and ionic balance are essential factors for stable management of nutrient solutions. Several methods for nutrient analysis and prediction are possible in maintaining appropriate ion concentration and ion balance. Previous techniques for EC-based closed-loop SCS-measured nutrient variations were mainly based on discrepancies between supplied nutrient concentrations and uptake concentrations, and through using data from controlled experiments.

Through nutrient management models and simulations, the problem of EC variation in closed-loop SCS can be analysed with much greater precision, making it possible to develop more advanced SCS. Trained deep learning models can be applied to different cultivation conditions or technologies such as plant factories, and other crops and/or varieties. Applications of theoretical modelling and artificial intelligence can be useful for the nutrient management of closed-loop SCS in greenhouses and vertical farms. These algorithms make it possible to predict and plan the management of nutrient solutions, reducing resource waste. Moreover, deep learning is a black-box model, so, in any situation with enough data, the deep learning approach can offer unprecedented information about the cultivation environment.

The need for research on sustainable resource management continues to grow. Fundamentally, this requires closing the loop in the supply chain for water and nutrients without significant losses in yields. However, the promotion of plant growth and improvement of nutrient-use efficiency can be conflicting priorities. Deep learning techniques such as RNN methodology can help reconcile these priorities, for example, by a better understanding of interactions in the root-zone environment. In particular, RNN can be used to analyse accumulative changes and make more effective predictions to ensure more precise nutrient measurement and supply.

## 6 Where to look for further information

The following articles provide a good overview of the subject:

- Ahn, T. I. and Son, J. E. (2019). Theoretical and experimental analysis of nutrient variations in electrical conductivity-based closed-loop soilless culture systems by nutrient replenishment method. *Agronomy* 9, 649.
- Gruda, N. S. (2019). Increasing sustainability of growing media constituents and stand-alone substrates in soilless culture systems. *Agronomy* 9, 298.
- Moon, T., Ahn, T. I. and Son, J. E. (2019). Long-short term memory for a model-free estimation of macronutrient ion concentrations of root-zone in closed-loop soilless cultures. *Plant Methods* 15, 59.
- Savvas, D. and Gruda, N. (2018). Application of soilless culture technologies in the modern greenhouse industry - A review. *Eur. J. Hortic. Sci.* 83, 280–293.

## 7 References

Adams, F. (1977). Ionic concentration and activity in soil solution. *Soil Sci. Soc. Amer. Proc.* 35, 420–426.

Ahn, T. I., Shin, J. W. and Son, J. E. (2010). Analyses of changes in ion concentration with time and drainage ratio under EC-based nutrient control in closed-loop soilless culture for sweet pepper plants (*Capsicum annuum* L. 'Boogie'). *J. Bio-Env. Cont.* 19, 298–304.

Ahn, T. I. and Son, J. E. (2019). Theoretical and experimental analysis of nutrient variations in electrical conductivity-based closed-loop soilless culture systems by nutrient replenishment method. *Agronomy* 9(10), 649.

Anpo, M., Fukuda, H. and Wada, T. (2018). *Plant Factory Using Artificial Light: Adapting to Environmental Disruption and Clues to Agricultural Innovation* (1st edn.). Elsevier Science, Amsterdam.

Baille, M., Baille, A. and Laury, J. C. (1994). A Simplified model for predicting evapotranspiration rate of nine ornamental species vs climate factors and leaf-area. *Sci. Hortic.* 59(3–4), 217–232.

Barber, S. A. (1995). *Soil Nutrient Bioavailability: A Mechanistic Approach*. John Wiley and Sons, New York.

Bassirirad, H. (2000). Kinetics of nutrient uptake by roots: responses to global change. *New Phytol.* 147(1), 155–169.

Beerling, E. A. M., Blok, C., van der Maas, A. A. and van Os, E. A. (2014). Closing the water and nutrient cycles in soilless cultivation systems. *Acta Hortic.* 1034(1034), 49–55.

Beltramo, T., Ranzan, C., Hinrichs, J. and Hitzmann, B. (2016). Artificial neural network prediction of the biogas flow rate optimised with an ant colony algorithm. *Biosyst. Eng.* 143, 68–78.

Bouchaaba, Z., Santamaria, P., Choukr-Allah, R., Lamaddalena, N. and Montesano, F. F. (2015). Open-cycle drip vs closed-cycle subirrigation: effects on growth and yield of greenhouse soilless green bean. *Sci. Hortic.* 182, 77–85.

Brunner, G., Wang, Y., Wattenhofer, R. and Weigelt, M. (2018). *Natural Language Multitasking: Analysing and Improving Syntactic Saliency of Hidden Representations. arXiv Preprint*. arXiv:1801.06024.

Carmassi, G., Incrocci, L., Maggini, R., Malorgio, F., Tognoni, F. and Pardossi, A. (2007). An aggregated model for water requirements of greenhouse tomato grown in closed rockwool culture with saline water. *Agric. Water Manag.* 88(1-3), 73-82.

Cho, K., Van Merriënboer, B., Gulcehre, C., Bahdanau, D., Bougare,S. F., Schwenk, H. and Bengio, Y. (2014). *Learning Phrase Representations Using RNN Encoder-Decoder for Statistical Machine Translation. arXiv Preprint*. arXiv:1406.1078.

Cho, W. J., Kim, H. J., Jung, D. H., Kim, D. W., Ahn, T. I. and Son, J. E. (2018). On-site ion monitoring system for precision hydroponic nutrient management. *Comput. Electron. Agric.* 146, 51-58.

Corwin, D. L., Rhoades, J. D. and Simunek, J. (2007). Leaching requirement for soil salinity control: steady-state versus transient models. *Agric. Water. Manag.* 90(3), 165-180.

Eggert, R. J. (2005). *Engineering Design*. Prentice Hall, Pearson, Upper Saddle River, NJ.

Ehret, D. L., Menzies, J. G. and Helmer, T. (2005). Production and quality of greenhouse roses in recirculating nutrient systems. *Sci. Hortic.* 106(1), 103-113.

Fischer,T. and Krauss, C. (2018). Deep learning with long short-term memory networks for financial market predictions. *Eur. J. Oper. Res.* 270(2), 654-669.

Gao, J., Ling, H., Hu, W. and Xing, J. (2014). Transfer learning based visual tracking with gaussian processes regression. In: *European Conference on Computer Vision*. Springer, Swiss, pp. 188-203.

Gieling, T. H., van Straten, G., Janssen, H. J. J. and Wouters, H. (2005). ISE and chemfet sensors in greenhouse cultivation. *Sens. Actuator B* 105(1), 74-80.

Glorot, X., Bordes, A. and Bengio, Y. (2011). Domain adaptation for large-scale sentiment classification: A deep learning approach. In: *Proceedings of the 28th International Conference on Machine Learning. ICML'11*. Institute of Museum and Library Services, Amherst, pp. 513-520.

Golnaraghi, F. and Kuo, B. C. (2010). *Automatic Control Systems* (9th edn.). Wiley, Hoboken, NJ.

Griffin, B. A. and Jurinak, J. J. (1973). Estimation of activity coefficients from the electrical conductivity of natural aquatic systems and soil extracts. *Soil Sci.* 116(1), 26-30.

Gruda, N. S. (2019). Increasing sustainability of growing media constituents and stand-alone substrates in soilless culture systems. *Agronomy* 9(6), 298.

Gutierrez, M., Alegret, S., Caceres, R., Casadesus, J., Marfa, O. and del Valle, M. (2008). Nutrient solution monitoring in greenhouse cultivation employing a potentiometric electronic tongue. *J. Agric. Food Chem.* 56(6), 1810-1817.

Hao, X. and Papadopoulos, A. P. (2002). Growth, photosynthesis, and productivity of greenhouse tomato cultivated in open or closed rockwool systems. *Can. J. Plant Sci.* 82(4), 771-780.

Heffernan, R., Yang, Y., Paliwal, K. and Zhou, Y. (2017). Capturing non-local interactions by long short-term memory bidirectional recurrent neural networks for improving prediction of protein secondary structure, backbone angles, contact numbers, and solvent accessibility. *Bioinformatics* 33(18), 2842-2849.

Hochreiter, S. and Schmidhuber, J. (1997). Long short-term memory. *Neural Comput.* 9(8), 1735-1780.

Howard, A. G., Zhu, M., Chen, B., Kalenichenko, D., Wang, W., Weyand, T., Andreetto, M. and Adam, H. (2017). *Mobilenets: Efficient Convolutional Neural Networks for Mobile Vision Applications*. arXiv Preprint. arXiv:1704.04861.

Incrocci, L., Malorgio, F., Della Bartola, A. and Pardossi, A. (2006). The influence of drip irrigation or subirrigation on tomato grown in closed-loop substrate culture with saline water. *Sci. Hortic.* 107(4), 365–372.

Jung, D. H., Kim, H. J., Choi, G. L., Ahn, T. I., Son, J. E. and Sudduth, K. A. (2015). Automated lettuce nutrient solution management using an array of ion-selective electrodes. *Trans. ASABE* 58, 1309–1319.

Kläring, H.-P. (2001). Strategies to control water and nutrient supplies to greenhouse crops. A review. *Agronomie* 21(4), 311-321.

Ko, M. T., Ahn, T. I. and Son, J. E. (2013). Comparisons of ion balance, fruit yield, water, and fertiliser use efficiencies in open and closed soilless culture of paprika (*Capsicum annuum* L.). Kor. *J. Hortic. Sci. Technol.* 31, 423–428.

Le Bot, J., Adamowicz, S. and Robin, P. (1998). Modelling plant nutrition of horticultural crops: a review. *Sci. Hortic.* 74(1–2), 47–82.

Lopez, G., Rubio, M. A., Martınez, M. and Batlles, F. J. (2001). Estimation of hourly global photosynthetically active radiation using artificial neural network models. *Agric. Forest Meteorol.* 107(4), 279–291.

Marti, H. R. and Mills, H. A. (1991). Nutrient-uptake and yield of sweet pepper as affected by stage of development and N-form. *J. Plant Nutr.* 14(11), 1165–1175.

Massa, D., Incrocci, L., Maggini, R., Bibbiani, C., Carmassi, G., Malorgio, F. and Pardossi, A. (2011). Simulation of crop water and mineral relations in greenhouse soilless culture. *Environ. Modell. Softw.* 26(6), 711–722.

Massa, D., Incrocci, L., Maggini, R., Carmassi, G., Campiotti, C. A. and Pardossi, A. (2010). Strategies to decrease water drainage and nitrate emission from soilless cultures of greenhouse tomato. *Agric. Water Manag.* 97(7), 971–980.

Meng, Z., Watanabe, S., Hershey, J. R. and Erdogan, H. (2017). *Deep Long Short-Term Memory Adaptive Beamforming Networks for Multichannel Robust Speech Recognition. arXiv Preprint*. arXiv:1711.08016.

Moon, T., Ahn, T. I. and Son, J. E. (2018). Forecasting root-zone electrical conductivity of nutrient solutions in closed-loop soilless cultures via a recurrent neural network using environmental and cultivation Information. *Front. Plant Sci.* 9, 859.

Moon, T., Ahn, T. I. and Son, J. E. (2019). Long-short term memory for a model-free estimation of macronutrient ion concentrations of root-zone in closed-loop soilless cultures. *Plant Methods* 15, 59.

Neocleous, D. and Savvas, D. (2016). NaCl accumulation and macronutrient uptake by a melon crop in a closed hydroponic system in relation to water uptake. *Agric. Water Manag.* 165, 22–32.

Olsen, C. (1950). The significance of concentration for the rate of ion absorption by higher plants in water culture. *Physiol. Plant.* 3(2), 152–164.

Olympios, C. M. (1999). Overview of soilless culture: advantages, constraints, and perspectives for its use in Mediterranean countries. *Cah. Options Méditerr.* 31, 307–324.

Pardossi, A., Malorgio, F., Incrocci, L., Carmassi, G., Maggini, R., Massa, D. and Tognoni, F. (2006). Simplified models for the water relations of soilless cultures: what they do or suggest for sustainable water use in intensive horticulture. *Acta Hortic.* 718(718), 425–434.

Pazuki, G. R. and Rohhani, A. A. (2006). A new model for the activity coefficients of individual ions in aqueous electrolyte solutions. *Fluid Phase Equilib.* 242(1), 65–71.

Putra, P. A. and Yuliando, H. (2015). Soilless culture system to support water use efficiency and product quality: a review. *Agric. Sci. Procedia* 3, 283–288.

Raviv, M., Krasnovsky, A., Medina, S. and Reuveni, R. (1998). Assessment of various control strategies for recirculation of greenhouse effluents under semi-arid conditions. *J. Hortic. Sci. Biotechnol.* 73(4), 485–491.

Raviv, M. and Lieth, J. H. (2008). *Soilless Culture: Theory and Practice* (1st edn.). Elsevier, Amsterdam, Boston and Heidelberg.

Rouphael, Y., Raimondi, G., Caputo, R. and De Pascale, S. (2016). Fertigation strategies for improving water use efficiency and limiting nutrient loss in soilless *Hippeastrum* production. *Hortscience* 51(6), 684–689.

Savvas, D. and Gizas, G. (2002). Response of hydroponically grown gerbera to nutrient solution recycling and different nutrient cation ratios. *Sci. Hortic.* 96(1–4), 267–280.

Savvas, D. and Gruda, N. (2018). Application of soilless culture technologies in the modern greenhouse industry - a review. *Europ. J. Hortic. Sci.* 83(5), 280–293.

Savvas, D. and Manos, G. (1999). Automated composition control of nutrient solution in closed soilless culture systems. *J. Agric. Eng. Res.* 73(1), 29–33.

Savvas, D., Stamati, E., Tsirogiannis, I. L., Mantzos, N., Barouchas, P. E., Katsoulas, N. and Kittas, C. (2007). Interactions between salinity and irrigation frequency in greenhouse pepper grown in closed-cycle hydroponic systems. *Agric. Water Manag.* 91(1–3), 102–111.

Shin, J. H., Park, J. S. and Son, J. E. (2014). Estimating the actual transpiration rate with compensated levels of accumulated radiation for the efficient irrigation of soilless cultures of paprika plants. *Agric. Water Manag.* 135, 9–18.

Shin, J. H. and Son, J. E. (2016). Application of a modified irrigation method using compensated radiation integral, substrate moisture content, and electrical conductivity for soilless cultures of paprika. *Sci. Hortic.* 198, 170–175.

Signore, A., Serio, F. and Santamaria, P. (2016). A targeted management of the nutrient solution in a soilless tomato crop according to plant needs. *Front. Plant Sci.* 7, 391.

Silberbush, M. and Ben-Asher, J. (2001). Simulation study of nutrient uptake by plants from soilless cultures as affected by salinity buildup and transpiration. *Plant Soil* 233(1), 59–69.

Snape, J. B., Dunn, I. J., Ingham, J. and Prenosil, J. E. (1995). *Dynamics of Environmental Bioprocesses: Modelling and Simulation*. VCH, New York.

Son, J. E. (1998). Precise control of nutrient solution and analysis of ionic contribution to EC reading. Kor. *J. Hortic. Sci. Technol.* 16, 362–354.

Son, J. E. and Okuya, T. (1991). Prediction of electrical conductivity of nutrient solution in hydroponics. *J. Agric. Meteorol.* 47(3), 159–163.

Son, J. E. and Takakura, T. (1989). A study on automatic control of nutrient solutions in hydroponics. *J. Agric. Meteorol.* 43, 147–151.

Ta, T. H., Shin, J. H., Noh, E. H. and Son, J. E. (2012). Transpiration, growth, and water use efficiency of paprika plants (*Capsicum annuum* L.) as affected by irrigation frequency. *Hortic. Environ. Biotechnol.* 53(2), 129–134.

Trejo-Perea, M., Herrera-Ruiz, G., Rios-Moreno, J., Miranda, R. C. and Rivasaraiza, E. (2009). Greenhouse energy consumption prediction using neural networks models. *Int. J. Agric. Biol.* 11, 1–6.

van Noordwijk, M. (1990). *Synchronisation of Supply and Demand Is Necessary to Increase Efficiency of Nutrient Use in Soilless Horticulture, Plant Nutrition-Physiology and Applications*. Kluwer Academic Publishers, Dordrecht, The Netherlands.

Wachinger, C., Reuter, M. and Klein, T. (2018). DeepNAT: deep convolutional neural network for segmenting neuroanatomy. *NeuroImage* 170, 434–445.

Wang, H., Sánchez-Molina, J. A., Li, M., Berenguel, M., Yang, X. T. and Bienvenido, J. F. (2017). Leaf area index estimation for a greenhouse transpiration model using external climate conditions based on genetics algorithms, back-propagation neural networks and nonlinear autoregressive exogenous models. *Agric. Water Manag.* 183, 107–115.

Xu, G., Levkovitch, I., Soriano, S., Wallach, R. and Silber, A. (2004). Integrated effect of irrigation frequency and phosphorus level on lettuce: P uptake, root growth and yield. *Plant Soil* 263(1), 297–309.

Zekki, H., Gauthier, L. and Gosselin, A. (1996). Growth, productivity, and mineral composition of hydroponically cultivated greenhouse tomatoes, with or without nutrient solution recycling. *J. Am. Soc. Hortic. Sci.* 121(6), 1082–1088.

# Chapter 11

## Advanced hydroponics design for plant cultivation in cities

*Giuseppina Pennisi, Alessandro Pistillo, Elisa Appolloni, Francesco Orsini and Giorgio Gianquinto, DISTAL – University of Bologna, Italy*



### 1 Introduction

Soilless culture systems (SCS), including hydroponics, can be defined as a soil-independent cultivation strategy where mineral nutrients and water are supplied directly onto plant roots utilizing a nutrient solution, with or without a mechanical root support media, for example, rock wool, perlite, vermiculite, peat, coconut fibre, and so on (Savvas, 2003; Savvas and Gruda, 2018). The attractiveness of SCS has increased considerably in recent years because of their advantage over soil cultures (Lee and Lee, 2015). Soilless cultures offer a solution to producing food in extreme climatic conditions and areas with limited land available (Lee and Lee, 2015) or where land is unsuitable for growing crops on soil (Treftz and Omaye, 2016). SCS applications are found in less-developed countries where they can improve food security through resource use optimization (Gianquinto et al., 2006). In contrast, highly technological SCS have been developed as part of space agencies research programmes (Resh, 2013), where they could contribute to the food security of long-term space missions. Technologies for SCS are also gaining a crucial role in urban environments, where their adoption in building-integrated agriculture or vertical farms is on the rise (Kozai, 2019).

http://dx.doi.org/10.19103/AS.2020.0076.13

The objective of urban agriculture (UA) is to grow crops for human consumption and ornamental use within an urban context (Orsini et al., 2013). This form of farming can be defined in different ways depending on product typology, primary function, market, production system and spatial allocation in the city area (Sanyé-Mengual et al., 2015a). It can be located inside or on the fringe of a town, city or metropolis, therefore distinguishing intra-urban from peri-urban agriculture (Mougeot, 2000; Van Veenhuizen, 2014; Opitz et al., 2016). Production type and scale can vary from small subsistence-oriented systems to large-scale businesses, involving people from different social and economic backgrounds and using productive systems with variable technological complexity (De Zeeuw et al., 2011). The main actors involved can be individuals, families, communities or companies, with a commitment ranging from a leisure activity to a full-time or part-time job.

The term 'urban' is related not only to its location but also to the strong connections that UA has with the urban ecosystem, using city resources and returning benefits to citizens (Mougeot, 2000). The benefits of SCS are determined by its multifunctional character (Orsini et al., 2013), which is not only limited to food supply but also includes social, economic, educational and environmental aspects. UA benefits include contributing to urban food security and nutrition, health, local economic development, social inclusion and gender equality as well as urban environmental management (Van Veenhuizen, 2014).

Food supply is the primary function of UA. Poorer sections of the population can rely on this form of agriculture to feed their families, both in developing and developed countries (Orsini et al., 2013). Its contribution has been and still is fundamental during wars and natural disasters, helping to create resilient cities (Mok et al., 2014) and to help refugees or migrants in camps and slums (Dehnavi and Süß, 2019). Benefits of UA are also related to the consumption of fresh vegetables and fruits, helping urban dwellers to access a balanced diet and therefore preventing illnesses. The easier access to affordable and nutritious food in low-income neighbourhoods can help to reduce the phenomenon of the so-called food deserts (Cummins and Macintyre, 2002), contributing to the decline of obesity issues caused by a diet rich in fats and carbohydrates (Centrone Stefani et al., 2018). The health function of UA is both connected to physical and psychological wellness, as shown by the role of therapeutic gardens built up in hospitals or other institutions to help people to deal with physical, mental or cognitive problems (Jarrott et al., 2002; Meneghello et al., 2016; Righetto et al., 2016).

UA may also have an economic role, providing an important form of income to alleviate urban poverty. Since urban dwellers spend between 60% and 80% of their income for food (Mougeot, 2005), self-production of horticultural goods can help families to save money and even earn small incomes from the selling of surplus products, farming outputs (e.g. compost from farming wastes), processed food or services (De Zeeuw et al., 2011; Van Veenhuizen, 2014). From a social standpoint, farming in a city context can help people on the margins of society such as immigrants, HIV-AIDS affected disabled, unemployed or elderly people to achieve greater integration, better livelihoods and community support (Van Veenhuizen, 2014). The environment can also benefit from UA, thanks to the reduction of negative urban outputs and increase of adaptation to (and mitigation of) climate change (De Zeeuw et al., 2011; Gruda et al., 2019a). Increasing urban vegetation can benefit urban microclimate control, biodiversity and stormwater flow management (Dubbeling et al., 2009). The reduction of transportation miles and energy for cooling, stocking and packaging can contribute to decreasing $CO_2$ emissions and the ecological footprint associated with the urban food systems (Specht et al., 2016). Protected cultivation can help in the reduction of greenhouse gas emissions and the overall environmental impact of food production (Gruda et al., 2019b).

Despite the benefits listed above, poor application of agriculture within city fringes can lead to negative impacts on urban communities and the environment, especially in cases of limited awareness and education among urban farmers, consumers and policymakers (Orsini et al., 2013). The main risks are related to public and individual health. Pathogens' contamination due to polluted wastewater, unhygienic post-harvest treatment or the presence of livestock represents the main hazards together with excessive use of agrochemicals and heavy metal contamination (Säumel et al., 2012; Van Veenhuizen, 2014). Environmental problems can also emerge due to the contamination of local water sources from leakage of fertilizers or pesticides (Van Veenhuizen, 2014). Because of these risks, authorities have often been reluctant to completely accept urban farming (De Zeeuw et al., 2011). However, the adoption of SCS in potentially contaminated sites could help reduce heavy metal accumulation in urban-grown vegetable products (Pennisi et al., 2016).

The most traditional way to undertake UA is using abandoned or underexplored urban soil. However, competition for high land costs in cities have forced urban farmers to find new spaces to grow their products, including exploring unused rooftop, external surfaces or inner spaces of residential or office buildings (Hui, 2011; Gasperi et al., 2016). Two forms have generally been identified:

- soil-based urban agriculture; and
- building-based urban agriculture.

In soil-based UA, the main types are (Opitz et al., 2016):

- self-organized *community gardens* managed by groups of people;

- individual allotments officially organized by municipalities;
- private backyard gardens for residents;
- urban farms managed by entrepreneurs or charity organizations to provide goods and services; and
- peri-urban agricultural holdings for food production business.

Building-based agriculture or zero-acreage farming (Z-farming) (Specht et al., 2014; Thomaier et al., 2015) refers to every kind of UA inside or on the top of urban buildings, including:

- monofunctional structures expressly constructed for food production, also named plant factories (Kozai and Niu, 2016a); and
- multifunctional structures with different uses beside horticultural products cultivation, including rooftop open-air farms or rooftop greenhouses found on top of residential or commercial buildings (Sanyé-Mengual et al., 2015b; Buehler and Junge, 2016).

As for the growing systems used, building-integrated agricultural systems generally make use of soilless cultures, either adopting high- or low-tech systems, in the form of edible walls or balconies, indoor farms or rooftop farms (Specht et al., 2014). Advanced SCS for building-integrated agriculture are described below.

## 2 Rooftop farming systems

Plant cultivation on the rooftop of city buildings is also referred to as rooftop agriculture (RA) (Orsini et al., 2017). RA is a form of urban farming that has been gaining attention among researchers and the public in the last decade. It can be defined as building-based agriculture applied on top of different typologies of urban buildings both in protected (rooftop greenhouse) and non-protected (open-air rooftop farm) conditions (Sanyé-Mengual et al., 2015a), with commercial, social, aesthetic or living quality objectives (Thomaier et al., 2015) using soil-based (e.g. soil-filled containers) or soil-less (e.g. hydroponics, aquaponics) systems (Nasr et al., 2017) (Fig. 1).

Rooftop cultivation has the same limits and benefits as UA conducted at soil level. However, due to its particular location, this type of agriculture faces particular issues not experienced in ground-based farming. Structural loading, rooftop access, accessibility of farming resources, high solar radiation and temperature are just some of the specific challenges that rooftop farms have to deal with (Hui, 2011; Caputo et al., 2017). Despite these limits, rooftop farming can also offer specific benefits for a building, helping to save up to 15% of annual energy consumption, thanks to the insulating effect provided by the green

**Figure 1** SCS used in rooftop farms. Plant cultivation on substrate growing systems in open-air rooftop gardens and rooftop greenhouses. (a) Experimental rooftop garden using box cultivation systems filled with the substrate at AgroParisTech (Paris, France); (b) substrate-based community rooftop garden at Post office La Chapelle (Paris, France); (c) kitchen rooftop garden on rockwool slabs in Barcelona at the residence of Joan Rieradevall Pons (Barcelona, Spain); (d) therapeutical rooftop garden using boxes filled with a substrate (Barcelona, Spain); (e) Simplified-substrate and (f) simplified-NFT vegetable production in the rooftop garden of the social housing buildings of Via Gandusio (Bologna, Italy). (g) NFT cultivation system for leafy vegetable production in the rooftop greenhouse of UrbanFarmers (The Hague, the Netherlands); substrate system on coir for the cultivation of green beans (h) leafy vegetables (i) at the experimental integrated-Rooftop Greenhouse (i-RTG) at the Universitat Autonoma de Barcelona (Barcelona, Spain). (Photos F. Orsini, private collection).

cover (Wong et al., 2003; McIntyre and Snodgrass, 2010). The integration of a rooftop greenhouse with energy, water and $CO_2$ flows produced by buildings can enhance the sustainability of both systems involved while producing high-value crops and maintaining the right thermal conditions inside the structure (Sanyé-Mengual et al., 2015a).

Open-air rooftop farms or gardens that use low-tech systems such as raised beds filled with soil are the most common forms of rooftop agriculture (Thomaier et al., 2015). These kinds of installation are less expensive to create compared to a rooftop greenhouse, therefore making them applicable to small-scale subsistence conditions in developing countries (Bryld, 2003). However, due to the absence of physical barriers, crops may be exposed to air pollution more easily than those in a rooftop greenhouse, even though both rooftop elevation (Säumel et al., 2012; Vittori Antisari et al., 2015) and soilless cultures (Pennisi et al., 2016) can considerably help in reducing risks.

Rooftop greenhouses are more often developed with sophisticated technologies and farming techniques (Thomaier et al., 2015), which can increase production capacity and resource-use efficiency (Sanyé-Mengual et al., 2015a). High-tech commercial rooftop farms are business-oriented start-ups with economic profitability as the first aim which have developed particularly in North America, Asia and Europe (Specht et al., 2015; Thomaier et al., 2015; Buehler and Junge, 2016). Depending on the level of integration of greenhouse technologies, productive capacity, range and quality of cultivable products are often high (Buehler and Junge, 2016). Despite the involvement of commercial businesses, rooftop farms are often also associated with non-profit aims, generally targeting the amelioration of urban living quality or the involvement of communities in social, recreational and educational activities (Thomaier et al., 2015). Rooftop cultivation can also be undertaken by hotels and restaurants, offering fresh, self-produced products to customers both to supply food ingredients and to enhance their image (Thomaier et al., 2015).

Rooftop farms may be integrated both into new and existing buildings (Sanyé-Mengual et al., 2015a; Buehler and Junge, 2016). While newly built constructions offer greater flexibility in planning to anticipate structural and management problems, older buildings may lead to higher costs of adaptation, less rational use of the rooftop surface and restricted range of applicable cultivation techniques (Caputo et al., 2017). Despite these problems, refitting still represents the most common situation for rooftop farming cases worldwide, applied on a variety of structures with functions including housing, warehouse and manufacturing, research and education, retail, hotel and restaurant, transportation facility, office, community service, food business or health uses (Thomaier et al., 2015).

Agriculture integration with buildings is not confined to rooftops. SCS combined with artificial lighting technologies and advanced systems for the control of temperature, $CO_2$ and relative humidity are making applications possible also within structures (Thomaier et al., 2015) or underground. Numerous examples of indoor farming are developing worldwide, showing particular suitability for commercial purposes, thanks not only to the great optimization of resources but also to the efficiency of space through multilayer cultivations (Specht et al., 2014; Kozai and Niu, 2016b). The indoor farming system may be of different sizes, ranging from a small-size system such as InFarm to industrialized systems such as Aerofarm, may be found underground as Growing Underground, where microgreens are cultivated in air-raid tunnels, or within containers as in Freight Farm. Advanced SCS for indoor farming are described in the next section.

## 3 Vertical farming with artificial lighting

One of the most technologically advanced solutions for growing crops in cities is the use of plant factories with artificial lighting, also called vertical farms. Vertical farming is a cultivation system where environmental factors (e.g. air temperature and humidity, light, $CO_2$ concentration) are controlled by minimizing exchanges between the indoor and outdoor environment, thanks to insulated cultivation rooms where a minimum amount of air and heat is exchanged with the outside environment (Kozai and Niu, 2016a; Kozai, 2019). This insulation from the external environment also gives the system a higher resilience to extreme weather events and offers easier control of pest and pathogens, as well as year-round productivity, as compared to more traditional soil-based, outdoor cultivation systems (Kozai, 2019). By developing plant cultivation vertically using multilayers systems with artificial lighting devices, this technique reduces the pressure on traditional agricultural land (Beacham et al., 2019). Pennisi et al. (2019a) reported that the land-surface-use efficiency of a 10-layer vertical growing system used for lettuce production was up to 20 times higher when compared with the productivity of traditional soil-based production systems. An air conditioning system is usually integrated into plant factories to eliminate the heat produced by the lamps and the water vapour produced by the plants, also guaranteeing a homogeneous redistribution of air through the various cultivation layers (Kozai and Niu, 2016a). Among most common plant species grown in plant factories are leafy vegetables (e.g. lettuce, basil, microgreens), medicinal plants (e.g. cannabis or other crops used in the preparation of pharmaceutical, herbal or cosmetic products), small fruits (e.g. berries), edible flowers and seedlings (e.g. grafted vegetables) (Kozai, 2013).

SCS is usually adopted in plant factories because it makes it possible to automate and optimize irrigation and fertilization management, to save labour, to avoid the use of substrates (thus helping to keep the system and products clean) and to allow vertical cultivation tiers to improve land-use efficiency (Wada, 2019). Indeed, thanks to the possibility to recirculate the drained nutrient solution and make it available to plants again, and to recover the water transpired by the plants, and vertical farms are also considered a nutrient and water-saving cultivation system (Pennisi et al., 2019b,c). Kozai (2013) reported water-use efficiency values in plant factories up to 30–50 times higher than those observed in open-field or greenhouse cultivation.

The most common SCS used are the (Fig. 2):

- nutrient film technique (NFT) (e.g. in AgricolaModerna®, GrowX® and Planty Cube®);
- ebb and flow system (e.g. in AgricolaModerna®, Grönska® and BrightBox®); and
- aeroponics system (e.g. in AeroFarms®)

**Figure 2** Vertical, multi-tier hydroponic cultivation system combining deep water culture and ebb and flow cultivation (images from Brightbox, Venlo, The Netherlands) (Photo F. Orsini).

An alternative growing solution for plants production in plant factories are cylindrical growth units (e.g. in ZipGrow®), where plants are grown above one another on the surface of upright cylindrical units (Beacham et al., 2019). The cylindrical units are generally filled with a substrate, such as perlite or rock wool, to enable nutrient solution retention and to support plants. The main limits of these growing systems are related to the shading effect of the upper plants on those below and a non-uniform distribution of nutrient solution (Linsley-Noakes et al., 2006), resulting in heterogeneous plant growth. When comparing the performance of lettuce cultivation in a vertical column and horizontal hydroponic systems, the former was shown to increase the biomass production per unit area (Touliatos et al., 2016).

In vertical farms, SCS are generally equipped with sensors and controllers enabling control of the root-zone environment, pH and electrical conductivity of the nutrient solution, macro- and micro-elements concentration, dissolved oxygen and temperature (Son et al., 2016). Linking these data with data measured at canopy level (e.g. temperature, relative humidity) and to morphological plant development by using Internet of Things (IoT) solutions, plant growth can be monitored to help growers maintaining the right environment, thus optimizing production and the quality of products (Wu et al., 2016). An IoT-based autonomous control android/iOS mobile application for an easy management of a soilless culture unit was successfully implemented by Peuchpanngarm et al. (2016) and included different typologies of sensors used for monitoring water level, air temperature and humidity, light intensity, connected to an Arduino platform to dynamically control the hydroponic environment. A mobile application was also created to guide grower cultivation activities. Gosavi (2017) developed an IoT-based soilless culture prototype in which a set of sensors and a microcontroller guide the photo periodical cycles of the lighting system. Mehra et al. (2018) developed an intelligent IoT-based SCS that incorporates pH, temperature, humidity, water level and light-sensing features and uses a deep neural network to ensure control with an accuracy of 88%. In the future of plant factories, these advanced technologies will be implemented to increase the productivity and resource-use efficiency of the system and optimize management costs (Kozai, 2019).

## 4 Innovative SCS for urban agriculture: aquaponics

A particular form of SCS is aquaponics. It is the result of the integration between fish farming (aquaculture) and soilless plant production (hydroponics) (Greenfeld et al., 2019). In the 1960s, aquaponics was introduced as a water-saving food production system, since water could be used for both fish and plants. Waste products from fish farming, such as fish faeces and decomposed fish feed, become the nutrients input for cultivation. Plants grown in the hydroponic unit filter and clean the water, which can then be moved back to the fish tank. The closed water cycle is possible, thanks to the action of nitrifying bacteria (*Nitrosomonas* and *Nitrobacter* spp.) that oxidize the toxic ammonia produced by the nitrates in the aquaculture system, which are then easily absorbed by plants roots (Calone et al., 2019). In an aquaponic system, water used for rearing fish is, first, treated to reduce organic matter and suspended solids. Water flows through a biofilter to remove ammonia and then passes to the soilless culture unit where roots absorb nitrates and finally is collected in a sump and returned to the fish-rearing tank (Rakocy et al., 2006). According to Palm et al. (2018), the majority of nutrients (> 50%) for optimal plant growth derive from wastes originating from fish farming. Several studies (Calone et al., 2019; Endo, 2019; and Kaburagi et al., 2020) have reported that water coming from the aquaculture unit needs the addition of ammonium, potassium, iron, manganese, and zinc to achieve optimal plant growth.

SCS that can be integrated into aquaponics are floating systems, NFT, aeroponics, ebb and flow and medium-based systems. While in medium-based systems the substrate can act as a surface for bacteria proliferation, in all the other systems an external biofilter is generally included (Nelson, 2008). Maucieri et al. (2018), reviewing 122 published papers on aquaponics, highlighted that the most frequently used systems were medium-filled growth beds followed by floating rafts. Medium-based soilless culture was used for a higher number of plant species, followed by the floating system and NFT.

The most common fish species used in aquaponics are Perciformes species, such as blue tilapia (*Sarotherodon aurea*), hybrid tilapia (*Oreochromis mossambicus* × *Oreochromis niloticus*), red tilapia (*Oreochromis* spp.) together with members of Siluriformes species such as the African catfish (*Clarias gariepinus*) and hybrid catfish (*Clarias microcephalus* × *C. gariepinus*) (Knaus and Palm, 2017). Leafy vegetables, such as lettuce (*Lactuca sativa)*, Chinese cabbage or Pak choi (*Brassica rapa* subsp. *pekinensis* and *chinensis*), herbs such as basil (*Ocimum basilicum*) and some fruity vegetables such as tomato (*Solanum lycopersicum)* and cucumber (*Cucurbita sativus*) are the major plant species cultivated in aquaponics (Knaus and Palm, 2017). Finding the best fish-plant combination to optimize productivity and resource-use efficiency is still a challenge for aquaponics (Calone et al., 2019).

There have been a number of studies on the economic feasibility of aquaponic production (Baker, 2010; Chaves et al., 2008; Rupasinghe and Kennedy, 2010). Quagrainie et al. (2018) performed a comparative study on the economic sustainability of hydroponics and aquaponics. The study found that the economics for aquaponics was worse than hydroponics due to lower vegetable production even if, with a cost/benefit ratio higher than 1, aquaponics is a feasible business. However, if we consider an increase in prices (+19.7%) due to organic production, aquaponics resulted in higher revenues than hydroponics. This means that product prices are fundamental to the economic sustainability of the system.

Current innovation in aquaponics is targeting the decrease in freshwater availability for agricultural activities, focussing on the use of fish and plant products adapted to saline water (Endo, 2019). Waller et al. (2015) reported that an aquaculture plant for sea bass (Dicentrarchus labrax) may be combined with a hydroponic unit for *Ripolium pannonicum* (sea aster), *Plantago coronopus* (buck's horn plantain) and *Salicornia dolichostachya* (long spiked glasswort) cultivation. The interest in these applications comes from the high commercial value of the associated food products. For example, sea bass can reach a market price of €9 $kg^{-1}$, and some edible halophytes (e.g. *Salicornia* spp.) are sold in the supermarket at prices up to €18 $kg^{-1}$ (Kotzen et al., 2019). These prices are significant compared with market prices of tilapia and lettuce, respectively, of \$4.42 $kg^{-1}$ (about €4.07 $kg^{-1}$) and \$0.57 $kg^{-1}$ (about €0.5 $kg^{-1}$), reported by Quagrainie et al. (2018).

## 5 Case study: innovative urban farming systems through the international student competition: UrbanFarm

Recently, UA projects have proliferated around the world, both promoted by top-down government support and by bottom-up community-based initiatives. In order to facilitate a wider uptake of UA, it is crucial to create awareness among both institutional actors and the civil society through innovative and interdisciplinary approaches. The international student challenge UrbanFarm aims at tackling the current need for cooperation between different disciplines by bringing together students from different fields of study into international teams addressing the regeneration of vacant urban spaces. The challenge has become an opportunity to link viewpoints and approaches while integrating state-of-the-art technologies and design for urban farming and urban planning. Starting from these project ideas, local administrators and urban planners gain the tools to foster sustainability in their cities. A major achievement of the UrbanFarm competition is the geographical distribution and expertise covered by the participating teams. The 2019 and 2020 UrbanFarm competitions involved 35 and 25 teams, respectively, with a total of 269 students from all over the world (Orsini et al., 2020). (For further information: https://site.unibo.it/urban-farm/en).

## 6 Summary and future trends

Innovative SCS applications for urban agriculture are being developed and applied in different regions. Building-integrated rooftop agriculture makes use of either basic or advanced technologies, with growing systems and greenhouse technologies developed in the rural environment adapted to fit the urban context. Recent advances in LED lighting technologies have led to the spread of indoor vertical farms. Here agriculture takes place on vertical tiers, with precisely tuned environmental and agronomic crop management that adapts dynamically to meet plant requirements. New areas of research include strategies to deal with the benefits and constraints associated with both urban resources and stressors. The emergence of integrated and interdisciplinary studies of urban SCS in the last few decades makes it possible to fulfil multiple goals such as the promotion of food security (Orsini et al., 2013) and safety (Pennisi et al., 2016), while also ensuring sustainability (Martin and Molin, 2019; Pennisi et al., 2019c) and eco-efficiency (Pennisi et al., 2019a). The sector however still suffers from a general lack of regulation and policies as well as standardized commercial systems and protocols. Simplified strategies for crop management (e.g. in small-scale systems or by non-trained operators) are needed (Sanyé-Mengual et al., 2018). Plant-growing techniques allowing for reducing costs or increasing product value (e.g. in indoor farming systems, where the costs associated with artificial lighting reduce competitiveness compared to rural farming) need to be identified (Kozai, 2019). Finally, UA needs to benefit from fuller integration within the urban infrastructure (Sanyé-Mengual et al., 2018).

## 7 Where to look for further information

Further information may be found in monographs concerning the main themes addressed in the article. Suggested books are Aquaponic gardening: a step-by-step guide to raising vegetables and fish together, Bernstein, 2011; Urban Agriculture, eds. Aubry, Duchemin, Nasr, 2012; Rooftop Urban Agriculture, eds. Orsini, Dubbelling, De Zeeuw, Gianquinto, 2017; LED lighting for urban agriculture, Kozai, 2016; Plant factory: an indoor vertical farming system for efficient quality food production, eds. Kozai, Niu, Takagaki, 2020.

## 8 References

Baker, A. A. C. (2010). Preliminary development and evaluation of an aquaponic system for the American Insular Pacific. Doctoral dissertation, University of Hawaii at Manoa, Honolulu.

Beacham, A. M., Vickers, L. H. and Monaghan, J. M. (2019). Vertical farming: a summary of approaches to growing skywards, *J. Hortic. Sci. Biotech.* 94(3), 277–283. doi:10.1080/14620316.2019.1574214.

Bryld, E. (2003). Potentials, problems, and policy implications for urban agriculture in developing countries, *Agric. Hum. Values* 20(1), 79–86. doi:10.1023/A:1022464607153.

Buehler, D. and Junge, R. (2016). Global trends and current status of commercial urban rooftop farming, *Sustainability* 8(11), 1108. doi:10.3390/su8111108.

Calone, R., Pennisi, G., Morgenstern, R., Sanyé-Mengual, E., Lorleberg, W., Dapprich, P., Winkler, P., Orsini, F. and Gianquinto, G. (2019). Improving water management in European catfish recirculating aquaculture systems through catfish-lettuce aquaponics, *Sci. Total Environ.* 687, 759–767. doi:10.1016/j.scitotenv.2019.06.167.

Caputo, S., Iglesias, P. and Rumble, H. (2017). Design of rooftop agriculture systems – elements of rooftop agriculture design. In: Orsini, F., Dubbeling, M., De Zeeuw, H. and Gianquinto, G. (Eds), *Rooftop Urban Agriculture*. Springer, Cham, Switzerland, pp. 39–59. doi:10.1007/978-3-319-57720-3_4.

Centrone Stefani, M., Orsini, F., Magrefi, F., Sanyé-Mengual, E., Pennisi, G., Michelon, N., Bazzocchi, G. and Gianquinto, G. (2018). Toward the creation of urban foodscapes: case studies of successful urban agriculture projects for income generation, food security, and social cohesion. In: Nandwani, D. (Ed.), *Urban Horticulture: Sustainability for the Future*. Springer, Dordrecht, The Netherlands, pp. 91–106. doi:10.1007/978-3-319-67017-1_5.

Chaves, P. A., Sutherland, R. M. and. Larid, L. M. (2008). An economic and technical evaluation of integrating hydroponics in a recirculation fish production system, *Aquac. Econ. Manag.* 3(1), 83–91. doi:10.1080/13657309909380235.

Cummins, S. and Macintyre, S. (2002). "Food deserts"–evidence and assumption in health policy making, *Br. Med. J.* 325(7361), 436–438. doi:10.1136/bmj.325.7361.436.

De Zeeuw, H., Van Veenhuizen, R. and Dubbeling, M. (2011). The role of urban agriculture in building resilient cities in developing countries, *J. Agric. Sci.* 149(Suppl. 1), 153-163. doi:10.1017/S0021859610001279.

Dehnavi, S. and Süß, V. (2019). Urban agriculture towards food security of Syrian refugees and vulnerable Lebanese host communities, *Dev. Pract.* 29(5), 635–644. doi:10.1080/09614524.2019.1630369.

Dubbeling, M., Campbell, M. C., Hoekstra, F. and Van Veenhuizen, R. (2009). Building resilient cities. *Urban Agriculture Magazine* 22, 3–11.

Endo, M. (2019). Aquaponics in plant factory. In: Anpo, M., Fukuda, H. and Wada, T. (Eds), *Plant Factory Using Artificial Light: Adapting to Environmental Disruption and Clues to Agricultural Innovation*. Elsevier, Amsterdam, The Netherlands, pp. 339–352. doi:10.1016/B978-0-12-813973-8.00032-4.

Gasperi, D., Pennisi, G., Rizzati, N., Magrefi, F., Bazzocchi, G., Mezzacapo, U., Centrone Stefani, M., Sanyé-Mengual, E., Orsini, F. and Gianquinto, G. (2016). Towards regenerated and productive vacant areas through urban horticulture: lessons from Bologna, Italy, *Sustainability* 8(12), 1347. doi:10.3390/su8121347.

Gianquinto, G., Orsini, F., Michelon, N., Ferreira Da Silva, D. and Damasio De Faria, F. (2006). Improving yield of vegetables by using soilless micro-garden technologies in peri-urban area of north-east Brazil, *Acta Hortic.* 747, 57–65. doi:10.17660/ActaHortic.2007.747.4.

Gosavi, J. V. (2017). Water monitoring system for hydroponics agriculture, *Int. J. Res. Appl. Sci. Eng. Technol.* 5, 19–34.

Greenfeld, A., Becker, N., Bornman, J. F. and Angel, D. J. (2019). Identifying knowledge levels of aquaponics adopters, *Environ. Sci. Pollut. Res.* 27, 4536–4540. doi:10.1007/s11356-019-06758-8.

Gruda, N., Bisbis, M. and Tanny, J. (2019a). Influence of climate change on protected cultivation: impacts and sustainable adaptation strategies - a review, *J. Clean. Prod.* 225, 481–495. doi:10.1016/j.jclepro.2019.03.210.

Gruda, N., Bisbis, M. and Tanny, J. (2019b). Impacts of protected vegetable cultivation on climate change and adaptation strategies for cleaner production - a review, *J. Clean. Prod.* 225, 324–339. doi:10.1016/j.jclepro.2019.03.295.

Hui, S. C. (2011). Green roof urban farming for buildings in high-density urban cities. In: *The 2011 Hainan China World Green Roof Conference*, Hainan (Haikuo, Boao and Sanya), China, 18-21 March 2011, pp. 1-9.

Jarrott, S. E., Kwack, H. R. and Relf, D. (2002). An observational assessment of a dementia-specific horticultural therapy program, *HortTechnology* 12(3), 403–410. doi:10.21273/horttech.12.3.403.

Kaburagi, E., Yamada, M., Baba, T., Fujiyama, H., Murillo-Amador, B. and Yamada, S. (2020). Aquaponics using saline groundwater: effect of adding microelements to fish wastewater on the growth of Swiss chard (*Beta vulgaris* L. spp. *cicla*), *Agric. Water Manag.* 227, 105851. doi:10.1016/j.agwat.2019.105851.

Knaus, U. and Palm, H. W. (2017). Effects of the fish species choice on vegetables in aquaponics under spring-summer conditions in northern Germany (Mecklenburg Western Pomerania), *Aquaculture* 473, 62–73. doi:10.1016/j.aquaculture.2017.01.020.

Kotzen, B., Emerenciano, M. G. C., Moheimani, N. and Burnell, G. M. (2019). Aquaponics: alternative types and approaches. In: Goddek, S., Joyce, A., Kotzen, B. and Burnell, G. M. (Eds), *Aquaponics Food Production Systems*. Springer, Cham, Switzerland, pp. 301–330. doi:10.1007/978-3-030-15943-6_12.

Kozai, T. (2013). Resource use efficiency of closed plant production system with artificial light: concept, estimation and application to plant factory, *Proc. Jpn. Acad. Ser. B Phys. Biol. Sci.* 89(10), 447–461. doi:10.2183/pjab.89.447.

Kozai, T. (2019). Towards sustainable plant factories with artificial lighting (PFALs) for achieving SDGs, *Int. J. Agric. Biol. Eng.* 12(5), 28–37. doi:10.25165/j.ijabe.20191205.5177.

Kozai, T. and Niu, G. (2016a). Role of the plant factory with artificial lighting (PFAL) in urban areas. In: Kozai, T., Niu, G. and Takagaki, M. (Eds), *Plant Factory: An Indoor Vertical Farming System for Efficient Quality Food Production*. Academic Press, Cambridge, MA, pp. 7–33. doi:10.1016/b978-0-12-801775-3.00002-0.

Kozai, T. and Niu, G. (2016b). Plant factory as a resource-efficient closed plant production system. In: Kozai, T., Niu, G. and Takagaki, M. (Eds), *Plant Factory: An Indoor Vertical Farming System for Efficient Quality Food Production*. Academic Press, Cambridge, MA, pp. 69–90. doi:10.1016/b978-0-12-801775-3.00004-4.

Lee, S. and Lee, J. (2015). Beneficial bacteria and fungi in hydroponic systems: types and characteristics of hydroponic food production methods, *Sci. Hortic.* 195, 206–215. doi:10.1016/j.scienta.2015.09.011.

Linsley-Noakes, G., Wilken, L. and de Villiers, S. (2006). High density, vertical hydroponics growing system for strawberries, *Acta Hortic.* 708, 365–370. doi:10.17660/ActaHortic.2006.708.63.

Martin, M. and Molin, E. (2019). Environmental assessment of an urban vertical hydroponic farming system in Sweden, *Sustainability* 11(15), 4124. doi:10.3390/su11154124.

Maucieri, C., Nicoletto, C., Junge, R., Schmautz, Z., Sambo, P. and Borin, M. (2018). Hydroponic systems and water management in aquaponics: a review, *Ital. J. Agron.* 13, 1012. doi:10.21256/zhaw-3671.

McIntyre, L. and Snodgrass, E. C. (2010). *The Green Roof Manual: A Professional Guide to Design, Installation, and Maintenance*. Timber Press, Portland, OR. doi:10.5860/choice.48-4481.

Mehra, M., Saxena, S., Sankaranarayanan, S., Tom, R. J. and Veeramanikandan, M. (2018). IoT based hydroponics system using Deep Neural Networks, *Comput. Electron. Agric.* 155, 473–486. doi:10.1016/j.compag.2018.10.015.

Meneghello, F., Marcassa, G., Koch, I., Sgaravatti, P., Piccolomini, B., Righetto, C., Prosdocimi Gianquinto, G. and Orsini, F. (2016). Garden therapy in neurorehabilitation: well-being and skills improvement, *Acta Hortic.* 1121, 13–18. doi:10.17660/ActaHortic.2016.1121.3.

Mok, H. F., Williamson, V. G., Grove, J. R., Burry, K., Barker, S. F. and Hamilton, A. J. (2014). Strawberry fields forever? Urban agriculture in developed countries: a review, *Agron. Sustain. Dev.* 34(1), 21–43. doi:10.1007/s13593-013-0156-7.

Mougeot, L. J. (2000). *Urban Agriculture: Definition, Presence, Potentials and Risks, and Policy Challenges*. Cities Feeding People Series Report 31.

Mougeot, L. J. (2005). Neglected issues on form and substance of research on urban agriculture. In: Mougeout, L. (Ed.), *Agropolis: The Social, Political and Environmental Dimensions of Urban Agriculture*. Earthscan, London, pp. 267–279.

Nasr, J., Komisar, J. and De Zeeuw, H. (2017). The status and challenges of rooftop agriculture - a Panorama of rooftop agriculture types. In: Orsini, F., Dubbeling, M., De Zeeuw, H. and Gianquinto, G. (Eds), *Rooftop Urban Agriculture*. Springer, Cham, Switzerland, pp. 9–29. doi:10.1007/978-3-319-57720-3_2.

Nelson, R. L. (2008). Aquaponic equipment: the biofilter, *Aquaponics J.* 48(2), 22–23.

Opitz, I., Berges, R., Piorr, A. and Krikser, T. (2016). Contributing to food security in urban areas: differences between urban agriculture and peri-urban agriculture in the Global North, *Agric. Hum. Values* 33(2), 341–358. doi:10.1007/s10460-015-9610-2.

Orsini, F., Dubbeling, M., De Zeeuw, H. and Gianquinto, G. (2017). *Rooftop Urban Agriculture*. Springer, Cham, Switzerland. doi:10.1007/978-3-319-57720-3.

Orsini, F., Kahane, R., Nono-Womdim, R. and Gianquinto, G. (2013). Urban agriculture in the developing world: a review, *Agron. Sustain. Dev.* 33(4), 695–720. doi:10.1007/s13593-013-0143-z.

Orsini, F., Pennisi, G., D'Alessandro, A., Kratochvilova, D., Steffan, G., Paoletti, M., Sabbatini, G., D'Ostuni, M., Trombadore, A., and Gianquinto, G. (2020). UrbanFarm2019: Bridging interdisciplinary knowledge for sustainable urban landscapes: results from the international student competition, *Acta Hortic.* (in press).

Palm, H. W., Knaus, U., Appelbaum, S., Goddek, S., Strauch, S. M., Vermeulen, T., Haïssam Jijakli, M. and Kotzen, B. (2018). Towards commercial aquaponics: a review of systems, designs, scales and nomenclature, *Aquac. Int.* 26(3), 813–842. doi:10.1007/s10499-018-0249-z.

Pennisi, G., Orsini, F., Gasperi, D., Mancarella, S., Sanoubar, R., Vittori Antisari, L., Vianello, G. and Gianquinto, G. (2016). Soilless system on peat reduce trace metals in urban grown food: unexpected evidence for a soil origin of plant contamination, *Agron. Sustain. Dev.* 36(4), 1–11. doi:10.1007/s13593-016-0391-9.

Pennisi, G., Blasioli, S., Cellini, A., Maia, L., Crepaldi, A., Braschi, I., Spinelli, F., Nicola, S., Fernandez, J. A., Stanghellini, C., Marcelis, L. F. M., Orsini, F. and Gianquinto, G. (2019a). Unravelling the role of red: blue LED lights on resource use efficiency and nutritional properties of indoor grown sweet basil, *Front. Plant Sci.* 10, 305. doi:10.3389/fpls.2019.00305.

Pennisi, G., Orsini, F., Blasioli, S., Cellini, A., Crepaldi, A., Braschi, I., Spinelli, F., Nicola, S., Fernandez, J. A., Stanghellini, C., Gianquinto, G. and Marcelis, L. F. M. (2019b). Resource use efficiency of indoor lettuce (*Lactuca sativa* L.) cultivation as affected by red:blue ratio provided by LED lighting, *Nat. Sci. Rep.* 9(1), 14127. doi:10.1038/s41598-019-50783-z.

Pennisi, G., Sanyé-Mengual, E., Orsini, F., Crepaldi, A., Nicola, S., Ochoa, J., Fernandez, J. and Gianquinto, G. (2019c). Modelling environmental burdens of indoor-grown vegetables and herbs as affected by red and blue LED lighting, *Sustainability* 11(15), 4063. doi:10.3390/su11154063.

Peuchpanngarm, C., Srinitiworawong, P., Samerjai, W. and Sunetnanta, T. (2016). DIY sensor-based automatic control mobile application for hydroponics. In: *2016 Fifth ICT International Student Project Conference (ICT-ISPC)*, Nakhon Pathom, Thailand, pp. 57–60. doi:10.1109/ict-ispc.2016.7519235.

Quagrainie, K. K., Flores, R. M. V., Kim, H. J. and McClain, V. (2018). Economic analysis of aquaponics and hydroponics production in the US Midwest, *J. Appl. Aquac.* 30(1), 1–14. doi:10.1080/10454438.2017.1414009.

Rakocy, J. E., Masser, M. P. and Losordo, T. M. (2006). *Recirculating Aquaculture Tank Production Systems: Aquaponics - Integrating Fish and Plant Culture*. SRAC Publication, Stoneville, MS. doi:10.1002/9781118250105.ch14.

Resh, H. M. (2013). *Hydroponic Food Production: A Definitive Guidebook for the Advanced Home Gardener and the Commercial Hydroponic Grower*. CRC Press, Boca Raton, FL.

Righetto, C., Prosdocimi Gianquinto, G., Orsini, F., Meneghello, F., Marcassa, G., Koch, I., Sgaravatti, P. and Piccolomini, B. (2016). Realization of a neuro-rehabilitation therapeutic garden: design criteria and horticultural choices, *Acta Hortic.* 1121, 51-58. doi:10.17660/ActaHortic.2016.1121.9.

Rupasinghe, J. W. and Kennedy, J. O. (2010). Economic benefits of integrating a hydroponic lettuce system into a barramundi fish production system, *Aquac. Econ. Manag.* 1(4), 81–96. doi:10.1080/13657301003776631.

Sanyé-Mengual, E., Orsini, F., Oliver-Solà, J., Rieradevall, J., Montero, J. I. and Gianquinto, G. (2015a). Techniques and crops for efficient rooftop gardens in Bologna, Italy, *Agron. Sustain. Dev.* 35(4), 1477–1488. doi:10.1007/s13593-015-0331-0.

Sanyé-Mengual, E., Cerón-Palma, I., Oliver-Solà, J., Montero, J. I. and Rieradevall, J. (2015b). Integrating horticulture into cities: a guide for assessing the implementation potential of Rooftop Greenhouses (RTGs) in industrial and logistics parks, *J. Urban Technol.* 22(1), 87–111. doi:10.1080/10630732.2014.942095.

Sanyé-Mengual, E., Specht, K., Krikser, T., Vanni, C., Pennisi, G., Orsini, F. and Gianquinto, G. P. (2018). Social acceptance and perceived ecosystem services of urban agriculture in Southern Europe: the case of Bologna, Italy, *PLoS ONE* 13(9), e0200993. doi:10.1371/journal.pone.0200993.

Säumel, I., Kotsyuk, I., Hölscher, M., Lenkereit, C., Weber, F. and Kowarik, I. (2012). How healthy is urban horticulture in high traffic areas? Trace metal concentrations in vegetable crops from plantings within inner city neighbourhoods in Berlin, Germany, *Environ. Pollut.* 165, 124–132. doi:10.1016/j.envpol.2012.02.019.

Savvas, D. (2003). Hydroponics: a modern technology supporting the application of integrated crop management in greenhouse, *J. Food Agric. Environ.* 1(1), 80–86.

Savvas, D. and Gruda, N. (2018). Application of soilless culture technologies in the modern greenhouse industry — a review, *Eur. J. Hortic. Sci.* 83(5), 280–293. doi:10.17660/eJHS.2018/83.5.2.

Son, J. E., Kim, H. J. and Ahn, T. I. (2016). Hydroponic systems. In: Kozai, T., Niu, G. and Takagaki, M. (Eds), *Plant Factory: An Indoor Vertical Farming System for Efficient Quality Food Production*. Academic Press, Cambridge, MA, pp. 213-221. doi:10.1016/b978-0-12-801775-3.00017-2.

Specht, K., Siebert, R., Hartmann, I., Freisinger, U. B., Sawicka, M., Werner, A., Thomaier, S., Henckel, D., Walk, H. and Dierich, A. (2014). Urban agriculture of the future: an overview of sustainability aspects of food production in and on buildings, *Agric. Hum. Values* 31(1), 33–51. doi:10.1007/s10460-013-9448-4.

Specht, K., Siebert, R., Thomaier, S., Freisinger, U. B., Busse, M., Dierich, A., Henckel, D. and Busse, M. (2015). Zero-acreage farming in the city of Berlin: an aggregated stakeholder perspective on potential benefits and challenges, *Sustainability* 7(4), 4511–4523. doi:10.3390/su7044511.

Specht, K., Weith, T., Swoboda, K. and Siebert, R. (2016). Socially acceptable urban agriculture businesses, *Agron. Sustain. Dev.* 36(1), 17. doi:10.1007/s13593-016-0355-0.

Thomaier, S., Specht, K., Henckel, D., Dierich, A., Siebert, R., Freisinger, U. B. and Sawicka, M. (2015). Farming in and on urban buildings: present practice and specific novelties of Zero-Acreage Farming (ZFarming), *Renew. Agric. Food Syst.* 30(1), 43-54. doi:10.1017/S1742170514000143.

Touliatos, D., Dodd, I. C. and McAinsh, M. (2016). Vertical farming increases lettuce yield per unit area compared to conventional horizontal hydroponics, *Food Energy Secur.* 5(3), 184–191. doi:10.1002/fes3.83.

Treftz, C. and Omaye, S. T. (2016). Hydroponics: potential for augmenting sustainable food production in non-arable regions, *Nutr. Food Sci.* 46(5), 672–684. doi:10.1108/NFS-10-2015-0118.

Van Veenhuizen, R. (ed.) (2014). *Cities Farming for the Future: Urban Agriculture for Green and Productive Cities*. RUAF Foundation, IDRC and IIRR, Silang, Philippines.

Vittori Antisari, L. V., Orsini, F., Marchetti, L., Vianello, G. and Gianquinto, G. (2015). Heavy metal accumulation in vegetables grown in urban gardens, *Agron. Sustain. Dev.* 35(3), 1139–1147. doi:10.1007/s13593-015-0308-z.

Wada, T. (2019). Theory and technology to control the nutrient solution of hydroponics. In: Anpo, M., Fukuda, H. and Wada, T. (Eds), *Plant Factory Using Artificial Light: Adapting to Environmental Disruption and Clues to Agricultural Innovation*. Elsevier, Amsterdam, The Netherlands, pp. 5–14. doi:10.1016/B978-0-12-813973-8.00032-4.

Waller, U., Buhmann, A. K., Ernst, A., Hanke, V., Kulakowski, A., Wecker, B., Orellana, J. and Papenbrock, J. (2015). Integrated multi-trophic aquaculture in a zero-exchange recirculation aquaculture system for marine fish and hydroponic halophyte production, *Aquac. Int.* 23(6), 1473–1489. doi:10.1007/s10499-015-9898-3.

Wong, N. H., Cheong, D. K. W., Yan, H., Soh, J., Ong, C. L. and Sia, A. (2003). The effects of rooftop garden on energy consumption of a commercial building in Singapore, *Energy Build.* 35(4), 353–364. doi:10.1016/S0378-7788(02)00108-1.

Wu, T. H., Chang, C. H., Lin, Y. W., Van, L. D. and Lin, Y. B. (2016). Intelligent plant care hydroponic box using IoTtalk. In: *Proceedings of the IEEE International Conference on Internet of Things (iThings) and IEEE Green Computing and Communications (GreenCom) and IEEE Cyber, Physical and Social Computing (CPSCom) and IEEE Smart Data (SmartData)*, China. doi:10.1109/iThings-GreenCom-CPSCom-SmartData.2016.94.

# Chapter 12

## Optimizing product quality in soilless culture systems (SCS)

*Pietro Santamaria, Barbara De Lucia and Angelo Signore, University of Bari Aldo Moro, Italy*



## 1 Introduction

The definition of the concept of 'quality' is complex, since it can have multiple meanings depending on the context (Gruda, 2005, 2019). In addition, the concept of quality is not static, as it evolves in relation to consumer and lifestyle needs (Schreiner et al., 2013). The concept of the 'quality' of vegetables in a soilless culture system (SCS) is relatively new. Figure 1 shows a total of 1615 articles between 1996 and 2018 recorded by Scopus.

Recently, there has been a tendency to broaden the concept of quality beyond the intrinsic sensory characteristics of the product by also including the method of production as a quality parameter. Consumers choose their products based on taste preferences and nutritional needs, while also considering the production conditions the products are grown in (Gruda, 2009, 2019; Santamaria and Valenzano, 2001). There are three main categories relating to production quality (Santamaria and Valenzano, 2001):

1 elements linked to tradition and culture, particularly the geographical origin of products;
2 elements linked to the environmental impact of the production process, with an increasing focus on sustainability for example 'organic' products; and
3 elements linked to ethics, with a socio-cultural focus, linked, for example, to issues such as worker safety, the exploitation of child labour such as 'fair trade' products, and so on.

http://dx.doi.org/10.19103/AS.2020.0076.14

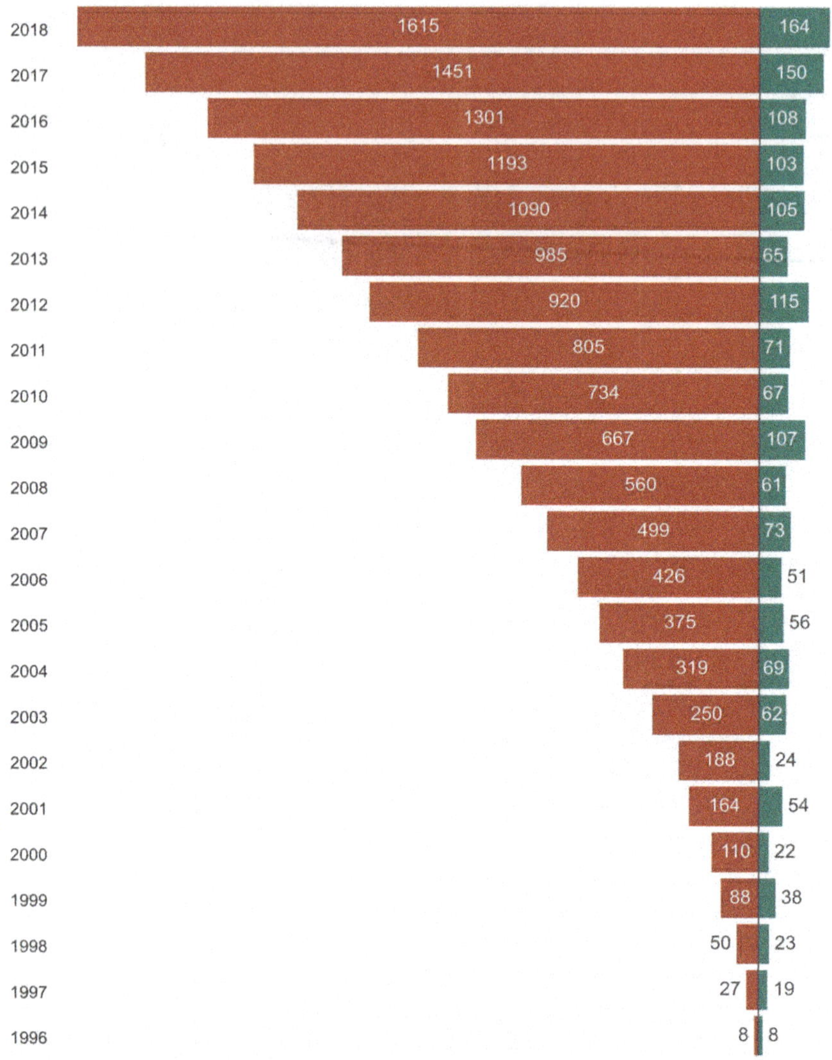


**Figure 1** Total number of articles (left, in red) and number of articles per year (right, in green) for the query "(TITLE-ABS-KEY(soil-less) OR TITLE-ABS-KEY(soilless) OR TITLE-ABS-KEY(hydroponic) AND TITLE-ABS-KEY(quality)) AND PUBYEAR > 1995 AND PUBYEAR < 2019 AND PUBYEAR <2019" in the title, abstract and keywords fields in Scopus between 1996 and 2018.

SCSs and greenhouse environments in general are artificial agricultural systems that can, in theory, be developed anywhere and do not need to be connected to a particular place. These systems can be seen by some as inconsistent with many consumers' preferences for 'organic' and 'natural' products (Schnitzler and Gruda, 2002; Santamaria and Valenzano, 2001). In some countries, an SCS can be considered 'organic'. For example, in the USA, Canada and some other countries, it is possible to market products obtained with an SCS as 'organic' while, in Europe, these products cannot be certified as 'organic' even when they are free of pesticide residues.

## 2 Quality of soilless vegetables and ornamentals

The spread of SCS raises new questions about the quality of the products that result from these systems (Santamaria and Valenzano, 2001). All SCSs have a common feature that is very important for their influence on product quality, namely the use of nutrient solutions (NS) with known concentration and composition. These solutions make it possible to control the process of moisture and mineral uptake by the plant. Therefore, SCSs are ideal for the optimization of root absorption as they regulate the temperature, aeration and nutrient composition of the NS for efficient plant growth and production. As a result, several produce characteristics may be favourably modified, due to a better control of environmental and technical factors, in comparison to conventional agricultural practices. Unfortunately, the cause-effect relations between quality and cultivation systems are not always clear. Furthermore, it is difficult to compare conventional and SCS cultivation systems accurately (Gruda, 2009). Using 30 different publications, these contrasting results are summarized using tomato as an example in Fig. 2.

An important aspect that positively affects the yield and quality of soilless grown vegetables is continuous water and nutrient supply to the plants using SCS. Selma et al. (2012) compared three genotypes of lettuce ('lollo rosso', red oak leaf and butterhead) cultivated simultaneously in soil and with an SCS in Valencia (Spain). The visual quality of ready-to-eat products of red-leafed genotypes cultivated with the SCS was better. The 'lollo rosso' genotypes cultivated with the SCS showed a higher content of bioactive compounds, including vitamin C and phenols, compared to those cultivated in soil. During the postharvest phase, the SCS allowed better control of microbial contamination. The crop grown in soil required a longer period of time for growth and maturity than in the SCS.

Manzocco et al. (2011) also reported that some types of green leafy vegetables grown in soil are more resistant to water loss than those grown with the hydroponic systems.

Valenzano et al. (2008) also reported being able to harvest products early when cultivated with an SCS. They compared this earliness of harvesting with the production of tomato using nutrient film technique (NFT), rockwool slabs and soil during two crop cycles (on average, 10 days in winter–spring and eight days in autumn–winter in the two SCSs). The results of research in this area are not definitive, due to the interaction between the cultivation system and other factors of production, as well as many other factors. For example, López et al. (2013) observed that, given more consistent cultivation conditions, peppers cultivated

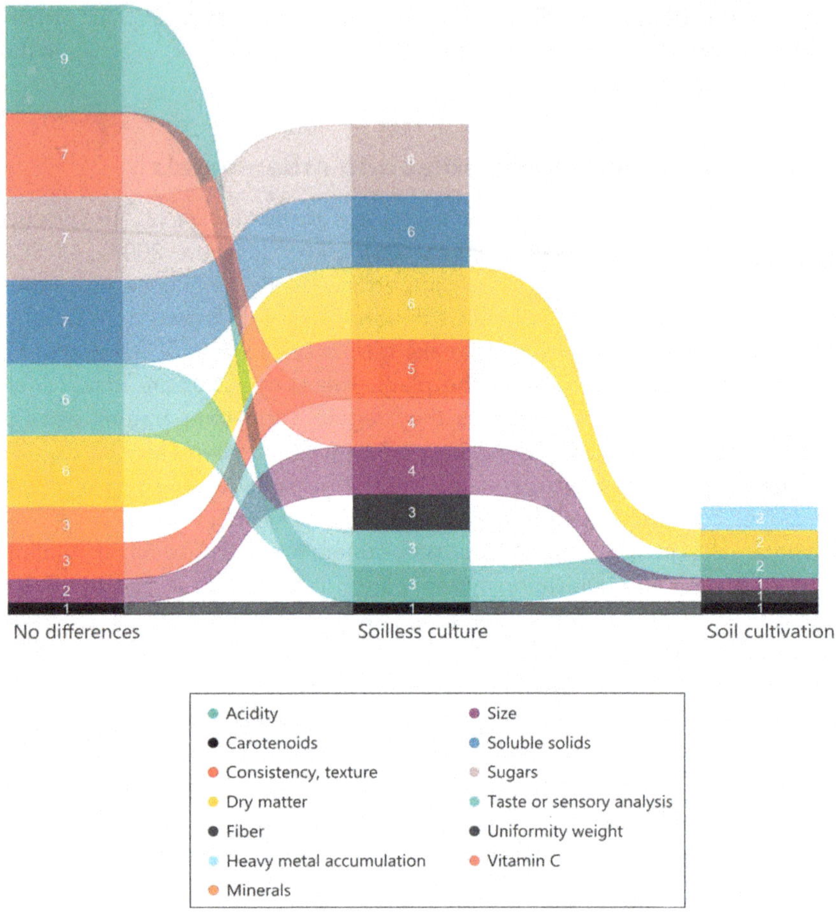


**Figure 2** Quality properties of tomato: a comparison between soil and an SCS according to the results obtained from 30 different articles. Numbers represent the number of scientific articles that report no differences or better results for soilless culture and soil cultivation. The colours represent different quality parameters. Based on Santamaria and Valenzano (2001) and Gruda (2009), modified and supplemented with results from recent studies.

in a SCS had higher nitrates and lower Ca concentration than those cultivated in soil. Cefola et al. (2011) found that table grape produced in containers with peat and perlite were superior to those cultivated in soil as the containers preserved the visual quality whilst reducing gilding of the cluster and weight loss.

Soilless culture systems allow producers to give a precise dosage of NS, which in turn provides crops with the right nutritional content in terms of chemical composition. This is important as some consumers are interested in products with well-defined nutritional composition due to their particular dietary needs. For example, La Malfa (1996) reported that baby food requires a very low level of nitrate content as well as high vitamin and mineral levels. This kind of food is also appropriate for individuals with particular metabolic disorders. These objectives can be achieved through the precise control of NS.

### 2.1 Quality in soilless ornamentals

The ornamental plant industry is one of the world's most dynamic and highly differentiated agricultural production sectors. Ornamental crops include a wide range of plant products: cut flowers, flowering and foliage potted plants, nursery stock, shrubs, bedding plants, and woody landscape ornamentals. External quality attributes widely cited in the literature include: design (shape and length, number of flowers and buds), colour of flowers and leaves, fresh and dry mass and absence of visual defects.

## 3 Examples of improving quality through soilless culture systems (SCS)

The quality of soilless products may be improved (or tailored) by controlling NS management in such areas as: composition, electrical conductivity, pH, and temperature. This section provides examples of using a soilless culture system to control these parameters and improve quality.

### 3.1 Biofortified vegetables

Humans require at least 22 mineral elements for their well-being (Graham et al., 2007). Increasing the number of bioavailable micronutrients in food is a major challenge for developing countries (White and Broadley, 2009; Graham et al., 2007; Welch and Graham, 1999). Recent data indicate an increase in world hunger: around 821 million people are chronically undernourished; they were 804 million in 2016 (FAO et al., 2018). Over 60% of the world's population suffers from iron (Fe) deficiency, over 30% from zinc (Zn) deficiency, 30% from iodine (I) deficiency and around 15% from selenium (Se) deficiency (Graham et al., 2007). The deficiencies of calcium (Ca), magnesium (Mg) and copper (Cu) are also common in several developed and developing countries. Such deficiencies may be solved through food diversification, mineral supplements, food fortification or by increasing the concentration and/or the bioavailability of mineral elements in food, for example, through biofortification. This process allows one or more elements to be incorporated into plants in the form of organic compounds (Smolen et al., 2019).

It is now possible to produce 'tailor-made', 'personalized' vegetables. Tailored food is intended for people suffering from particular metabolic disorders. Besides increasing nutritional factors, this process of tailoring foods can be achieved through the reduction of anti-nutritional factors in plants such as phytates, oxalates, nitrates, histamine, chromium, zinc and nickel, especially for those with allergies and/or metabolic disorders. This process also adds value to vegetables for other consumers.

The EU Regulation No. 1169/2011 of the European Parliament and of the Council of 25 October 2011 regulates the use of nutritional labels on food. For example, these labels can display the content of mineral salts if a 'significant' quantity is present. A quantity is considered significant, in terms of 100 g or 100 mL, when its concentration is at least 15% of the recommended dose (Table 1).

Due to its rapid oxidation in the soil and low mobility in the phloem, soluble iron-based fertilizers are inefficient in increasing the concentration of Fe in plants, even with foliar applications (Graham et al., 2000). The Fe content in lettuce, spinach and onion leaves grown in an NFT for about a month increased linearly over time using ammonium ferric citrate (AFC, 100 mg/L) or EDTA-Fe (50 mg/L) in the last 3–15 hours of the crop cycle (Inoue et al., 2000). Using AFC, the Fe content was increased from approx. 0.82 mg/100 g to 5–7 mg/100 g in fresh spinach after 9 hours, while it increased from 0.82 mg/100 g to 3.05 mg/100 g in lettuce and from 0.58 mg/100 g to 2.00 mg/100 g in onion after 6 hours.

During two studies, using NS with increasing concentrations of silicon (2, 50 and 100 mg/L), provided in the form of potassium metasilicate ($K_2SiO_3$) with the floating system (Fig. 3), the 'baby leaves' accumulated silicon, highlighting a close relationship with the provided dose, without colour substantial differences in the content of other nutrients (D'Imperio et al., 2016a,b). Furthermore, the bioaccessibility, that is, the amount of nutrient released by the plant matrix

**Table 1** Mineral salts that can be declared on the food label and their Recommended Dietary Allowances (RDA).

| Mineral salts | RDA (mg) |
|---|---|
| Calcium | 800 |
| Chloride | 800 |
| Phosphorus | 700 |
| Iron | 14 |
| Fluorine | 3.5 |
| Magnesium | 370 |
| Manganese | 2 |
| Molybdenum | 50 |
| Potassium | 2000 |
| Copper | 1 |
| Selenium | 55 |
| Zinc | 10 |
| Iodine | 0.15 |

**Figure 3** A silicon biorforticated leafy chicory to be used as a freshly cut product. The floating system is the simplest SCS for obtaining vegetables with a 'personalized' quality profile.

during an in vitro gastro-intestinal digestion process, was higher compared to plants that were fed without additional potassium metasilicate (D'Imperio et al., 2017). Equally promising results have also been obtained for green bean. Interestingly, the amount of silicon of biofortificated pods remained almost constant, even after the boiling or steaming processes (Montesano et al., 2016).

Biofortification treatments may also be used for other mineral elements (I, Mg, Zn, etc.) to obtain nutritionally enriched products for the market. These products can then be marked accordingly. Apart from biofortification with essential mineral elements, an SCS allows the modification of bioactive compound composition. For example, in purslane leaves the content of α-linolenic acid (ω-3) may be doubled when the molar ratio of $NH_4$:$NO_3$ is 0.5:0.5, rather than 1:0, 0.75:0.25 or 0.25:0.75 (Palaniswamy et al., 2000).

Developing protocols for the production of microgreens with low potassium content helps patients suffering from chronic renal failure, a disease that limits the consumption of rich-potassium foods. These vegetables are generally rich in this element. Renna et al. (2018) found that it was possible to obtain microgreens with K content around 100 mg/100 g fresh mass, more than 30% less than the average K content in vegetables, and market them as 'low potassium rate' vegetables, in compliance with the Regulation (EC) No. 1924/2006. Asaduzzaman et al. (2018) have shown that by reducing the amount of potassium nitrate in the NS, from anthesis to harvesting, it is possible to produce melon fruits with a lower K content (<20%) than the K content of a greenhouse-grown melon (340 mg/100 g fresh mass).

### 3.2 Nutrient solution (NS) composition and quality attributes

In many cases, especially in NFT systems, the growth and yield of vegetables in response to the concentration of nutrients is lower than expected. For example, Massey and Winsor (1980a) did not observe any effect on tomato yield when N-$NO_3$ was distributed in concentrations varying between 10 mg/L and 320 mg/L. Similar results were observed by varying the P concentration from 5 mg/L to 200 mg/L (Massey and Winsor, 1980b). When approaching the end of a crop cycle, a reduction in macroelement concentrations of NS (Gunes et al., 1998; Siddiqi et al., 1998), especially of N, will increase mineral element use efficiency without affecting yield and this reduces the environmental impact of SCS (Le Bot et al., 2001).

In a similar way to vegetables, the visual quality of ornamental crops is linked to an adequate balance of nutrients. The reduction of potassium (K) in the NS is an alternative method for chemical growth regulation in summer flowering bedding plants (Hansen and Petersen, 2004) and can lead to a darker green or red colouration in many species. Şirin (2011) showed that in gerbera 'Skyline', grown in an open SCS with peat and perlite, the N dose of 150 mg/L compared to 210 mg/L improved the quality of the capitulum diameter and the stalk's upper diameter; stalk length was not affected. In potted anthurium, Sonneveld and Voogt (1993) showed that nitrogen supply must not exceed 6 g/plant in a year, otherwise flower quality will decline.

For leafy vegetables, the increase of nitrogen concentration in the NS is accompanied by the accumulation of nitrates in the leaves. For example, by increasing nitrogen concentration from 1 mM to 8 mM, the concentration of nitrates increased from 5300 mg/kg to 8100 mg/kg of fresh mass on average for cultivated rocket [*Eruca vesicaria* (L.) Cav.] and 'wild' rocket [*Diplotaxis tenuifolia* (L.) DC.] (Santamaria et al., 2002). The content of $NO_3$ may be reduced by also increasing the molybdenum concentration in the NS (Moncada et al., 2019).

The availability of N in nitric form, rather than in ammoniac form, favours the accumulation of $NO_3$ in leafy vegetables (Santamaria, 2006; Blom-Zandstra, 1989), resulting in negative effects on commercial quality of the products. The response to ammonium depends especially on the species considered (Santamaria et al., 1998a, 1999). Providing ammonium as the sole chemical source of N, decreases the accumulation of nitrates in vegetables (Santamaria et al., 1998b), but most species exhibit toxic effects in such situations. Among these, it is necessary to point out the endive (Fig. 4), which is capable of producing heads free of nitrates and of weight similar to those obtained with $NO_3$-N fertilization (Bonasia et al., 2008; Santamaria and Elia, 1997).

Banijamali et al. (2018) showed that a dose of 2 mM $NH_4^+$ (36 mg/L) increased not only the number of flowers, but the length of pedicles and fresh

**Figure 4** Endive (*Cichorium endivia* L. var. *crispum* Hegi) grown in hydroculture with 100% nitrogen in nitric (left) or ammoniacal form (right). Endive tolerates ammoniacal nitrogen very well, but due to the reduction of the pH values, the roots are shorter and more compact.

weight in rose 'Vendetta', while 5 mM $NH_4^+$ (90 mg/L) caused a decrease in most of the quality parameters. A recent experiment showed that a 80:20 ($NO_3^-$:$NH_4^+$) ratio should be used for gerbera cut flower production to achieve the best quality in terms of the number of leaves per plant, fresh and dry weights of roots and shoots, number of flowers, flower stalk diameter and capitulum diameter (Khalaj et al., 2017).

Potassium (K) constitutes up to 10% of plant dry weight and is considered a macronutrient essential for plants, with fundamental effects on their health, growth, and quality. Sonntag et al. (2019) reported a positive impact of K on the parameters linked to the taste of cocktail tomatoes, namely, total soluble solids, titratable acidity, dry matter, colour and firmness. Ornamental plants normally demand K in higher quantities. Nutritional guidelines for flowering plants suggested that K concentration is in the 250–350 mg/L range. The optimum NS for anthurium may contain very wide ratios of K as long as Ca and Mg levels are maintained at a low proportion in the NS.

Several studies have reported that Ca is important in cut flower cultivation when using an SCS. In African daisy plants grown in volcanic rock, the best Ca dose for 'Amaretto' was 240 mg/L and for 'Darling' 180 mg/L, which produced a higher capitulum diameter and scape length compared to using a Ca dose of 120 mg/L (Albino-Garduño et al., 2007). Bernstein et al. (2018) showed that

110 mg/L Ca plus 10% N-$NH_4$ prevented flower abortion and leaf damage in *Anemone coronaria*, and enhanced the quality of flowers, compared to 60 mg/L.

One of the most important disorders of the tomato is the blossom-end rot (BER) determined by calcium deficiency (Hagassou et al., 2019). Some strategies to avoid BER are shown in Table 2.

Although silicon (Si) is not considered an essential element for higher plants, its beneficial effects have been proven in several species. In cucumber, as in other cucurbits, the addition of Si in the NS improves the structure of the cell walls, especially when its concentration in the water used for NS preparation is scarce. This makes the plant's tissues more resistant to fungal attacks, such as the powdery mildew. As a result, fewer fungicide treatments are required and it is possible to obtain products with lower residues of pesticide in the edible organs. Interesting results in this sense were obtained by Buttaro et al. (2009) and Gottardi et al. (2012). Buttaro et al. (2009) studied the effect of Si in the NS on yield, quality and sensitivity to powdery mildew in 'carosello' and 'barattiere' (*Cucumis melo* L.), two Italian local varieties of unripe melon, sensitive to this disease. The distribution of Si reduced the severity (3.3% vs. 29.5%) and the incidence (49.1% vs. 81.5%) of powdery mildew on both landraces with respect to the control plants grown without Si. Gottardi et al. (2012), on the other hand, studied the effect of the addition of Si in the NS on the quantitative and qualitative aspects of two cultivars of lamb's lettuce cultivated in a floating system: the supply of Si increased yield and product quality by reducing the concentration of nitrates into the tissues, and improved the post-harvest quality of the product.

An adequate amount of aluminium (Al) must be provided during the production of potted blue hydrangeas [*Hydrangea macrophylla* (Thunb.) Ser.] (Fig. 5) because Al binds with the anthocyanin pigment delphinidin-3 glucoside to produce blue sepals (Takeda et al., 1985). Absence of Al in the root medium produces pink sepals while inadequate amounts of Al result in an undesirable murky mauve colour. Repeated drenches of high concentrations of aluminium sulphate solution are typically applied during greenhouse production of blue hydrangeas due to the inability of growing media to retain Al. Opena

**Table 2** How to avoid blossom-end root (BER) and calcium deficiency in tomato by acting on the nutrient solution.

| Strategy | Parameters of nutrient solution |
|---|---|
| To optimize the Ca uptake | Ca: 120 mg/L;<br>Temperature: < 25 °C;<br>$O_2$: > 3 mg/L. |
| To maintain the balance of nutrients | N: 180 mg/L (<15% N-$NH_4$);<br>K: 300–400 mg/L;<br>P: > 5 mg/L. |

**Figure 5** In an acidic growing medium, the aluminium ion becomes water-soluble and is absorbed from the roots, followed by transport to the sepal vacuoles to give a blue color.

and Williams (2003) showed that Al nutrient retention can be improved by incorporating zeolites into a soilless root medium.

### *3.3 Electrical conductivity of the nutrient solution*

A key factor affecting the quality of tomato fruits is the electrical conductivity (EC) of the NS. An increase in EC increases dry matter content, total soluble solids, reducing sugars, titratable acidity, volatile compounds, minerals, carotenes and vitamin C. With an SCS it is also possible to use different nutrient concentrations, and therefore EC, during the day and at night. For example, Santamaria et al. (2004), using EC of 6 dS/m during the night and 2 dS/m during the day, obtained tomatoes with higher total soluble solids content when compared to the use of a NS with a constant CE of 2 dS/m all day long (7.0 vs 6.6 °Brix for 'Naomi' and 4.6 vs 4.3 for 'Diana') without decreasing the yield. Moreover, Li et al. (2001) showed that a shoot environment treatment that depresses potential transpiration of tomato plants can modify the effect of the root zone salinity, both by reducing the fraction of fruits that is to be discarded and reducing the decrease in fresh weight of each fruit.

The use of a lower concentration of NS than the standard gives quality advantages in floriculture. Zheng et al. (2005) and Rouphael et al. (2008) proved that when NS concentration was reduced by 50%, it did not show any detrimental effect on quality parameters, respectively, in gerbera and zonal pelargonium. Rose is traditionally categorized as a salt-sensitive species. More recent research had shown that variations in salt tolerance exist among cultivars with moderately salt-tolerant cultivars and rootstocks (Niu and Sun, 2019).

Gerbera has been classified as moderately sensitive to salinity (Sonneveld et al., 1999). High EC of NS reduces stem length and number of gerbera flowers (De Kreij et al., 1999); for example, an EC value between 2.1 and 3.7 dS/m in a closed system reduces the average flower weight by 6.2% (Savvas et al., 2002). In an open system, an EC of 2.4 dS/m provides a higher fresh weight, capitulum diameter, stem length and diameter higher than plants grown with 1.6 dS/m (Paradiso et al., 2003). Contradictory results have been highlighted in hydroponically grown *Lilium*: Morgan (2006) reported that an EC level higher than 1.8 dS/m resulted in a stunted plant and small flowers; on the other hand, Al-Hammouri et al. (2017) found that, in a zeolitic tuff closed SCS, the stem diameter and buds number were not affected with an EC level of up to 2.3 dS/m for Asiatic lily 'Brunello'.

### 3.4 pH of the nutrient solution

The pH influences the chemical behaviour of the essential elements and their availability. Most crops grow better with a pH value ranging from 5.5 to 6.5 in the root environment. For example, lettuce produces the best results at pH 6, while a higher or lower pH reduces both the nitrate concentration and the fresh weight of the leaves (Lee et al., 1998). Cut flower rose (*Rosa* x *hybrida*) had a length of 60 cm longer with an NS that had a pH of 6.0 compared to an NS with a pH of 8.0 (Kim et al., 2005). In rose 'Gran Gala', Roosta and Rezaei (2014) discovered that the increase in pH from 4.5 to 6.5 resulted in an ascending trend in flowering bud diameter; it was likely related to better uptake of nutrients, particularly P, Fe, Mn, Zn, and Cu.

### 3.5 Nutrient solution temperature

Due to solar radiation, the air temperature inside the greenhouse has high daily fluctuations, in contrast to what happens to the NS, because compared to air, water has a higher heating capacity and lower thermal transmittance. Low temperatures in SCS may affect the functionality of the radical system involved in the uptake of nutrients such as nitrate, iron and sulphate (Iacuzzo et al., 2011). As the temperature of the NS increases, the removal of water and all the macronutrients increases. Temperatures higher than 20°C can increase tomato growth, but can have a negative effect on fruit quality. In fast-growing crops such as cucumber these temperatures can have a positive effect on fruit quality (Adams, 1999). By increasing the temperature of the NS from 15°C to 30°C the nitrate content in lettuce increases, but not the yield, which is maximized at 20°C (Lee et al., 1998). However, the increase of the temperature of the NS causes a decrease in the concentration of dissolved oxygen and an increase in the consumption of oxygen (respiration increases with increasing temperature).

Results on the effects of root temperature in SCS rose plants are mainly dependant on a combination of factors such as root-air temperatures, cultivars and rootstock. Zeroni and Gale (1982) suggested that a root-zone temperature of 18°C was optimal for shoot growth of 'Better Times' and 'Sonia' roses grafted on the rootstock of *Rosa indica*. Zeroni and Gale (1987) also discovered that increasing root temperatures from 18°C to 25°C had no effects on stem length and flower production. Calatayud et al. (2008) showed that 'Gran Gala' plants were well adapted to 10°C without a decrease in dry biomass. Furthermore, a cold solution increased $NO_3^-$ uptake and thin-white root production but decreased water uptake.

### 3.6 Fertigation management

It is possible to reduce $NO_3$ content by eliminating N-$NO_3$ or by replacing it with Cl, $SO_4$ or $NH_4$ in the NS in the period immediately before harvesting. Some examples are summarized by Santamaria (2006) and, for rocket crop, one of the species with the higher accumulation of nitrates, by Santamaria et al. (2001). The frequency and volumes of fertigation play an important role in the qualitative and qualitative aspects of vegetable production. Since irrigation in SCS is typically controlled continuously by a computer, water stress is rarely a problem. However, the quality of the product may be improved by reducing the water regime to 80% of daily evapotranspiration, but this may have negative consequences on yield (Aydiner et al., 2016; Shinohara et al., 1995).

Water stress also negatively influences the quality of ornamentals. Raviv and Blom (2001) reported that low substrate water content reduced shoot length in cut rose 'Sonia' and 'Laser', since water stress caused stomata closures, reducing turgidity. Bolla et al. (2010) found a reduction in the production of marketable cut rose 'Eurored' stems, grown in three different substrates, under water deficit condition, using only 67% of the irrigation needs. In gerbera SCS, Tsirogiannis et al. (2010) showed that stem length and flower diameter were unaffected by the irrigation scheduling.

In ornamental nurseries, smaller leaves and shorter internodes, decrease of flowers number, size and quality occur due to drier irrigation regimes, but many ornamental crops show different responses to drought, making it possible to use moderate deficit irrigation whilst keeping a good overall quality. In container-grown carnation, quality characteristics such as foliage width, number of shoots and flowers per plant, colour of flowers were not reduced (Álvarez et al., 2009).

### 3.7 Harvesting methods

Another advantage of SCS is the possibility to sell leafy vegetables in containers with their growing media (Di Gioia et al., 2015) or submerged into NS (Fig. 6). In both cases, the vegetables maintain a longer shelf life and are harvested by the end-user. In the latter case, as shown by Kotsiras et al. (2016) with lettuce crops, the product is also cleaner and it has a better shelf life, relative fresh weight and visual quality in comparison with product harvested in the greenhouse. By moving the spinach in a SCS containing a NS without N, two or three days before harvesting, the nitrate concentration is reduced and vitamin C is increased (Savvas and Gruda, 2018; Mozafar, 1996).

**Figure 6** Endive plants sold with the roots (top) and a tray with *Brassica* microgreens in a container (bottom).

## 4 Conclusion and future trends

Using an SCS makes it possible to produce vegetables and ornamentals with improved quality, as well as vegetables with tailored concentrations of some elements for people with special dietary requirements. The main factors through which it is possible to improve the quality profile of soilless horticultural produce are the composition and management of the NS.

The numerous examples reported in this chapter demonstrate the potential of the SCS in producing vegetables and ornamental plants of "superior" quality. A new challenge in agriculture is the production of tailor-made foods specifically suited to target groups of people with special nutritional needs, which can be achieved through using an SCS. An emerging topic is that of the biofortification of vegetables obtained with SCS. In collaboration with the Institute of Sciences of Food Production of the National Research Council of Bari (Italy), after identifying the elderly as target of the population with specific nutritional requirements for silicon and calcium, we biofortified different leafy vegetables, such as purslane, chicory, Swiss chard, basil, tatsoi and mizuna (D'Imperio et al., 2016a,b), to be used in the ready-to-eat sector, using the floating system.

Future trends also include using sensors and computers for real-time greenhouse climate and for monitoring and management of NS, as well as interpretation algorithms based on machine learning logic to analyze such data and make management decisions (Sambo et al., 2019).

## 5 Where to look for further information

In this chapter, the authors have described the main factors through which it is possible to influence and improve the quality profile of soilless horticultural produce. For more information on the concept of the quality of products, it is useful to consult Gruda's works (2005, 2009, 2019; see in the References).

To explore the topic of nitrate content in vegetables - one of the most important traits for leafy vegetables and the only one to be regulated at EU level - a reference review from Santamaria (2006) is recommended.

Increasing the nutrients in food is an important challenge for developing countries. A recent review by Gruda et al. (2018) and Rouphael and Kyriacou (2018) provides an overview and a future perspective on scientific progress in SCS to improve the quality of vegetables through the NS management and biofortification applications.

SCSs are ideal for the optimization of root absorption, by regulating temperature, aeration and nutrient composition of the NS, for an efficient plant growth and production. Two reviews discussing limitations and opportunities of SCS in plant mineral nutrition process are those of Sambo et al. (2019) and Savvas and Gruda (2018).

## 6 References

Adams, P. (1999). Plant nutrition demystified. *Acta Hortic.* 481, 341–344.

Albino-Garduño, R., Zavaleta-Mancera, H. A., Ruiz-Posadas, L. M., Sandoval-Villa, M. and Castillo-Morales, A. (2007). Response of gerbera to calcium in hydroponics. *J. Plant Nutr.* 31(1), 91–101.

Al-Hammouri, A. A. N., Al-Ghawanmeh, K., Hani, N. B. and Karam, N. (2017). Effect of irrigation with nutrient solutions mixed with treated wastewater on Asiatic lily 'Brunello' grown in a closed soilless culture. *Acta Agric. Slov.* 109(1), 29–42.

Álvarez, S., Navarro, A., Bañón, S. and Sánchez-Blanco, M. J. (2009). Regulated deficit irrigation in potted Dianthus plants: effects of severe and moderate water stress on growth and physiological responses. *Sci. Hortic.* 122(4), 579–585.

Asaduzzaman, M., Talukder, M. R., Tanaka, H., Ueno, M., Kawaguchi, M., Yano, S., Ban, T. and Asao, T. (2018). Production of low-potassium content melon through hydroponic nutrient management using perlite substrate. *Front. Plant Sci.* 9, 1382.

Aydiner, E., Tuzel, Y., Tuzel, I. H., Tunali, U. and Oztekin, G. B. (2016). Effects of irrigation based on different moisture levels of growing medium on soilless grown greenhouse tomatoes. *Acta Hortic.* 1142, 93–98.

Banijamali, S. M., Feizian, M., Bayat, H. and Mirzaei, S. (2018). Effects of nitrogen forms and calcium amounts on growth and elemental concentration in *Rosa hybrida* cv. 'Vendentta'. *J. Plant Nutr.* 41(9), 1205–1213.

Bernstein, N., Sacks, M., Snir, P. and Rosenberg, R. (2018). Prevention of flower abortion and leaf damages in *Anemone coronaria* by Ca supplements. *Isr. J. Plant Sci.* 65(01–02), 97–108.

Blom-Zandstra, M. (1989). Nitrate accumulation in vegetables and its relationship to quality. *Ann. Appl. Biol.* 115(3), 553–561.

Bolla, A., Koukourikou-Petridou, M., Voyiatzis, D. and Chimonidou, D. (2010). Physiological responses associated to substrate water availability of Rosa 'Eurored' plants grown in soilless greenhouse conditions. *Sci. Hortic.* 121(1), 80–83.

Bonasia, A., Conversa, G., Gonnella, M., Serio, F. and Santamaria, P. (2008). Effects of ammonium and nitrate nutrition on yield and quality in endive. *J. Hortic. Sci. Biotechnol.* 83(1), 64–70.

Buttaro, D., Bonasia, A., Minuto, A., Serio, F. and Santamaria, P. (2009). Effect of silicon in the nutrient solution on the incidence of powdery mildew and quality traits in carosello and barattiere (*Cucumis melo* L.) grown in a soilless system. *J. Hortic. Sci. Biotechnol.* 84(3), 300–304.

Calatayud, Á., Gorbe, E., Roca, D. and Martínez, P. F. (2008). Effect of two nutrient solution temperatures on nitrate uptake, nitrate reductase activity, $NH_4^+$ concentration and chlorophyll a fluorescence in rose plants. *Environ. Exp. Bot.* 64(1), 65–74.

Cefola, M., Pace, B., Buttaro, D., Santamaria, P. and Serio, F. (2011). Postharvest evaluation of soilless-grown table grape during storage in modified atmosphere. *J. Sci. Food Agric.* 91(12), 2153–2159.

De Kreij, C., Voogt, W. and Baas, R. (1999). Nutrient solutions and water quality for soilless cultures. Research Station for Floriculture and Glasshouse Vegetables (PBG), Naaldwijk, The Netherlands, Brochure 196.

Di Gioia, F., Mininni, C. and Santamaria, P. (2015). How to grow microgreens. In: *Microgreens* (Di Gioia, F. and Santamaria, P., Eds). Eco-logica editore, Bari, Italy, 51–80. ISBN 978-88-909289-3-2.

D'Imperio, M., Renna, M., Cardinali, A., Buttaro, D., Santamaria, P. and Serio, F. (2016a). Silicon biofortification of leafy vegetables and its bioaccessibility in the edible parts. *J. Sci. Food Agric.* 96(3), 751–756.

D'Imperio, M., Renna, M., Cardinali, A., Buttaro, D., Serio, F. and Santamaria, P. (2016b). Calcium biofortification and bioaccessibility in soilless "baby leaf" vegetable production. *Food Chem.* 213, 149–156.

D'Imperio, M., Brunetti, G., Gigante, I., Serio, F., Santamaria, P., Cardinali, A., Colucci, S. and Minervini, F. (2017). Integrated in vitro approaches to assess the bioaccessibility and bioavailability of silicon-biofortified leafy vegetables and preliminary effects on bone. *In Vitro Cell. Dev. Biol. Anim.* 53(3), 217–224. ISSN: 1071-2690.

FAO, IFAD, UNICEF, WFP and WHO (2018). The state of food security and nutrition in the world 2018. Building climate resilience for food security and nutrition. FAO, Rome. Available at: http://www.fao.org/3/I9553EN/i9553en.pdf (Accessed 15 July 2019).

Gottardi, S., Iacuzzo, F., Tomasi, N., Cortella, G., Manzocco, L., Pinton, R., Römheld, V., Mimmo, T., Scampicchio, M., Dalla Costa, L. and Cesco, S. (2012). Beneficial effects of silicon on hydroponically grown corn salad (*Valerianella locusta* (L.) Laterr) plants. *Plant Physiol. Biochem.* 56, 14–23.

Graham, R. D., Welch, R. M. and Bouis, H. E. (2000). Addressing micronutrient malnutrition through enhancing the nutritional quality of staple foods: principles, perspectives and knowledge gaps. *Adv. Agron.* 70, 77–142.

Graham, R. D., Welch, R. M., Saunders, D. A., Ortiz-Monasterio, I., Bouis, H. E., Bonierbale, M., de Haan, S., Burgos, G., Thiele, G., Liria, R., Meisner, C. A., Beebe, S. E., Potts, M. J., Kadian, M., Hobbs, P. R., Gupta, R. and Twomlow, S. (2007). Nutritious subsistence food systems. *Adv. Agron.* 92, 1–74.

Gruda, N. (2005). Impact of environmental factors on product quality of greenhouse vegetables for fresh consumption. *Crit. Rev. Plant Sci.* 24(3), 227–247.

Gruda, N. (2009). Do soilless culture systems have an influence on product quality of vegetables? *J. Appl. Bot. Food Qual.* 82, 141–147.

Gruda, N. (2019). Assessing the impact of environmental factors on the quality of greenhouse produce. In: *Achieving Sustainable Greenhouse Cultivation* (Marcelis, L. and Heuvelink, E., Eds). Burleigh Dodds Science Publishing Limited, Sawston (Cambridge), UK.

Gruda, N., Savvas, D., Colla, G. and Rouphael, Y. (2018). Impacts of genetic material and current technologies on product quality of selected greenhouse vegetables – a review. *Eur. J. Hortic. Sci.* 83(5), 319–328.

Gunes, A., Alpaslan, M. and Inal, A. (1998). Critical nutrient concentrations and antagonistic and synergistic relationships among the nutrients of NFT-grown young tomato plants. *J. Plant Nutr.* 21(10), 2035–2047.

Hagassou, D., Francia, E., Ronga, D. and Buti, M. (2019) Blossom end-rot in tomato (*Solanum lycopersicum* L.): A multi-disciplinary overview of inducing factors and control strategies. *Sci Hortic.* 249, 49–58.

Hansen, C. W. and Petersen, K. K. (2004). Reduced nutrient and water availability to hibiscus rosa-sinensis 'Cairo Red' as a method to regulate growth and improve post-production quality. *Eur. J. Hortic. Sci.* 69(4), 159–166.

Iacuzzo, F., Gottardi, S., Tomasi, N., Savoia, E., Tommasi, R., Cortella, G., Terzano, R., Pinton, R., Dalla Costa, L. and Cesco, S. (2011). Corn salad (*Valerianella locusta* (L.) Laterr.) growth in a water-saving floating system as affected by iron and sulfate availability. *J. Sci. Food Agric.* 91(2), 344–354.

Inoue, K., Kondo, S., Adachi, A. and Yokota, H. (2000). Production of iron enriched vegetables: effect of feeding time on the rate of increase in foliar iron content and foliar injury. *J. Hortic. Sci. Biotechnol.* 75(2), 209–213.

Khalaj, M. A., Kiani, S., Khoshgoftarmanesh, A. H. and Amoaghaie, R. (2017). Growth, quality, and physiological characteristics of gerbera (*Gerbera jamesonii* L.) cut flowers in response to different $NO_3^-$:$NH_4^+$ ratios. *Hortic. Environ. Biotechnol.* 58(4), 313–323.

Kim, H. J., Cho, Y. S., Kwon, O. K., Cho, M. H., Hwang, J. B., Bae, S. D. and Jeon, W. T. (2005). Effect of pH and EC of hydroponic solution on the growth of greenhouse rose. *Asian J. Plant Sci.* 4(4), 348–355.

Kotsiras, A., Vlachodimitropoulou, A., Gerakaris, A., Bakas, N. and Darras, A. I. (2016). Innovative harvest practices of butterhead, lollo rosso and bataviagreen lettuce (*Lactuca sativa* L.) types grown in floating hydroponic system to maintain the quality and improve storability. *Sci. Hortic.* 201, 1–9.

La Malfa, G. (1996). La qualità degli ortaggi nella coltivazione fuori suolo (Vegetable quality in soilless culture). *Tecnica Agricola*, 48(4), 35–46.

Le Bot, J., Jeannequin, B. and Fabre, R. (2001). Growth and nitrogen status of soilless tomato plants following nitrate withdrawal from the nutrient solution. *Ann. Bot.* 88(3), 361–370.

Lee, E. H., Lee, B. Y., Kim, K. D., Lee, J. W. and Kwon, Y. S. (1998). Nitrate content and activities of nitrate reductase and glutamine synthetase as affected by temperature and pH of nutrient solution in leaf lettuce and water dropwort. *J. Korean Soc. Hortic. Sci.* 39, 157–160.

Li, Y. L., Stanghellini, C. and Challa, H. (2001). Effect of electrical conductivity and transpiration on production of greenhouse tomato (*Lycopersicon esculentum* L.). *Sci. Hortic.* 88(1), 11–29.

López, A., Fenoll, J., Hellín, P. and Flores, P. (2013). Physical characteristics and mineral composition of two pepper cultivars under organic, conventional and soilless cultivation. *Sci. Hortic.* 150, 259–266.

Manzocco, L., Foschia, M., Tomasi, N., Maifreni, M., Dalla Costa, L., Marino, M., Cortella, G. and Cesco, S. (2011). Influence of hydroponic and soil cultivation onquality and shelf life of ready-to-eat lamb's lettuce (*Valerianella locusta* L. Laterr). *J. Sci. Food Agric.* 91(8), 1373–1380.

Massey, D. M. and Winsor, G. W. (1980a). Some responses of tomato to nitrogen in recirculating solutions. *Acta Hort.* 98, 127–137.

Massey, D. M. and Winsor, G. W. (1980b). Some responses of tomato to phosphorus concentration in nutrient film culture. Proceedings of the 5th International Conference Soilless Culture, Wageningen, The Netherlands, 205–214.

Moncada, A., Miceli, A., Sabatino, L., Iapichino, G., D'Anna, F. and Vetrano, F. (2019). Effect of molybdenum rate on yield and quality of lettuce, escarole, and curly endive grown in a floating system. *Agronomy* 8(9), 171.

Montesano, F. F., D'Imperio, M., Parente, A., Cardinali, A., Renna, M. and Serio, F. (2016). Green bean biofortification for Si through soilless cultivation: plant response and Si bioaccessibility in pods. *Sci Rep.* 6, 31662.

Morgan, L. (2006). Grow gorgeous lilies hydroponically. *Growing Edge* 17(4), 36–40.

Mozafar, A. (1996). Decreasing the NO3 and increasing the vitamin C contents in spinach by a nitrogen deprivation method. *Plant Food Hum. Nutr.* 49(2), 155–162.

Niu, G. and Sun, Y. (2019). Salt tolerance in roses-a review. *Acta Hortic.* 1232, 143–150.

Opena, G. B. and Williams, K. A. (2003). Use of precharged zeolite to provide aluminium during blue hydrangea production. *J. Plant Nutr.* 26, 1825–1840.

Palaniswamy, U. R., McAvoy, R. J. and Bible, B. B. (2000). Omega-3-fatty acid concentration in *Portulaca oleracea* is altered by nitrogen source in hydroponic solution. *J. Am. Soc. Hortic. Sci.* 125(2), 190–194.

Paradiso, R., De Pascale, S., Aprea, F. and Barbieri, G. (2003). Effect of electrical conductivity levels of nutrient solution on growth, gas exchanges and yield of two gerbera cultivars in soilless system. *Acta Hortic.* 609, 165–171.

Raviv, M. and Blom, T. J. (2001). The effect of water availability and quality on photosynthesis and productivity of soilless-grown cut roses. *Sci. Hortic.* 88(4), 257–276.

Renna, M., Castellino, M., Leoni, B., Paradiso, V. M. and Santamaria, P. (2018). Microgreens production with low potassium content for patients with impaired kidney function. *Nutrients* 10(6), 675.

Roosta, H. R. and Rezaei, I. (2014). Effect of nutrient solution pH on the vegetative and reproductive growth and physiological characteristics of rose Cv 'Grand Gala'in hydroponic system. *J. Plant Nutr.* 37(13), 2179–2194.

Rouphael, Y., Cardarelli, M., Rea, E. and Colla, G. (2008). The influence of irrigation system and nutrient solution concentration on potted geranium production under various conditions of radiation and temperature. *Sci. Hortic.* 118(4), 328–337.

Rouphael, Y. and Kyriacou, M. C. (2018). Enhancing quality of fresh vegetables through salinity eustress and biofortification applications facilitated by soilless cultivation. *Front. Plant Sci.* 9, 1254.

Sambo, P., Nicoletto, C., Giro, A., Pii, Y., Valentinuzzi, F., Mimmo, T., Lugli, P., Orzes, G., Mazzetto, F., Astolfi, S., Terzano, R. and Cesco, S. (2019). Hydroponic solutions for soilless production systems: issues and opportunities in a smart agriculture perspective. *Front. Plant Sci.* 10, 923.

Santamaria, P. (2006). Nitrate in vegetables: toxicity, content, intake, and EC regulation. *J. Sci. Food Agric.* 86(1), 10–17.

Santamaria, P., Cantore, V., Conversa, G. and Serio, F. (2004). Effect of night salinity level on water use, physiological responses, yield and quality of tomato. *J. Hortic. Sci. Biotech.* 79(1), 59–66.

Santamaria, P. and Elia, A. (1997). Producing nitrate free endive heads: effect of nitrogen form on growth, yield and ion composition of endive. *J. Am. Soc. Hortic. Sci.* 122(1), 140–145.

Santamaria, P., Elia, A., Papa, G. and Serio, F. (1998a). Nitrate and ammonium nutrition in chicory and rocket salad plants. *J. Plant Nutr.* 21(9), 1779–1789.

Santamaria, P., Elia, A., Parente, A. and Serio, F. (1998b). Fertilization strategies for lowering nitrate content in leafy vegetables. Chicory and Rocket Salad Cases. *J. Plant Nutr.* 21(9), 1791–1803.

Santamaria, P., Elia, A. and Serio, F. (2002). Effect of solution nitrogen concentration on yield, leaf element content, and water and nitrogen use efficiency of three hydroponically-grown rocket salad genotypes. *J. Plant Nutr.* 25(2), 245–258.

Santamaria, P., Elia, A., Serio, F., Gonnella, M. and Parente, A. (1999). Comparison between nitrate and ammonium nutrition in fennel, celery, and Swiss chard. *J. Plant Nutr.* 22(7), 1091–1106.

Santamaria, P., Gonnella, M., Elia, A., Parente, A. and Serio, F. (2001). Ways of reducing rocket salad nitrate content. *Acta Hortic.* 548, 529–536.

Santamaria, P. and Valenzano, V. (2001). La qualità degli ortaggi allevati senza suolo (The quality of vegetables grown in soilless systems). *Italus Hortus* 8(6), 31–38.

Savvas, D. and Gruda, N. (2018). Application of soilless culture technologies in the modern greenhouse industry – a review. *Eur. J. Hortic. Sci.*, 83(5), 280–293.

Savvas, D., Manos, G., Kotsiras, A. and Souvaliotis, S. (2002). Effects of silicon and nutrient-induced salinity on yield, flower quality, and nutrient uptake of gerbera grown in a closed hydroponic system. *J. Appl. Bot.* 76, 153–158.

Schnitzler, R. V. H. and Gruda, N. (2002). Quality issues of greenhouse production. *Acta Hortic.* 614, 663–674.

Schreiner, M., Korn, M., Stenger, M., Holzgreve, L. and Altmann, M. (2013). Current understanding and use of quality characteristics of horticulture products. *Sci. Hortic.* 163, 63–69.

Selma, M. V., Luna, M. C., Martínez-Sánchez, A., Tudela, J. A., Beltrán, D., Baixauli, C. and Gil, M. I. (2012). Sensory quality, bioactive constituents and microbiological quality of green and red fresh-cut lettuces (*Lactuca sativa* L.) are influenced by soil and soilless agricultural production systems. *Postharv. Biol. Technol.* 63(1), 16–24.

Shinohara, Y., Akuba, K., Maruo, T. and Ito, T. (1995). Effect of water stress on the fruit yield, quality and physiological condition of tomato plants using the gravel culture. *Acta Hortic.* 396, 211–218.

Siddiqi, M. Y., Kronzucker, H. J., Britto, D. T. and Glass, A. D. M. (1998). Growth of a tomato crop at reduced nutrient concentrations as a strategy to limit eutrophication. *J. Plant Nutr.* 21(9), 1879–1895.

Şirin, U. (2011). Effects of different nutrient solution formulations on yield and cut flower quality of gerbera (*Gerbera jamesonii*) grown in soilless culture system. *Afr. J. Agric. Res.* 6(21), 4910–4919.

Smolen, S., Kowalska, I., Kovácik, P., Halka, M. and Sady, W. (2019). Biofortification of six varieties of lettuce (*Lactuca sativa* L.) with iodine and selenium in Combination with the application of salicylic acid. *Front. Plant Sci.* 10, 143.

Sonneveld, C., Baas, R., Nijssen, H. M. C. and De Hoog, J. (1999). Salt tolerance of flower crops grown in soilless culture. *J. Plant Nutr.* 22(6), 1033–1048.

Sonneveld, C. and Voogt, W. (1993). The concentration of nutrients for growing *Anthurium andreanum* in substrate. *Acta Hortic.* 342, 61–68.

Sonntag, F., Naumann, M., Pawelzik, E. and Smit, I. (2019). Improvement of cocktail tomato yield and consumer-oriented quality traits by potassium fertilization is driven by the cultivar. *J. Sci. Food Agric.* 99(7), 3350–3358.

Takeda, K., Kariuda, M. and Itoi, H. (1985). Blueing of sepal colour of *Hydrangea macrophylla*. *PhytoChemistry* 24(10), 2251–2254.

Tsirogiannis, I., Katsoulas, N. and Kittas, C. (2010). Effect of irrigation scheduling on gerbera flower yield and quality. *HortScience* 45(2), 265–270.

Valenzano, V., Parente, A., Serio, F. and Santamaria, P. (2008). Effect of growing systems and cultivar on yield and water use efficiency of greenhouse-grown tomato. *J. Hortic. Sci. Biotec.* 83(1), 71–75.

Welch, R. M. and Graham, R. D. (1999). A new paradigm for world agriculture: meeting human needs: productive, sustainable, nutritious. *Field Crops Res.* 60(1–2), 1–10.

White, P. J. and Broadley, M. R. (2009). Biofortification of crops with seven mineral elements often lacking in human diets – iron, zinc, copper, calcium, magnesium, selenium and iodine. *New Phytol.* 182(1), 49–84.

Zeroni, M. and Gale, J. (1982). The effect of root temperature on the development, growth and yield of "Sonia" roses. *Sci. Hortic.* 18(2), 177–184.
Zeroni, M. and Gale, J. (1987). The effect of root temperature on rose plants in relation to air temperature. *Plant Soil* 104(1), 93–98.
Zheng, Y., Graham, T. H., Richard, S. and Dixon, M. (2005). Can low nutrient strategies be used for pot gerbera production in closed-loop subirrigation? *Acta Hortic.* 691, 365–372.

# Part 3

## Case studies

# Chapter 13

## Advances in soilless cultivation of tomatoes and other fruit vegetables

*V. Truffault, Futura Gaïa, France; and E. Legast, Centre Technique Interprofessionnel des Fruits et Légumes (CTIFL), France*



### 1 Introduction

Soilless culture systems (SCS) have become a primary production technique in the agricultural industry, notably for high-value crops like fruit and vegetables. In order to manage the crop microclimate and management accurately, basic principles of plant physiology must be integrated into cultivation systems. The present chapter gives an overview of current scientific and applied advances of SCS for tomatoes and other fruit vegetables.

Tomato is cultivated worldwide over an area of 4.7 million ha and is one of the most popular vegetable crops. According to data from FAOSTAT, 182 million tons of tomatoes were produced in 2018 leading to an average yield of 3.83 kg $m^{-2}$. In Europe, northern countries rely on heated glasshouses whereas Mediterranean countries use plastic greenhouses; the yield associated with heated glasshouse production could reach up to 65 kg $m^{-2}$, and could reach up to 8.5 kg $m^{-2}$ for plastic greenhouses. Sweet pepper and cucumber are the two other major crops cultivated in SCS. In 2018, sweet pepper

http://dx.doi.org/10.19103/AS.2020.0076.16

production reached 37 million tons on 2 million ha (average yield was 1.85 kg $m^{-2}$); cucumber production reached 75 Million tons on 2 million ha (average yield was 3.75 kg $m^{-2}$).

Tomato fruit contains 95% water. It has a low caloric intake (3% carbohydrates, 1% protein and trace amounts of lipids). Tomato contains fibres and minerals (potassium, magnesium, calcium). Sugars (glucose, fructose) and acids (malic and citric acids) are also major constituents (Fig. 1). The colouring of tomato is due to the accumulation of carotenoids, mainly lycopene and beta-carotene (Davies and Hobson, 1981). Tomatoes also contain other molecules with antioxidant properties, such as flavonoids and hydroxycinnamic acid derivatives, and vitamin C. Taste assessment by the consumer is complex, combining texture, juiciness, flavours, sweetness and acidity. The historical indicator for taste was the sugar:acid ratio (Grierson and Kader, 1986) but a more recent study (Tieman et al., 2012) has revealed the crucial contributions of 6 volatiles (2-butylacetate, *cis*-3-hexen-1-ol, 3-methyl-1-butenol, 2-methylbutanal, 1-octen-3-one and *trans*,*trans*-2,4-decadienal), in combination with fructose and citric acid contents. Sweet pepper is 90% water, is a rich source of vitamins C, B6, B9, E and A and contains a high amount of polyphenols and carotenoids. Cucumber is 96% water, has a low caloric intake and is a source of potassium. All the metabolic processes involved in the final

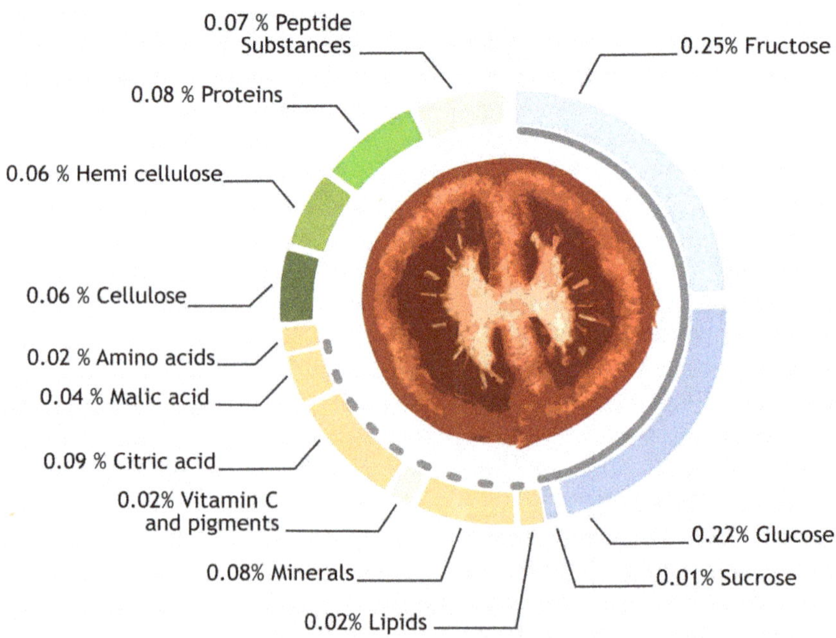


**Figure 1** Tomato fruit internal quality descriptors. Adapted from Davies and Hobson (1981).

composition of fruit vegetables depend on many factors interacting with each other during the growth and development of the fruit: climate, fertigation and cultivation practices can change the quality potential of each variety.

## 2 Soilless culture systems

### 2.1 Basic principles

The production of fruit vegetables is divided into three major stages:

1. stage 1 is planting and establishment of the rooting system;
2. growth stage 2 establishes carbon sources (vegetative) and ensures early fruiting; and
3. growth stage 3 ensures a well-balanced crop.

In the planting stage (1), growers must avoid extreme temperature and increase relative humidity over the very first day as plants are in the juvenile phase. Plant density is crucial to establish a well-balanced, productive crop, with a need to consider the high light requirements of fruit vegetables. Planting density of cucumber is 1.25–3 stems $m^{-2}$; tomato is 2.5–4.4 stems $m^{-2}$ (higher density for cherry tomatoes); sweet pepper is 3–4 stems $m^{-2}$. In relation to the planting date, evolution density should follow solar radiation. Crop performance also relies on rooting system, so that grafting is a usual practice in tomato and sweet pepper. No obvious gain has been reported in cucumber-grafted crops.

Stage 2 focusses on producing a robust vegetative plant with healthy roots as well as fast shoot and leaf growth to support future yield potential. In addition to gentle climatic settings (low day-night difference), fruit and axillary bud pruning are essential cultural operations. In cucumber, the first 7–8 fruits produced are removed to promote root growth.

Stage 3 covers the major part of the crop cultivation cycle and relies on the equilibrium between source and sink to ensure a regular yield. Most of the time in European conditions without artificial lighting, source activity is limited in comparison to sink needs. The transport of assimilates to fruits is relatively tricky in fruit vegetables as it promotes vegetative or generative phenological aspects of the plant. In cases where photosynthesis is not limited, the plant accumulates dry matter in its vegetative parts and reduces its harvest index. A limitation of photosynthesis can affect phloem flow to fruit and penalize yield. However, the imbalance can also be linked to too high a fruit load: the fruits then compete with each other, leading to a decrease in size. In large-fruited tomatoes, 5–7 fruits must be kept on every truss. In cucumber, 50% of the fruits are removed (one in, one out). Crop architecture indicators, easy to follow like leaf area index, could help the growers.

## 2.2 Production system

Modern SCS climate and fertigation are computer controlled. Suitable techniques for soilless fertigation are reviewed by Savvas and Gruda (2018). SCS with standard equipment and management for heating, ventilation and dehumidification are widely used in northern countries. $CO_2$ enrichment is ensured either by condensation of fumes coming from the boiler during winter or by pure liquid $CO_2$ during summer.

To cope with climate change effects, developments through equipment design and management have to take into consideration high water-use efficiency, heat control during summer, artificial lighting during winter, sufficient pest control and efficient humidity control (Gruda et al., 2019). Notably, the use of fossil fuels leading to $CH_4$, $N_2O$ and $CO_2$ released in the atmosphere will increase the occurrence of climate changes. Summer conditions will be warmer, increasing the pressure from pests and decreasing water availability, while winter conditions will be characterized by high humidity and low irradiance.

For year-round production, adaptations such as semi-closed greenhouses and/or artificial lighting are required. Semi-closed greenhouses allow for active temperature and humidity control, high $CO_2$ content and good pest control by closing windows. A semi-closed greenhouse has a limited number of roof vents and is equipped with a climatic chamber for inside-air recirculation and forced outside-air intake. Cooling is ensured by pads and fans and heating by heat exchangers, coupled with a distribution of the mixture of air through ducts under each crop gutter. Several studies have shown 16–40% energy savings associated with yield increases in tomato of 9–20% in temperate climates (de Gelder et al., 2012a; Truffault et al., 2017). In Mediterranean climates, the potential of semi-closed greenhouses is even higher: in association with a decrease of the gas consumption of over 70%. Grisey et al. (2013) reported an increase in tomato yield of 20%.

Sources of energy must evolve to favour renewable energies (de Zwart et al., 2019). The use of a Life Cycle Assessment (LCA) calculation provides an indicator of the carbon footprint of production. A French study comparing sources of energy, and greenhouse design (including semi-closed greenhouse), reported an LCA-decrease of 46% using renewable energy, and of 52% using a semi-closed greenhouse in comparison to a modern greenhouse heated with natural gas (Grisey et al., 2020). The feasibility of using solar thermal energy to heat, cool and dehumidify heated vegetable greenhouses has been confirmed (Grisey et al., in press) in regions with high irradiance and a warm climate, such as Mediterranean region, if coupled with efficient energy storage. Solar panels installed on greenhouse roofs can reduce non-renewable energy consumption, but this conflicts with high-light-loving crops, such as tomato and other fruit vegetables (Yano and Cossu, 2019).

## 3 Climatic requirements

### 3.1 Light

Tomato, cucumber and sweet pepper are high-light demanding crops; tomato plants bearing fruits require a minimum of 20 moles $day^{-1}$ (daily light integral, DLI; Spaargaren, 2001). In Northern countries, light requirements are not fulfilled from October to March (outside DLI < 20 moles $day^{-1}$). Considering that, in commercial greenhouses, 70–80% of the outside solar radiation reaches the plants, and growers have the possibility to use additional lighting through, for example, high-pressure-sodium (HPS) lamps. Nowadays, Light Emitting Diodes (LED) are increasingly used because of cost-saving. According to Zhang et al. (2017a), the energy consumption of LED is reduced compared to HPS. LED is used by plant scientists to control the effect of the light spectrum on plant physiology (Fanwoua et al., 2019; Kaiser et al., 2019). Results from studies highlight the immense potential of this technology. Challenges linked to suitable cultivars (Ouzounis et al., 2016), optimal light spectra (Gómez and Mitchell, 2016), LED positioning (Jiang et al., 2017) or light management (Iersel and Gianino, 2017) are still not fully resolved for tomatoes and other fruit vegetables. For a review, see Bantis et al. (2018).

Studies with so-called photosynthetic wavelengths (classically 80% red–20% blue) showed no negative or positive effect on tomato fruit quality (Dzakovich et al., 2015). In another trial using different light spectra, Dzakovich et al. (2017) found that additional light from LEDs could marginally influence tomato fruit quality in a greenhouse, in relation to the prevalence of sun light (even in winter). In contrast, tomato fruits specifically illuminated at 263 μmol $m^{-2}$ $s^{-1}$ of white light for 13 days showed their vitamin C content increased by 32% compared to fruits subjected to low light intensities (Ntagkas et al., 2016). LED monochromatic light in the blue region might enhance defence compounds of plants (Huché-Thélier et al., 2016; Taulavuori et al., 2017), while red-wavelength rich environments could increase lycopene content (Grozeff et al., 2016). For an example of fruit illuminated with specific wavelengths, see Section 7.

### 3.2 Carbon dioxide ($CO_2$)

The effect of $CO_2$ on plant physiology is related to photosynthetic activity: increasing $CO_2$ content above the ambient level will significantly increase tomato, cucumber or sweet pepper photosynthesis (Nederhoff and Vegter, 1994). In greenhouses, a confined environment, plant photosynthetic activity reduces the ambient $CO_2$ content, thus $CO_2$ enrichment is common. However, $CO_2$ enrichment has to be correctly managed, accorded to plant needs and greenhouse design. Kuroyanagi et al. (2014) studied the efficiency of $CO_2$

enrichment in an unventilated plastic greenhouse: the authors concluded that an average of 45.5% of the daily $CO_2$ supply was efficiently absorbed by tomato crops. $CO_2$ efficiency increased up to 50% under high solar radiation and low wind conditions. de Gelder et al. (2012b) aimed to reduce $CO_2$ injection while limiting the impact on the final yield of the crop: a 50% reduction in the quantities supplied over the production period reduced tomato yield by only 1 kg $m^{-2}$. The optimization of $CO_2$ management is a current economic and environmental issue. The main challenges are reviewed by Li et al. (2018). $CO_2$ also influences quality establishment in tomatoes: vitamin C, soluble sugars increased respectively by 18.5% and 14.7% under high $CO_2$ concentrations (Dong et al., 2018).

### 3.3 Temperature and humidity

Temperature and humidity are closely related in greenhouses. With the exception of semi-closed greenhouses, temperature and humidity controls are jointly managed by the same actuators (roof vents, fans, etc.). When possible, humidity setting should allow complete stomata opening. Vapour pressure deficit is a good indicator of plant activity facing temperature and humidity environment. Looking at Table 1, to keep plant transpiration activity in the right zone (highlighted in grey), we observe that the higher the temperature, the narrower the range of optimal humidity while, conversely, the lower the temperature, the wider the range of optimal humidity.

Tomato adapts well to different temperature ranges, tolerating from 5°C to 40°C for a few hours. Optimum day temperature is from 18°C to 24°C; night temperature is from 16°C to 20°C (24 h-average from 17°C to 22°C). The difference between day and night temperature control the vegetative or generative status of the plant: if the plant is too generative, differences should be reduced, and vice versa. Twenty-four-hour temperature and organogenesis are correlated (Turcotte et al., 2015). Cucumber grows optimally at average 24 h-temperature around 22°C, and is sensitive to temperature below 16–17°C. In general, greenhouse settings are 18°C at night and 24°C during the day. First stages of the vegetative growth (until flowering) are optimal using a 20/19°C day/night temperature regime. It is then considered advantageous to close roof openings to increase temperature regarding light intensity up to 28°C. Targets for humidity control should be set to 75–85% in cucumber, 65–75% in tomato and sweet pepper. Sweet pepper growth optimum temperature is 20–25°C, day/night temperatures of 25°C/21°C are ideal. Pepper is vulnerable below 15°C and higher than 33°C. Sweet pepper is very sensitive to sunscald.

However, it is clear that each organ of a tomato, cucumber or sweet pepper plant (apical bud, leaf, stem, fruit and root) has a specific need in terms of temperature. Recording apical bud temperature of cucumber in North-West

**Table 1** Target (highlighted in grey) air vapour pressure deficit (g cm$^{-3}$) depending on leaf temperature (°C) and air relative humidity (%), adapted from Portree, 1993

| | | Leaf temperature (°C) | | | | | | | | | | | | | | | |
|---|---|---|---|---|---|---|---|---|---|---|---|---|---|---|---|---|---|
| *VPD (g cm$^{-3}$)* | | 15 | 16 | 17 | 18 | 19 | 20 | 21 | 22 | 23 | 24 | 25 | 26 | 27 | 28 | 29 | 30 |
| **Relative humidity (%)** | **95** | 0.5 | 0.6 | 0.6 | 0.7 | 0.7 | 0.8 | 0.8 | 0.9 | 0.9 | 1.0 | 1.0 | 1.1 | 1.2 | 1.3 | 1.4 | 1.4 |
| | **90** | 1.1 | 1.2 | 1.3 | 1.3 | 1.4 | 1.5 | 1.6 | 1.7 | 1.8 | 2.0 | 2.0 | 2.2 | 2.4 | 2.5 | 2.7 | 2.8 |
| | **85** | 1.7 | 1.8 | 1.9 | 2.0 | 2.1 | 2.2 | 2.4 | 2.6 | 2.7 | 3.0 | 3.0 | 3.3 | 3.6 | 3.7 | 4.1 | 4.2 |
| | **80** | 2.2 | 2.3 | 2.5 | 2.7 | 2.9 | 3.0 | 3.3 | 3.5 | 3.7 | 3.9 | 4.1 | 4.4 | 4.7 | 5.0 | 5.3 | 5.7 |
| | **75** | 2.8 | 2.9 | 3.1 | 3.3 | 3.6 | 3.8 | 4.1 | 4.3 | 4.6 | 4.9 | 5.2 | 5.5 | 5.9 | 6.3 | 6.7 | 7.1 |
| | **70** | 3.3 | 3.5 | 3.7 | 4.0 | 4.3 | 4.5 | 4.9 | 5.2 | 5.5 | 5.8 | 6.2 | 6.6 | 7.1 | 7.5 | 8.0 | 8.5 |
| | **65** | 3.9 | 4.1 | 4.3 | 4.6 | 5.0 | 5.3 | 5.7 | 6.0 | 6.4 | 6.8 | 7.2 | 7.7 | 8.3 | 8.7 | 9.3 | 9.9 |
| | **60** | 4.4 | 4.7 | 5.0 | 5.3 | 5.7 | 6.1 | 6.5 | 6.8 | 7.4 | 7.8 | 8.2 | 8.8 | 9.4 | 9.9 | 10.8 | 11.3 |
| | **55** | 5.5 | 5.3 | 5.6 | 5.9 | 6.4 | 6.8 | 7.3 | 7.7 | 8.3 | 8.8 | 9.2 | 9.9 | 10.6 | 11.2 | 11.9 | 12.7 |
| | **50** | 5.5 | 5.8 | 6.2 | 6.6 | 7.1 | 7.5 | 8.1 | 8.6 | 9.2 | 9.7 | 10.3 | 11.0 | 11.7 | 12.4 | 13.2 | 14.0 |

France (CTIFL, Carquefou) led to the conclusion that its transpiration rate was little understood. Over the period analysed, we observed an average difference between the air temperature and the apical bud temperature of 0.7°C during the day and 2.8°C at night. The maximum difference (day and night records) was 4.9°C when the air was warmer than the apical bud. The maximum deviation was 6.6°C when the bud is warmer than the air, a situation that only occurs during the day (Truffault, unpub. data).

A similar observation has been published by Savvides et al. (2013). The authors reported that apical bud temperature deviated from air temperature in a range between −4.1 °C and 3.0 °C for cucumber, between −2.6°C and 3.8°C for tomato (lower transpiration rate). Savvides et al. (2017) set up an experiment to explore the impact of the apical bud temperature on the whole performance of a cucumber plant. They isolated the apical bud and applied a temperature treatment: either the apical bud was at the same temperature as the cucumber plant (22°C), or it was warmer or colder (26°C and 18°C). Three different ideotypes could be described. If the apical bud temperature is lowered below the plant temperature, not only does the plant form fewer leaves, but the leaves are thicker: this penalizes photosynthesis. By selectively heating the apical bud, independently of the other plant organs, the development of the plant will be faster. The sink force should be increased but the apical bud should attract the assimilates more effectively, to the detriment of fruit growth. As stated by Boulard et al. (2017), temperature and humidity are not distributed homogenously in the greenhouse, so these situations occur frequently. During the winter, as heating pipes are on the ground, apical buds could be colder than the plant. On the contrary, in the summer, sun radiation warms mostly the apical buds. Thus, it is necessary to maintain proper ventilation inside the greenhouse.

The distribution of temperature also affects fruit growth and quality (Fig. 2; Legast et al., in press). Considering climatic distribution and crop management (high-wire training system), fruit growth and development experience various climates. Indeed, to favour truss initiation and high cell division rates, for example, when the fruits are in the upper part of the plant, temperatures should be maintained between 14°C and 26°C. To optimize cell growth phases and water influx, for example, when the fruits are in the middle part of the plant, set points should be 18°C–23°C. Fruit ripening is promoted by high temperatures leading to secondary metabolite accumulation and high sink strength, which are good for both fruit quality and agronomical performance. Therefore, Legast et al. (in press) proposed an ideal fruit temperature ranging from 23°C to 28°C at the bottom part of the plant. The vertical gradient must be controlled to reach these optimums for tomato fruit quality, as such a gradient seems to have no effect on tomato yield on a yearly basis (Kawasaki and Yoneda, 2019; Qian et al., 2015). However the situation might be different for other species (Kawasaki

**Figure 2** Fruit and temperature heterogeneity along the greenhouse vertical gradient. Adapted from Legast et al. (in press).

and Yoneda, 2019). It is noteworthy that applying a cold pre-night temperature is a simple way to increase yield and quality (Zhang et al., 2010). Indeed, at the end of the day, the leaves are normally 'full' of sugars. A significant drop in air temperature causes a preferential allocation of photo-assimilates to the fruit. This can be explained by the fact that the fruit cools more slowly than the leaves. The photo-assimilates will then be preferentially directed in the plant towards the warmer organs.

As for roots, most studies have focussed on pathogen resistance in relation to root temperature (Kawasaki and Yoneda, 2019). Controlling root temperature is more crucial in an NFT system, where the external climate has a strong influence on the floating root temperature, rather than in rock wool. It has been reported that root temperatures lower than 16°C reduce yield, while root heating could be useful when air temperature is not optimal (Kawasaki and Yoneda, 2019). He et al. (2019) set up an experiment controlling root temperature at 10°C using different air temperatures (winter: 18°C–23°C; summer: 21–29°C). Yield was largely reduced in winter but not significantly in summer. More interestingly, root cooling was reported to increase overall fruit quality, in terms of sugars, vitamin C, carotenoids contents (He et al., 2019).

There are three important functions at the leaf level:

i photosynthesis which, within a wide range, is not dependent on temperature (Schwarz et al., 2014);
ii respiration, which increases with temperature; and
iii transpiration which is very dependent on leaf temperature.

The transpiration rate results from a sum of factors such as humidity, heating and solar radiation. According to Stanghellini (1987), between 15°C and 30°C and above 60% relative humidity the stomata are 100% open in greenhouses. The transpiration rate and vapour pressure deficit of the ambient air (Amitrano et al., 2019) or between the leaf and the ambient air ($VPD_{leaf\text{-}air}$; Truffault et al., 2016) are closely correlated. As greenhouse actuators should be managed according to plant needs, recordings of $VPD_{leaf\text{-}air}$ are of significant interest to maintain a high transpiration rate. $VPD_{leaf\text{-}air}$ is calculated by the following equation (1):

$$VPD_{leaf\text{-}air} = VPLs - VPA$$

where VPLs is the saturated vapour pressure of the leaf estimated by the equation (2):

$$VPLs = 0.61078 * e^{(17.2624 * \text{leaf temperature})/(\text{leaf temperature} + 238.3)}$$

and VPA is the actual vapour pressure of the greenhouse air calculated by the equation (3) including the current relative humidity (RH):

$$VPA = 0.61078 * e^{(17.2624 * \text{air temperature})/(\text{air temperature} + 238.3)} * \frac{RH}{100}$$

Calculation of $VPD_{leaf\text{-}air}$ relies on the temperature of the canopy and the air temperature and the relative humidity in greenhouse. Optimal hygrometry is 65–80% (Turcotte et al., 2015). However, studying $VPD_{leaf\text{-}air}$ may give growers another view of the theoretical optimum. Manipulating the calculation methods of $VPD_{leaf\text{-}air}$ highlights that to increase by 0.1 kPa $VPD_{leaf\text{-}air}$, growers may:

- increase the air temperature by 2.5°C;
- decrease relative humidity by 4%; and
- raise the plant temperature by 0.7°C.

Therefore, it is obvious that there is a greater advantage to focus on the plant than the air temperature. Relocation of the heating pipes closer to the plant points in that direction (Albert et al., 2013; Bournet et al., 2017; Kempkes et al., 2000).

## 4 Nutrient solutions and substrates

Hoagland and Arnon (1950) established the first well-balanced nutrient solution, enabling soilless crop production. Adaptation of nutrient recipes to tomato crops has been carried out according to its average mineral content. An average tomato leaf consists of 30.1 g of nitrate (N) per kg of dry matter, 3.9 g of phosphorus (P), 30.0 g of potassium (K), 60 g of calcium (Ca) and 9.6 g of magnesium (Mg). An average tomato fruit is 25.2 g of N per kg of dry matter, 3.9 g of P, 50.7 g of K, 2.0 g of Ca and 1.9 gof Mg. Nutrient recipes are adapted to the developmental stage of the plant in such a way that the starting recipe is promptly changed according to plant growth and development. The main purpose is to follow the variability between organ needs depending on fruiting or vegetative stages. The nutrient recipe of de Kreij et al. (1997) is the one that is mostly used by growers. As highlighted in Table 2, slight adaptations to Mediterranean conditions (Savvas and Gruda, 2018) and to temperate regions (Truffault et al., 2019) have been proposed in the literature. To avoid any deficiency, it is important to monitor the ratio of K to the sum of Ca and Mg, due to antagonistic reactions between those elements (Sambo et al., 2019). In practice for tomatoes in SCS, fertigation is managed according to the drainage percentage, drainage electric conductivity and pH, aiming for 30% drainage, EC = 3–4 dS $m^{-1}$ and pH = 55.8. These aims ensured correct oxygenation of the substrate to maintain root respiration, thus absorption of mineral elements.

Different nutrients are not taken up by the plant in equivalent proportions, leading to depletion or accumulation of one element in the drainage solution (Kempen, 2015). Among mineral elements, P is the least mobile and the least bioavailable for plants, due to the high reactivity of phosphate ions with the substrate (Hinsinger et al., 2011). On the other hand, a significant part of the world's phosphate resources is non-renewable, and it is well established that a shortage of phosphate fertilizers from rock phosphate could threaten food security (Cordell et al., 2009). In SCS, irrigation and fertilization are managed conjointly – fertigation, and the loss of water by plant transpiration is replaced by water supply. Irrigation schedules are managed according to solar radiation. However, nutrient uptake does not exactly follow irradiance, being driven by growth rate and plant needs. Especially when solar radiation is high, the water

**Table 2** Recommended nutrient recipes for tomato in closed soilless culture system, adapted from de Kreij et al., 1997, Savvas and Gruda, 2018, Truffault et al., 2019.

| | | Savvas and Gruda, 2018 | | de Kreij et al., 1997 | | | | | Truffault et al., 2019 | | | |
|---|---|---|---|---|---|---|---|---|---|---|---|---|
| | | Vegetative stage | Reproductive stage | Start | Flower 1st truss | Flower 3rd truss | Flower 5th truss | Flower 10th truss | Start | Flower 3rd truss to 6th truss | Flower 6th truss to harvesting 2nd truss | Harvesting 3rd truss |
| mmol $L^{-1}$ | **$NO_3^-$** | 13.7 | 12.8 | 10.8 | 10.8 | 10.8 | 10.8 | 10.8 | 15.0 | 15.0 | 13.0 | 12.0 |
| | **$H_2PO_4^-$** | 1.4 | 1.3 | 1.3 | 1.3 | 1.3 | 1.3 | 1.3 | 2.0 | 2.0 | 1.5 | 1.5 |
| | **$K^+$** | 8.2 | 8.0 | 6.5 | 5.3 | 7.5 | 10.0 | 7.5 | 7.5 | 9.0 | 9.5 | 7.5 |
| | **$Ca^{2+}$** | 3.5 | 3.0 | 5.5 | 5.8 | 5.3 | 4.0 | 5.3 | 6.0 | 5.5 | 4.0 | 4.0 |
| | **$Mg^{2+}$** | 1.8 | 1.4 | 2.0 | 2.3 | 1.8 | 1.5 | 1.8 | 2.0 | 2.0 | 1.5 | 1.5 |
| | **$SO_4^{2-}$** | 2.4 | 1.8 | 3.0 | 3.0 | 3.0 | 3.0 | 3.0 | 3.0 | 3.0 | 2.5 | 2.0 |
| µmol $L^{-1}$ | **Fe** | 15.0 | 15.0 | 15.0 | 15.0 | 15.0 | 15.0 | 15.0 | 15.0 | 15.0 | 15.0 | 15.0 |
| | **Mn** | 10.0 | 10.0 | 10.0 | 10.0 | 10.0 | 10.0 | 10.0 | 10.0 | 10.0 | 10.0 | 10.0 |
| | **Zn** | 7.0 | 6.0 | 4.0 | 4.0 | 4.0 | 4.0 | 4.0 | 4.0 | 4.0 | 4.0 | 4.0 |
| | **B** | 20.0 | 20.0 | 20.0 | 20.0 | 20.0 | 20.0 | 20.0 | 20.0 | 20.0 | 20.0 | 20.0 |
| | **Cu** | 1.0 | 0.8 | 0.8 | 0.8 | 0.8 | 0.8 | 0.8 | 0.8 | 0.8 | 0.8 | 0.8 |
| | **Mo** | 0.5 | 0.5 | 0.5 | 0.5 | 0.5 | 0.5 | 0.5 | 0.5 | 0.5 | 0.5 | 0.5 |
| ratio | $K^+$ / $Ca^{2+}$+$Mg^{2+}$ | 0.8 | 0.9 | 0.4 | 0.3 | 0.5 | 0.9 | 0.5 | 0.5 | 0.6 | 0.9 | 0.7 |

demand is often higher than the nutrient demand (Le Bot et al., 1998). Over-fertilization is common, leading to 30–80% of nutrients being actually used by crops (Rácz, 2007), impacting yield or quality (Truffault et al., 2019).

It is important to note that a large percentage of nutrients are taken up during the night when no fertigation is traditionally scheduled. Kempen et al. (2017) highlighted this precisely for tomato crops: 35% of the total daily $NO_3^-$ uptake occurred during the night, whatever the irradiance sum of the day. In comparison, up to 45% of total daily $H_2PO_4^-$ and from 48% (under low irradiance) to 41% (under high irradiance) of total daily $Ca^{2+}$ (calcium is passively absorbed by plants, following water needs) were taken up during the night. Using a nutrient film technique (NFT) system, a similar approach by Nomiyama et al. (2017) strengthened the conclusion that night-time transpiration was not the only driver for nutrient absorption: while 18% of the daily transpiration occurred during the night, 24% of $NO_3^-$, 26% of $H_2PO_4$ and 24% of $Ca^{2+}$ daily ion absorption were recorded.

SCS relies on precision horticulture. Inorganic substrates have been favoured over time, which are biologically inactive. However, roots contribute to the pool of organic nutrients in the substrate: dead roots are decomposed by micro-organisms and living roots exude sugars, amino acids and carboxylic acids which help in maintenance of the rhizosphere microbiome (Pieterse et al., 2016). In a symbiotic reaction, beneficial micro-organisms enhance nutrient availability for plant absorption and protect the root from pathogens.

Conversa et al. (2013) reported that arbuscular mycorrhizal fungi (AMF) selected for tomato reduced the need for P supply in soil. AMF solubilized the inorganic phosphate which became more easily absorbed by roots. A single strain of *Glomus* or AMF mixture has proved their efficacy in field tomato increasing yield up to 26% (Baum et al., 2015). Result reproducibility is difficult in field conditions. Obviously, such a close connexion is much smaller in SCS in comparison to soil-type (Alsanius and Wohanka, 2019). It also means that competition between micro-organisms for substrate colonization is absent, leaving the first micro-organism to appear as 'winning' the battle. In tomato, a prime example is *Agrobacterium rhizogenes* (hairy roots syndrome) which finds perfect conditions for growth in soilless crops (Grunert et al., 2014), notably due to the lack of competitive micro-organisms at a pH of around 5.5–5.8. Indeed, a less-acidic environment (pH 6.8–7) is more advantageous for plant growth, as highlighted by Husson, (2013). Maintaining pH at 5.5–5.8 is only related to fertilizer providing manganese and boron which have low availabilities at pH > 6.5 (Lucas and Davis, 1961). Nowadays, chelates of those elements are available (Chelal, BMS Micro-Nutrients NV, Bornem, Belgium).

The next step is then to guide the structure of soil microbial communities to act to the detriment of plant pathogens (integrative rootzone management, IRM). Grunert et al. (2019) used phytotrons to analyse how the microbial community, fertilizers and tomato root metabolism/catabolism affected the composition of the rhizosphere. The authors revealed that tomato root exudates had a stronger impact on microbiome establishment, regardless of the fertilizer used. Tomato plants modulated the nitrification activity in the rhizosphere to shape microbial communities (Grunert et al., 2019). Further studies are awaited. Furthermore, SCS growers are moving towards organic substrates first to improve LCA scores, and secondly to switch to more sustainable production technique (Gruda, 2019). The carbon footprint of coconut coir and rock wool were respectively from 162 g to 244 g of $CO_2$ $m^{-2}$ and from 640 g to 1120 g of $CO_2$ $m^{-2}$ on a yearly basis (Grasselly et al., 2009). Substrates with the least impact on global warming were wood-based substrates of local species requiring little transport. These substrates are currently little used on tomatoes. Coconut coir or wood-based substrates have the advantage of being renewable, unlike substrates of mineral origin, which contribute to the depletion of natural resources.

## 5 Managing climate and fertigation according to plant needs

The performance of SCS tomato crops suffers from the lack of knowledge of the plant's needs in real time and challenges in predicting climatic conditions. Cultivation environments are also becoming increasingly technical and more and more complex for growers to master, which hinders technical innovation in farms. Nowadays, the management of climate and fertigation is carried out through corrective measures, to get as close as possible to the needs of the plant according to the visual observations of the grower. There is an increasing shift to the sensor-based approaches and modelling tools which are reviewed in the sections that follow.

### *5.1 Sensors-based approaches*

Except for adjustments of the EC, nutrient recipes are not adapted to a plant's actual consumption. Growers have combined readings from substrate sensors (EC, temperature and hygrometry of the substrate are recorded in real-time) with the visual appearance of the crop (Thompson et al., 2018) to apply corrections to the nutrient recipe. Fertigation is managed through corrective approaches (Incrocci et al., 2017), whether or not it contains a laboratory analysis of the nutrients remaining in the drainage solution. The optimization of fertigation will need reliable sensors or models, and most of the available ones rely on plant transpiration/water status of the substrate.

The dynamics of plant weight over a day or a period could be monitored using automatic weighting systems. A decrease in weight (except from harvesting operations) is used as a signal to schedule an irrigation. An increase in weight is an estimation of plant growth. Such systems are already available for growers (Aquabalance®, Hoogendoorn, Vlaardingen, The Netherlands; Prodrain®, Ridder Growing Solutions, Maasdijk, The Netherlands; Root Optimizer®, Priva Horticulture, De Lier, The Netherlands), and validated by scientists (Lee and Son, 2019). High-precision systems face a number of disadvantages that hinder their application, such as high costs, the inability to move a device once installed, the need for regular monitoring and data that is not easy to interpret (Thompson et al., 2018).

An infra-red camera offers a cheaper and easy-to-interpret alternative to manage irrigation schedules: it provides canopy temperature in real time which is used to calculate $VPD_{leaf-air}$. Transpiration rate and $VPD_{leaf-air}$ are closely correlated, making it possible to adjust the irrigation cycle based on the plant's needs (PlantVoice®, Hoogendoorn, Vlaardingen, The Netherlands; TopCrop Monitor®, Priva Horticulture, De Lier, The Netherlands). The 'plant speaking' approach provided by those sensors is particularly suitable for cherry tomatoes, which are sensitive to fruit cracking, and has been validated in cucumber (Truffault et al., 2016).

Sap flow rate and stem diameter recordings are indicators of a plant's reaction to its environment. These types of sensor are powerful when used in parallel (Fig. 3). A sap flow sensor measures the flow of xylem sap that comes from the root to the plant's aerial parts. It provides water flux measurements linked to plant transpiration (De Swaef and Steppe, 2010). This information is very useful for monitoring the activity of the plant and its water potential (Liu et al., 2007). A stem diameter sensor accurately measures the variation of stem size, which is linked to phloem unloading. Dynamics of sap flow rate and stem size is inversely correlated to each other. On a daily basis, it is possible to

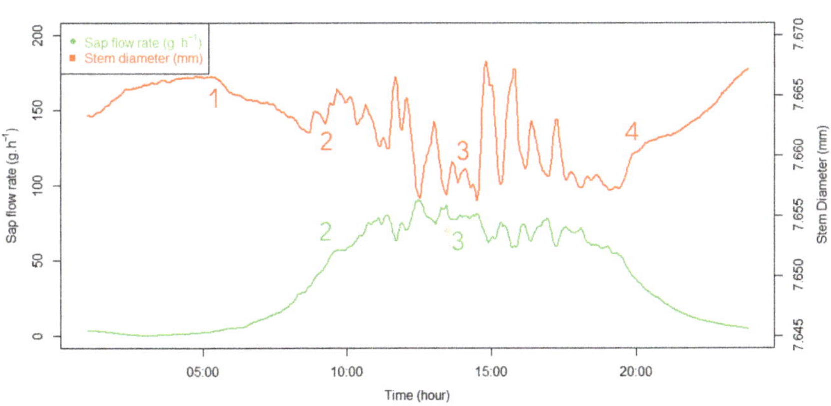


**Figure 3** Sap flow rate (in green; $g.h^{-1}$) and stem diameter (in orange; mm) dynamics throughout a typical day. 1 - heating pipes switched on. 2 - first irrigation schedule. 3 - succession of 'ups' and 'downs' illustrating generative or vegetative activities of the plant. 4 - recovery phase (sunset).

characterize all the decisions, such as greenhouse management, climate and fertigation, made by growers.

In Fig. 3 - [1], greenhouse heating starts that activates phloem unloading generating a reduction in the diameter of stem. In Fig. 3 - [2], first irrigation of the day stimulates the activation of the plant and leads to a significant influx of sap towards the plant. Irrigations are noticeable as being the only events showing dual increase of both sap flow rate and stem diameter. In addition to crop management events, changes to the environment perceived by plants are recorded through a succession of increases and drops (Fig. 3 - [3]) which illustrates the physiological activity and 'break' phases of the plant. The correlation between sap flow rate and solar radiation is close to 80%, which means that the sap flow measurements are also indicators of light use efficiency (transpiration and photosynthesis). At sunset, the recovery phase starts, increasing stem diameter (Fig. 3 - [4]). The slope of the curve gives an indication of the impact of climate on plant carbon status. Stem diameter is affected by the stem cell growth and declining cell elasticity (lignin synthesis): thus, it increases during the night, without any plant activity. Using the balance between plant growth and sink (mostly fruits) strength on the plant reserves gives a good index of plant vegetative/generative status (De Swaef and Steppe, 2010). Sap flow rate and stem diameter measurements are very useful tools for growers but can be tricky to analyse.

Nutrient uptake is not equivalent and does not follow water influx exactly (Kempen, 2015). Thus, the accuracy and durability of crop nutrition is low when only based on transpiration rates. On-line ion-sensing could be a valuable option to monitor a plant's real consumption. Kim et al. (2013) reviewed ion-sensing technology, but this is quite expensive and may not be reliable over the long term. Nowadays, studies have been carried out by measuring one or only a few plant macronutrients at a time (Vardar et al., 2015) and available sensors do not include micronutrients. Advances in real-time N fertigation based on plant growth have been proposed (Adamowicz et al., 2012). Further technological developments are awaited.

## 5.2 Modelling-based approaches

Instrumentation of greenhouses with multiple sensors must include a decision support system for growers. To monitor the physiological activities of each plant organ is a real challenge in SCS for fruit vegetables. Currently, definition of set points is crucial but may be inefficient if effective prediction is not possible. One of the major challenges in modelling is linked to the energy crisis. Modelling tools including weather forecasting have allowed energy savings over short periods of time (in achieving forecasting robustness). Shen et al. (2018) succeeded in optimizing the daily average temperature over 7 days to save 9%

of energy costs in winter conditions. However, optimal set points (temperature, light, $CO_2$) for tomato crop growth were considered constants in this study.

The large number of variables involved in crop growth and development makes it necessary to use a tool for simplifying systems: modelling. Top-down (from phenotype to genotype) or bottom-up (from genotype to phenotype) approaches are studied by scientists but yield prediction of tomato still suffers from a lack of accuracy. In contrast, yield prediction of sweet pepper has been shown to be reliable (de Gelder et al., 2006), which has led to the generation of new set points able to reduce yield fluctuations and energy consumption (by 10%; Van Henten et al., 2006).

Lin et al. (2019) improved tomato yield prediction through the combination of two different models (TOMGRO, Jones et al., 1991); Vanthoor model, Vanthoor, 2011). TOMGRO, one of the most studied and robust tomato models, is based on the source-sink approach. The Vanthoor model is a greenhouse environmental system. Relative mean square error (RMSE) was used as an indicator of model performance. RMSE between actual yield and Lin, TOMGRO and Vanthoor outputs were respectively 2.6, 17.2 and 17.6 (Lin et al., 2019). The fruit vegetables' supply chain requires a better match between supply and market demand to decrease the pressure on product pricing. Such an improvement on yield prediction will enable better sales prices. Focussing on nutrient and water management and integrating weather forecasts, generation of optimized settings over a period of 3 days led to an increase of 10% of tomato fruit weight and the enhancement of nutrient uptake efficiencies (Elings et al., 2004).

As nutrient and water uptakes may not be managed conjointly, Juárez-Maldonado et al. (2014) developed six regression models for N, P, K, Ca, Mg and S for tomato. Simulated nutrient concentrations within each organ (leaf, fruit, root, stem) showed a good fit with destructive measurements over time: the accuracy of the model is higher than 95% for N, P, K and S, 75% for Ca and 89% for Mg. Integrating these approaches could lead to significant savings in fertilizer: the HortSyst model (Martínez-Ruiz et al., 2019) has led to a 50% saving of N fertilizer in soilless tomato crops.

All the above examples rely on mathematical models. Artificial intelligence (AI) techniques are well adapted to complex systems like greenhouses as they benefit from cloud computing technology and can manage large volumes of data. Continuous learning and adaptation to historical data are likely to be more suitable than mathematical models in generating optimized set points for growers. With the help of AI and with the addition of new data each year, tomato yield predictions reached an accuracy of between 83% and 93% (*LetsGrow.com*, 2020).

Such an integrative approach has recently been investigated (Hemming et al., 2019). In an initiative entitled 'Autonomous greenhouse challenge', five

96 m² compartments were given to several AI specialists between August and December 2018 (http://www.autonomousgreenhouses.com/). Each team was instructed to grow cucumbers in one compartment using AI, without going into the greenhouse or 'seeing' the crop (except from cameras). One team was able to beat the grower reference in terms of crop performance (+3.5%) and net profit (+16.7%; Hemming et al., 2019). The response time to changing physiological activities of plants can be measured in seconds or minutes using AI techniques (Kaiser et al., 2018), a time interval not achievable by growers.

## 6 Pre-harvest and post-harvest factors affecting product quality management

### *6.1 Pre-harvest influences on fruit quality*

Fruit quality of SCS and soil-based tomato crops do not differ in terms of the parameters that are most investigated (Gruda, 2009). In a recent study, 12 tomato cultivars produced in soil and soilless culture at the same time of the year were described using sensory and biochemical analyses. Minor biochemical differences were measured but sensory panels were unable to distinguish between soilless and soil tomatoes (Cottet et al., 2019).

Fruit colour and size are the only criteria available for the consumer at purchase. Tomato colour is mainly driven by carotenoid synthesis. Lycopene is responsible for the red colour of tomatoes. Lycopene synthesis is sensitive to potassium supply (Ramírez et al., 2009) and strongly dependent on the fruit temperature (Dumas et al., 2003). Fruit size is highly dependent on genetics and the carbon status of the plant: high photosynthetic activity and appropriate carbon partitioning to the fruit are the main drivers. It is assumed that sources (mainly leaves) are suppliers of carbon to sinks (fruits, roots). In the 1940s–1950s, breeders introduced the 'uniform ripening' *u* mutation to ensure better visual quality (no green shoulders, easier to harvest), but this was also related to a decrease in chloroplast number of fruits. Fruits developed to their normal size and ripened at the same rate as wild type, but the accumulation of sugars in the red fruit was repressed by 10–15% (Cocaliadis et al., 2014; Powell et al., 2012). Therefore, fruit sugar content might be enhanced by genotypes with high chloroplast number and appropriate crop management. Independent of genetics, deleafing at the basal level of the plants ensured high irradiance to ripening fruits, which has been reported to increase vitamin C (Gautier et al., 2009) and shorten ripening time (Johkan et al., 2013) but no positive effects on the sugar content have been reported (Johkan et al., 2013; Ntagkas et al., 2019).

Fruit cracking is an important disorder leading to low fruit marketability and shelf life (Gruda et al., 2018). It may occur when internal pressure on the fruit epidermal cuticle is high. Cherry tomatoes are particularly affected because fruit growth continues after the mature green stage (Domínguez et al., 2012), unlike large fruited tomatoes. Ensuring adequate leaf transpiration is key to avoid excess turgor pressure at the fruit level. Irrigation management using VPD is useful for cherry tomatoes, probably even more than for large-fruited genotypes (Vurro et al., 2019; Zhang et al., 2017b), as it not only takes into account the canopy transpiration in real-time but also solar radiation. Since greenhouse tomato plants are indeterminate with different stages of fruit development occurring at the same time on an individual plant (cultivation period is 10 months), it may be difficult to adopt a water deficit in practice without any adverse effect on a ripening stage or on crop performance (Wang and Xing, 2017). Fertilization management is a valuable strategy to achieve better fruit quality in greenhouses. The positive effect of EC on tomato quality has been well reported (Gruda et al., 2018; Krauss et al., 2006; Rouphael and Kyriacou, 2018).

Above all, nitrogen supply is crucial for yield establishment and fruit quality as stated by Benard et al. (2009) and Truffault et al. (2019) on modern cultivars. Plant metabolism shifts from carbon-containing compounds (sugars, acids, vitamin C) to nitrogen-containing compounds (proteins) as a function of nitrogen supply (Albornoz, 2016). The impacts of other macroelements are well documented (Gruda et al., 2018). Blossom end-rot is a major disorder for tomato fruit quality. It is caused by a calcium deficiency in the distal part of the fruit which has a lower density of xylem-stained vascular bundles. As calcium is transported within the plant following the water gradient, when leaf transpiration is high, almost all calcium is directed to and remains in leaves. Thus, the calcium supply of tomato requires special attention. Bertin and Genard (2018) have reviewed the establishment of tomato fruit quality and the influence of environment and crop management.

## *6.2 Post-harvest influences on fruit quality*

Pre-harvest conditions influence the ability of tomato to increase shelf life, notably the time of harvest, ripening stage and production systems (Beckles, 2012). Adding far-red LED-light during cultivation resulted in an increase in cold tolerance during storage of tomato fruits; higher fruit firmness at harvest was also associated with lower post-harvest weight loss (Affandi et al., 2020). However, post-harvest conditions are still the major influence on fruit quality. Recently (and in relation to LED developments) efforts have been devoted to the light environment at the post-harvest stage. 10 days red-rich wavelengths at the post-harvest stage enhanced 2-methylbutanal and 3-methyl-1-butanol levels in tomato fruits, two strong aroma markers of tomato (Colquhoun et al., 2013).

Tomato is sensitive to chilling injury. Therefore, storage conditions are critical to extend shelf life while maintaining good overall quality. Storing tomatoes at 4°C, 10°C, 12.5°C or 20°C induced different quality disorders. At 4°C for 30 days, volatiles content dramatically decreased by 66% compared to the initial stage (Renard et al., 2013). At 10°C, unappreciated odours appeared and consumers could perceive a lack of freshness (Ponce-Valadez et al., 2016). Consumers were not able to distinguish between tomatoes stored at 12.5°C and at 20°C (Ponce-Valadez et al., 2016). Guidelines for proper tomato storage can be found (Suslow and Cantwell, 2009).

## 7 Case study: adapting light spectrum to apical bud and fruit requirements using light-emitting diode (LED) lighting

The organogenesis, morphogenesis and photosynthetic activity of plants are fundamentally adapted to the climatic environment. As mentioned in the earlier discussion of 'Climatic requirements', the vertical temperature distribution affects physiological activities of plants, notably apical buds and fruit. Light distribution over the plant canopy, either in quantity (Buck-Sorlin et al., 2009) or quality (Dieleman et al., 2019) has the potential to affect the physiological activities of plant organs. The quantity of light in the environment near to apical buds was correlated to the leaf initiation rate in cucumber and tomato (Savvides et al., 2014). Authors have suggested that the photosynthate availability in the apical bud itself affects leaf appearance. Light signals might also affect regulation, through photoreceptors of the red:far red ratio or the blue spectrum. Unlike under the natural light circadian rhythm, greenhouse crops under additional light (in winter for the following example) are subjected to approximately 8 h of LED spectrum and 8 h of solar input complemented by LED spectrum. It is not clear whether or not this dynamic light environment affects the organogenesis and morphogenesis of the plant. Specific wavelengths dedicated to fruit quality could also be a valuable tool to increase fruit nutritive content, notably in the blue region (Taulavuori et al., 2017).

### *7.1 Plant material and growing conditions*

Soilless tomato cultivation took place in a Venlo-type heated greenhouse compartment (CTIFL Carquefou, France), 5.5 m height and with a 320 m$^2$ ground surface. Tomato plants (*Solanum lycopersicum* var. 'Climston' (bunch tomatoes), Syngenta, Bâle, Switzerland) were two-head grafted on the rootstock DRO 141 (De Ruiter™ Seeds, Bergschenhoek, the Netherlands). Sowing was carried out on October 19, 2018, and transplantation on December 13, 2018, with a planting density of 3.05 stems m$^{-2}$. To follow the irradiance increase, leaving the axillary shoot from the axil just below the 5th truss led to a final stem density of 4.06 stems m$^{-2}$. Growing substrates were rock wool in slabs (Cultilene EXACT® AIR, Saint-Gobain Cultilene, Rijen, the Netherlands);

integrated pest management was carried out with Biobest (Biobest Group NV, Westerlo, Belgium). A MultiMa® climate computer (Ridder Growing Solutions, Maasdijk, The Netherlands) was used to record and control the greenhouse climatic conditions. Climate management led to the establishment of a 24-h average temperature of 22.7°C; day temperature 24.2°C; night temperature 19.1°C. The air saturation deficit was set to 6.7 g of water per $kg^{-1}$ of dry air during the day, and 3.9 g $kg^{-1}$ during the night. The injection of $CO_2$ led to a $CO_2$ content of 526 ppm.

## 7.2 Light treatment

LED lamps were positioned at three levels (Fig. 4): one at the top to study the influence of light on the apical bud, one intra-canopy (near leaf number 12, counting from the top) and the last one near the ripening fruits. LED devices

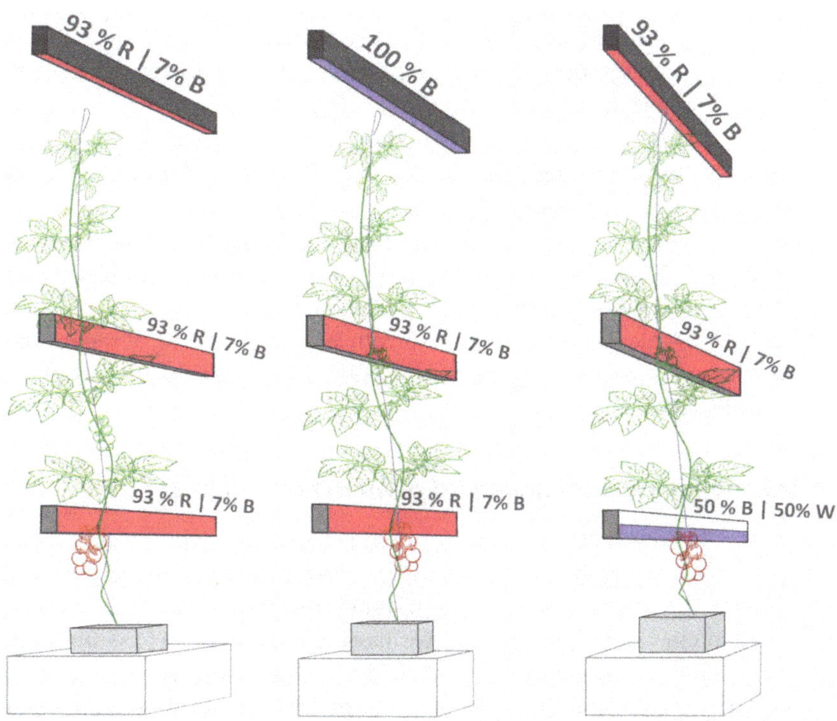


**Figure 4** Design of the LED experiment in CTIFL greenhouse compartment (Carquefou, France). Control treatment is illuminated during 18 h using 93% red 7% blue at the top, middle and bottom levels. apexLED treatment is illuminated with 100% blue for 2 h when switched on, then 93% red 7% blue for the next 14 h at the top level. Middle and bottom levels were similar to control treatment. fruitLED treatment illuminated ripening fruits (bottom level) using with 50% blue 50% white. Top and middle levels were similar to control treatment.

were designed and operated by VGD-LED (Vegetal Grow Development, Chateaurenard, France). Each LED lamp (counts: 192) could be managed independently from each other in terms of light intensity and spectrum, with six wavelengths available for intra-canopy devices (red 660 nm, blue 450 nm, far-red 730 nm, purple 415 nm, green 525 nm, white 4200 K) and four wavelengths for uppermost LED (red 660 nm, blue 450 nm, far-red 730 nm, white 4200 K). Maximum intensity per wavelength was 100 μmol $m^{-2}$ $s^{-1}$ for intra-canopy LEDs, (except for green and far-red, respectively limited at 75 μmol $m^{-2}$ $s^{-1}$ and 30 μmol $m^{-2}$ $s^{-1}$) and 200 μmol $m^{-2}$ $s^{-1}$ for the uppermost LED fixture. LED intensity and spectrum were adjusted in relation to the recorded solar radiation to decrease energy consumption, as reported by Pinho et al. (2013). Although it was not the purpose of the experiment, energy consumption was reduced by 30% in comparison to normal LED practice. Such equipment is quite unique in Europe.

The photoperiod was 18 h. Except for the light spectrum (see below), all the parameters were similar between treatments (daily light integral, climatic and fertigation management, cultivation practices). In addition to sun light, the following light management was undertaken (Fig. 4):

- control: 108 plants were illuminated with 93% red 7% blue at the top, middle and bottom levels.
- apexLED: 162 plants were illuminated with 100% blue for 2 h at the switch on, then 93% red 7% blue for the next 14 h at the top level. Middle and bottom levels were similar to control treatment.
- fruitLED: 54 plants were illuminated with 93% red 7% blue at the top and middle levels. At the bottom level, ripening fruits were illuminated with 50% blue 50% white.

## 7.3 Agronomic performance and plant measurements

Tomatoes were harvested bi-weekly (ripening stage as defined by commercial practices). Total marketable fruit yield per square metre is expressed as kg $m^{-2}$. Plant architecture was evaluated through weekly measurements of stem growth (expressed in cm $week^{-1}$), length and width of leaf number 10 (counting from the top of the plant; first leaf was larger than 6 $cm^2$. Data expressed in cm). Image analysis (ImageJ software, US National Institutes of Health, Bethesda, MD, USA) was carried out to estimate terminal leaflet surface area (expressed in $cm^2$). Data are the averages of 10 plants per treatment. Sap flow and stem diameter sensors (2Grow, Destelbergen, Belgium) were fixed to two independent plants per treatment. Vitamin C content was assayed using HPLC on 3 replicates of 10 fruits per treatment: fruits were harvested in April 2019, and frozen immediately at −80°C using liquid nitrogen before grinding.

## 7.4 Blue light dose at the apical bud level influences plant water status decreasing yield

LED treatments showed little impact on the architecture of the plants (Table 3), apart from the blue effect on cumulative growth. Chlorophyll fluorescence, maximum photosynthetic activity, stomatal conductance and chlorophyll content of leaves were assayed repeatedly during the experiment, but data showed no consistency and no statistical difference could be noted (data not shown). Harvesting started on week 9 and ended on week 35 with a final yield of 40.3, 36.9 and 38.1 kg $m^{-2}$ for control, apexLED and fruitLED respectively. fruitLED yield was not statistically different from other treatments.

A difference of 8.3% between control and apex LED was significant ($p<0.05$). On week 21 (halfway through the experiment), differences between those two treatments reached 13.7% in favour of the control. In a comparable study using 3 h of 85 μmol $m^{-2}$ $s^{-1}$ of blue (447 nm) when lights were turned on followed by monochromatic red (670 nm) lighting, tomato yield increased by 8% in comparison to the reference 100% red (Dieleman et al., 2014). Our study revealed a detrimental effect of 100% blue lighting when this treatment was applied lights switching on.

Sap flow measurements highlighted this negative impact. Indeed, sap flow rate is representative of plant transpiration activity throughout the

**Table 3** Agronomical performance, plant architecture and fruit vitamin C content of tomato plants subjected to different wavelengths (see Figure 4).

| | Control | apexLED | fruitLED |
|---|---|---|---|
| **Final stem length (m)** | 10.1 | 9.8 | 8.9 |
| *p<0.05* | a | b | c |
| **Average leaf length (cm)** | 43.1 | 43.7 | 43.0 |
| *p<0.05* | a | a | a |
| **Average leaf width (cm)** | 41.4 | 42.3 | 42.0 |
| *p<0.05* | a | a | a |
| **Average terminal leaflet area ($cm^2$)** | 510.1 | 538.0 | 542.0 |
| *p<0.05* | a | a | a |
| **Average dry matter (%)** | 11.0 | 10.8 | 10.7 |
| *p<0.05* | a | a | a |
| **Yield (kg $m^{-2}$)** | 40.3 | 36.9 | 38.1 |
| *p<0.05* | a | b | ab |
| **Fruit vitamin C content (mg $100g^{-1}$ juice)** | 17.2 | 17.0 | 18.9 |
| *p<0.05* | b | b | a |

day, according to its environment. Sap flow rate is strongly correlated with the radiation receive by the plant (Jones, 2004): higher radiation leads to higher sap flow rates. In our experiment (Fig. 5a), we calculated the ratio between the sap flow rate and the DLI (daily light integral), which provided reliable information on stomatal functioning (which are bonds between the

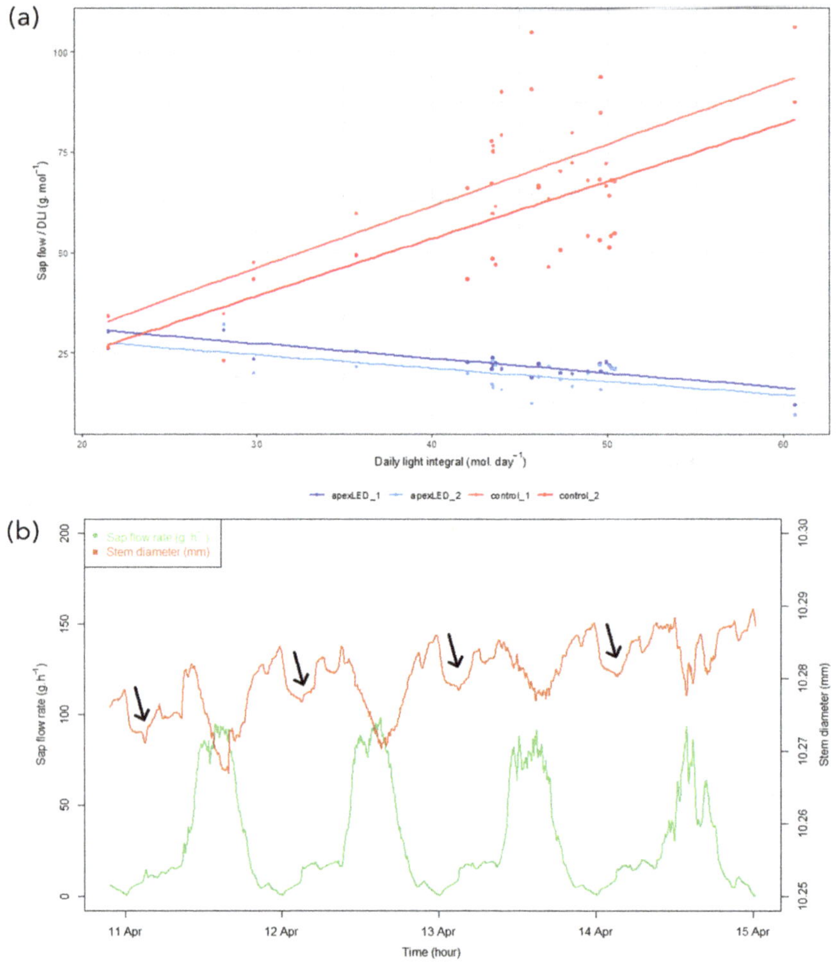


**Figure 5** (a) Sap flow rate and LED treatments. Ratio of the cumulative daily sap flow rate (gram of sap per mole of light) and the DLI (daily light integral; moles.day$^{-1}$) according to the DLI. Two plants were instrumented and used as replicates (control_1 and control_2; apexLED_1 and apexLED_2). Control = red dots and line. apexLED = blue dots and line. The lines are the linear regression of each treatment. (b) Sap flow rate (in green; g.h$^{-1}$) and stem diameter (in orange; mm) measurements over several days (April 2019). Black arrows indicate the cavitation phenomenon on the stem diameter as described by Vermeiren et al. (2017).

environment and transpiration activity). By taking the cumulative daily flow and the DLI, we were able to observe the activity of the plant in relation to the light received during the day. Figure 5a shows a contradictory pattern between control plants and apexLED plants. Although under medium radiation, no difference could be noted, apexLED led to a limitation of plant transpiration activity under high light radiation (DLI>30 moles.day$^{-1}$). However, this limitation had no influence on maximum photosynthetic activity. It will be interesting to push investigations forward to find out which processes were impacted.

The stem diameter's dynamic throughout the day raise our attention (Fig. 5b). The stem diameter dropped suddenly after LED lights were switched on. This is called the cavitation phenomenon. Vermeiren et al. (2017) reported that using artificial lighting, the stem diameter drops after the lamps were switched on and then returned to normal dynamics one to two hours later. Under natural sunrise, light increases in a gradual way, as does the transpiration rate. Lamps switching on lead to a very fast increase of the transpiration rate which creates a cavitation phenomenon if root water uptake cannot keep up the demand. Moreover, after a cavitation phenomenon occurred, we observed (Fig. 5b) that stem diameter increased again, sap flow rate dropped, showing that the activity of the plant is stagnating. Therefore, plant transpiration activity remained low throughout the night-time period (where LEDs are the only light providers). We are putting some efforts to avoid these detrimental effects. To repress the cavitation phenomenon, delaying the switch on of intra-canopy lighting by one hour after the toplighting started in addition to a gradual increase of the light intensity of the toplighting LED (to mimic the sunrise) has been evaluated as a suitable solution (data not shown).

## 7.5 Blue illumination leads to fruit quality increase

Fruit quality improvement is required by consumers. Vitamin C content increased by 10% in the fruit LED treatment in comparison to the control. The blue-rich environment stimulated antioxidant synthesis (Taulavuori et al., 2013), such as vitamin C. Therefore, fruit-specific illumination using LED light could increase nutritional quality, but economic aspects must be taken into account. Furthermore, in our experiment, we were not able to detect a similar improvement in quality later in the season when solar radiation increased (data not shown). In the context of out-of-season production, issues linked to fruit quality have to be solved. In conclusion, LED lighting could be adjusted to plant and organ needs if specifically positioned. Adaptation of the climatic environment and fertigation management are crucial in performing LED trial, especially when fine-tuning the light spectrum. Photobiology is very complex and further studies are awaited.

## 8 Conclusions and future trends

Fruit vegetable production in SCS is a competitive sector known for its high-quality production and performance. It contributes to the nutrition and to the economy of whole regions, as it is also a labour-intensive sector. Scientific research is key to improve the sustainability of SCS, both economically and environmentally, to meet requirement in terms of agronomic performance and quality. Fruit and vegetable cultivation practices rely on experience and technical skills of the grower: sensors and modelling advances help to improve crop performance and reliability in the face of climate change and society expectations, taking into account real-time plant physiology. SCS technology has started a new revolution in data on strategy and systems, automating previously manual tasks. The reduction of energy costs, and globally of input-output fluxes, is needed to achieve energy neutral self-supporting greenhouses, as society's expectations in lowering the ecological footprint of greenhouse productions are high. Innovation transfer between research and growers will ensure enhancement of the vegetable production sector.

Pressure on arable land, climatic change and demography evolving to increase the urban population set many future challenges. Fruit vegetables contribute substantially to the human diet and the need for tasty and healthy food will raise the standards for modern breeding and integration of innovative technologies. Although SCS technology will be important, new ways of growing fruit vegetable crops in vertical farms will also require specific cultivars (O'Sullivan et al., 2020).

The integration of SCS into a circular economy is a step towards increasing the sustainability of the system. Whether for energy, water or $CO_2$, synergies with industries using these elements as waste are to be found. This will reduce the economic and ecological costs of fruit vegetables production. Waste from tomato, sweet pepper and cucumber crops should find outlets to become by-products with a market value. During the growing period (through deleafing) and at the end of the production cycle (leaves and stems), large quantities of residual biomass with no added value are discarded. In relation to carbon partitioning, approximately 66% of the total carbon will be directed into leaves and stems (33% in fruits) during the growing period, meaning that for 1 kg of tomatoes harvested and sold, 1.6 kg of carbon won't find any economic opportunities. After the last harvest, Junker-Frohn et al. (2019) suggested exposing tomato plants to abiotic stress (nitrogen deficiency) to increase rutin (flavonoid) content in leaves. In association with appropriate extraction methods, tomato leaves became a novel source of secondary metabolites for cosmetics or food industries. However, at the end of the cultivation period, 70% of the residual biomass will be stem material, which has not yet been intensively studied for its composition. As it is a crucial economical point for growers, initiatives like the Tomato Lab® in France (*Conservatoire de la tomate*, 09/19) should be encouraged.

# 9 Where to look for further information

## 9.1 Further reading

Application of soilless culture technologies in the modern greenhouse industry – A review by Savvas and Gruda (2018).

Guidelines to use tomato in experiments with a controlled environment by Schwarz et al. (2014).

The reference book 'Tomatoes' by Heuvelink (2018).

## 9.2 Key journal/conferences

GREENSYS (International Symposium on Advanced Technologies and Management for Innovative Greenhouses) holds regular meetings specifically on controlled environment agriculture (greenhouse and vertical farming).

EUCARPIA (European Association for Research on Plant Breeding) holds regular meetings, including section on Vegetables (and a specific working group on tomato).

# 10 References

Adamowicz, S., Le Bot, J., Huanosto Magaña, R. and Fabre, J. (2012). Totomatix: a novel automatic set-up to control diurnal, diel and long-term plant nitrate nutrition. *Annals of Botany* 109(1): 309–319.

Affandi, F. Y., Verdonk, J. C., Ouzounis, T., Ji, Y., Woltering, E. J. and Schouten, R. E. (2020). Far-red light during cultivation induces postharvest cold tolerance in tomato fruit. *Postharvest Biology and Technology* 159: 111019.

Albert, B., Brajeul, E., Le Quillec, S., Lesourd, D. and Loda, D. (2013). Using heating relocation and air dehumidification/cooling in soilless cucumber crop in greenhouse: application in the West of France. In: *International Symposium on New Technologies for Environment Control, Energy-Saving and Crop Production in Greenhouse and Plant*, vol. 1037, pp. 741–748.

Albornoz, F. (2016). Crop responses to nitrogen overfertilization: a review. *Scientia Horticulturae* 205: 79–83.

Alsanius, B. W. and Wohanka, W. (2019). Root zone microbiology of soilless cropping systems. *Soilless Culture* 2019: 149–194.

Amitrano, C., Arena, C., Rouphael, Y., De Pascale, S. and De Micco, V. (2019). Vapour pressure deficit: the hidden driver behind plant morphofunctional traits in controlled environments. *Annals of Applied Biology* 175(3): 313–325.

Bantis, F., Smirnakou, S., Ouzounis, T., Koukounaras, A., Ntagkas, N. and Radoglou, K. (2018). Current status and recent achievements in the field of horticulture with the use of light-emitting diodes (LEDs). *Scientia Horticulturae* 235: 437–451.

Baum, C., El-Tohamy, W. and Gruda, N. (2015). Increasing the productivity and product quality of vegetable crops using arbuscular mycorrhizal fungi: a review. *Scientia Horticulturae* 187: 131–141.

Beckles, D. M. (2012). Factors affecting the postharvest soluble solids and sugar content of tomato (*Solanum lycopersicum L.*) fruit. *Postharvest Biology and Technology* 63(1): 129–140.

Benard, C., Gautier, H., Bourgaud, F., Grasselly, D., Navez, B., Caris-Veyrat, C., Weiss, M. and Génard, M. (2009). Effects of low nitrogen supply on tomato (*Solanum lycopersicum*) fruit yield and quality with special emphasis on sugars, acids, ascorbate, carotenoids, and phenolic compounds. *Journal of Agricultural and Food Chemistry* 57(10): 4112–4123.

Bertin, N. and Genard, M. (2018). Tomato quality as influenced by preharvest factors. *Scientia Horticulturae* 233: 264–276.

Boulard, T., Roy, J.-C., Pouillard, J.-B., Fatnassi, H. and Grisey, A. (2017). Modelling of micrometeorology, canopy transpiration and photosynthesis in a closed greenhouse using computational fluid dynamics. *Biosystems Engineering* 158: 110–133.

Bournet, P. E., Brajeul, E., Truffault, V., Chantoiseau, E. and Pinoit, C. (2017). Impact of heating location and mechanical ventilation on climate distribution inside a greenhouse cucumber crop. In: *International Symposium on New Technologies for Environment Control, Energy-Saving and Crop Production in Greenhouse and Plant*, vol. 1227, pp. 339–346.

Buck-Sorlin, G. H., Hemmerling, R., Vos, J., and de Visser, P. H. (2009). Modelling of spatial light distribution in the greenhouse: description of the model. In: *Third International Symposium on Plant Growth Modeling, Simulation, Visualization and Applications*, Beijing, China, vol. 2009, pp. 79–86.

Cocaliadis, M. F., Fernández-Muñoz, R., Pons, C., Orzaez, D. and Granell, A. (2014). Increasing tomato fruit quality by enhancing fruit chloroplast function. A double-edged sword? *Journal of Experimental Botany* 65(16): 4589–4598.

Colquhoun, T. A., Schwieterman, M. L., Gilbert, J. L., Jaworski, E. A., Langer, K. M., Jones, C. R., Rushing, G. V., Hunter, T. M., Olmstead, J., Clark, D. G. and Folta, K. M. (2013). Light modulation of volatile organic compounds from petunia flowers and select fruits. *Postharvest Biology and Technology* 86: 37–44.

Conservatoire de la tomate (09/19) Tomato Lab, incubateur d'innovations. Available at: https://www.conservatoiredelatomate.fr/portfolio/tomato-lab-incubateur-dinnovations-autour-de-la-tomate/ (Accessed 10 February 2020).

Conversa, G., Lazzizera, C., Bonasia, A. and Elia, A. (2013). Yield and phosphorus uptake of a processing tomato crop grown at different phosphorus levels in a calcareous soil as affected by mycorrhizal inoculation under field conditions. *Biology and Fertility of Soils* 49(6): 691–703.

Cordell, D., Drangert, J.-O. and White, S. (2009). The story of phosphorus: global food security and food for thought. *Global Environmental Change* 19(2): 292–305.

Cottet, V., Navez, B., Jost, M., Causse, M. and Fontanet, L. (2019). Modes de production et variétés en tomate traditionnelle. La diversité révélée par analyse sensorielle. *Infos-Ctifl* 348: 15–22.

Davies, J. N. and Hobson, G. E. (1981). The constituents of tomato fruit–the influence of environment, nutrition, and genotype. *Critical Reviews in Food Science and Nutrition* 15(3): 205–280.

de Gelder, A., Bontsema, J., Hemming, J., Buwalda, F., and van Henten, E. J. (2006). Toward an optimal control strategy for sweet pepper cultivation-1. A dynamic crop model. In: *III International Symposium on Models for Plant Growth, Environmental Control and Farm Management in Protected Cultivation*, vol. 718, pp. 367–374.

de Gelder, A., Dieleman, J. A., Bot, G. P. A. and Marcelis, L. F. M. (2012a). An overview of climate and crop yield in closed greenhouses. *The Journal of Horticultural Science and Biotechnology* 87(3): 193–202.

de Gelder, A., Warmenhoven, M. G., Kromdijk, J., Meinen, E., de Zwart, H. F., Stolker, H. and Grootscholten, M. (2012b). *Gelimiteerd CO2 en het nieuwe telen Tomaat*. Rapport Wageningen UR Glastuinbouw 1159: 1–76.

de Kreij, C., Voogt, W., Van den Bos, A., and Baas, R. (1997). Voedingsoplossingen voor de teelt van tomaat in gesloten teeltsystemen. Brochure VG Tomaat, The Netherlands.

De Swaef, T. and Steppe, K. (2010). Linking stem diameter variations to sap flow, turgor and water potential in tomato. *Functional Plant Biology* 37(5): 429–438.

de Zwart, F., Vanthoor, B. and Koreneef, S. (2019). Tuinbouw zonder fossiele energie. Wageningen Plant Research, Business Unit Glastuinbouw 853, pp. 1–64.

Dieleman, A., Weerheim, K. and Dueck, T. A. (2014).Report on final specifications with an empirical proof on claims for high productivity horticulture. Available at: http://www.hi-led.eu/public-documents/ (Accessed 11 February 2020).

Dieleman, J. A., De Visser, P. H. B., Meinen, E., Grit, J. G. and Dueck, T. A. (2019). Integrating morphological and physiological responses of tomato plants to light quality to the crop level by 3D modeling. *Frontiers in Plant Science* 10: 839.

Domínguez, E., Fernández, M. D., Hernández, J. C. L., Parra, J. P., España, L., Heredia, A. and Cuartero, J. (2012). Tomato fruit continues growing while ripening, affecting cuticle properties and cracking. *Physiologia Plantarum* 146(4): 473–486.

Dong, J., Gruda, N., Lam, S. K., Li, X. and Duan, Z. (2018). Effects of elevated $CO_2$ on nutritional quality of vegetables: a review. *Frontiers in Plant Science* 9: 924.

Dumas, Y., Dadomo, M., Di Lucca, G. and Grolier, P. (2003). Effects of environmental factors and agricultural techniques on antioxidant content of tomatoes. *Journal of the Science of Food and Agriculture* 83(5): 369–382.

Dzakovich, M. P., Gómez, C. and Mitchell, C. A. (2015). Tomatoes grown with light-emitting diodes or high-pressure sodium supplemental lights have similar fruit-quality attributes. *HortScience* 50(10): 1498–1502.

Dzakovich, M. P., Gómez, C., Ferruzzi, M. G. and Mitchell, C. A. (2017). Chemical and sensory properties of greenhouse tomatoes remain nnchanged in response to red, blue, and far red supplemental light from light-emitting diodes. *Hortscience* 52(12): 1734–1741.

Elings, A., De Visser, P. H. B., Marcelis, L. F. M., Heinen, M., van den Boogaard, H. A. G. M., Gieling, T. H. and Werner, B. E. (2004). Feed-forward control of water and nutrient supply in greenhouse horticulture: development of a system. *Acta Horticulturae* 654: 195–202.

Fanwoua, J., Vercambre, G., Buck-Sorlin, G., Dieleman, J. A., de Visser, P. and Génard, M. (2019). Supplemental LED lighting affects the dynamics of tomato fruit growth and composition. *Scientia Horticulturae* 256: 108571.

Gautier, H., Massot, C., Stevens, R., Sérino, S. and Génard, M. (2009). Regulation of tomato fruit ascorbate content is more highly dependent on fruit irradiance than leaf irradiance. *Annals of Botany* 103(3): 495–504.

Gómez, C. and Mitchell, C. A. (2016). In search of an optimized supplemental lighting spectrum for greenhouse tomato production with intracanopy lighting. In: *VIII International Symposium on Light in Horticulture*, vol. 1134, pp. 57–62.

Grasselly, D., Hamm, F., Quaranta, G., and Vitrou, J. (2009). Carbon footprint of coconut fibre (coir) substrates. *Infos-Ctifl* 249: 55–59.

Grierson, D. and Kader, A. (1986). Fruit ripening and quality. In: Atherton, J. G. and Rudich, J. (Eds), *The Tomato Crop*, Chapman and Hall, London, pp. 241–280.

Grisey, A., Rosso, L., Auger, E., Tisiot, R. and d'Amaral, F. (2013). « Sustainable energy greenhouse » results of the project: reduction of energy consumption and energy storage in aquifer. In: *International Symposium on New Technologies for Environment Control, Energy-Saving and Crop Production in Greenhouse and Plant*, vol. 1037, pp. 147–154.

Grisey, A., Grasselly, D., Le Quillec, S., Brajeul, E., and Tisiot, R. (2020). Survey of semi-enclosed greenhouses: a new concept under development in France. *Infos-Ctifl* 359: 38–43.

Grisey, A., Levaillant, C., Martin, D., et al. (2020). Using solar energy to heat, cool and dehumidify tomato greenhouses in France: a pre-feasibility study. In: *International Symposium on Advanced Technologies and Management for Innovative Greenhouses,* Acta Horticulturae 1296, 731–738.

Grozeff, G. E. G., Senn, M. E., Alegre, M. L., Chaves, A. R. and Bartoli, C. G. (2016). Nocturnal low irradiance pulses improve fruit yield and lycopene concentration in tomato. *Scientia Horticulturae* 203: 47–52.

Gruda, N. (2009). Do soilless culture systems have an influence on product quality of vegetables? *Journal of Applied Botany and Food Quality* 82(2): 141–147.

Gruda, N., Savvas, D., Colla, G. and Rouphael, Y. (2018). Impacts of genetic material and current technologies on product quality of selected greenhouse vegetables–A review. *European Journal of Horticultural Science* 83(5): 319–328.

Gruda, N., Bisbis, M. and Tanny, J. (2019). Impacts of protected vegetable cultivation on climate change and adaptation strategies for cleaner production – a review. *Journal of Cleaner Production* 225: 481–495.

Gruda, N. S. (2019). Increasing sustainability of growing media constituents and stand-alone substrates in soilless culture systems. *Agronomy* 9(6): 298.

Grunert, O., Hernandez-Sanabria, E., Perneel, M., Van Labeke, M. C., Reheul, D. and Boon, N. (2014). Molecular insights on the functional microbial community from organic and mineral growing media and its interaction with *Agrobacterium rhizogenes*. *Communications in Agricultural and Applied Biological Sciences* 79(3): 345–356.

Grunert, O., Robles-Aguilar, A. A., Hernandez-Sanabria, E., Schrey, S. D., Reheul, D., Van Labeke, M. C., Vlaeminck, S. E., Vandekerckhove, T. G. L., Mysara, M., Monsieurs, P., Temperton, V. M., Boon, N. and Jablonowski, N. D. (2019). Tomato plants rather than fertilizers drive microbial community structure in horticultural growing media. *Scientific Reports* 9(1): 9561.

He, F., Thiele, B., Watt, M., Kraska, T., Ulbrich, A. and Kuhn, A. J. (2019). Effects of root cooling on plant growth and fruit quality of cocktail tomato during two consecutive seasons. *Journal of Food Quality* 2019: 1–15.

Hemming, S., de Zwart, F., Elings, A., Righini, I. and Petropoulou, A. (2019). Remote control of greenhouse vegetable production with artificial intelligence—greenhouse climate, irrigation, and crop production. *Sensors* 19(8): 1807.

Heuvelink, E. (2018). *Tomatoes* (2nd edn.). CABI Publishing, Wallingford, UK.

Hinsinger, P., Brauman, A., Devau, N., Gérard, F., Jourdan, C., Laclau, J., Le Cadre, E., Jaillard, B. and Plassard, C. (2011). Acquisition of phosphorus and other poorly mobile nutrients by roots. Where do plant nutrition models fail? *Plant and Soil* 348(1-2): 29-61.

Hoagland, D. R. and Arnon, D. I. (1950). *The Water-Culture Method for Growing Plants without Soil*. Circular. California agricultural experiment station, 347. Berkeley: College of Agriculture, University of California.

Huché-Thélier, L., Crespel, L., Gourrierec, J. L., Morel, P., Sakr, S. and Leduc, N. (2016). Light signaling and plant responses to blue and UV radiations–perspectives for applications in horticulture. *Environmental and Experimental Botany* 121: 22-38.

Husson, O. (2013). Redox potential (Eh) and pH as drivers of soil/plant/microorganism systems: a transdisciplinary overview pointing to integrative opportunities for agronomy. *Plant and Soil* 362(1-2): 389-417.

Iersel, M. W. and Gianino, D. (2017). An adaptive control approach for light-emitting diode lights can reduce the energy costs of supplemental lighting in greenhouses. *Hortscience* 52(1): 72-77.

Incrocci, L., Massa, D. and Pardossi, A. (2017). New trends in the fertigation management of irrigated vegetable crops. *Horticulturae* 3(2): 37.

Jiang, C., Johkan, M., Hohjo, M., Tsukagoshi, S., Ebihara, M., Nakaminami, A. and Maruo, T. (2017). Photosynthesis, plant growth, and fruit production of single-truss tomato improves with supplemental lighting provided from underneath or within the inner canopy. *Scientia Horticulturae* 222: 221-229.

Johkan, M., Ishii, M., Maruo, T., Na, L., Tsukagoshi, S., Hojoh, M., Nakaminami, A. and Shinohara, Y. (2013). Improved light conditions at the fruit truss accelerate harvest time and enhance ascorbic acid concentration in a low-truss, high-density tomato production system. *Journal of the Japanese Society for Horticultural Science* 82(4): 317-321.

Jones, H. G. (2004). Irrigation scheduling: advantages and pitfalls of plant-based methods. *Journal of Experimental Botany* 55(407): 2427-2436.

Jones, J. W., Dayan, E., Allen, L., et al. (1991). A dynamic tomato growth and yield model (TOMGRO). *Transactions of the ASAE* 34(2): 663-0672.

Juárez-Maldonado, A., Benavides-Mendoza, A., de-Alba-Romenus, K. and Morales-Díaz, A. B. (2014). Dynamic modeling of mineral contents in greenhouse tomato crop. *Agricultural Sciences* 05(2): 114–123.

Junker-Frohn, L. V., Lück, M., Schmittgen, S., Wensing, J., Carraresi, L., Thiele, B., Groher, T., Reimer, J. J., Bröring, S., Noga, G., Jupke, A., Schurr, U., Usadel, B., Wiese-Klinkenberg, A. and Wormit, A. (2019). Tomato's green gold: bioeconomy potential of residual tomato leaf biomass as a novel source for the secondary metabolite rutin. *ACS Omega* 4(21): 19071-19080.

Kaiser, E., Morales, A. and Harbinson, J. (2018). Fluctuating light takes crop photosynthesis on a rollercoaster ride. *Plant Physiology* 176(2): 977-989.

Kaiser, E., Weerheim, K., Schipper, R. and Dieleman, J. A. (2019). Partial replacement of red and blue by green light increases biomass and yield in tomato. *Scientia Horticulturae* 249: 271-279.

Kawasaki, Y. and Yoneda, Y. (2019). Local temperature control in greenhouse vegetable production. *The Horticulture Journal* 88(3): 305-314.

Kempen, E. (2015). Nutrient and water use of tomato (Solanum Lycopersicum) in soilless production systems. Ph.D. Thesis. Stellenbosch University, Stellenbosch, South Africa, 177pp.

Kempen, E., Agenbag, G. A. and Deckers, S. (2017). Variations in water and macronutrient uptake of soilless tomato as affected by the nutrient solution composition. *South African Journal of Plant and Soil* 34(2): 139–148.

Kempkes, F. L. K., Van de Braak, N. J. and Bakker, J. C. (2000). Effect of heating system position on vertical distribution of crop temperature and transpiration in greenhouse tomatoes. *Journal of Agricultural Engineering Research* 75(1): 57–64.

Kim, H.-J., Kim, W.-K., Roh, M.-Y., Kang, C., Park, J. and Sudduth, K. A. (2013). Automated sensing of hydroponic macronutrients using a computer-controlled system with an array of ion-selective electrodes. *Computers and Electronics in Agriculture* 93: 46–54.

Krauss, S., Schnitzler, W. H., Grassmann, J. and Woitke, M. (2006). The influence of different electrical conductivity values in a simplified recirculating soilless system on inner and outer fruit quality characteristics of tomato. *Journal of Agricultural and Food Chemistry* 54(2): 441–448.

Kuroyanagi, T., Yasuba, K., Higashide, T., Iwasaki, Y. and Takaichi, M. (2014). Efficiency of carbon dioxide enrichment in an unventilated greenhouse. *Biosystems Engineering* 119: 58–68.

Le Bot, J., Adamowicz, S. and Robin, P. (1998). Modelling plant nutrition of horticultural crops: a review. *Scientia Horticulturae* 74(1–2): 47–82.

Lee, J. W. and Son, J. E. (2019). Nondestructive and continuous fresh weight measurements of bell peppers grown in soilless culture systems. *Agronomy* 9(10): 652.

Legast, E., Brajeul, E. and Truffault, V. (2020). Effect of temperature on tomato fruit growth: a modelling-based proposal for optimal temperature distribution within heated greenhouse. In: *International Symposium on Advanced Technologies and Management for Innovative Greenhouses,* Acta Horticulturae 1296, 49–56.

LetsGrow.com (2020). Moving from green fingers to growing without hands. Available at: https://www.letsgrow.com/en/news/moving-from-green-fingers-to-growing-without-hands/ (Accessed 10 February 2020).

Li, Y., Ding, Y., Li, D. and Miao, Z. (2018). Automatic carbon dioxide enrichment strategies in the greenhouse: a review. *Biosystems Engineering* 171: 101–119.

Lin, D., Wei, R. and Xu, L. (2019). An integrated yield prediction model for greenhouse tomato. *Agronomy* 9(12): 873.

Liu, H. F., Génard, M., Guichard, S. and Bertin, N. (2007). Model-assisted analysis of tomato fruit growth in relation to carbon and water fluxes. *Journal of Experimental Botany* 58(13): 3567–3580.

Lucas, R. E. and Davis, J. F. (1961). Relationships between pH values of organic soils and availabilities of 12 plant nutrients. *Soil Science* 92(3): 177–182.

Martínez-Ruiz, A., López-Cruz, I. L., Ruiz-García, A., Pineda-Pineda, J. and Prado-Hernández, J. V. (2019). HortSyst: a dynamic model to predict growth, nitrogen uptake, and transpiration of greenhouse tomatoes. *Chilean Journal of Agricultural Research* 79(1): 89–102.

Nederhoff, E. and Vegter, J. (1994). Photosynthesis of stands of tomato, cucumber and sweet pepper measured in greenhouses under various $CO_2$ concentrations. *Annals of Botany* 73(4): 353–361.

Nomiyama, R., Yasutake, D., Sago, Y. and Kitano, M. (2017). Application of transpiration integrated model to simulation of dynamics of ion absorption by tomato roots growing in soil-less culture. *Environment Control in Biology* 55(1): 29–35.

Ntagkas, N., Min, Q., Woltering, E. J., Labrie, C., Nicole, C. C. S. and Marcelis, L. F. M. (2016). Illuminating tomato fruit enhances fruit vitamin C content. In: *VIII International Symposium on Light in Horticulture*, vol. 1134, pp. 351–356.

Ntagkas, N., Woltering, E., Nicole, C., Labrie, C. and Marcelis, L. F. M. (2019). Light regulation of vitamin C in tomato fruit is mediated through photosynthesis. *Environmental and Experimental Botany* 158: 180–188.

O'Sullivan, C. A., McIntyre, C. L., Dry, I. B., Hani, S. M., Hochman, Z. and Bonnett, G. D. (2020). Vertical farms bear fruit. *Nature Biotechnology* 38(2): 160–162.

Ouzounis, T., Heuvelink, E., Ji, Y., Schouten, H. J., Visser, R. G. F. and Marcelis, L. F. M. (2016). Blue and red LED lighting effects on plant biomass, stomatal conductance, and metabolite content in nine tomato genotypes. In: *VIII International Symposium on Light in Horticulture*, vol. 1134, pp. 251–258.

Pieterse, C. M. J., de Jonge, R. and Berendsen, R. L. (2016). The soil-borne supremacy. *Trends in Plant Science* 21(3): 171–173.

Pinho, P., Hytönen, T., Rantanen, M., Elomaa, P. and Halonen, L. (2013). Dynamic control of supplemental lighting intensity in a greenhouse environment. *Lighting Research and Technology* 45(3): 295–304.

Ponce-Valadez, M., Escalona-Buendía, H. B., Villa-Hernández, J. M., de León-Sánchez, F. D., Rivera-Cabrera, F., Alia-Tejacal, I. and Pérez-Flores, L. J. (2016). Effect of refrigerated storage (12.5°C) on tomato (*Solanum lycopersicum*) fruit flavor: A biochemical and sensory analysis. *Postharvest Biology and Technology* 111: 6–14.

Portree, J. (1993). *Greenhouse Vegetable Production Guide for Commercial Growers*. Ministry of Agriculture, Fisheries and Food, Province of British Columbia, Ext. Systems Branch, Victoria, B. C., Canada.

Powell, A. L., Nguyen, C. V., Hill, T., Cheng, K. L., Figueroa-Balderas, R., Aktas, H., Ashrafi, H., Pons, C., Fernández-Muñoz, R., Vicente, A., Lopez-Baltazar, J., Barry, C. S., Liu, Y., Chetelat, R., Granell, A., Van Deynze, A., Giovannoni, J. J. and Bennett, A. B. (2012). Uniform ripening encodes a golden 2-like transcription factor regulating tomato fruit chloroplast development. *Science* 336(6089): 1711–1715.

Qian, T., Dieleman, J. A., Elings, A., de Gelder, A. and Marcelis, L. F. M. (2015). Response of tomato crop growth and development to a vertical temperature gradient in a semi-closed greenhouse. *The Journal of Horticultural Science and Biotechnology* 90(5): 578–584.

Rácz, I. (2007). Study on nutrient conditions in soilless vegetable production. *Cereal Research Communications* 35(2): 965–968.

Ramírez S., L. F., Muro E., J. and Sánchez G., P. (2009). Potassium affects the lycopene and β-carotene concentration in greenhouse tomato. *Acta Horticulturae* 821: 223–228.

Renard, C. M., Ginies, C., Gouble, B., Bureau, S. and Causse, M. (2013). Home conservation strategies for tomato (*Solanum lycopersicum*): storage temperature vs. duration- Is there a compromise for better aroma preservation? *Food Chemistry* 139(1–4): 825–836.

Rouphael, Y. and Kyriacou, M. C. (2018). Enhancing quality of fresh vegetables through salinity eustress and biofortification applications facilitated by soilless cultivation. *Frontiers in Plant Science* 9: 1254.

Sambo, P., Nicoletto, C., Giro, A., Pii, Y., Valentinuzzi, F., Mimmo, T., Lugli, P., Orzes, G., Mazzetto, F., Astolfi, S., Terzano, R. and Cesco, S. (2019). Hydroponic solutions for soilless production systems: issues and opportunities in a smart agriculture perspective. *Frontiers in Plant Science* 10: 923.

Savvas, D. and Gruda, N. (2018). Application of soilless culture technologies in the modern greenhouse industry: a review. *European Journal of Horticultural Science* 83(5): 280–293.

Savvides, A., Van Ieperen, W., Dieleman, J. A. and Marcelis, L. F. (2013). Meristem temperature substantially deviates from air temperature even in moderate environments: is the magnitude of this deviation species-specific? *Plant, Cell and Environment* 36(11): 1950–1960.

Savvides, A., Ntagkas, N., van Ieperen, W., Dieleman, J. A. and Marcelis, L. F. M. (2014). Impact of light on leaf initiation: A matter of photosynthate availability in the apical bud? *Functional Plant Biology: FPB* 41(5): 547–556.

Savvides, A., van Ieperen, W., Dieleman, J. A. and Marcelis, L. F. (2017). Phenotypic plasticity to altered apical bud temperature in *Cucumis sativus*: more leaves-smaller leaves and *vice versa*. *Plant, Cell and Environment* 40(1): 69–79.

Schwarz, D., Thompson, A. J. and Kläring, H. P. (2014). Guidelines to use tomato in experiments with a controlled environment. *Frontiers in Plant Science* 5: 625.

Shen, Y., Wei, R. and Xu, L. (2018). Energy consumption prediction of a greenhouse and optimization of daily average temperature. *Energies* 11(1): 65.

Spaargaren, J. (2001). Supplemental lighting for greenhouse crops. Available at: https://agris.fao.org/agris-search/search.do?recordID=NL2001004039.

Stanghellini, C. (1987). Transpiration of greenhouse crops: an aid to climate management. Ph.D. Thesis. Wageningen University, Wageningen, The Netherlands, 150 pp.

Suslow, T. and Cantwell, M. (2009). Tomato: recommendations for maintaining postharvest quality. In: Kader, A. A. (Ed.), *Produce Facts*. Postharvest Technology Research & Information Center, Davis.

Taulavuori, E., Taulavuori, K., Holopainen, J. K., Julkunen-Tiitto, R., Acar, C. and Dincer, I. (2017). Targeted use of LEDs in improvement of production efficiency through phytochemical enrichment. *Journal of the Science of Food and Agriculture* 97(15): 5059–5064.

Taulavuori, K., Julkunen-Tiitto, R., Hyöky, V. and Taulavuori, E. (2013). Blue mood for superfood. *Natural Product Communications* 8(6): 791–794.

Thompson, R., Delcour, I., Berckmoes, E., et al. (2018). The fertigation bible. Available at: http://www.fertinnowa.com/the-fertigation-bible/.

Tieman, D., Bliss, P., McIntyre, L. M., Blandon-Ubeda, A., Bies, D., Odabasi, A. Z., Rodríguez, G. R., van der Knaap, E., Taylor, M. G., Goulet, C., Mageroy, M. H., Snyder, D. J., Colquhoun, T., Moskowitz, H., Clark, D. G., Sims, C., Bartoshuk, L. and Klee, H. J. (2012). The chemical interactions underlying tomato flavor preferences. *Current Biology: CB* 22(11): 1035–1039.

Tomato Genome Consortium (2012). The tomato genome sequence provides insights into fleshy fruit evolution. *Nature* 485(7400): 635–641.

Truffault, V., Albert, B., Schuppe, J., Le Quillec, S. and Brajeul, E. (2016). Improvement of irrigation and sanitary risk control in a cucumber greenhouse crop using vapour pressure deficit and fruit temperature sensor. In: *V. International Symposium on Models for Plant Growth, Environment Control and Farming Management in Protected Cultivation*, vol. 1182, pp. 129–136.

Truffault, V., Le Quillec, S. and Brajeul, E. (2017). Insights into the potential of semi-closed greenhouses and future perspectives for tomato crops. In: *International Symposium on New Technologies for Environment Control, Energy-Saving and Crop Production in Greenhouse and Plant*, vol. 1227, pp. 141–150.

Truffault, V., Marlene, R., Brajeul, E., Vercambre, G. and Gautier, H. (2019). To stop nitrogen overdose in soilless tomato crop: A way to promote fruit quality without affecting fruit yield. *Agronomy* 9(2): 80.

Turcotte, G., Larouche, R., Carrier, A., et al. (2015). Production de la tomate de serre au Québec. In: *Ministère de l'agriculture, des pêcheries et de l'alimentation du Québec (MAPAQ) 2015*. Available at: http://www.agrireseau.net/legumesdeserre/documents/95590/production-de-la-tomate-de-serre-au-quebec-guide-technique/.

Van Henten, E. J., Buwalda, F., De Zwart, H. F., de Gelder, A., Hemming, J. and Bontsema, J. (2006). Toward an optimal control strategy for sweet pepper cultivation-2. optimization of the yield pattern and energy efficiency. In: *III International Symposium on Models for Plant Growth, Environmental Control and Farm Management in Protected Cultivation*, vol. 718, pp. 391–398.

Vanthoor, B. H. (2011). A model-based greenhouse design method. Ph.D. Thesis. Wageningen University, Wageningen, The Netherlands, 307 pp.

Vardar, G., Altıkatoğlu, M., Ortaç, D., Cemek, M. and Işıldak, İ (2015). Measuring calcium, potassium, and nitrate in plant nutrient solutions using ion-selective electrodes in hydroponic greenhouse of some vegetables. *Biotechnology and Applied Biochemistry* 62(5): 663–668.

Vermeiren, J., Fabri, S., Wittemans, L., Moerkens, R., Vanlommel, W., Marien, H. and Steppe, K. (2017). Revised sap flow driven stem diameter model for tomatoes grown under assimilation lighting. *X International Workshop on Sap Flow*, vol. 1222, pp. 27–34.

Vurro, F., Janni, M., Coppedè, N., Gentile, F., Manfredi, R., Bettelli, M. and Zappettini, A. (2019). Development of an *in vivo* sensor to monitor the effects of vapour pressure deficit (VPD) changes to improve water productivity in agriculture. *Sensors* 19(21): 4667.

Wang, X. and Xing, Y. (2017). Evaluation of the effects of irrigation and fertilization on tomato fruit yield and quality: a principal component analysis. *Scientific Reports* 7(1): 1–13.

Yano, A. and Cossu, M. (2019). Energy sustainable greenhouse crop cultivation using photovoltaic technologies. *Renewable and Sustainable Energy Reviews* 109: 116–137.

Zhang, H., Burr, J. and Zhao, F. (2017a). A comparative life cycle assessment (LCA) of lighting technologies for greenhouse crop production. *Journal of Cleaner Production* 140: 705–713.

Zhang, D., Du, Q., Zhang, Z., Jiao, X., Song, X. and Li, J. (2017b). Vapour pressure deficit control in relation to water transport and water productivity in greenhouse tomato production during summer. *Scientific Reports* 7(1): 43461.

Zhang, L., Hao, X., Li, Y. and Jiang, G. (2010). Response of greenhouse tomato to varied low pre-night temperatures at the same daily integrated temperature. *HortScience* 45(11): 1654–1661.

# Chapter 14

## Advances in soilless culture strawberry production

*Mike Nichols, formerly Massey University, New Zealand*



### 1 Introduction

Strawberries are the most popular berry fruits (soft fruits) in the world, with some 9 million tonnes being produced on just less than 400000 ha. Table 1 (FAO, 2018) demonstrates that PR China produces 40% of the world's strawberries and that the United States produces a further 16%. In PR China, production is spread throughout the country (Zhang et al., 2019) and includes a considerable area of strawberries grown in greenhouses, whereas in the United States most of the production is grown in the open field, and restricted to one small area–southern California which produces 91% of their crop. Hydroponics is a relatively new technology for strawberries, and currently, few (if any) statistics exist regarding areas or levels of production. It is well established that there are considerable areas of greenhouse strawberries in PR China, but these are predominantly in 'solar' greenhouses and grown in the soil. There are some modern greenhouses (e.g. near Shanghai) where hydroponics is used, but these are the exception rather than the rule. The use of hydroponics as a means to produce strawberries is very much a European technology, which slowly

http://dx.doi.org/10.19103/AS.2020.0076.17

**Table 1** Major producers of strawberries

| Country | Area (ha) | Production (t) | t/ha |
|---|---|---|---|
| PR China | 133114 | 3717283 | 27.90 |
| USA | 21327 | 1449280 | 46.27 |
| Mexico | 13650 | 658436 | 48.24 |
| Egypt | 11072 | 407240 | 36.78 |
| Turkey | 15392 | 400167 | 26.00 |
| Spain | 6819 | 360416 | 52.85 |
| Belgium | 1976 | 47500 | 24.04 |
| The Netherlands | 1691 | 66100 | 39.09 |
| WORLD | 395884 | 9223815 | |

FAO, 2018 www.fao.org/faostat/en

became established in the late 1960s/1970s with the development of Nutrient Film Technique (NFT) by Cooper (1979) and other researchers using solid media such as rockwool and peat (Lieten, 2014).

## *1.1 History of strawberry production*

The present-day strawberry (*Fragaria X anannasa* Duchesne) is of a fairly recent origin. The native European strawberry (*F.vesca*) has very small insignificant red fruit, but the early European explorers returned from the Americas and brought back with them many new plants, including *F. chiloense* (from Chile), and *F. virginiana* (from North America). By chance, a hybrid was produced probably in France in the 18th century (Darrow, 1966). Neither of these strawberries is particularly impressive, and the hybrid would never have occurred naturally, because of the physical distance between North America and Chile. The large red aromatic fruit was immediately popular, and from this hybrid has developed by selection and breeding the varieties, which dominate the present-day strawberry industry.

Currently, there are three groups of commercially grown strawberries, namely:

1. Short-day types which initiate flowers under short days (less than 12 hours photoperiods). These are also called June bearers.
2. Day-neutral types, which will initiate flowers irrespective of day length, although flower initiation may be reduced under high temperatures.
3. A much less important group of long-day types, which initiate flowers only under long days.

Hydroponic strawberry production probably owes its origin to the 19th-century development by the rich and famous to produce out-of-season fruit and vegetables (such as pineapples and peaches) in walled gardens using greenhouses. Strawberries were one of the choice crops, and this involved obtaining new runners as early as possible (mid-summer) and growing the plants in large pots until the autumn when the pots were then placed on their sides and left outside in the cold. In the New Year, the pots would then be brought into a heated greenhouse and fed with liquid manure. The Waterperry Horticultural School (near Oxford, UK), regularly grew strawberries (cultivar Royal Sovereign) in this manner to exhibit at the Chelsea Flower Show in the 1950s.

It is unclear when hydroponic strawberry production as we know it today commenced, but it was probably the development of the NFT by Dr Allen Cooper in England in the 1960s, (Cooper, 1979) along with the use of media-based hydroponic systems using rockwool or peat in Belgium and the Netherlands which provided added impetus to the development of hydroponics in general, and strawberries in particular.

Of course, it is not necessary to use protected cropping systems for hydroponics (Fig. 1), but in the case of strawberries the weather-sensitive nature of the fruit, along with the high capital cost of the hydroponic infrastructure makes the additional investment more logical.

## 2 Soilless culture – strawberry production systems

Because of the structure of the plant, the two main systems used for hydroponic strawberry production are based on structures, which allow the fruit to hang

**Figure 1** Outdoor NFT hydroponic strawberries, New Zealand 2006.

down clear of the ground. This means that the plants are grown in gulleys or gutters, generally suspended at a convenient height above the ground to make harvesting more efficient and to reduce the risk of fruit rots.

There are two main systems used:

1 Solid growing media.

This involves a solid media, either naturally occurring organic media such as coir or peat, or manufactured media such as rockwool, or horticultural foam. Although there have been numerous experiments with these media little reliable information exists. This is primarily because the physical characteristics of the organic media are not stated (e.g. Massetani et al., 2017). Coir or peat should be defined at the very minimum in terms of their water-holding and aeration characteristics.

2 Liquid growing systems.

These are either NFT or aeroponic systems. The Floating Raft /Deep flow system (popular for leafy greens) is not considered appropriate for a fruiting crop such as strawberry. Initially, the systems used for hydroponic strawberry production were NFT and the use of rockwool or peat as solid media.

Belgium and the Netherlands have a long tradition of using local peat as a growing medium for greenhouse crops, but environmental concerns are likely to reduce the use of this medium in the future.

## *2.1 Solid media*

Media can play a major role in the performance of strawberries. We are only now starting to formulate the relevant questions to ask in order to establish how the choice of media can influence strawberry performance. There have been, over the years, innumerable studies with different media for example Braet et al. (2017), Massetani et al. (2017), Taghavi et al. (2017), Beyene et al. (2014), and Palha et al. (2019), but in almost none of the projects are the physical characteristics of the media defined with amendments, such as sand or perlite. There is a range of different coirs and also a wide range of commercial peats. The characterising of the physical characteristics of such media appears to have been disregarded by researchers when comparing the influence of media on strawberries (with the exception of Taghavi et al., 2017), who defined their media in terms of drainage, aeration, water retention, and gas diffusion, but not with any physical measurements. They emphasise the importance of fertigation and irrigation strategies as being important in determining the potential productivity of any particular substrate. Duggan-Jones et al. (2013)

are some of the few researchers to have examined the effect of coir particle size and particle size mix (the major drivers for aeration, moisture holding in coir) on the productivity of any crop (tomatoes). With the exception of Belgium and Netherlands, coir would appear to be the preeminent medium in which to grow hydroponic strawberries, probably because of the environmental sensitivity of using peat, and yet we have still to define the ideal physical characteristics of the medium for growing strawberries. There is a need to determine what is the most satisfactory combination of aeration and moisture holding requirements for hydroponic strawberry production not only for coir, but also for peat, rockwool, or any other solid medium.

Nichols et al. (2015) compared the yield of greenhouse strawberries when grown in coir and when grown in horticultural foam. The results were inconclusive, however, because the volume of foam (per plant) was slightly smaller than the volume of coir (per plant). Foam has several advantages over coir (and rockwool), namely that it could be produced on site, it is clean and after use can be disposed of by burning. Conversely, rockwool has disposal problems, and the coir has further uses after strawberry production.

### 2.1.1 Rockwool

Rockwool also has environmental concerns, as energy is required for its production, and it does also have disposal problems away from where it is manufactured. Its major advantage is that being a manufactured product it is possible to achieve considerable standardisation in relation to the important horticultural physical characteristics–aeration and water holding capacity. The challenge is to establish the optimum irrigation and nutrition strategy for growing strawberries in this medium–as has been achieved with tomatoes.

### 2.1.2 Peat

Peat, being a naturally occurring product, can vary considerably in physical characteristics and therefore in horticultural value. It is difficult to wet when dry and is the medium of choice in Belgium/the Netherlands, where there is a large amount of experience on its use.

### 2.1.3 Coir

Coir, produced from the outside of the coconut fruit is (relatively speaking) the new boy on the block. There is a real need for this renewable resource to be standardised. Coir is a totally underutilised resource–with most of the horticultural product coming from Sri Lanka and India. The world's major producer of coconuts, Indonesia, produces very little horticultural coir at present. It has major advantages over rockwool–as its production involves very little energy, and because it is compressed its freight costs are considerably less than that for the equivalent volumes of rockwool, which is freighted around the world in its final modular form. A major problem with coir-based media is the lack of standardisation and certification. Variation in aeration and moisture holding characteristics occur both within and between the different suppliers.

One marked advantage of using coir-based, or other naturally based growing media, is that they have a value after growing strawberries. For instance, the coir used for growing strawberries is then used as a substrate to grow out-of-season greenhouse asparagus in New Zealand (Nichols, 2020).

### 2.1.4 Horticultural foam

A novel growing medium is foam, which is currently manufactured from petrochemicals, No doubt it could be produced from plant-derived chemicals if necessary. It can be manufactured on the site (in scale) but currently is normally shipped around the world in a finished form rather like rockwool. It can be disposed of easily after horticultural use simply by burning. Horticultural foam has a number of major advantages over rockwool as it can be manufactured locally.

Experience has suggested that the productivity of strawberries is influenced by the type of growing media used, with fine peat preforming in a superior manner to other solid media, and with rockwool preforming worst of many other solid media. This poses an essential consideration of why? Modern hydroponic methods enable water and/or nutrients to be applied on demand, and media with specific aeration/water holding characteristics can be developed, so clearly, some other factors (s) have come into play? Why (for example) did rockwool perform poorly compared to peat for Jansen (1997)? Was the nutrition incorrect, or was the frequency of irrigation inadequate when compared to peat?

Essentially we need to understand the requirements of the strawberry in terms of aeration and moisture retention, and to ensure that a medium with these characteristics is available. The question also arises as to why such a standardised material such as rockwool does not perform as well as some natural organic products.

## 2.2 Liquid systems

### 2.2.1 The nutrient film technique (NFT)

NFT has the advantage of simplicity. There is no problem of deciding when to water, or how much water to apply, as the plant roots are slightly immersed in a continuous film of nutrient solution, which is pumped to the end of the gulleys, which are at a slight slope, so that the solution flows down the gulley, where it is collected, and returned to the reservoir. The pH and nutrient status, generally on the basis of conductivity, is automatically controlled in the reservoir. Provided that the gulleys are not excessively long then the nutrient status is reasonably consistent along the length of the gulley, and there should be no differences occurring during the day, unlike solid media systems, where moisture and nutrient status can vary over time due to the intermittent application of nutrients. NFT could be used without any greenhouse protection for strawberry production (Fig. 1), but this is a risky proposition as it provides the plants with no protection against inclement weather.

### 2.2.2 Aeroponics

Few studies have been undertaken with strawberries. In addition to providing lighter weight structures, aeroponics has the potential to modify root temperatures (He & Lee, 1998) and root aeration. The downside is the need to have an extremely reliable delivery system of fine droplets of nutrient solution to ensure that the plants suffer no unwanted moisture stress, as the buffer due to any power disruption is small. Kanechi et al. (2017) have shown that very fine droplets (called dry fog spray) reduces the temperature of the rhizosphere and results in enhanced flower production during the summer for day-neutral varieties in Japan. Of course increased flower numbers do not necessarily result in increases in productivity. Droplet size is important in aeroponics but the smaller (more desirable) the droplet, the higher the risk of nozzle blocking, and the need for improved filtration. With other horticultural crops, Nichols et al. (2002) have shown that tomato and cucumber seedling growth rates are increased by using up to 80% oxygen in the root zone, while He et al. (2003) have demonstrated that root zone carbon dioxide increases productivity with aeroponic lettuce. Clearly similar research is required for strawberries. A possible disadvantage of the aeroponic system is that the roots tend to clump together, rather than spread out.

As with NFT the moisture status and nutrient status at the root surface do not vary over time cf solid media. The only concern must be in terms of gas exchange.

### 2.2.3 Nutrient solution control

There are two approaches to the control of the nutrient solution status.

One may use either:

**Table 2** Nutrient solution used by Melis et al. (2017)

| **Macro elements** | **K** | **Mg** | **Ca** | **$NH_4$** | **$NO_3$** | **P** | **Cl** | **$SO_2$** | **$HCO_3$** |
|---|---|---|---|---|---|---|---|---|---|
| (mmol.L$^{-1}$) | 4.50 | 1.25 | 3.50 | 1.00 | 10.00 | 2.00 | 0.50 | 1.25 | 0.00 |
| **Minor elements** | **Fe** | **Mn** | **Cu** | **Zn** | **B** | **Mo** | | | |
| (μmol.L$^{-1}$) | 30.00 | 18.00 | 0.75 | 7.00 | 12.00 | 0.50 | | | |

1 a re-circulating system in which the drainage water is returned to the reservoir directly or via some form of sterilization, the so-called closed-loop system, or
2 a water to waste system, in which the drainage water is disposed of elsewhere, the so-called open-loop system.

The first essential is to ensure that the pH of the solution is appropriate for strawberries, and this is normally considered to be between 5.8 and 6.3. This is usually adjusted by using nitric acid to reduce pH, or potassium hydroxide to increase pH. The basic hydroponic solution is obtained by mixing two concentrated solutions together in a significant volume of water. One of the solutions contains calcium nitrate and the other potassium phosphate, but mixing the two in a dilute solution does not result in the precipitation of calcium phosphate, which would occur in a concentrated solution. (Table 2).

Information on the appropriate conductivity for the nutrient solution can best be described as varied. With NFT or aeroponics, the chemical status of the solution is virtually constant for plants, but for any solid media system (e.g. rockwool, peat, or coir) the conductivity within the medium can vary over time and space considerably. There are a number of approaches to this problem:

1 to measure the electric conductivity of the drain water and to irrigate when the drain water exceeds 2.5 ds.m$^{-1}$.
2 to irrigate to ensure that about 25% of the irrigation water is 'lost' as drainage, to leach out excess salts in the medium.
3 measure the conductivity and so on in random modules, using a handheld conductivity probe.
4 use a weighing lysimeter to record water loss. This system has potential problems if the plants in the lysimeter are not typical of the general population.

Future development will be the introduction of specific ion electrodes, which will enable more precise changes to the chemical content of the nutrient solution elements.

There have been numerous research projects on the hydroponic nutrition of strawberries over the years. The main conclusions that can be derived are that strawberries require full nutrition (Table 2) throughout their life (Sarraf

**Figure 2** Tip runner production (Courtesy Berryworld, New Zealand).

et al., 2017). The appropriate nutrition of tray plants can have an influence on the later productivity (Melis, et al., 2017), and that for some day-neutral varieties (Gallace et al., 2017) they are complex productivity implications. Although conductivities of 1.6, 1.2, and 0.7 ds.m$^{-1}$had no influence on the number of flower trusses per plant, the higher conductivities resulted in more quaternary and quinary fruits, and a reduction in the yield of larger (>31 mm) fruits.

## 3 Propagation

Strawberries can be propagated either from seed or vegetatively. Seed is the standard method for producing new varieties and has also been used as a method to develop (via $F_1$ hybrid seed production) a simpler method of mass production of planting material. Seed, however, has the inherent downside of not providing the genetic stability of vegetative propagation.

Vegetative production has potential problems, because of the increased risk of the transmission of pathogens between generations. The vegetative propagation of high health strawberry plants has developed since the middle of the 20th century. Strawberries can suffer from a wide range of systemic diseases, and the elimination of these has been a major challenge. The current method is to use tissue culture, combined with heat attenuation. This involves the growing of the strawberry plants at a high temperature followed by the removal of the meristem, and its tissue culture in a sterile medium. A critical part of producing high-health planting material is the regular testing of the plant material for pathogens. This may be through a range of different technologies, including the use of molecular biological methods. The high health plants are then used for the mass production of commercial plants for fruit production.

The problem with any form of tissue culture is the risk of mutations, which may not be apparent until fruit production.

The conventional method of producing this planting material is via runners grown in the field in the soil which has been fumigated for disease, pest and weed control. The healthy mother plants are planted in the field during spring at approximately 1m × 1m spacing, and the runners are produced during the long days of summer. These are dug up in the autumn as bare root transplants. No fruit is permitted to grow on the mother plants. Runner plants with a range of physiological ages (and sizes) will be produced, as the first runners will develop as early as mid-summer, and runner production will continue well into the autumn,

The lack of efficient 'soil fumigants' has led to the development of the 'tip runner' propagation method, in which the runners are isolated from the soil either by a polythene film or else by being grown from mother plants hydroponically 1-2 m above the ground, and the runners which develop are harvested before they reach the ground. The unrooted runners are then inserted into a sterile rooting medium and placed in a mist propagation unit until roots develop (usually 3-5 days). These plants are usually described as plug plants or tray plants.

This is usually a summer operation as runner production occurs under long day conditions, but for year round strawberry production young plants are required throughout the year. Recent studies have shown that they can also be successfully stored at -1.5°C for nearly 12 months without any ill effects. Runners can be produced year round in a greenhouse with supplementary lighting to provide continuous long days. Because of the runners hanging down, it is possible to produce many runners in a very small area and this means that a plant factory might provide an alternative method for producing year-round runners.

## 4 Strawberry plant architecture

In simple terms the strawberry is a herbaceous, determinate, rosette plant, which terminates in a flowering structure when vegetative growth turns into generative growth under specific thermoperiods, depending on genotype; however other environmental factors, for example agronomic and nutritional, can also be of major significance particularly during the propagation period (Massetani et al., 2011).

Most of the studies on strawberry plant architecture have been undertaken in Belgium on the cv. 'Elsanta', and it is now common for plants to be examined to evaluate the development in terms of floral initiation (Van Delm, et al., 2014). A number of limited studies on strawberry plant architecture have emphasised the flowering aspects of the plant and has paid less attention to two other potentially important aspects, namely runner production, and the development of 'daughter' crowns.

It is clear that the growing medium and the supply of nutrients play an important role in determining not only when floral initiation begins, but the level of flower production. However, flower production does not necessarily equate with yield or economic productivity. For example, Gallace et al. (2017) demonstrate a complex interaction between three-day neutral cultivars, and three nutrient solution conductivities namely 0.7, 1.2 and 1.5 $ds.m^{-1}$. Increasing fruit number may increase total yield but could compromise marketable yield by producing smaller fruits.

One of the major costs of growing hydroponic strawberries is the expense of fresh plants for each growing cycle, along with the cost of removing the old plants and replacing the growing medium.

Most of the plant architecture studies have been associated with the variety 'Elsanta'–essentially a short-day variety. There are obvious limits to the long-term productivity of short-day strawberry varieties, and there is a need to obtain information on possible strategies to control flowering in the day-neutral varieties. Ideally, we should be able to plant a young runner, knowing that it would produce a fixed number of fruits of a 'roughly' specified size range over a reasonably long time period. This might be achieved by the control of fruit number per truss and the number of new crowns allowed to develop. The fruit size would be determined by market demand, and the very important relationship between the source, such as plant photosynthetic capacity, and sinks, such as fruit number and size.

Provided the propagation phase is undertaken under controlled conditions (i.e. in a plant factory) the examination of the apical meristem might be unnecessary. Beyene et al. (2014) suggested that primary runners produce more flowers per plant than tertiary runners (but provide no fruit yield data), whereas D'Anna and Iapichino (2002) suggest that runner position on the stolon is not important in terms of fruit yield. Currently, most tip runners are produced in large numbers (Fig. 2). However, ideally, for precise crop timing, these runners should be harvested when they are all at the same physiological stage, as the later produced plants are unlikely to be as physiologically advanced as the earlier developed plants. Nutrition plays a major role in floral initiation in strawberries, and floral induction is greatly enhanced by nutrient stress, particularly nitrogen (Lieten, 2002). However, following induction, a higher level of nitrogen is required to develop the floral organs. Thus some compromise will need to be established, in which stolon production is enhanced by adequate nutrition; floral induction is stimulated by nutritional stress, and flower development is then enhanced by adequate nutrition.

It is becoming increasingly important to balance the vegetative with the generative part of the plant. This is already done with tomatoes but is a fairly new concept for strawberries. We need to balance the photosynthetic capacity of the plant with the fruit load, with particular reference to the season. There is also the complication that it costs much more to harvest small fruits compared to large fruits because for every 'gram' increase in the mean fruit weight a harvester will pick 1 extra kg/hour.

## 5 Pest and disease management

Strawberries can suffer from a range of different pathogens, from minute virus particles to large caterpillars. The majority of the systemic diseases that afflict strawberries are normally controlled using heat attenuation, tissue culture, and various techniques to test the plant material for freedom from systemic pathogens.

Other pests and diseases are normally controlled either by hygienic measures, the use of naturally occurring predators, naturally occurring biological agents (biopesticides), or by the use of 'conventional' pesticides.

As hydroponic strawberry normally is carried out in greenhouses, *Botrytis cinerea* (grey mould) is not problem but powdery mildew (*Podosphera aphanis*) can be a problem in greenhouse production. It appears that the use of Ultra Violet Radiation (actually UV-C) with the appropriate dose and treatment interval will provide some control.

The use of conventional pesticides may well become necessary if the biological methods fail. However, the use of conventional pesticides pose problems in terms of residues on a fruit crop which matures over several weeks, and cannot be easily washed. Therefore, Integrated Pest management (IPM) is the logical and pragmatic solution.

The major soil-borne pathogens are essentially controlled by ensuring that the young plants are propagated from healthy plants in a hygienic manner. With the disappearance of methyl bromide and no satisfactory alternative, this means that tip runners are the best alternative. With any hydroponic system, the risk of spreading soil-borne pathogens is high, and this risk is increased if a recirculating hydroponics system is used unless some form of sterilization (e.g. UV) of the drain water is carried out. In any case, all fresh water introduced into the system should be pathogen free by using either UV or ozone. The alternative is to use a sand filter and to rely on obtaining some form of equilibrium between the various microorganisms.

The range of predators used to control pests is determined by the local biota. In some geographically isolated countries (e.g. New Zealand/Australia) this is limited due to strict biosecurity rules. For example in Australia, Bumble bees are not present on the mainland (only on Tasmania) and this valuable pollinator of greenhouse crops is not permitted on the mainland. In New

Zealand, many valuable biological predators are not present and their possible introduction is an extremely tedious bureaucratic, often futile operation.

The key factor in biological control is to introduce the predators before the pest has built up. In the case of white fly and red spider mites, the trigger may be only 1 pest per leaf, and this is very difficult to observe, even by experts. However, systems are being developed to scan greenhouses and transmit pictures back to a base computer for analysis of the leaf spots which can establish the possible presence and intensity of these pests, Nieuwenhuizen et al. (2020).

## 6 Supplementary lighting

It is extremely difficult to produce ripe strawberry fruit in the winter months at high latitude due to the short photoperiod and the low levels of photosynthetically active radiation from the sun and sub optimum temperatures. In Belgium (latitude 51 N) ripe fruit production in January and February is virtually impossible without the use of supplementary lighting, even though greenhouse temperatures can be optimised. The initial use of metal-halide lighting was eventually replaced by High Pressure Sodium (HPS) lights, and more recently by the more efficient LEDs (Light Emitting Diode), which offer the potential of monochromatic lights (Van Delm et al., 2017).

At lower latitudes (e.g. Portugal) (40 N) the value of supplementary lighting for winter greenhouse strawberry production is less certain. Palha et al. (2019) found no improvement in productivity from the use of supplementary LED lighting; however, their fruit production period was from mid-February and not the more difficult period of January/February.

In Belgium currently, there is interest in producing ripe fruit virtually free from solar radiation, by relying almost entirely on supplementary lighting from LEDs (Vanderbruggen, 2019).

## 7 Case study

This case study refers to 'The Wee Red Barn' which is situated near Masterton, which is some 100 km NE of Wellington, New Zealand's capital city. Masterton has a mild temperate climate tending towards the Mediterranean, with a fairly equable rainfall of just over 900 mm per year. The Wee Red Barn is owned by Allan and Dot Bissett, who also are the Australasian agents for Haygrove Tunnels. On their property, currently, they not only grow strawberries, but also gooseberries, raspberries, blueberries, blackberries, cherries and asparagus all under high plastic tunnels. Their main crop is strawberries, and these are grown entirely hydroponically using a table-top system, using imported coir modules as the growing medium.

The strawberry runners are obtained in late May from a commercial high health runner producer as bare rooted plants, and these are immediately planted into new 1m-long coir modules which have been fully hydrated and then buffered (using calcium nitrate solution) immediately prior to planting. The three-row table top system is an A frame shape, with the central row being some 30–40 cm above the two adjacent rows. The gutters for the coir modules are produced on the property by means of a specialised rolled steel shaper. Each 1 long coir modules has seven plants, and three pressure regulated drippers are used to supply the plants with all the water and fertilizers required. There are only three table tops per house (Fig. 3) and the Bissetts prefer to leave enough space between the table tops to spray pesticides by means of a tractor additional. A frame might possibily could be fitted into a house by using a different spraying technique.

The Bissetts grow predominately day-neutral varieties from California, namely Albion, Monterey and San Andreas. They also grow a small area of a short-day variety (Camerosa) under shelter netting, just for the pre-Christmas market. This is not harvested after Christmas, and not watered, and the plants are removed when they have dried off. The coir modules are re-used for one future Camerosa crop the following year; in this case, whereas with all the other strawberries coir modules are replaced every season.

The plants are regularly sprayed with pesticides, particularly to control cyclamen mite and red spider mite. No biological methods are used. Pollination is by any natural bumble bee population and by honey bees, which enter the houses when the sides are raised for ventilation.

The hydroponic nutrient solution is not re-circulated but is collected and sprayed onto the nearby pasture.

An extremely small labour force maintains the high tunnels, but during the harvest season, temporary workers (usually backpackers) do the harvesting. This is facilitated by using trolleys between the rows and is not a particularly arduous task with the A frame table tops. The large fruit achieved with good irrigation and nutrition provides a good opportunity to earn a good income. The moisture status (and conductivity) of the nodules is regularly checked by means of a hand monitor.

Yields in excess of 100t/ha are regularly achieved, as the harvesting normally commences in early October, and continues until the end of May. As the season progresses there is a tendency for fruit size to fall, and with the variety Albion this can be overcome to some extent by cutting the plants hard back in mid-January–forgoing some crop, but getting the opportunity to use a harsh miticide against cyclamen mites (for which Albion is particularly sensitive), knowing that no fruit would be harvested for 4–5 weeks.

Most of the fruits are sold at the gate, but if there is an oversupply the surplus fruit is sold through toe market system to supermarkets.

**Figure 3** Three-row strawberry system–Wee Red Barn, New Zealand.

One of the advantages of using coir as a growing medium is that when the strawberry crop is removed there is still a valuable organic material. At the Wee Red Barn, this is not immediately used outside in the field but is used to construct 900 mm high beds for growing out of season asparagus in high tunnels (Nichols, 2020). Asparagus is normally harvested from September to December in the New Zealand (spring), but by using high tunnels and coir beds harvesting can commence in late July, and it may be possible to obtain 3 crops every 2 years, always avoiding the 'low-price' normal harvest period. Peat would probably work just as well, but rockwool would not.

This system of strawberry production (from early October to late May) appears to suit New Zealand, although there is an increasing interest in year round production in order to reduce the biosecurity risk of introducing new pests and diseases from imported fruit from Australia and United States during the winter off season.

## 8 Future trends in research

Fifty years ago, strawberry was only available in the early summer from short-day varieties, also called 'June bearers', which initiated flowers in the autumn, experienced the winter chill, and then flowered in the spring. About this time also an interest developed in extending the production season for strawberries by using greenhouses. This is particularly important because an extended harvest season could be possible using a single variety, e.g. 'Elsanta' in Belgium.

It is almost certain that in the future strawberry production will move from the field, initially into greenhouses, but inevitably into completely environmentally controlled plant factories, where many of the tasks (such as fruit picking) will be carried out by robots. Two recent developments suggest that the manner with which we grow strawberries is going to change dramatically in the near future. This move is already starting, with strawberries being grown at Proftein Hoogstraten in gutters suspended one above the another, and beneath each gutter is LED lighting. Vanderbruggen (2019) (Fig. 4). A Dutch group (Anon, 2019) has established a strawberry innovation cluster (Fragaria innova) and is linked with Signify (formerly Phillips Lighting) with an interest not only in using plant factories, but also in improving the taste of the strawberry fruit, presumably by using selected wavelengths of LED lighting. Plant factories in which the plants are grown in a fully insulated building with all the light (radiation) being provided artificially is almost a reality. When this is combined with robotic harvesting of the ripe fruit produced with the assistance of Artificial Intelligence (AI)-controlled systems the future trend is very clear.

Labour for harvesting is one of the industries' major problems, and the Spanish developed Agrobot (2019) which allows multiple rows of strawberries to be harvested 24/7 with minimal labour is clearly the future trend.

A big problem with hand harvesting of strawberries is the need to have a reasonable distance between the rows for the harvesters to operate. This is not an efficient use of expensive greenhouse space; because of the structure of the strawberry plant a considerable amount of solar radiation is not absorbed by the plant. With vine crops (such as tomatoes the depth of the canopy ensures efficient light interception. This has been an ongoing problem in the past, with expensive designs developed which move alternative rows up or down, or sideways to permit ease of crop training and harvesting, and thus make more efficient use of the greenhouse space. As the robotic harvester requires less-working distance between the rows than people, then more can be grown in the greenhouse or plant factory.

Light Emitting Diode (LED) lights are monochromatic, with two major wavelengths (red and blue) which provide the major radiation for photosynthesis, but some of the other wavelengths may also have an important role to play, in providing the radiant energy of the appropriate wavelength to develop desirable aroma and taste chemicals within the fruit. The quality of strawberries is not only Brix but also a range of aromatic chemicals along with texture.

Hydroponic strawberry production will demand plants which are pre-programmed to perform in a predetermined manner. Planting field-grown runners will not provide the appropriate degree of certainty required for precision production. A clear understanding of the treatments required to provide such plants will be a critical part of the fruit-production process.

**Figure 4** Multilayer greenhouse strawberry planting. (Courtesy Proftein Hoogstraten, Belgium)

The combination of plant factory production and robotic harvesting should enable strawberries to be grown anywhere on the planet to provide ripe fruit on any day of the year. Seasonality is already rapidly disappearing due to air freight and will disappear entirely in the future.

## 9 Conclusion

Strawberry is still primarily field grown, but the demand by consumers for consistent quality, and availability of fresh fruit over a longer season, is slowly moving production under protected cultivation. Soilless cultivation is an important part the investment for improving the growing environment for the crop leading to marked increases in productivity.

## 10 Where to look for further information

On the world scene, strawberries are a minor crop. The International Society for Horticultural Science (www.ishs.org) has a Workgroup called 'Strawberry Culture and Management', which arranges international symposia on strawberry production every 2–3 years and the papers are published in Acta Horticulturae after the meeting, following peer review. Other relevant presentations are to be found within the ISHS network in the generic division Protected Cultivation and Soilless Culture. Other possible sources of information are to be found in CAB abstracts, and research journals such as *HortResearch*, *Scientia Horticulturae* and a wide range of industry publications. There is also numerous industry conferences held worldwide (though mainly in Western Europe), which provide a valuable interaction between scientists and industry.

## 11 References

Agrobot (2019). Available at: https://www.agrobot.com/.

Anon (2019). Innovative cultivation concept for lit strawberry cultivation. Available at: www.FreshPlaza.com. Accessed 18 November 2019.

Beyene, G. T., Kehoe, F., Mac Sturtain, M. and Hunter, A. (2014). Effect of different substrates on rooting and flower development of primary and tertiary runners of strawberry trayplants cultivar "Elsanta". *Acta Horticulturae* 1049, 581-585.

Braet, E., Simoen, J. and Hanssens, J. (2017). Strawberry cultivation on substrate: comparison of a continuous cultivation and a fresh spring planting. *Acta Horticulturae* 1156, 229-234.

Cooper, A. (1979). *The ABC of NFT. Nutrient Film Technique*. pp. 181. Grower Books, London.

Darrow, G. M. (1966). The Strawberry. History, Breeding and Physiology. Holt, Rinehart and Winston, New York.

Duggan-Jones, D. I., Nichols, M. A. and Woolley, D. J. (2013). Effect of physical characteristics of coir on the productivity of greenhouse tomatoes. *Acta Horticulturae* 1004, 101-106.

FAO (2018). Available at: www.fao.org/faostat/en.

Gallace, N., Boonen, M., Lieten, P. and Bylemans, D. (2017). Electrical conductivity of the nutrient solution: implications for flowering and yield in day neutral cultivars. *Acta Horticulturae* 1156, 223-228.

He, J., Austin, P. T., Nichols, M. A. and Lee, S. K. (2003). Effect of root zone (RZ) $CO_2$ on productivity and photosynthesis in aeroponically grown lettuce plants. *Acta Horticulturae* 648, 38-45.

He, J. and Lee, S. K. (1998). Growth and photosynthetic responses of three aeroponically grown lettuce cultivars (*Lactuca sativa* L.) to different rootzone temperatures and growth irradiances under tropical aerial condition. *The Journal of Horticultural Science and Biotechnology* 73(2), 173-180.

Jansen, W. A. G. M. (1997). Growing media and lant densities for strawberry tray plants. *Acta Horticulturae* 439, 457-470.

Kanechi, M., Hikosaka, Y., Fukuda, C. and Uno, Y. (2017). Ever-bearing strawberry culture using a new aeroponic system with dry fog spray fertigation during the summer season. *Acta Horticulturae* 1176, 37-44.

Lieten, P. (2002). The effect of nutrition prior to and during flower differentiation on phyllody and plant production on short day strawberry 'Elsanta'. *Acta Horticulturae* 567, 145-148.

Lieten, P. (2014). The strawberry nursery industry in the Netherlands an update. *Acta Horticulturae* 1049, 99-106.

Massetani, F., Gangatharam, R. and Neri, D. (2011). Plant architecture of strawberry in relation to abiotic stress, nutrient application and type of propagation system. *Genes, Genomes and Genomics* 5(Special Issue 1), 13–23.

Massetani, F., Savini, G. and Neri, D. (2017). Effect of substrate and container type in the strawberry soilless cultivation. *Acta Horticulturae* 1156, 295–300.

Melis, P., Stiffels, K. Vervoort, M. and Van delm, T. (2017). Programming production potential and harvest spread through fertilisation in "Elsanta" trayplants. *Acta Horticulturae* 1156, 315–321.

Nichols, M. A., Woolley, D. J. and Christie, C. B. (2002). Effect of oxygen and carbon dioxide concentration in the root zone on the growth of vegetables. *Acta Horticulturae* 578, 119–122.

Nichols, M. A., Duggan-Jones, D. and Christie, C. B. (2015). Greenhouse strawberry variety trial. *Practical Hydroponics and Greenhouses* 161, 38–43.

Nichols, M. A. (2020). Year round asparagus production. *NZ Grower* 75(2), 38–39.

Nieuwenhuizen, A. T., Kool, J., Suh, H. K. and Hemming, J. (2020). Automated spider mite detection on tomato leaves in greenhouses. *Acta Horticulturae* 1268, 165–172.

Palha, M. G., Pestana, F. and Oliveira, C. M. (2019). Plant growth, yield and fruit quality of *Fragaria ananassa* genotypes under supplementary LED lighting systems and substrate cultivation. *Acta Horticulturae* 1265, 91–97.

Sarraf, E., Gauthier, L., Gosselin, A. and Desjardin, Y. (2017). Effects of fertigation on yield of day neutral strawberry (fragaris X anannasa) cultivars grown hydroponically in high tunnels. *Acta Horticulturae* 1156, 323–326.

Taghavi, T., Fortin, J. -P., Hughes, B. R., Zandstra, J., Dale, A. and Wright, B. (2017). Developing substrate culture strategies for the production of day-neutral strawberries. *Acta Horticulturae* 1156, 277–282.

Van Delm, T., Melis, P., Stoffels, K., Van De Vyver, F. and Baets, W. (2014). Strawberry plant architecture and flower induction in plant production and strawberry cultivation. *Acta Horticulturae* 1049, 489–494.

Van Delm, T., Stoffels, K., Melis, P., Vervoort, M. and Vanderbuggen, R. (2017). Overcoming climatic limitations: cultivation systems and winter production under assimilation lighting resulting in year-round strawberries. *Acta Horticulturae* 1156, 517–528.

Vanderbruggen, R. (2019). Belgium: cultivating strawberries in a four-layered cultivation system. *Hortidaily* November 11, 2019.

Zhang, Y., Wang, G., Dong, J. and Chang, L. (2019). Strawberry culture and breeding studies in China. *Chronica Horticulturae* 59(2), 29–36.

# Chapter 15

## Advances in soilless culture of ornamentals

*Erik van Os and Arca Kromwijk, Wageningen University & Research, The Netherlands*



### 1 Introduction

The cultivation of ornamentals, whether flowers and pot plants, is less widespread than vegetable production. The area used for ornamental cultivation is much less compared to the area used for vegetable production in most countries. Most production takes place conventionally in soil and soilless culture systems (SCS) are less often used. Research on the subject is limited, so specialists from different countries are consulted, depending on their expertise. For instance, Columbia is a leading producer of various types of flowers, whereas Ecuador specializes in the production of rose and exports products to the United States and Europe.

Where ornamentals are cultivated in soils, investment level is often low and traditional cultivation methods are only adapted slightly, depending on the exporter's preference in terms of variety and colour (pers. Comm. Nieves Garcia, WUR GH). As an example, carnation is sometimes grown in substrate beds. In North America, the United States shows developments in upscaling and increasing efficiency in cultivation, as well as using soilless growing methods, but many farms still grow in soil (pers. Comm. Chris Higgins, 2019). The key factor is the level of investment needed. In Kenya, Ethiopia and Uganda, for example, production mainly takes place in soil. The knowledge gap between growth in soil and in soilless conditions for staff is often too great and results

http://dx.doi.org/10.19103/AS.2020.0076.19

in innovations in SCS being unsuccessful and in a return to traditional soil-growing methods (pers. Comm. Nieves Garcia, WUR GH).

Fernández et al. (2018) has examined current trends in protected cultivation in Europe and highlighted how floriculture has begun to adapt new methods. European growers, for example, in the Netherlands, have developed SCS. In Spain, for example, SCS have been used to produce roses. SCS use 50% less water and 60% less fertilizer than soil (Van Os, 1998). However, the use of SCS is still limited by higher installation costs as well as by the greater technical skills and expertise required. Ornamental production is scattered throughout Europe. Begonia and azaleas are grown in container fields in Belgium (https://lv.vlaanderen.be/nl/voorlichting-info/publicaties/praktijkgidsen/water/duurzaam-watergebruik-de-sierteelt.). In Denmark, pot plant production in greenhouses is popular and in Norway small greenhouses use artificial light to produce pot plants for the local market. Germany mainly produces pot plants and cut flowers for local markets. Italy also has large areas dedicated to producing container plants for export.

Countries such as Canada have developed governmental programmes for growers to be more efficient in the use of water. Deloitte (2009) and Zheng (2018) found that larger-scale growers use more ebb/flood and drip-irrigation techniques, and recirculate surplus water, compared to smaller-scale growers who use sprinkler irrigation. In Europe, traditional systems show the use of (movable) sprinkler irrigation while others have introduced the ebb and flood system for irrigation (Van Os, 2017). Water is sometimes recirculated but irrigation systems are mainly open systems. In the Netherlands, the area of ornamental production in greenhouses has always been comparable to that of vegetable production (Raaphorst, 2017). The Netherlands has 1000 ha in soil for cultivation of cut flowers such as chrysanthemums, lisianthus, lily and freesia, and about 2000 ha using pot plants. The area for roses has declined from about 900 ha to 250 ha. The area for ornamental cultivation is slowly decreasing, whilst the number of growers/companies is strongly decreasing (Raaphorst, 2017).

The change from soil to soilless cultivation started in the late 1970s with the introduction of the stone wool and nutrient film technique (NFT). Higher yields and better quality were achieved in vegetable cultivation. Problems with soil-borne diseases also stimulated a change to SCS. Roses suffered from nematodes and fungi in soils requiring soil to be disinfected via steam sterilization between cultivation cycles. Increasing energy prices between 1980s and 1990s also encouraged rose growers to change to soilless cultivation. The shift to SCS for flower crops was accelerated by the development of techniques for recirculation of nutrient solution along with the disinfection of recirculated drain water (Van Os, 1998, 2009).

In 2000, the EU Water Framework Directive (WFD, 2000) was introduced, which emphasized the importance of clean ground and surface water. In the Netherlands, it became a requirement for nutrient emissions and plant protection products (PPPs) to be reduced in surface water. Dutch governmental bodies and horticultural organizations agreed on reaching a zero nitrogen and phosphate emission rate by 2027 (Fig. 1). Southern European countries focussed on losses of water from greenhouses due to large levels of water consumption, while northern European countries were more aware of the polluting properties of nutrients and PPPs. All European countries are looking for more efficient use of water, partly because the supply of good quality water is limited and partly because pollution of ground and surface water needs to be avoided.

The discharge of nutrients in surface water (e.g. nitrates and phosphates) is seen as a problem but PPPs are also increasingly seen as a concern (Vermeulen et al., 2010; Van Ruijven et al., 2019; Beerling et al., 2014). In countries such as Belgium, Sweden and the United Kingdom, measures have been taken or considered to minimize the discharge of a nutrient solution with large amounts of nutrients and PPPs into surface water. New Dutch legislation requires growers to purify discharge water from soilless greenhouse cultivation to eliminate PPPs by 95% (Van Ruijven et al., 2019). The idea of using a zero-liquid-discharge system has also been introduced (Van Os et al., 2019, 2020). A concern was why growers were discharging between 5% and 15% of their total annual water supply when the water was being recirculated (Fig. 2). The main reasons for discharges were the fear of a high sodium concentration and an unbalanced composition of the nutrient solution. More and more growers are trying to become a 'zero-discharger'. If a grower can prove he is not discharging any nutrient solution containing PPPs, he will be approved as a zero-discharger and therefore does not need to invest in purification equipment.

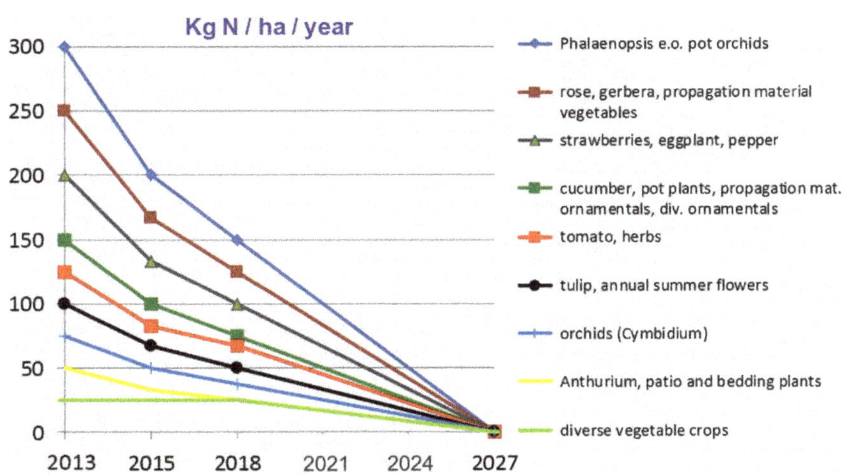


**Figure 1** Nitrogen emission standards for soilless grown crops in the Netherlands (Activiteitenbesluit Landbouw, 2013 and 2018).

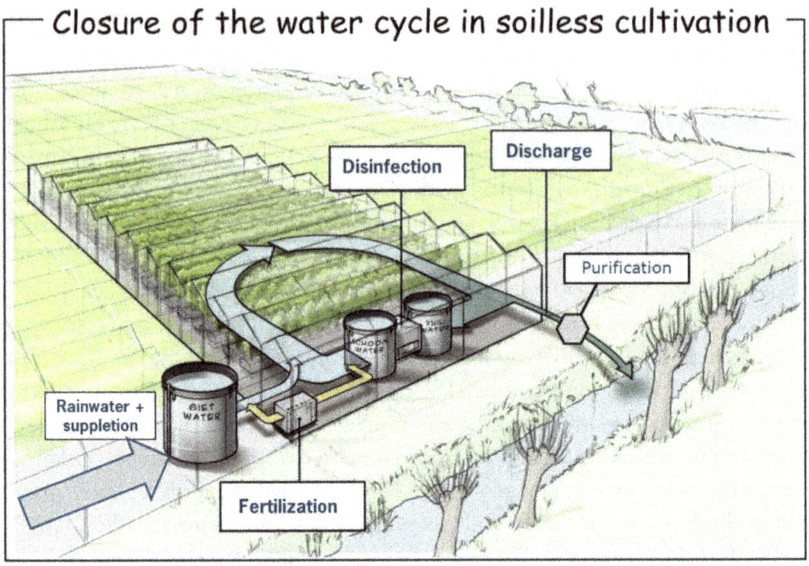


**Figure 2** Schematic overview of the water flows at a Dutch farm.

## 2 Growing systems for cut flowers: growing in rows

Various growing systems are used in soilless ornamental production. This section provides an overview of the systems used, adaptations and challenges for the future, starting with cut flowers. The large variation in cut flowers also resulted in many types of systems. Basic groups are (Van Os et al., 2019):

- Rose, gerbera and cymbidium growing in rows; 5–10 plants/m$^2$.
- Anthurium, amaryllis, gypsophila, bouvardia, carnation, alstroemeria and lily growing in beds of 120–140 cm wide, 20–40 plants/m$^2$.
- Chrysanthemum, lisianthus, freesia and aster mainly growing in soil, 40–80 plants/m$^2$.
- Forced bulb flowers, such as tulip, hyacinth and iris grown in ebb and flood systems without any substrate.

This section discusses growing in rows. Other systems are discussed in the following sections.

Rose- and gerbera-growing row systems are comparable with vegetable systems (Fig. 3). The system is based on a cultivation period of more than one year (rose 3–5 years; gerbera 2–3 years) and the almost daily need to check plants to harvest the flowers. Rose cuttings are rooted in stone wool or coir blocks and transplanted to slabs of the same material. Slabs lie in long rows behind each other on about 0.2% sloped troughs. They are placed in greenhouses in a support system in the soil. The height above the soil made

**Figure 3** Gerbera in a rack system to fix containers filled with coir or rockwool cubes (left). Rose growing in a bent cane system (right).

the bent cane system possible, producing more roses of better quality (Fig. 3). Irrigation takes place by drip lines with pressure-compensated drippers of 2 L/h. Surplus supply water (drain water; 30–50% of the supply) is collected in troughs and flows on gravity to a drain water collection tank to be pumped out of the greenhouse to be disinfected and reused again. The distance between the troughs depends on plant density requirements of the crop as well as the stands and span width of the greenhouse construction.

Young gerbera plants are placed on slabs of stone wool or coir or in containers of about 20 cm diameter filled with stone wool cubes or coir (Fig. 3). In the same way as roses, slabs are placed in troughs with drip irrigation and the drain water is recirculated. If planted in containers, a rack system is often used to keep the containers well-balanced and to avoid pulling the plants out of the substrate during harvesting. A trough is placed below the containers. In the case of both rose and gerbera, growth in separate acclimatized compartments has been shown to improve yields (Van Weel, 2005). However, high annual costs and problems with the irrigation system have not made these systems commercially feasible (Van Tuijl et al., 2005). New varieties and the use of artificial lighting, as well as zero-liquid discharge systems, have also improved the efficiency of SCS.

## 3 Growing in beds

Many flower crops are cultivated in beds, as the plant density is much higher (20–40 plants/m$^2$), compared to rose and gerbera (5–10 plants/m$^2$).

### *3.1 Anthurium*

Anthurium for cut flower production has been grown in soilless systems for a few decades. The system uses a bed with containers filled with a substrate with a high bulk density (>500 kg/m$^3$). Clay or lava granulate is widely used. A lighter substrate causes turnover of the plants after a few years of cultivation; long and heavy shoots develop outside the container. It is also possible to grow the plants directly in a bed filled with lava or clay granulates. In the middle of each 120 cm wide bed lies a drainage pipe to collect the surplus of nutrient solution to be recirculated.

### 3.2 Amaryllis (Hippeastrum)

Amaryllis (Hippeastrum) cultivation uses two systems. The growing of bulbs takes mainly place in soil, but the forcing of flowers is now changing to a soilless system, primarily due to an increasing problem with *Fusarium and nematodes* (Kromwijk et al., 2016). Perlite and clay granulates are the main substrates (Fig. 4). As bulbs are big, a loose substrate layer of 5-15 cm is popular. Cultivation takes 3–4 years after which bulbs are renewed. Bulbs are planted in beds of 120–140 cm wide, with a drain pipe in the middle. Growers were reluctant to recirculate the nutrient solution because of the release of toxic plant root exudates.

Kromwijk et al. (2016) have compared disinfection of drain water with UV light, with combined UV light and hydrogen peroxide (advanced oxidation) for break-down of the root exudate lycorine. During the three-year trial there were no differences in production and crop growth, although growth of the roots in the control was slightly less. Van Marrewijk (2012) found lycorine as potential toxic root exudate with high concentrations of 68 mg/kg fresh weight in the bulb itself and lower concentrations in the roots, with an average 4.5 mg/L. Further experiments showed no presence of lycorine in drain water in soilless cultivation of amaryllis. Lycorine is not broken down by heat treatment (95°C), but a high UV light dosage (240 mJ/cm$^2$) is effective. It means that, with reuse of drain water, UV light is the better disinfection method.

After cultivation for 3–4 years, the substrate has to be renewed or disinfected. Normally steam sterilization is used in soil. In substrate beds, which are rather dry, steam sterilization is also possible if all synthetic materials are resistant to temperatures of more than 90°C. Another option is to remove the

**Figure 4** Amaryllis in perlite (left), and Freesia in sand beds (right) with a polystyrene cover for light reflection.

substrate and to refill the system with a fresh, unused substrate, which is not very much done. Both are being investigated in practical trials by the growers.

### 3.3 Freesia

Freesia is a tuber crop which produces flowers in 5-7 months and where after the corms stay in the soil to grow further for a short period. After digging them up, corms are dried and receive a high temperature treatment during storage. They are then replanted and after 2-3 weeks soil temperature is changed to about 16 to induce flowering and produce new flowers. During the last ten years, a few experiments have taken place to a soilless system with sand as a substrate (Fig. 4). Flower production was improved (Blok & Kromwijk, 2014a) but corm quality was not sufficient to be reused in new cultivations.

Better results are achieved if the sand layer has a uniform thickness. An anti-root foil is used below the sand. Below that a drainpipe is used to collect the drain water, lying in a layer of clay granulates. All foils must be resistant to steam sterilization (90–100°C). Blok and Kromwijk (2014a) compared 10 substrates for growing freesia and concluded that coarse coir or coarse peat resulted in better growth, possibly due to high air content at saturation level. However, too many roots are formed in coir while, in substrates as perlite and lava granulates, it is difficult to separate tubers (1–3 cm) from substrate granules with the same size.

### 3.4 Chrysanthemum and lisianthus

Chrysanthemum is still mainly grown in soil in the Netherlands. Van Os (1980) reported a fully mechanized system using a NFT with flat troughs of 3–5 cm wide as the rooting block. While cultivation was successful, the system did not prove economically feasible. Later the 'mobysant' project was launched in which bare-rooted cuttings were placed in coir in 6 cm troughs (Van Os, 2006). The troughs were on a 0.5% slope and drain water was collected and reused. The main problem was the combination of a substrate and NFT which caused too many dry and wet spots in the trough and substrate and, consequently, a much higher risk for *Pythium*. This meant that fewer than 90% flowers from the planted cuttings could be harvested and, consequently, the system was not economical. Similar results were reported by Vermeulen (2009). The combination of substrate and a hydroponic system such as NFT and DFT is harder to manage and causes higher risks of diseases such as *Pythium*. Another development is bare rooting of the cuttings (Van Winsen, 2015; Dodde, 2017; Glastuinbouw Nederland, 2019). Results are promising after 4 years though economic feasibility has to be proven and a solution has to be found for transport of bare-rooted cuttings to final planting. Bare-rooted cuttings cannot survive a long time out of the solution. Either propagation needs to take place at the grower stage or transport has to be reorganized. Overviewing all the experiments in producing chrysanthemums, it clearly takes a number of years to become familiar with a system to get maximum growth. It is also important that the system should always be economical for the grower. Economic feasibility is challenging, particularly given further improvements in soil cultivation (e.g. faster growing cycles, higher yield and quality).

Lisianthus has been grown on NFT systems on a semi-practical scale, for example, in coir with sub-irrigation (ebb/flood) (Van Velden, 2014). Drip irrigation did not work because of uneven water content of the coir substrate with more than 40 plants per $m^2$. The high sensitivity for fungal diseases requires frequent disinfection of circulating water and substrate (Sleegers, 2013). The system is not yet feasible technically or economically, but progress is being made.

In general it is difficult to make a substrate system with more than 40 plants per $m^2$ economic. The use of drip irrigation for each plant is too expensive and handling at planting and harvesting stages is not practical. Applying one dripper for 4 or 8 plants mostly gives uneven growth. These types of crops often have a short rotation with 4–6 plantings per year. Drippers have to be removed after harvest and replaced. The substrate also has to be tilled, otherwise a heterogenic substrate develops because cuttings are rooted in substrate (mainly peat) blocks which stay behind after each harvest.

## 4 Forcing of bulbs for cut flower production

Tulips, hyacinth and iris are examples of crops where bulb production takes place in the open field, while forcing takes place in a one-layer or a multi-layer SCS on NFT or ebb/flood in a greenhouse from January to May. Bulbs grow for the entire season in the soil and are dug up in autumn. They are cooled under specific temperature conditions after which they can be forced. Bulbs are placed in special trays (Fig. 5) to keep them straight up, especially needed when

**Figure 5** Tulip bulbs are pricked in trays by hand (left); empty trays for pricking bulbs on the ebb/flood irrigation tables (middle); forced tulips in greenhouse (right).

the heavy flower appears and turnover during transport of the trays is a risk. The trays are placed on tables for a period of about 3 weeks. One system consists of three stacked layers. First, the lowest layer without artificial light is planted with the first signs of the flowers emerging from the bulbs. The final ripening of the flowers takes place at the upper layer at the top of the greenhouse. Growing tables can be automatically moved from the working shed to the greenhouse and back. The total water system is a recirculating system with disinfection of the nutrient solution with UV or ozone. In another system, only one layer of tables is used in the greenhouse. Growing tables are also planted and harvested in the work shed.

## 5 Growing systems for pot plants

Small flowering pot plants (e.g. kalanchoe, cyclamen, pot chrysanthemum, begonia, saintpaulia, phalaenopsis and many others) grow in 8–12 cm pots on ebb/flood tables or benches, have a short growing cycle and require substantial labour (Fig. 6). The tables can be fixed to one location, rolling from left to right in the same span or automatically transported from the work shed to the greenhouse and back. Large foliage pot plants, such as palms, ficus, monstera, dracaena and others, grow often on concrete floors in pots of 15–30 cm (Fig. 6). Machinery to place and collect the containers can operate on concrete floors. Plants require little labour but need a long cultivation time. Irrigation is mostly done by ebb/flood. Nutrition of the plants is done via the fertigation system, adapting to the growing needs of plants. Alternatively, controlled release fertilizers can be used directly when young plants are potted up. There are big differences in automation. Some growers place the plants by hand on the tables. Others use specific rakes to pick up one row of 10–15 plants. Such rakes are also able to space the plants from about 3 cm to 10–15 cm. Large-scale growers automate all handling from filling the pots and placing on the tables in the work shed to transport of the tables to the greenhouse. In some countries, such as The Netherlands, image analysis is already standard to grade plants by size or colour. In addition, automated detection of insects and fungi will be commercially available in the next few years.

**Figure 6** Pot plants on ebb/flood tables (left and middle) and on a concrete floor in clay granulates (right).

Outdoor production of container plants, such as azalea, heath (Erica species), nursery stock and begonia, also takes place on floors. Sprinkler irrigation from above is normal, with ebb/flood the exception. To recirculate surplus water the water needs first to be filtered and preferably disinfected. An additional problem is rain and snow. Rainwater should be collected from a closed floor to be used first, but the installation needs adaptation to discharge a certain volume of water to surface water.

## 6 Phalaenopsis: a case study

Phalaenopsis is an orchid grown in pots. It is one of the most important pot plants in the Netherlands (Kromwijk et al., 2019) with a turnover in 2016 of €500 million which is 33% of the total annual pot plant turnover (Royal Flora Holland, 2016). The production area was more than 250 ha in 2016 (Raaphorst, 2017). The plants were grown on open bottom tables and irrigated from above with sprinklers. Dutch legislation followed the EU Water Framework Directive (WFD, 2000) which favoured phalaenopsis cultivation. Recirculation was not needed because of sodium sensitivity and very limited use of fertilizer (Van Os, 1998; Beerling et al., 2017). Once production techniques were well-established, there was a dramatic increase in production area which was followed by a fall in prices.

A phalaenopsis pot plant requires a vegetative period of 28 weeks with high temperatures (28°C), followed by a generative period of about 18 weeks with lower temperatures (about 19–22°C). At this point the use of fertilizers is increased from an electric conductivity (EC) of 0.5 or less in the past to about 1.5 mS/cm now. Under new environmental legislation (Fig. 1) phalaenopsis was placed in the group where most emission of nitrogen was allowed compared to other crops. It also became clear that growers had to change their growing system towards a recirculation system. An enquiry among 30 growers highlighted their main concerns (Schoenmakers, 2012). These included: presence of sodium, dispersal of root-borne pathogens and imbalances in nutrient composition.

Traditionally phalaenopsis has been grown in an open-loop system. Based on Baas (2010) the estimated transpiration is on average 0.6–1.1 mm/day. This is about 2000–4000 $m^3$/ha/year. As water supply is once in 4–6 days 10–15 L/ $m^2$ with a drain percentage of 70% it was calculated that between 4300 and 9600 $m^3$/ha/year was supplied to the crop (Van Os and Kromwijk, 2014a). When the nutrient solution contains 15 mmol/L it can be calculated that between 900 and 2000 kg N per ha per year is discharged to ground or surface water or the sewage system. This is much more than the earlier expected discharge of 300 kg/ha/year, which is comparable with 1400 $m^3$/ha/year (Van Os and Kromwijk, 2014b). This means a closed-loop system is required. Growers have had to invest in collecting drain water below the mesh tables, such as drain water storage, filters, disinfection equipment, pumps and pipework. It would also be useful to separate the water circuits in the vegetative and generative growth periods to be able to supply a different EC of the nutrient solution.

Systems used are mostly open mesh tables with galvanized wire bottoms on which 9–15 cm pots are placed. The pots contain a bark substrate (Fig. 7). The surplus water drains through and between the pots and through the wire mesh to the soil. Several growers have placed a 1.0–1.5 mm liner beneath the tables. However, zinc concentrations became too high. Zinc is required in the nutrient solution (2–5 μmol/L). In most crops values of more than 25 μmol/L give toxicity symptoms which often looks like iron deficiency (yellowness in the young leaves) (Blok and Kromwijk, 2014b). The cause of this problem is the use of galvanized mesh growing tables. Now growers are gradually transitioning to a closed-loop system. A watertight floor is used with a drainpipe lying in lava granulates to collect the surplus solution. Mesh tables are placed on this. A further future development could be the use of aluminium mesh tables or a closed polyester-profiled bottom, similar to pot plants. The surplus can then be collected, disinfected and reused.

In SCS it is mostly recommended to avoid the use of urea, $CO(NH_2)_2$, because urea does not increase the EC value of the nutrient solution but does decrease its pH-value. In open-loop systems, growers usually use a nutrient solution with a concentration of 1.0 dS/m and 3.5 mmol/L urea which is similar to 7 mmol N and 0.7 EC. An EC of 1.7 mS/cm is supplied to the plants (Blok and Kromwijk, 2015). To realize the same ion concentration in the root environment, the bivalent ions Ca and Mg should be lowered and the monovalent ion K has to be increased when changing from an open-loop system to a closed-loop system. Current advice is to use an EC of 1.2 mS/cm in the open-loop system and 1.3 mS/cm in the recirculation system in the vegetative stage, which changes to 1.1–1.3 mS/cm (open/closed) in the generative stage. K/Ca changes from 1.5 in open system to 2.7 in closed system in the vegetative stage, while in the generative stage it changes from 2.1 to 3.0.

**Figure 7** Phalaenopsis on open-growing tables (left) and after sprinkler irrigation (right) in the vegetative growth period. Bark is used as substrate.

Kromwijk et al. (2019) compared the reuse of the nutrient solution with an open run-to-waste system on traditional mesh tables and sprinkler irrigation. Results showed 70% drainage in open systems, compared to 30–50% in closed systems of vegetable and cut flower production. The increase in EC in drain water was limited (0.2 mS/cm). Before recirculation, drain water was disinfected by UV treatment. In the open system >1500 kg N per ha was discharged. In the two recirculation treatments 182 kg N/ha was discharged at a sodium concentration of 1.2 mmol/L and 96 kg N/ha was discharged due to an unbalanced nutrient solution. In all treatments the same production and quality was achieved. It could be concluded that if good-quality rainwater is used, recirculation is possible and discharge of nitrogen is dramatically decreased. Sodium does not accumulate very rapidly (e.g. a threshold value of 2 mmol/L looks possible) (Kromwijk et al.,2017), but has to be proven for phalaenopsis.

## 7 Conclusions and future trends

The need to be more efficient with water supply is an important step in greenhouse horticultural development. There is a change from soil to SCS and, within soilless systems, from open, run-to-waste systems to closed-loop systems. The first step is to move from a closed system with a minor 5–10% discharge to a zero liquid discharge. This means that all water used stays in the system and there is no discharge of water, nutrients and PPPs to sewage, ground or surface water. However, this requires significant investment in both technology and expertise.

In ornamental production there is a slow change to SCS. In some European countries, such as in the Netherlands, legislation is the steering factor. The established crops grown using soilless systems (rose, gerbera) are looking to zero liquid discharge, while the traditionally soil-grown crops are looking for economically feasible systems to avoid discharge to the environment. For amaryllis (Hippeastrum) the change looks promising but, for other crops such as freesia, chrysanthemum, and lisianthus, there is still a long way to go.

Phalaenopsis, the leading pot plant crop in the Netherlands, had to make a big change. Initially legislation did not require recirculation and an open system was accepted. During the last 15 years improvements in cultivation led to increased use of fertilizers resulting in unacceptably high discharges of nitrogen, requiring a transition to a closed system. The fear of sodium accumulation became reality and, consequently, clean fresh and/or rainwater had to be used, while large investments were required to recirculate the nutrient solution.

As the world faces increasing shortages of water, water-use efficiency in the cultivation of ornamentals needs to be further improved. In several European countries there is the desire to avoid discharge of nitrogen and PPPs. In the Netherlands a zero liquid discharge of nitrogen should be achieved by 2027

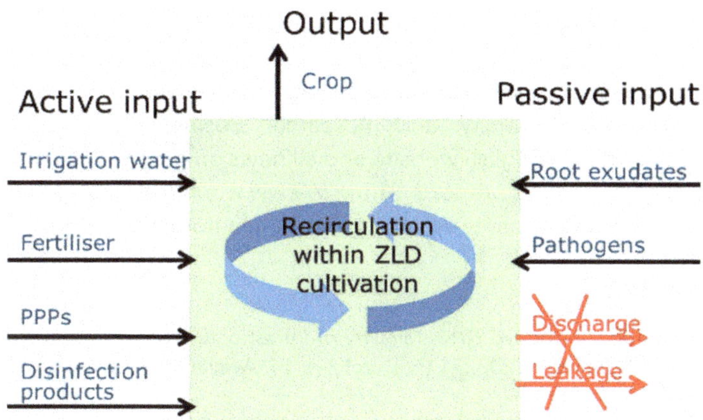


**Figure 8** Keeping the balance in a zero-liquid discharge growing system.

while, since 2018, growers have had to purify discharge water to eliminate PPPs by 95%. Gerbera and pot plant growers have shown how to achieve a zero-liquid discharge (Van Ruijven et al., 2019; Van Os et al., 2019, 2020). As shown in Fig. 8 a good balance in the water system is essential. Inputs in the system have to be taken up by the plant or broken down. This applies to fertilizers and to pipework-cleaning products such as bleach, chlorine products and hydrogen peroxide, as well as so-called growth improvers to improve plant resilience against diseases. However, it is most essential to start with sodium-free and pathogen-free water supply.

The change to SCS is likely to increase because it is easier to prevent disease and water- and fertilizer-use efficiencies are much better and, consequently, yield and quality of the produce are higher. However, higher initial investment and lack of knowledge remain as barriers to change. Another tendency is the upscaling of the sector as small companies are acquired by larger businesses. Because of the larger scale, the investment level can be increased, for example, in better irrigation systems to increase water-use efficiency (Zheng, 2018; Raaphorst, 2017). Governmental bodies can steer these improvements by legislation but they also need to offer subsidies for environmentally friendly investments in technology.

## 8 Where to look for further information

- International Society for Horticultural Science, ISHS: www.ishs.org. The leading organization for all horticultural sciences which are reported in symposia and the Acta Horticulturae proceedings.
- The technical magazine for growers ('Onder Glas') has also an English issue: https://www.ingreenhouses.com/. Magazines can be read online for

free, but with a delay of months, or a paid subscription. It gives information about the latest results in research and commercial development for the entire greenhouse sector, especially meant for growers.

- Wageningen University and Research, Business Unit Greenhouse Horticulture with English website and all news on greenhouse horticulture including scientific reports. https://www.wur.nl/en/Research-Results/Research-Institutes/plant-research/greenhouse-horticulture.htm

## 9 References

Activiteitenbesluit Landbouw, 2013. Ministry of Infrastructure and Environment, 01-01-2013. IenM/BSK-2012/233271 (in Dutch), p. 14. Available at: www.rijksoverheid.nl (revision 2018).

Baas, R., 2010. Verdampingsmodel Phalaenopsis. Report Productschap Tuinbouw, 33p (in Dutch).

Beerling, E., Van Os, E. A., Van Ruijven, J., Janse, J., Lee, A. and Blok, C., 2017. Water-efficient zero-emission greenhouse crop production: a preliminary study. *Acta Hortic.* 1170, 1133–1140.

Beerling, E. A. M., Blok, C., Van der Maas, A. A. and Van Os, E. A., 2014. Closing the water and nutrient cycles in soilless cultivation systems. *Acta Hortic.* 1034, 49–56.

Blok, C. and Kromwijk, J. A. M., 2014a. *Freesia op substraat: praktijkproeven 2013 en 2014.* (Rapport / Wageningen UR Glastuinbouw; No. 1309), 22p. Bleiswijk: Wageningen UR Glastuinbouw (in Dutch).

Blok, C. and Kromwijk, J. A. M., 2014b. Recirculatie potorchidee. Flyer no. 5, Zink. Available at: https://www.glastuinbouwwaterproof.nl/content/3Onderzoek/Recirculatie_potorchidee/5_Flyer_Zink.pdf (in Dutch).

Blok, C. and Kromwijk, A., 2015. Recirculatie potorchidee. Flyer no. 9, Nieuw voedingsschema phalaenopsis (in Dutch).

Deloitte, 2009. The impact of ornamental horticulture on Canada's economy. Report Deloitte and Touche, 138p.

Dodde, H. 2017. Chrysanten laten zich goed telen op water. *Nieuwe Oogst*, 24 October 2017. Available at: https://www.nieuweoogst.nl/nieuws/2017/10/24/chrysanten-laten-zich-goed-telen-op-water (in Dutch).

Fernández, J. A., Orsini, F., Baeza, E., Oztekin, G. B., Muñoz, P., Contreras, J. and Montero, J. I., 2018. Current trends in protected cultivation in Mediterranean climates. *Eur. J. Hortic. Sci.* 83(5), 294–305. ISSN 1611-4426 print, 1611-4434 online | © ISHS 2018.

Glastuinbouw Nederland, 2019. Chrysant op water is ook nieuw leren telen. Available at: https://www.glastuinbouwnederland.nl/nieuws/chrysant-op-water-is-ook-nieuw-leren-telen/ (in Dutch)

Kromwijk, A., Blok, C., van Os, E. A. and Beerling, E., 2019. Reusing drainwater to lower the emission of nutrients in the cultivation of phalaenopsis. *Acta Hortic.* 1262, 101–108. DOI: 10.17660/ActaHortic.2019.1262.15

Kromwijk, A., Voogt, W., Steenhuizen, J., van Winkel, A. and van Mourik, N., 2017. *Natriumgevoeligheid en recirculatie bij Cymbidium in 3 teeltjaren: Behoud plantgezondheid en voorkomen groeiremming bij hergebruik drainwaterdrain water.* Bleiswijk: Wageningen University & Research, BU Greenhouse Horticulture. (Rapport WPR; No. 735), 54p. doi: 10.18174/433587.

Kromwijk, J. A. M., van der Burg, R., Nijs, L., Overkleeft, J., Eveleens-Clark, B. A., Blok, C., Woets, F., 2016. *Recirculatie bij snij-amaryllis (Hippeastrum) in drie teeltjaren (2013-2015): Behoud plantgezondheid en voorkomen groeiremming bij hergebruik drainwater*. (Rapport GTB; No. 1398), 88p. Bleiswijk: Wageningen UR Glastuinbouw (in Dutch).

Raaphorst, M. G. M., 2017. *Kwantitatieve Informatie Glastuinbouw 2016-2017*. Bleiswijk: Wageningen University & Research, BU Greenhouse Horticulture (in Dutch).

Royal Flora Holland, 2016. Jaarverslag 2016. Available at: https://viewer.onlinepublisher.nl/flora-holland-jaarverslag-2016-nl (in Dutch).

Schoenmakers, M., 2012. Inventarisatie knelpunten emissiebeperking pot-orchideeën. Eindrapportage LTO Groeiservice PT 14263-28. Available at: https://www.glastuinbouwwaterproof.nl/content/3Onderzoek/Knelpunten_emissiebeperking_potorchidee.pdf (in Dutch).

Sleegers, J. 2013. Lisianthus uit de grond heeft nog eind te gaan. *Vakblad voor de Bloemisterij*, 12, 34–35. Available at: https://proeftuinzwaagdijk.nl/publicaties/Lisianthus%20uit%20de%20grond%20heeft%20nog%20eind%20te%20gaan.pdf

Van Marrewijk, I., 2012. Amaryllis, gaat recirculatie samen met een gezond gewas? Report Horticultural Board, PT 146660 (in Dutch).

Van Os, E. A., 1980. Complete mechanization of the growing of cut chrysanthemums in nutrient film. In: ISOSC Proceedings Fifth Congress, Wageningen, the Netherlands, pp. 187–196.

Van Os, E. A., 1998. Closed soilless growing systems, the finishing touch. *Acta Hortic.* 458, 279–289.

Van Os, E. A., 2006. Mobysant: meer dan 28% energiebesparing per tak is mogelijk. *Onder Glas*. 3(1), 44–45 (in Dutch).

Van Os, E. A., 2009. Comparison of some chemical and non-chemical treatments to disinfect a recirculating nutrient solution. *Acta Hortic.* 843, 229–234.

Van Os, E. A., 2017. Recent advances in soilless culture in Europe. *Acta Hortic.* 1176, 1–8.

Van Os, E. A., Beerling, E. A. M., Blok, C., Janse, J., Leyh, R., van Ruijven, J. P. M., van der Staaij, M. and Kaarsemaker, R., 2019. Zero discharge of nutrients and pesticides to the environment in hydroponic production. *Acta Hortic.* 1266, 443–450.

Van Os, E. A., Beerling, E. A. M., Blok, C., Leyh, R., van Ruijven, J. P. M., van der Staaij, M., Janse, J., Kaarsemaker, R. and Roosen, W., 2020. Zero liquid discharge in soilless greenhouse horticulture: solutions to save water and the environment while ensuring an optimal production. *Acta Hortic.* 1273, 129–135.

Van Os, E. A. and Kromwijk A., 2014a. Recirculatie potorchidee. Flyer 7, Watervoorziening. Available at: https://www.glastuinbouwwaterproof.nl/content/3Onderzoek/Recirculatie_potorchidee/7_Flyer_Watervoorziening.pdf (in Dutch).

Van Os, E. A. and Kromwijk A., 2014b. Recirculatie potorchidee. Flyer 10. Scenarioberekeningen stikstof emissie. Available at: http://www.glastuinbouwwaterproof.nl/fileadmin/user_upload/waterproof/Projecten/doc/Potorchidee/2014_08_20_flyer_recirculatie_potorchidee_10_scenarioberekeningen_emissie.pdf (in Dutch).

Van Ruijven, J. P. M., van Os, E. A., Beerling, E., Blok, C., Janse, J., 2019. Towards zero-liquid discharge in hydroponic cultivation. *Acta Hortic.* 1242, 863–872.

Van Tuijl, B. A. J., van Henten, E. J. and van Os, E. A., 2005. Mobiele gerberateelt en toekomstige ontwikkelingen. A&F rapport 330, 42p (in Dutch).

Van Velden, P., 2014. Alles is gericht op een ziektevije teelt met hogere opbrengst. *Onder Glas* April 2014, 18–19. Available at: https://www.onderglas.nl/magazines/#dflip-df_830/18/ (in Dutch).

Van Weel, P. A., 2005. Movable gutters role in rose production. *FlowerTech* 8(7), 30–32.

Van Winsen, J., 2015. Teelt chrysant op water onder de loep. *Nieuwe Oogst* 11 April 2015. Available at: https://www.proeftuinzwaagdijk.nl/publicaties/Chrysant%20op%20water%2011-4-2015%20Nieuwe%20Oogst.pdf (in Dutch).

Vermeulen, T., 2009. Literatuurstudie chrysant los van de grond. Wageningen UR Glastuinbouw. Report 233, 60p (in Dutch).

Vermeulen, T., van der Linden, A. M. A. and van Os, E. A., 2010. Emissions of plant protection products from glasshouses to surface water in The Netherlands. (Rapport / Wageningen UR Greenhouse Horticulture/RIVM rapport; No. GTB-1002/607407001). Bleiswijk: Wageningen UR Greenhouse Horticulture.

WFD, 2000. EU Water Framework Directive 2000/60/EC. Available at: http://ec.europa.eu/environment/water/water-framework/index_en.html

Zheng, Y. 2018. Current nutrient management practices and technologies used in North American greenhouse and nursery industries. *Acta Hortic.* 1227, 435–441.

# Index

CPSIA information can be obtained
at www.ICGtesting.com
Printed in the USA
BVHW010004200221
600633BV00001B/2

9 781786 764355